Aero- and Hydro-Acoustics

International Union of Theoretical and Applied Mechanics

Aero- and Hydro-Acoustics

IUTAM Symposium
Ecole Centrale de Lyon
3–6 July, 1985

Editors
G. Comte-Bellot and **J. E. Ffowcs Williams**

With 276 Figures

Springer-Verlag
Berlin Heidelberg NewYork Tokyo

Prof. Geneviève Comte-Bellot
Département Acoustique
Ecole Centrale de Lyon
36, av. Guy de Collongue, BP 163
69131 Ecully Cedex

Prof. John E. Ffowcs Williams
Dept. of Engineering
University of Cambridge
Trumpington Street
GB Cambridge CB2-1PZ
England

ISBN-13 : 978-3-642-82760-0 e-ISBN-13 : 978-3-642-82758-7
DOI : 10.1007 / 978-3-642-82758-7

Library of Congress Cataloging-in-Publication Data
Aero- and hydro-acoustics.
At head of title: International Union of Theoretical and Applied Mechanics.
1. Fluid dynamics–Congresses.
2. Underwater acoustics–Congresses.
3. Aerodynamic noise–Congresses.
I. Comte-Bellot, Geneviève.
II. Ffowcs Williams, John E.
III. International Union of Theoretical and Applied Mechanics.
TA357.A33 1986 620.2'5 86-11862

Softcover reprint of the hardcover 1st edition 1986

2161/3020-543210

Scientific Committee

Prof. L. BJORNO, Danmark
Prof. G. COMTE-BELLOT, France, Chairman
Prof. D.G. CRIGHTON, UK
Prof. J.E. FFOWCS WILLIAMS, UK, Chairman
Dr M.E. GOLDSTEIN, USA
Prof. M.S. HOWE, UK
Prof. M.J. LIGHTHILL, UK
Prof. L.M. LIAMSHEV, USSR
Prof. E.A. MULLER, Germany
Dr. A. POWELL, USA

Local Organizing Committee

Dr. D. JUVE
Prof. C. LESUEUR
Dr. G. ROBERT, Secretary
Dr. M. ROGER
Prof. J. SABOT
Prof. M. SUNYACH, Chairman

Preface

The Symposium "Aero-and Hydroacoustics" was held at the Ecole Centrale de Lyon, France, on 3rd - 6 July 1985, at the initiative of the International Union of Theoretical and Applied Mechanics (IUTAM).

The research topics discussed concerned the acoustics of unstable and turbulent flows and the noise generated by movable surfaces. Other subjects of interest were thermoacoustics, diffraction, and acoustic propagation through turbulence. Applications of numerical and experimental techniques to source diagnostics were discussed, as was active noise control. In Hydroacoustics the emphasis was on noise due to flow induced vibrations.

There were 8 general invited lecturers and 43 other contributors, with 150 participants form 14 countries. Sufficient time was provided for discussions and social gatherings. For example, there was the opening of the high speed anechoic wind tunnel of the Centre Acoustique, and a welcome address given by the Mayor of Lyon in his historic City Hall.

The organizing Committee of the Symposium wish to express their thanks to la Direction des Recherches Etudes et Techniques du Ministère des Armées (DRET), le Centre National de la Recherche Scientifique (CNRS), le Conseil Général du Rhône, l'Association Universitaire de Mécanique (AUM), le Ministère de l'Environnement, for financial support. Le Groupement des Acousticiens de langue française (GALF) helped advertise the meeting. The organizers are also indebted to the International Union of Theoretical and Applied Mechanics for providing travel subsidies to young scientists.

Lyon, October 1985

G. COMTE-BELLOT

J.E. FFOWCS WILLIAMS

Participants

ABADIE P. (F)/ CERDAN.DCAN Toulon, BP 77, 83800 TOULON NAVAL.

ACTON E. (GB)/ TOPEXPRESS LTD, 13.14 Round Church St., Cambridge, CB4 4TQ.

ALI H.B. (I)/ SACLANT ASW Res. Center, Viale San Bartolomeo 400, 1-19026 San Bartolomeo -SP-, La Spezia.

ALLORY G. (F)/ Société NOVATOME, G.I.E. Arny, 91680 BRUYERES LE CHATEL.

ALLOUCHE J.L. (F)/ I.R.C.N., 47 Rue de Marceau, 78008 PARIS.

ARBEY H.S. (F)/ I.N.R.S., Avenue de Bourgogne, 54501 VANDOEUVRE CEDEX.

ATTIACH J.C. TSI (F)/ TSI France Inc. 68 Rue de Paris, 93804 EPINAY SUR SEINE CEDEX.

AUDOLY Ch. (F)/ G.E.R.D.S.M., Le Brusc, Ch de la Gardiole, 83140 SIX FOURS LES PLAGES.

AUGER J.M. (F)/ U.T.C., Div Acoustique et Vibrations Industrielles, B.P. 23, 60206 COMPIEGNE CEDEX.

BATAILLE J. (F)/ E.C.L., Mécanique des Fluides, 69131 ECULLY CEDEX.

BECHERT D. (RFA)/ D.F.V.L.R., Institut fur Turbulenzforschung Muller-Breslau-Strass 8, D1000 BERLIN 12.

BENARROUS E. (F)/ Société METRAFLU, 64 Chemin des Mouilles, 69134 ECULLY CEDEX.

BIRON D. (F)/ ONERA-CERT-DERMES, 2 Av E. Belin, 31055 TOULOUSE CEDEX.

BJORNO L. (DK)/ Industrial Acoustics Lab, Tech Univ Denmark, Building 352, DK-2800 Lyngby.

BLAISE A. (F)/ INSA de Lyon, LVA-Bat 303, 20 Av Albert Einstein, 69621 VILLEURBANNE CEDEX.

BLAKE W.K. (USA)/ David Taylor Naval Ship R and D Center, Bethesda, MARYLAND 20084.

BLANC-BENON Ph. (F)/ E.C.L., Centre Acoustique, 69131 ECULLY CEDEX.

BOISRAYON M.(F)/ G.E.R.D.S.M., Le Brusc, Ch de la Gardiole, 83140 SIX FOURS LES PLAGES;

BOREL M. (F)/ DCAe/STPA/MO.O, 4 Av de la Porte d'Issy, 75996 PARIS ARMEES.

BORRELLO G. (F)/ C.N.E.S., 4 Rue Charles Beaudelaire, 91000 EVRY.

BOUCHARD P. (F)/ TECHNICATOME, Serv Mesure Vibration Informatque, Et de Cadrache, 13115 Saint Paul lez Durance.

BOVIS A. (F)/ Bassin d'Essais des Carènes de Paris, 8 Bd Victor, 75732 PARIS CEDEX 15.

BRAJOU F. (F)/ G.E.R.D.S.M., Le Brusc, 83140 SIX FOURS LES PLAGES.

BRENOT D. (F)/ O.N.E.R.A., BP 72, 92322 CHATILLON CEDEX.

BROCHER E. (F)/ I.M.F.M., 1 Rue Honnorat, 13003 MARSEILLE.

BRUGGEMAN J.C. (NL)/ Lab for Fluid Dynamics and Heat Transfer, P.O. Box 513, 5600 MD EINDHOVEN.

BRUNEAU M. (F)/ Lab d'Acoustique, Univ du Maine, BP 535, 72017 LE MANS CEDEX.

CANDEL S. (F)/ O.N.E.R.A., BP 72, 92322 CHATILLON CEDEX.

CANARD S. (F)/ O.N.E.R.A., BP 72, 92322 CHATILLON CEDEX.

CARON A. (F)/ Avions Marcel Dassault-BA, Div Acoustique, 78 Quai Carnot, 92214 SAINT CLOUD CEDEX.

CHABAS F. (F)/ O.N.E.R.A., BP 72, 92322 CHATILLON CEDEX.

CHAPMAN C.J. (GB)/ I.S.V.R., Univ of Southampton, S09 5NH.

CHATEL G. (F)/ SECAV, 74 Bd de Marseil Leveyre, 13008 MARSEILLE.

CHING G. (F)/ THOMSON CSF, Div. Activités Sous Marine, Chemin des Travails, 06802 CAGNES SUR MER.

CHOSSON M. B et K (F)/ Bruel et Kjaer France, 16 Av Pierre Allard, 69500 BRON.

COMTE-BELLOT G. (F)/ E.C.L., Centre Acoustique, 69131 ECULLY CEDEX.

CONTINO M.Ch. (F)/ MATRA, Service "Calcul Espace", 52/54 Av Dutartre, 78150 LE CHESNAY.

CRANCE C. (F)/ Bassin d'Essais des Carènes de Paris, 8 Bd Victor, 75732 PARIS CEDEX 15.

CRIGHTON D. (GB)/ School of Mathematics, Univ of Leeds, Leeds LS2 9JT.

D'AMBRA F.N. (F)/ Aerospatiale, Div Hélicoptères, H.DE/ERA, BP 13, 13725 MARIGNANE CEDEX.

DAMONGEOT A. (F)/ Aerospatiale, Div Hélicoptères, BP 13, 13725 MARIGNANE CEDEX.

DELCAMBRE J. (F)/ E.D.F. DER, Serv AEE, Départ Acoustique, 1 Av du Général de Gaulle, 92141 CLAMART CEDEX.

DI NICOLA G. (F)/ DRET-Gr VI, 26 Boulevard Victor, 75996 PARIS.

DOWLING A.P. (GB)/ Cambridge Univ, Engng Dept, Trumpington Street, Cambridge CB2 1PZ.

DRUGEON J.L. (F)/ E.D.F. DER, Serv AEE, Départ Acoustique, 1 Av du Général de Gaulle, 92141 CLAMART CEDEX.

EATWELL G. (GB)/ TOPEXPRESS LTD, 13 Round Church Street, Cambridge CB5 8AD.

ELIAS G. (F)/ O.N.E.R.A., BP 72, 92322 CHATILLON CEDEX.

ESPARCIEUX Ph. (F)/ CERDAN.DCAN Toulon, BP 77, 83800 TOULON NAVAL.

ESPOSITO P. (F)/ E.D.F. DER, Serv AEE, Départ Acoustique, 1 Av du Général de Gaulle, 92141 CLAMART CEDEX.

FARASSAT F. (USA)/ NASA Langley Research Center, MS 461, NASA LARC HAMPTON VA. 23665.

FERNANDEZ G. (F)/ DRET.SDR.G63, 26 Bd Victor, 75015 PARIS.

FFOWCS WILLIAMS J.E. (GB)/ Univ Engng Depart, Trumpington Street, Cambridge CB2 1PZ.

FILIPPI P.J.T. (F)/ L.M.A. - CNRS Marseille, 31 Ch Joseph Aiguier, 13277 MARSEILLE CEDEX 09.

FITZPATRICK J.A. (IRL)/ Trinity College, Dept Mech Engng, Dublin 2.

FORESTIER B.E. (F)/ I.M.F.M., LA 03, 1 Rue Honnorat, 13003 MARSEILLE.

FUMOUX Ch. (F)/ BERTIN et Cie, 3 Allée Gabriel Voisin, BP 3, 78373 PLAISIR CEDEX.

GARCIA A. (F)/ THOMSON CSF, DASM, Lab LTS, BP 53, 06802 CAGNES SUR MER.

GAY B. (F)/ E.C.L., Mécanique des Fluides, 69131 ECULLY CEDEX.

GEORGE A. (USA)/ Cornell University, 105 Upson Hall, Ithaca, N.Y.

GINSBERG J. (USA)/ School of Mech. Engng., Georgia Inst. of Tech., Atlanta, Georgia 30332.

GROSCHE F.R. (RFA)/ DFVLR Gottinger, IVK, Chaussée de Waterloo 72, B1640 RHODE ST GENESE - BELGIQUE

GUERIN B. (F)/ INSA de Lyon, LVA-Bat 303, 20 Av Albert Einstein, 69621 VILLEURBANNE CEDEX.

GUO Y.P. (GB)/ Cambridge Univ, Engng Depart, Trumpington Street, CAMBRIDGE CB2 1PZ.

HAERTIG J. (F)/ I.S.L., 12 Rue de l'Industrie, 68300 Saint LOUIS.

HAMDI M.A. (F)/ U.T.C., Div Acoustique et Vibrations Industrielles, B.P. 233, 60206 COMPIEGNE CEDEX.

HANDLER R.A. (USA)/ Naval Research Lab, Code 5844, Washington DC 20375.

HANSON D. (USA)/ Hamilton Standard, Div of United Tech Corporation, Windsor Locks, Connecticut 06096.

HARDIN J. (USA)/ NASA Langley Research Center, Hampton Va 23665.

HAYEK S.I. (USA)/ Applied Res Lab, Penn State Univ, P.O. Box 30, State College, PA 16801.

HECKL M. (GB)/ Univ of Cambridge, Depart Engng, Trumpington Street, Cambridge CB2 1PZ.

HECKL M. (RFA)/ Institut fur Tech Akustik Sekr. TA 7, Tech Univ Berlin, Einstreinufer 27, D1000 BERLIN 10.

HILL Ch. (GB)/ Cambridge Univ, Engng Dept, Emmanuel College, Cambridge CB2 3AP.

HO C.M. (USA)/ Dept of Aerospace Engng, Univ of Southern California, LA. CAL 90089.

HOCQUET Ph. (F)/ Bassin d'Essais des Carènes de Paris, 8 Bd Victor, 75732 PARIS CEDEX 15

HODGES Ch. (GB)/ TOPEXPRESS LTD, 13.14 Round Church St., Cambridge CB5 8AD.

HUERRE P. (USA)/ Univ. of Southern California, Dept of Aerospace Engng, Los Angeles, California 90007.

HUGHES I. (GB)/ Cambridge Univ, Engng Dept, Trumpington Street, Cambridge CB2 1PZ.

JOHANNESEN N. (GB)/ Dept of Aeronautical Engng, Univ of Manchester, Manchester M13 9PL.

JOUAILLEC F. (F)/ O.N.E.R.A., 29 Av de la Division Leclerc, 92320 CHATILLON SOUS BAGNEUX.

JULIENNE A. (F)/ O.N.E.R.A., BP 72, 92322 CHATILLON CEDEX.

JUVE D. (F)/ E.C.L., Centre Acoustique, 69131 ECULLY CEDEX.

KAMBE T. (JAPAN)/ Dept of Physics, Faculty of Science, Univ Tokyo, Hongo, Bunkyo-ku, Tokyo 113.

KERSCHEN E.J. (USA)/ Univ of Arizona, Aerospace and Mech Engng Depart, Tucson, Arizona 85721.

KO S.H. (USA)/ Dept of the Navy, Naval Underwater Systems Center, New London CT 06320.

LAAKE A. (RFA)/ Max-Planck-Institut fur Stromungsforschung, Bottingerstr. 4.8, D.3400 Gottingen.

LAGIER M. (F)/ THOMSON CSF, Div. Activités Sous Marines, Chemin des Travails, 06802 CAGNES SUR MER.

LAM K.T. (HONG KONG)/ Acoustics Lab, Dept Mech Engng, Univ of Hong Kong.

LARSEN P.N. (DK)/ Industrial Acoustics Lab, Tech Univ Denmark, Building 352, DK-2800 LYNGBY.

LAVERDANT A.M. (F)/ O.N.E.R.A., 29 Av de la Division Leclerc, 92322 CHATILLON SOUS BANGEUX.

LEEHEY P. (GB)/ Office of Naval Research, London, 223 Old Marylebone Road, London NW1 5TH.

LEGOFF J.P. (F)/ D.R.E.T. SDR/G6, 26 Bd Victor, 75996 PARIS ARMEES.

LE GUET P-L. (F)/ Bassin d'Essais des Carènes de Paris, Lab d'Hydrodynamique Navale, 8 bd Victor, 75732 PARIS CEDEX 15.

LESUEUR C. (F)/ I.N.S.A. de LYON, LVA-Bat.303, 69621 VILLEURBANNE CEDEX.

LEVINE H. (USA)/ Dept of Math, Stanford Univ, Stanford California 94305.

LUZZATO E. (F)/ E.D.F. DER, Serv A.E.E., Départ Acoustique, 1 Av du Général de Gaulle, 92141 CLAMART CEDEX.

LYAMSHEV L.M. (URSS)/ Joint Scientific Council on Physical and Tech Acoustics, USSR Academy of Sciences, Shvernik 4, Moscow V-36.

MAHAN J.R. (USA)/ College of Engng, Virginia Polytechnic Inst. and State Univ., Depart of Mech Engng, Blacksburg, Virginia 24061.

MAIDANIK G. (USA)/ David Taylor Naval Ship R and D Center, Bethesda, Maryland 20085-5000.

MARQUETTE G.R.B. (F)/ BERTIN et Cie, 3 Allée Gabriel Voisin, BP 3, 78373 PLAISIR CEDEX.

MATHIEU J. (F)/ E.C.L., Mécanique des Fluides, 69131 ECULLY CEDEX.

MEIER G.E.A. (RFA)/ Max-Planck-Institut fur stromungsforschung, Bottingerstr. 4.8, D.3400 Gottingen.

METCALFE R. (USA)/ Flow Industries Inc., 21414, 68th Ave So. Kent, WA 98032.

MICHALKE A. (RFA)/ Hermann-Fottinger-Institut fur Thermo-und Fluiddynamik Technische Univ. Berlin, Strasse des 17. Juni 135, D1000 BERLIN 12.

MOHRING W. (RFA)/ Max-Planck-Inst. Stromungsforsch., Bottingerstr. 4-8, D3400 Gottingen.

MULLER E.A. (RFA)/ Max-Planck-Institut fur stromungsforschung, Bunsenstrass 10, D3400 Gottingen.

NARAYANAN S. (RFA)/ Institut fur Technische Akustik, TU Berlin, Einsteinufer 27, 1000 BERLIN 10.

NELSON Ph.A. (GB)/ I.S.V.R., Univ of Southampton, S09 5NH.

OBERMEIER F. (RFA)/ Max-Planck-Institut fur Stromungsforschung, Bottingerstr. 4.8, D3400 Gottingen.

OPPENEAU J.C. (F)/ Ministère de l'Environnement - Cabinet, 14 Bd du Général Leclerc, 92524 NEUILLY SUR SEINE CEDEX.

PARKER R. (GB)/ Swansea Univ, Mech Engng, Singleton Park, Swansea W. Glam 5AZ 8PP.

PATRICK W.P. (USA)/ United Tech Research Center, East Hartford, Connecticut 06108.

PAUZIN S. (F)/ ONERA-CERT-DERMES, 2 Av E. Belin, 31055 TOULOUSE CEDEX.

PERRAUD J.C. (F)/ O.N.E.R.A., BP 72, 92322 CHATILLON CEDEX.

PERRET R. (F)/ NEYRTEC, Etablissement Grenoble, ALSTHOM-ATLANTIQUE, BP 61 X, 38041 GRENOBLE CEDEX.

PETITJEAN A. (F)/ E.D.F., DER. Serv A.E.E. Dept Acoustique, 1 Av du Général de Gaulle, 92141 CLAMART CEDEX.

PIRAUX J. (F)/ L.M.A. - CNRS Marseille, 31 Ch Joseph Aiguier, 13277 MARSEILLE CEDEX 09.

POWELL A. (USA)/ David Taylor Naval Ship R and D Center, Bethesda, Maryland 20084.

PURSHOUSE M. (SCOTLAND)/ YARD Ltd, Charing Cross Tower, Glasgow G24PP.

RAILLON L. (F)/ THOMSON CSF, DASM, Lab LTS, BP 53, 06802 CAGNES SUR MER

REBER J.D. (F)/ TECHNICATOME, Serv Mesure Vibration Informatique, Et d Cadarache, 13115 Saint Paul Lez Durance.

REBILLAT J.C. (F)/ C.E.A.T. Poitiers, 43 Rte de l'Aérodrome, 86000 POITIERS.

REETHOF G. (USA)/ Noise Control Lab, The Penn State Univ, 213 Engng Unit E, Univ Park, PA 16802.

REISCHMAN M.M. (USA)/ Office of Naval Research, Code 432F, Arlington, VA 22217.

RICHARZ W. (CAN)/ Dept Mech Aeronautical Engng, Carleton Univ, Ottawa Ontario.

RIENSTRA S.W. (NL)/ National Aerospace Lab, P.O. Box 90502, 1006 BM Amsterdam.

ROBERT G. (F)/ E.C.L., Centre Acoustique, 69131 ECULLY CEDEX.

ROEBUCK I. (GB)/ Admiralty Research Establishment, Portland, Durset DT5 2JS.

ROGER M. (F)/ Centre Acoustique, Ecole Centrale de Lyon, 36 Av Guy de Collongue, BP 163, 69131 ECULLY CEDEX.

RONNEBERGER D. (RFA)/ 3. Physikalisches Institut, Burgerstr. 42.44, D3400 Gottingen.

ROSS C. (GB)/ TOPEXPRESS LTD, 13 Round Church Street, Cambridge CB5 8AD.

SCHAAF K. (RFA)/ 3. Physikalisches Institut, Burgerstr. 42.44, D3400 Gottingen.

SCOTT J. (GB)/ TOPEXPRESS LTD, 13 Round Church Street, Cambridge CB5 8AD.

SEVIK M. (USA)/ David Taylor Njaval Ship R and D Center, Bethesda, Maryland 20084.

SILHOUETTE L. (F)/ CERDAN.DCAN Toulon, BP 77, 83800 TOULON NAVAL.

SMADJA P. (F)/ E.D.F., DER, Serv AEE, Départ Acoustique, 1 Av du Général de Gaulle, 92141 CLAMART CEDEX.

SMITH D.J. (GB)/ Mech of Fluids Lab, Univ of Manchester, Barton Airport, ECCLES, Manchester M30 7RU.

STOKES A.N. (GB)/ Dept of Mech Engng, Univ College of Swansea SA 28PP SWANSEA-WALES UK.

SUNYACH M. (F)/ E.C.L., Centre Acoustique, 69131 ECULLY CEDEX.

SZUMONWSKI A. (PL)/Politechnika Warszawska, ul. Nowowiejska 24, PL.00665 Warszawa.

TOCQUET B. (F)/ G.E.R.D.S.M., Le Brusc, Ch de la Gardiole, 83140 SIX FOURS LES PLAGES.

TOURRET J. (F)/ C.E.T.I.M., BP 67, 60304 SENLIS.

TRAN VAN NHIEU M. (F)/ SINTRA, 1 Av Aristide Briand, 94117 ARCUEIL.

TRUMPFF B. (F)/ THOMSON CSF, Div. Activités Sous Marines, Chemin des Travails, 06802 CAGNES SUR MER.

VASUDEVAN M.S. (GB)/ Institute of Sound and Vibration Research, The University, Southampton SO9 5NH

VILLE J.M. (F)/ U.T.C., Div Acoustique et Vibrations Industrielles, B.P. 233, 60206 COMPIEGNE CEDEX.

VILLOUVIER V. (F)/ E.D.F. DER, Serv AEE, Départ Acoustique, 1 Av du Général de Gaulle, 92141 CLAMART CEDEX.

WELSH M.C. (AUST)/ CSIRO, Div of Energy Tech, P.O. Box 26, Highett, 3190.

WETTA P. (F)/ METRAVIB, Section Acoustique, 64 Ch des Mouilles, BP 182, 69132 ECULLY CEDEX.

YOSHIDA H. (JAPAN)/ Dept of Mech Engng, Chiba Univ, Yayoi-Cho 1.33, Chiba-Shi 260.

ZHAO Keyong (CHINE)/ Institut d'Acoustique, Academia Sinica, 5 Zhongguancunlu, P.B. 2712, Beijin.

Contents

SESSION IV - "ROTOR NOISE"
Chairman : Dr. F. D'AMBRA, FRANCE

SESSION V - "THERMO ACOUSTICS"
Chairman : Prof. S. CANDEL, FRANCE

SESSION VI - "COMPUTATIONAL ACOUSTICS"
Chairman : Prof. H. LEVINE, U.S.A.

SESSION VII - "HYDROACOUSTICS"
Chairman : Prof. P. LEEHEY, U.S.A.

SESSION VIII - "HYDROACOUSTICS (C' nd)" :
Chairman : Dr. G. DI NICOLA, FRANCE

SESSION IX - "FLOW NOISE"
Chairman : Prof. F.OBERMEIER, R.F.A.

Participants of the Symposium

Diffraction

Diffraction and Radiation

Harold LEVINE

Department of Mathematics
Stanford University

The subject and its evolution

Diffraction relates, in general terms, to a characteristic wave phenomenon within regions of space not directly insonified or illuminated by a given source < e.g., the shadow of an object interposed in a beam of light > ; and is manifest by a non-zero level of excitation in such regions. This, and the kindred phenomena of reflection and refraction, is explicable by means of superposed wave functions that are separately linked with the primary or given source and the secondary effect of its environment < resulting from the presence, say, of a material object nearby > . The classical development of diffraction theory envisaged configurations without any relative motion between the primary source and its surrondings ; and invoked the abstractions inherent in time-periodic or aperiodic point sources and plane waves for purposes of mathematical simplicity. Thus, Sommerfeld obtained the first explicit and exact analytical solution for a diffraction problem in linear wave theory, relating to a plane wave incident upon a thin opaque semi-infinite screen with a straight edge ; afterwards, Macdonald, Herglotz and others derived the corresponding solution for a point source and a wedge formed by two semi-infinite intersecting plane surfaces.

The analysis of excitations due to moving point and distributed sources, without regard for any irregularities or structures in their vicinity, was initiated in the context of electromagnetic fields by Lienard and Wiechert and drew less attention when a

Aero- and Hydro-Acoustics
IUTAM Symposium Lyon 1985
Editors: G. Comte-Bellot and J. E. Ffowcs Williams

relativistic modelling of extended charge distributions proved difficult. After the discovery of Cerenkov radiation in dielectrics, the analogue of a characteristic large scale disturbance due to a small aerodynamic source in uniform supersonic motion, the study of interactions between moving sources and their environment greatly intensified. The quantitative description, by Tamm and Frank, of Cerenkov radiation from a moving point charge inspired numerous Russian publications on moving source problems which introduced new terminologies, such as diffraction and transition radiation. Diffraction radiation is attributable, in particular, to the electric current induced on the surface of a highly conducting body as a consequence of an external charge in motion < or current > nearby ; unsteadiness in time < and variability with location > typifies the surface current distribution, regardless of the manner in which the external charge moves. Since a point charge of fixed strength in uniform motion at a constant speed less than that of light in its surroundings merely excites a local field and does not transfer energy to distant locations it is the indirect effect of diffraction which accounts, in this instance, for the whole of the radiation. There are various prototypes of transition radiation, associated with the passage of a source < mass, force, thermal,... > through an interface between differently constituted media, across a region of turbulence or at a variable distance from a boundary surface.

In the matter of adapting Lighthill's acoustic analogy so as to account for the effect of structures on a primary source < namely, a region of turbulent motion > , Ffowcs Williams and Hall considered a fixed point quadrupole close by the edge of a thin plate and predicted, on the basis of a locally incompressible and inviscid flow near the edge, an energy output considerably above that which the same quadrupole would generate as an isolated source. Ffowcs Williams later encouraged the author to analyse the coupling between a moving quadrupole and the edge, which constitutes a problem of diffraction radiation ; and this has been followed by numerous investigations that bear on moving sources and passive or active environments in the context of fluid/acoustic configurations. Progress is evident as regards the features of diffraction radiation for streamlined bodies, rather than point sources, and especially for aeroacoustic source models of unsteady flows, wherein line vortices and quadrupoles generate sound due to their interaction with a solid surface. Thus, Howe presented the theory of diffraction radiation which pertains to a line vortex of fixed strength --the primary source-- in uniform translatory motion, at speed v_1, parallel to a rigid semi-infinite plate ; a parallel mean flow, at speed v_2, on both sides of the plate is taken into account, with an anticipated difference in the resultant features according as $v_1 \gtrless v_2$. The application of a Kutta condition at the trailing edge, which augments the secondary < image > source in the plate by a distribution of < vortical > sources along the downstream extension of the plate, is shown to significantly alter the magnitude of the forward radiation field.

The literature on diffraction radiation, surveyed and created some years ago by Bolotovskii and Voskresenskii, provides analytical results for several configurations, involving a single half plane, a pair of parallel half planes and a regular lattice or comb of such planes ; the primary source is either a uniformly charged line filament or a point charge in steady non-accelerated motion. Yates has developed the analogous theory for line and point charges in the exterior of a perfectly conducting wedge. Evidently, the acoustical counterparts of these problems may be described with the same detail and precision, which is to say, for the entire subsonic range of source speeds. Shegurova adopted a finite scale model for the primary source of an acoustic disturbance, namely a slender and rigid axisymmetric body ; and assumed that the body moves steadily, in the direction of its axis, past a semi infinite plane. There is no essential difference, as regards the pattern and frequency spectrum of diffraction radiation in this instance, between point and finite scale sources, apart from a shape factor which relates to the Fourier transform of the cross-sectional area of the body.

Some theoretical results which concern sound radiation in fluids wherein a primary source travels along a curved path and interacts with a solid surface are available ; thus, Crighton considered a single < line > vortex whose trajectory loops around the edge of a semi-infinite plane while Cannel and Ffowcs Williams gave attention to vortex pairs that exit from a parallel-sided semi-infinite duct. Away from the edge or duct opening the vortices follow virtually straight line paths parallel to rigid boundaries and remain acoustically silent ; the radiation emitted while the vortices proceed along curved paths in the vicinity of the edges may be estimated for slightly compressible fluids by matching an inner incompressible flow to an outer wave field.

The analytical techniques

To sketch some mathematical aspects of diffraction radiation and indicate efficient means for the relevant analysis, consider a general monopolar source whose prescribed orbit takes it past a fixed body ; and suppose that the pressure in the homogeneous medium of density ρ_0 satisfies the linear wave equation

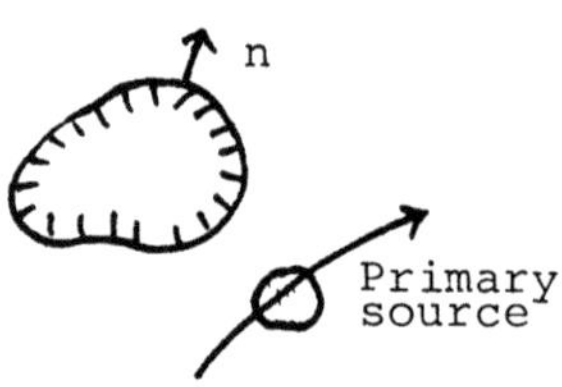

$$\left(\nabla^2 - \frac{1}{c^2}\frac{\partial^2}{\partial t^2}\right) p(\underline{r},t) = -\frac{\partial Q(\underline{r},t)}{\partial t}$$

where $Q(\underline{r},t)$ describes the mobile and extended source. Let

$$p(\underline{r},t) = p_P(\underline{r},t) + p_S(\underline{r},t)$$

with the designations P and S for the primary and secondary components, the former being characteristic of the given source as an isolated entity and the latter, specified by the homogeneous wave equation

$$(\nabla^2 - \frac{1}{c^2}\frac{\partial^2}{\partial t^2})\, p_S(\underline{r},t) = 0$$

being that which represents the effect of the body on the exterior field. If the body is a rigid one, these components of the total pressure are mutually involved in the boundary condition

$$\frac{\partial}{\partial n}\,(p_P(\underline{r},t) + p_S(\underline{r},t)) = 0$$

at the surface ; hence, given $p_P(\underline{r},t)$, the determination of p_S $(\underline{r},t)$becomes a matter of finding that solution of the homogeneous wave equation which has a prescribed normal derivative at the boundary surface and an appropriate far field behavior.

Taking account of the identity

$$p(\underline{r},t)\,\frac{Q(\underline{r},t)}{\rho_0} = \nabla\,.\,(p\underline{v}) + \frac{\partial}{\partial t}\,(\frac{1}{2}\,\rho_0 v^2 + \frac{1}{2}\,\frac{p^2}{\rho_0 c^2})$$

Where $\underline{S} = p\underline{v}$ describes the acoustical energy flux vector, it appears that an alternative to flux integration over a remote control surface for specifying the total amount of radiated energy, E, is provided by the integral

$$E = \frac{1}{\rho_0}\int_{-\infty}^{\infty} p(\underline{r},t)\; Q\;(\underline{r},t),\; d\underline{r}\; dt$$

within the primary source domain itself.

An explicit realization of the secondary or response function $p_S(\underline{r},t)$ is made difficult, in particular, by the relatively intricate form of the functions $p_p(\underline{r},t)$ which describe moving sources, as compared to the plane and spherical wave forms which typify the primary sources in classical diffraction theory. The solutions found thus far rest on especially simple selection of source characteristics and shapes of diffracting surfaces ; point-like sources with fixed strength and thin semi-infinite diffracting screens offer a manageable schematization for detailed analysis. Two or three dimensional problems involving a plane screen may be usefully recast in terms of an integral equation thereupon, and the resources of complex analysis are available for its solution. A precise calculation of the total energy radiated on passage of an otherwise silent monopolar source by a rigid screen is feasible ; here

$$Q(\underline{r},t) = q \quad \delta(\underline{r} - \underline{r}_P\ (t)), \ \frac{d^2\underline{r}_P}{dt^2} = 0 \quad , \quad v = \left|\frac{d\underline{r}_P}{dt}\right| < c$$

and the single integral representation

$$E = \frac{q}{\rho_0} \int_{-\infty}^{\infty} p_S\ (\underline{r}_P(t),\ t)\ dt$$

relates to the unsteady secondary pressure at the instantaneous primary source location. When the source trajectory lies in a plane normal to the edge of the screen, the radiated energy varies inversely with respect to the impact distance d from the edge and is independent of the angle of the trajectory with respect to the plane of the screen.

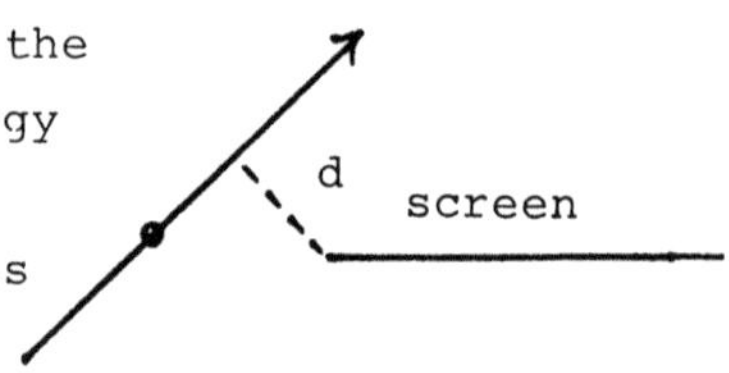

A simple Doppler factor of the form

$$\frac{f(v/c)}{(1-\ v^2/c^2)^n}$$

is applicable, after specifying the integer exponent n and the polynomial f, to the monopolar, dipolar,... source types.

Yates has successfully dealt with a more complicated boundary value problem which concerns diffraction radiation from a right-angled wedge ; in this configuration a coupled pair of integral equations at the respective wedge faces present themselves and the requisite complex analysis is non-standard. He obtains a definite integral representation for the total energy radiated,

$$E = \frac{q^2}{2\pi d \cosh\beta} \int_{\beta}^{\infty} \frac{\cosh\left(\frac{2}{3}\xi\right)\left[\cosh 2\xi - \cosh^2\beta\right]}{\sinh^3\xi \sqrt{\cosh^2\xi - \cosh^2\beta}}\, d\xi$$

where

$$\cosh\beta = \frac{c}{v},$$

on the assumption that the linear trajectory of a point charge lies in a plane normal to the vertex of a wedge whose faces are perfectly conducting. It may thus be noted that the measures of radiated energy for screen and wedge configurations share a common dependence on the impact distance and an insensitivity to the orientation of the trajectory for the primary point source ; however, a contrasting feature appears with respect to the dependence on source speed, inasmuch as a simple and explicit Doppler screen factor is replaced by a complicated function of the speed in the corresponding wedge factor. Asymptotic representations for the radiation field, which apply at large distances from the screen edge or wedge vertex and describe both the directivity patterns and their frequency variations, are known and prove to conform with the anticipated geometrical optics pattern at high source speeds.

Approximation procedures need to be invoked at an early stage when the diffracting surfaces are neither plane or wedge-shaped and the technicalities that stand in the way of a complete analysis are manifest ; for example, in their study of acoustical diffraction radiation generated by the motion of a point source near a rough surface, Pavlov and Sukhorukov employ a perturbation technique that rests on the hypothesis of a surface profile with but small deviations from an ideally smooth plane. The same authors investigate the transition radiation resulting from the passage of a point heat source across a spherical interface between regions with different sound speeds ; here it is the restriction to long or short wave lengths compared with the sphere diameter, respectively, that precedes the deduction of approximate analytical results.

Recent studies whose aim is to calculate and interpret, for model radiating systems, the effect of changes in source or material parameters may eventually be incorporated into a more comprehensive description of source interaction problems. Thus, Bolotovskii, Davydov and Rok have already derived some exact results in cases of radiation from point sources with a continuously variable velocity or dipole moment and contrasted the effects which accompany rapid or slow changes in these parameters ; another example pertains to transition radiation at a diffuse, rather than sharp, boundary between two media. The joint specification of a variable lifetime and velocity for the primary source which enters into a composite radiating system and, eventually, the determination of self-consistent motions and field reactions on moving sources offer challenging problems for future consideration.

References

1. Bolotovskii , B.M. and Voskresenskii , G.V. Diffraction radiation, Sov. Phys. Usp., 9 (1966) 73.

2. Bolotovskii , B.M., Davydov , V.A., Rok , V.E. Radiation of electromagnetic waves on instantaneous change of the state of the radiating system, Sov. Phys. Usp., 21 (1978) 865-872.

3. Bolotovskii , B.M., Davydov , V.A., Rok , V.E. The emission of electromagnetic waves in the case of a smooth variation of parameters of a radiating systems, Sov. Phys. Usp., 25 (1982) 167-175.

4. Cannell , P. and Ffowcs Williams , J.E. Radiation from line vortex filaments exhausting from a two-dimensional semi-infinite duct. J. Fluid Mech. 58 (1973) 65-80.

5. Crighton , D.G. Radiation from vortex filament motion near a half-plane, J. Fluid Mech. 51 (1972) 357-362.

6. Howe , M.S. The effect of forward flight on the diffraction radiation of a high speed jet, Journal of Sound and Vibration, 50 (1977) 183-193.

7. Levine , H. Acoustical diffraction radiation, Philips Research Reports 30 (1975) 240-276.

8. Pavlov , V.I. and Sukhorukov , A.I. Transition radiation of sound by a mass source moving over a rough surface, Sov. Phys. Acoust. 29 (1983) 397-399.

9. Pavlov , V.I. and Sukhorukov , A.I. Transition radiation of sound by a source crossing a spherical interface, Sov. Phys. Acoust. 30 (1984) 146-148.

10. Shegurova , G.I. Radiation of sound by extended streamlined bodies moving past a half-plane. Sov. Phys. Acoust. 17 (1971) 237-240.

11. Yates , R.A. Some problems in electromagnetic diffraction by a right angle wedge, Thesis, Stanford University, (1976).

Influence of Airfoil Mean Loading on Convected Gust Interaction Noise

by

E.J. KERSCHEN and M.R. MYERS

Aerospace and Mechanical Engineering Department
University of Arizona
Tucson, Arizona 85721, U.S.A.

Abstract

A theoretical model is developed for the noise generated when a convected vortical or entropic gust encounters an airfoil at non-zero angle of attack. The analysis is based on Rapid Distortion Theory. High frequency gusts, whose wavelengths are short compared to the airfoil chord, are considered. The noise generation is shown to be concentrated near the airfoil leading edge. The level of the generated noise is increased by airfoil mean loading, with the appropriate scaling parameter being the local leading edge incidence angle. The trailing edge simply scatters the leading edge sound field, and here the mean loading effects scale on the airfoil total lift. Calculations are presented which illustrate that, at high frequencies, moderate levels of airfoil steady loading can dramatically increase the noise produced by airfoil convected gust interactions.

Introduction

The first attempts to include steady loading effects in airfoil convected gust interaction noise models were due to Ffowcs Williams and Hawkings[1,2]. They developed an acoustic analogy formulation in which the interaction between the convected disturbances and the mean flow gradients appeared in the form of a quadrupole source term. A number of authors used this formulation to predict mean loading effects on turbomachinery noise; a typical model is presented in Ref. 3.

The acoustic analogy approach is formally exact, but its use requires specification of the unsteady velocity fields which appear in the source terms. This information is not generally available, and thus approximate expressions must be utilized. In the acoustic analogy models discussed above, the mean flow disturbance created by the airfoil was represented by that due to a point force. This approximation may be useful for the case of gust wavelengths that are long compared to the airfoil chord, but for the case of short wavelength gusts the details of the flow in the

Aero- and Hydro-Acoustics
IUTAM Symposium Lyon 1985
Editors: G. Comte-Bellot and J. E. Ffowcs Williams

vicinity of the airfoil must be considered.

The quadrupole acoustic analogy formulation was a significant development in airfoil noise prediction methods, and through it information has been obtained for a large number of applications. However, for noncompact situations, the source term modeling is difficult and thus the accuracy of the results is less certain. The present analysis involves highly noncompact sources, since the gust wavelength is assumed small compared to the airfoil chord.

Recently, Goldstein[4] has developed an alternative formulation for the interaction of aerodynamic bodies with convected disturbances. He utilizes the fact that the problem can be viewed as an inviscid linear disturbance of the nonuniform steady mean flow past the body. In terms of the present application, there are two advantages to Goldstein's approach. The first is that the source terms which arise in Goldstein's formulation are known functions, in contrast to the acoustic analogy approach where heuristic modeling is required. In addition, since the linearization is about the nonuniform mean flow, the acoustic propagation effects appear explicitly in the wave operator rather than as additional source terms.

We utilize Goldstein's formulation to analyze the far field noise generated by the interaction of a convected gust with a flat plate airfoil at angle of attack, α, to a subsonic stream. A complete theory, which includes the first order perturbations to the noise field produced by the effects of nonuniform mean flow, is developed. The theory is a systematic asymptotic expansion which contains no heuristic modeling. Many features of the earlier acoustic analogy models are recovered, confirming the physical mechanisms identified by Ffowcs Williams. New features of the physics are also found, illustrating the importance of propagation effects which were ignored by the acoustic analogy models.

Analysis

The analysis considers linear, three-dimensional disturbances of a two-dimensional, inviscid, irrotational, compressible mean flow. Convected velocity and entropy disturbances are assumed to be present in the uniform flow far upstream of the airfoil. Since the disturbances are linear, we represent arbitrary upstream disturbances as a superpostition of harmonic components. Thus, the solution will be developed for a single harmonic component.

Goldstein[4] showed that the complete disturbance velocity can be expressed as $\mathbf{u} = \nabla G + v$, where $\mathbf{v}$ is a known function of the upstream convected gust and contains the additional vorticity generated by the interaction of the upstream velocity and entropy disturbances with the mean flow gradients. Substituting the expression for $\mathbf{u}$ into the continuity equation produces the following convected wave equation for G:

$$\frac{D_0}{Dt}\left(\frac{1}{c_0^2}\frac{D_0 G}{Dt}\right) - \frac{1}{\rho_0}\nabla\cdot(\rho_0\nabla G) = \frac{1}{\rho_0}\nabla\cdot\rho_0\mathbf{v} \qquad (1)$$

On the airfoil surface, the disturbance velocity potential satisfies the boundary condition $\partial G/\partial n = -\mathbf{v}\cdot\mathbf{n}$, where $\mathbf{n}$ is the unit normal vector. The radiation condition applies far from the airfoil. The pressure fluctuations are given by $p = -\rho_0 D_0 G/Dt$.

The above equation is linear, but for a general compressible mean flow the variable coefficients must be determined numerically. To allow the development of approximate closed form solutions, we utilize Prantl-Glauert theory for the two-dimensional mean flow description. We remove the harmonic dependence of G on time and the spanwise coordinate, and introduce the mean flow streamfunction and velocity potential as coordinates. Thus, the problem is reduced to solving a second order partial differential equation for the modified disturbance velocity potential $h(\phi,\psi)$. The reduced system is given in Refs. 5 and 6.

This reduced system is analyzed by a combination of singular perturbation and complex variable techniques. We assume that the wavelength of the disturbance is small compared to the airfoil chord ($k \gg 1$). The various regions involved in the asymptotic analysis are illustrated in Figure 1. Local regions, whose size scales on the gust wavelength, are present at the leading and trailing edges. Away from the airfoil edges, there is an outer region in which the solution has a multiple scales behavior. In addition, behind the airfoil, a "transition" region extends over an angular domain of $O(1/\sqrt{k})$. In this transition region, the geometric acoustics outer solution gives way to a diffraction type solution.

The solution in the local leading edge region is developed utilizing Fourier transform techniques. Since the mean flow Mach number is O(1), wave propagation effects are important in this local region. However, we have the simplification that the airfoil appears semi-infinite. Thus the Wiener-Hopf technique is applicable. An additional assumption, that the shift of the mean flow stagnation point from the leading edge is small compared to the gust wavelength ($\alpha^2 k \ll 1$), is introduced in analyzing the leading edge region. The solution can then be developed in a perturbation series of the form

$$h(\phi,\psi) = 1/k\,(h_0 + \alpha\sqrt{k}\,(h_1 + h_2) + O(\alpha^2 k)) \tag{2}$$

The function h_0 is the standard leading edge field for the case of zero mean loading. We have separated the $O(\alpha\sqrt{k})$ correction terms into two functions h_1 and h_2. The function h_1 includes two physical effects. The first is the interaction between the field h_0 and the local variations in mean flow velocity and speed of sound. Since these local variations have the same length scale as the acoustic wave, this interaction results in additional far field sound radiation. In addition, h_1 also includes the effect of airfoil surface boundary condition modifications due to vorticity distortion by the mean flow gradients. The physical effects represented by h_1 have been neglected in models based on the acoustic analogy.

The field h_2 is generated by volume source terms that are similar to the acoustic analogy quadrupoles. However, our volume source terms include effects of density gradients which have been neglected in acoustic analogy models. The scattered quadrupoles considered by Ffowcs Williams and Hall[7] are also included in h_2.

The solution in the outer region has two components. The first is due to a distribution of "quadrupole" type volume sources. These sources generate local fluctuations, but make negligible contributions to the far field sound. The second component is the ray acoustic continuation of the asymptotic behavior of the local leading edge solution. This is the true sound field. The mean flow gradients produce significant sound field phase distortions which are captured quite naturally in our analysis. In order to obtain these phase distortions from an acoustic analogy model, quadrupole sources in addition to $\rho u_i u_j$ would have to be accounted for.

The ray acoustic outer solution is then scattered by the trailing edge. This process is analyzed by a local trailing edge solution developed via the Wiener-Hopf technique. The O(1) phase distortions discussed above significantly modify the scattered field produced by the trailing edge. The asymptotic behavior of the local trailing edge solution matches with an additional ray acoustics solution for directions outside the "transition" region, while the matching is with a diffraction type field in the transition region. Thus, the far field sound consists of the direct ray field generated by the leading edge and the scattered field from the trailing edge. Details of the analysis may be found in Refs. 5 and 8.

Numerical Results and Disucssion

Numerical results which illustrate the dominant features of the far field sound are presented below. It turns out that, except for extremely high frequencies, the scattered noise from the trailing edge noticeably modifies the leading edge fields. Thus, in examining the relative importance of the terms h_0, h_1 and h_2, we combine the relevant leading and trailing edge contributions. For all the calculations presented here, we have set the spanwise wavenumber k_z and the amplitude of the entropy disturbance equal to zero. The angle $\Lambda = \tan^{-1}(A_n/A_t)$ gives the direction of the gust velocity fluctuations with respect to the uniform flow at upstream infinity.

In Figure 2, we compare the relative contributions to the far field noise from the terms h_0, h_1 and h_2. The quantity A_i is equal to $(2\pi r)^{-1}\int|p_i(\theta)|\,d\theta$, where the integral is evaluated over a large circle of radius r and p_i is the pressure associated with h_i and its scattered field from the trailing edge. Thus, the ratios A_1/A_0 and A_2/A_0 provide an approximate measure of the relative size of the coefficients of $\alpha\sqrt{k}$ in the expansion for the far field pressure. It can be seen that both ratios are significantly greater than one, indicating that the far field sound is sensitive to airfoil angle of attack. It is also clear that the relative importance of h_1 and h_2 depends strongly on Mach number. In general, h_1 is somewhat larger at lower Mach numbers, while h_2 increases rapidly and dominates at near sonic conditions.

Figure 3 presents the variation in total acoustic power as a function of angle of attack for three different gust wavelengths. If the trailing edge contributions were neglected, our perturbation theory would predict a linear variation with a slope proportional to $1/\sqrt{\bar{K}}$. The curves in Figure 3 exhibit this general behavior, with mild modulations due to the scattering by the trailing edge. The magnitude of the increases are quite striking. The far field acoustic power for $\alpha = 6$ degrees is approximately three times that for $\alpha = 0$.

The analysis has considered a flat plate airfoil, but the general features are also applicable to a cambered airfoil without thickness. For high frequency gusts, the mean flow parameter which controls the acoustic power level is not the airfoil lift force, as assumed implicitly in the earlier acoustic analogy models, but rather the shift of the mean flow stagnation point from the leading edge. This result lends theoretical support to the experiments of Ginder and Newby[9], who found that aircraft fan broadband noise levels correlate much better on leading edge incidence angle than on fan total loading.

Acknowledgement

The work described herein was funded by NASA Lewis Research Center under grant NAG3-357, with Mr. John F. Groeneweg as grant monitor. We gratefully acknowledge this support. We also thank Prof. T. F. Balsa and other colleagues at the University of Arizona for many stimulating and productive discussions.

References

1. Ffowcs Williams, J.E. and Hawkings, D.L. Sound generation by turbulence and surfaces in arbitrary motion. Phil. Trans. R. Soc. **A 1151** (1969) 264-321.
2. Ffowcs Williams, J.E. and Hawkings, D.L. Theory relating to the noise of rotating machinery. J. Sound Vib. **10** (1969) 10-21.
3. Goldstein, M.E., Rosembaum, B.M. and Albers, L.U. Sound radiation from a high-speed axial-flow fan due to the inlet turbulence quadrupole interaction. NASA TN D-7667 (1974).
4. Goldstein, M.E. Unsteady vortical and entropic disturbances

round arbitrary obstacles. J. Fluid Mech. **89** (1978) 433-468.

5. Kerschen, E.J. and Myers, M.R. Effect of airfoil mean loading on convected gust interaction noise. AIAA Paper No. 84-2324 (1984).
6. Kerschen, E.J. and Myers, M.R. Equation governing linearized disturbances of a perfect-gas base flow. AIAA J. (to be submitted) (1985).
7. Ffowcs Williams, J.E. and Hall, L.H. Aerodynamic sound generation by turbulent flow in the vicinity of a scattering half plane. J. Fluid Mech. 40 (1970) 657-670.
8. Kerschen, E.J. and Myers, M.R. Effect of incidence angle on noise generated by airfoils interacting with high-frequency convected gusts. J. Fluid Mech. (to be submitted) (1985).
9. Ginder, R.B. and Newby, D.R. An improved correlation for the broadband noise of high speed fans. J. Aircraft **14** (1977) 844-849.

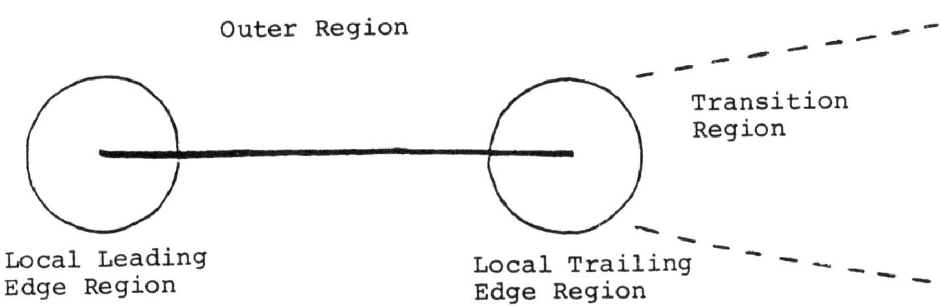

Figure 1. Flat plate airfoil with various regions illustrated

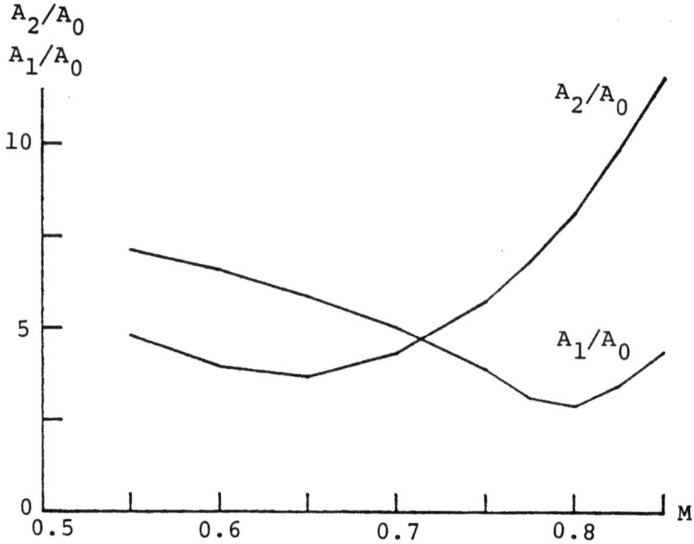

Fig.2. Magnitudes of h_1 and h_2 relative to h_0 as a function of Mach number. Λ = 75 deg.

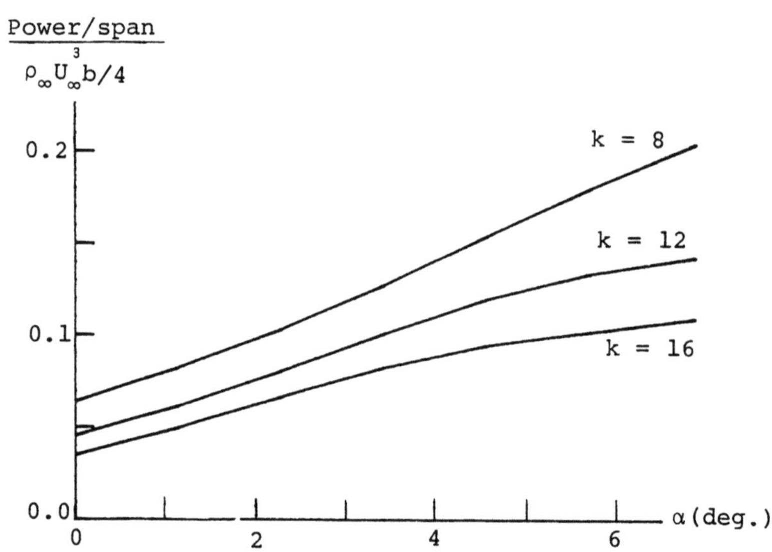

Fig.3. Far field acoustic power per unit span as a function of airfoil incidence angle. M = 0.7, Λ = 45 deg.

Acoustic Waves Emitted by Vortex-Body Interaction

T. KAMBE,* T. MINOTA** and Y. IKUSHIMA**

* Department of Physics, Faculty of Science, University of Tokyo
Hongo, Bunkyo-ku, Tokyo 113, Japan
** Department of Applied Science 36, Kyushu University
Hakozaki, Fukuoka 812, Japan

Summary
Acoustic emission by a vortex ring moving near a solid body is investigated both experimentally and theoretically for three typical configurations of the vortex-body system. In the first case, a vortex ring passes near by a circular cylinder whose axis is set at right angles to the vortex motion. In the remaining two cases, a vortex ring passes by a sharp edge of a flat plate, its width being comparable with the vortex size for the one case and semi-infinite for the other. This work presents a basis of comparison between the observations and the theoretical predictions.

1. An introduction and the related theories

A recent investigation of the acoustic emission from head-on collision of two vortex rings [1] implies that the intensity of the sound is proportional to U^8, where U is the translation velocity of a single vortex ring. This is consistent with the Lighthill's theory of the aerodynamic sound generation by turbulence in which U denotes a representative velocity of the turbulence. However, the sound emission by vortex-body interaction is more powerful at low Mach number flows. Influence of the solid bodies in a flow on the emission was first studied by Curle [2], who showed that the intensity is proportional to U^6. Later, Ffowcs Williams & Hall [3] and Crighton & Leppington [4] investigated the effect of a sharp edge and showed that the edge-flow interaction yields the intensity law proportional to U^5.

Our present study confirms these power laws (with some experimental dispersions) by using a vortex ring which moves near a solid body with or without a sharp edge. In addition, this study presents the wave profiles of the emission as well as the directionality pattern. For a body of a finite cross-section

Aero- and Hydro-Acoustics
IUTAM Symposium Lyon 1985
Editors: G. Comte-Bellot and J. E. Ffowcs Williams

such as a circular cylinder or a flat plate of a finite width, the directionality of the emission has a dipole character, as is known in the theory. However, in the present study which uses a vortex ring, the dipole axis rotates as the vortex position changes relative to the body. In the third case of the semi-infinite plate, the intensity exhibits a cardioid directionality pattern (without rotation).

Based on the Lighthill's equation, Curle [2] obtained the acoustic pressure at a point $\mathbf{x}=(x_i)$ in the far field, generated by the flow-body interaction at a time t, as

$$p(\boldsymbol{x}, t+\frac{r}{c}) = -\frac{1}{4\pi c}\frac{x_i}{r^2}\frac{d}{dt}F_i(t) \quad , \qquad r=|\boldsymbol{x}| \; , \tag{1}$$

where F_i is the total resultant force exerted on the body by the surrounding fluid which is characterized by the uniform density ρ_0 and the sound speed c. Thus the fluctuating force exerted on the fluid by the body, i.e. $-\mathbf{F}$, generates a dipole wave, whose axis is in the direction of the force.

Suppose that there exists a vortex motion near a cylindrical body of an arbitrary cross-section. According to Obermeier [5], the same acoustic pressure is expressed differently as

$$p(\boldsymbol{x}, t+\frac{r}{c}) = \frac{\rho_0}{4\pi c}\sum_{1,2}\frac{x_i}{r^2}\left(\frac{d}{dt}\right)^2 \int \psi_i(\boldsymbol{y})\,\omega_3(\boldsymbol{y},t)\,d\boldsymbol{y} \tag{2}$$

(in dimensional form) with the coefficient corrected, where the x_3-axis is taken in the direction of cylinder axis and ψ_i (i=1,2) are two-dimensional streamfunctions of potential flows in the i-th direction around the cylinder. When the vortex has a loop structure of strength Γ (like a vortex ring) with the curve C representing the loop, it can be shown that

$$\int \psi_i\,\omega_3\,d\boldsymbol{y} = \Gamma\,\Phi_i(t) \, , \qquad \Phi_i(t) = \int_S [\,\mathrm{rot}(\psi_i\mathbf{k})]_n\,dS \; ,$$

where $\mathbf{k}=(0,0,1)$ and S is an open surface having the curve C as its boundary. Here Φ_i is the volume flux of the potential flow ψ_i through the loop C. Thus the pressure (2) is reduced to

$$p(\boldsymbol{x}, t+\frac{r}{c}) = \frac{\rho_0\Gamma}{4\pi c}\,\frac{\sin\phi}{r}\left(\frac{d}{dt}\right)^2 \Phi_c \; , \tag{3}$$

where $\Phi_c = (\Phi_1 x_1 + \Phi_2 x_2)/(x_1^2 + x_2^2)^{1/2}$ and $\sin\phi = (x_1^2 + x_2^2)^{1/2}/r$.

When the vortex ring moves near the edge of a semi-infinite plate, it is found [6] that the acoustic pressure takes the form

$$p(x, t + \frac{r}{c}) = \frac{\rho_0 \Gamma}{\sqrt{2\pi^3}\, c} \frac{\sin\frac{1}{2}\theta \sqrt{\sin\phi}}{r} \left(\frac{d}{dt}\right)^{3/2} \Phi_e , \qquad (4)$$

where Φ_e is the volume flux (through the ring) of an imaginary potential flow around the edge, and the angles θ and ϕ are defined in Fig. 10.

2. Experimental setup

Vortex ring is produced by the same method as in [1,6]. When a shock wave emerges out of a nozzle (diameter 6mm), a vortex ring (its radius R about 4.7mm) is formed at the nozzle exit by the shock impulse. The vortex ring thus formed moves forward with a self-induced velocity U and interacts with a body placed near its orbit. An acoustic wave is excited by the interaction. Observation of the wave is made in an anechoic chamber of inner size 1.8m on all sides with four microphones set in the far field. Experimental arrangements of the nozzle and body are shown in Figs. 1, 7 and 10. The data acquisition and processing are made similarly as in [1,6].

3. Cylinder-vortex interaction

A perspective profile of the wave pressure generated by the vortex-cylinder interaction (Fig.1) is shown in Fig.2 for the vortex speed U=27m/s. This profile is obtained as follows.

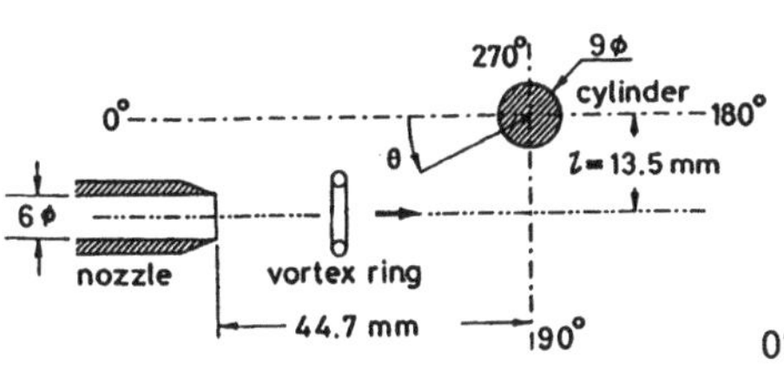

Fig.1.

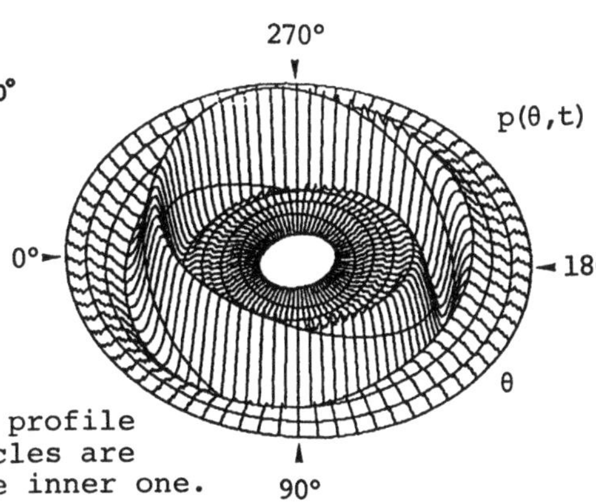

Fig.2. Perspective wave profile p(θ,t). The outer circles are earlier in time than the inner one.

First the wave pressure $p(t,\theta)$ observed at a fixed r is decomposed into Fourier series of cos $k\theta$ and sin $k\theta$ (k is an integer and θ is the angle of the polar coordinates in the plane $\phi=90°$ perpendicular to the cylinder axis). The calculation was done up to $k=2$, and it has been found that the principal components are the terms with $k=1$, namely the pressure is well approximated by the form

$$p(t,\theta) = a_1(t)\cos\theta + b_1(t)\sin\theta \qquad (5)$$

representing a dipole wave with a rotating axis. This is clearly seen in Fig.3 A-C (the radial coordinate being the pressure p at the direction θ), in which the vortex position backed by the retarded-time distance is depicted together. One observes a spiral wave which is related to the rotating dipole axis in Fig.2 (radial axis= t, vertical axis= the pressure p in the direction). Figure 4 shows the comparison in the absolute scale between the observed wave profiles (solid) and the predicted ones (broken) of the formula (2), where $\delta/R=0.15$ is assumed and a correction to the measured l is made, δ being the radius of the vortex core.

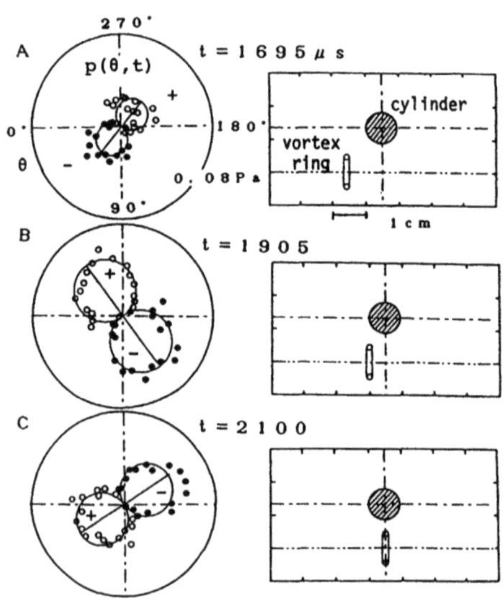

Fig.3.

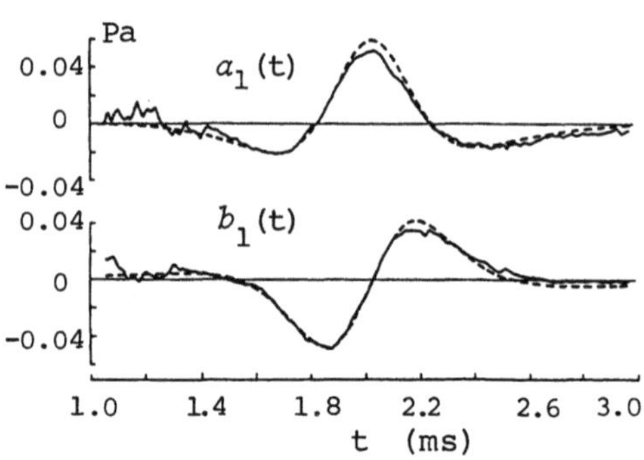

Fig.4. Dominant Fourier coefficients a_1, b_1 and theoretical prediction (broken curve).

Figure 5 gives the coefficient $-b_1(t)$ for various values of the vortex speed U. The pressure amplitudes Δp are plotted against U for $\theta = 70°$ ($\phi = 90°$) in figure 6, in order to see the scaling law by the relation

$$\log \Delta p = \beta(\theta) \log U + \gamma(\theta) \quad , \tag{6}$$

where the parameters β and γ are determined by the least square method. The value of β is found to be 3.2. This behavior is not far from the theoretical scaling law: $p \propto U^3$.

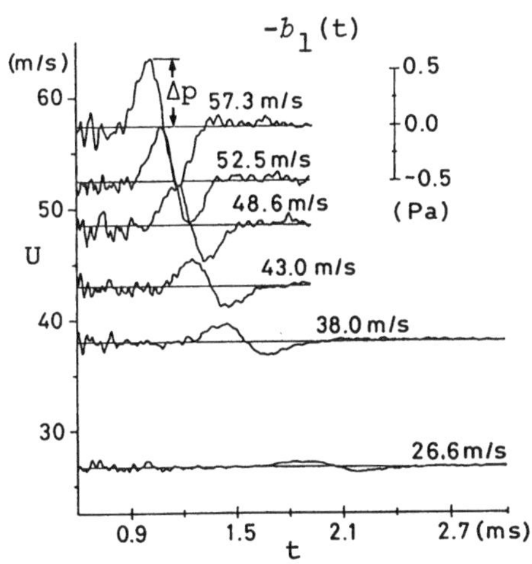

Fig.5. Fourier coefficient $-b_1(t)$ for various values of U.

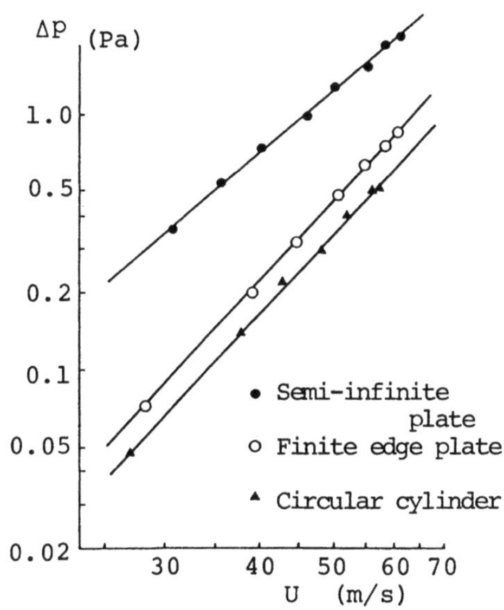

Fig.6. Scaling law: log(Δp) vs. log U.

4. An edge plate

Figure 8 shows a perspective wave generated by the vortex moving near the finite edge plate (Fig. 7), with U=27m/s. The coefficients $a_1(t)$ and $b_1(t)$ in the decomposition (5) for this case are given in Fig.9. It is remarkable that the $\sin\theta$ -

component is very small in contrast with the previous case. The acoustic pressure is observed for various values of U from 28 to 61 m/s. The pressure amplitudes Δp are plotted against U for $\theta = 140°$ ($\phi = 90°$) in Fig.6. The value of the slope β in the form (6) is found to be 3.3.

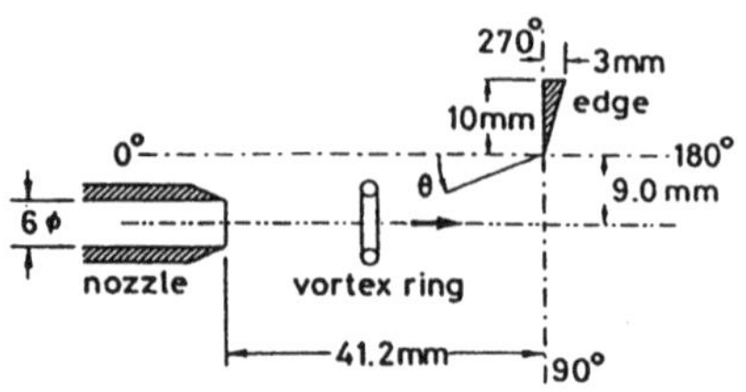

Fig.7.

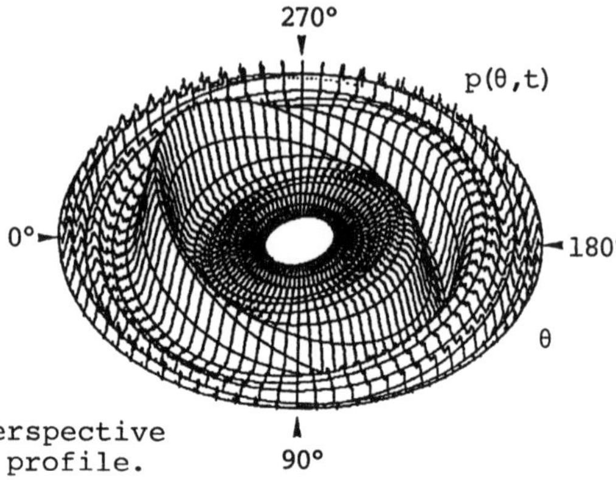

Fig.8. perspective wave profile.

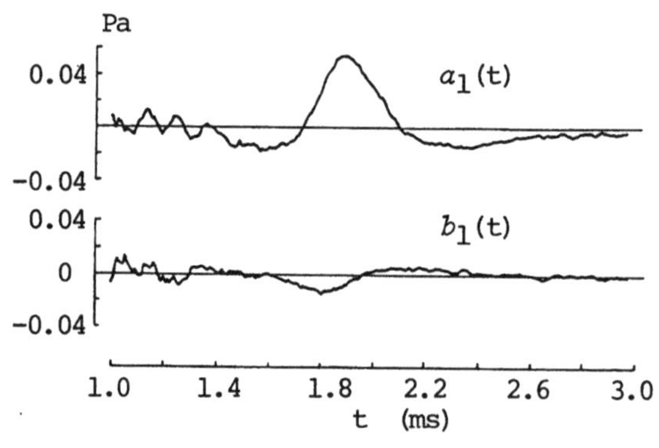

Fig.9. Fourier coefficients a_1 and b_1.

5. Semi-infinite plate [6]

Figure 11 is a perspective profile when the vortex moves perpendicularly to the semi-infinite plate (Fig.10). The main component of the observed wave is given by $p(t,\theta,\phi=90°) = c(t)\sin(\theta/2)$. Figure 12 compares the observed profile c(t) in the absolute scale (Pa) with the theoretical ones obtained from (4) with U=30m/s ($\phi=90°$, r=634mm) for $\delta/R=0.2$ (thin broken curve) and 0.3 (thick), δ being the radius of the vortex core.

An instantaneous directivity of the wave pressure is illustrated in Fig. 13, as in Fig. 3 for the dipole directivity. The open (positive) and filled (negative) circles show the observation, while the solid curve is $c(t)$ of Fig. 11 at t=1146 µs. The intensity shows the cardioid curve, $I \propto p^2 \propto \sin^2(\theta/2)$. In order to see the power law, pressure amplitude Δp is plotted against U for $\theta = 140°$ ($\phi = 90°$) in Fig. 6. The value of the slope β in the form (6) is found to be 2.53. Theoretical scaling law is $p \propto U^{2.5}$, which leads to the intensity law $I \propto U^5$.

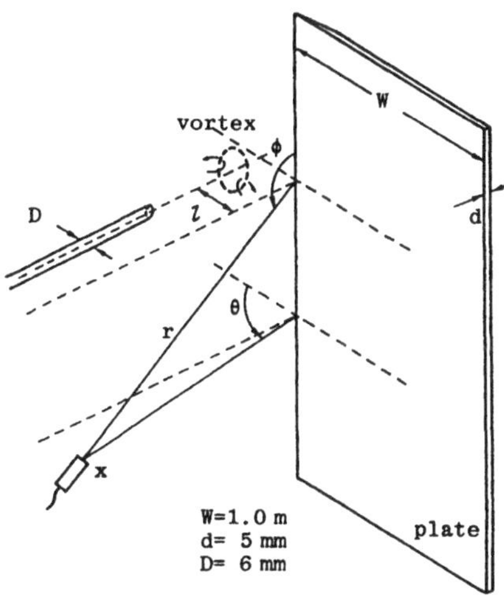

Fig.10.

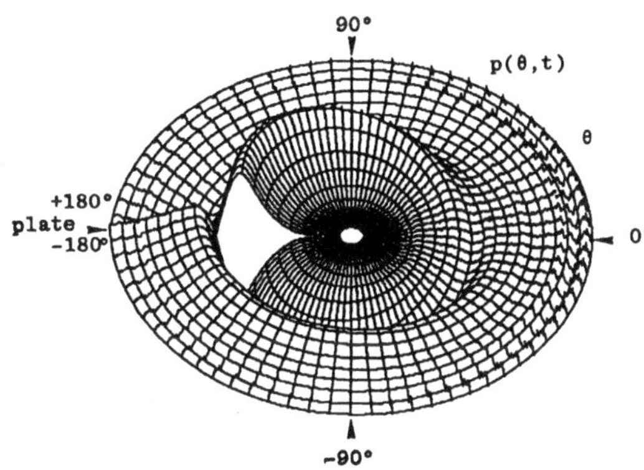

Fig.11. Perspective wave profile.

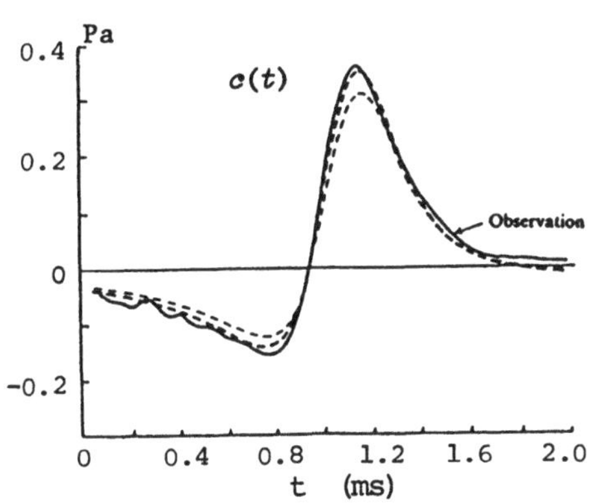

Fig.12. Comparison between observed $c(t)$ and predicted ones (two broken curves).

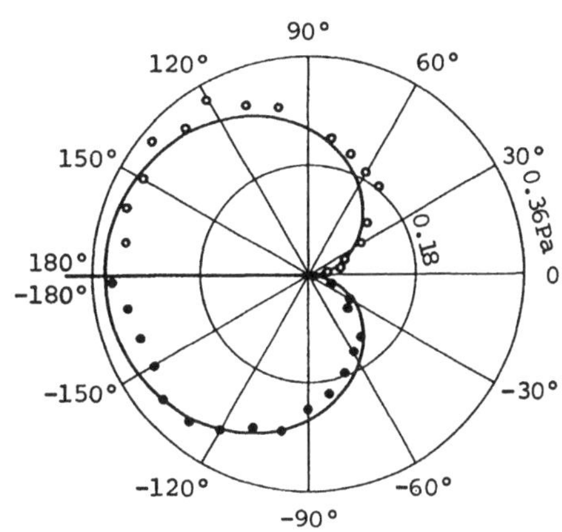

Fig.13. Directivity plot at t = 1146 µs.

References

[1] Kambe, T. & Minota, T. (1983) Proc. R. Soc. Lond. A **386**, 277-308.

[2] Curle, N. (1955) Proc. R. Soc. Lond. A **231**, 505-514.

[3] Ffowcs Williams, J.E. & Hall, L.H. (1970) J. Fluid Mech. **40**, 657-670.

[4] Crighton, D.G. & Leppington, F.G. (1970) J. Fluid Mech. **43**, 721-736.

[5] Obermeier, F. (1980) J. Sound Vib. **72**, 39-49.

[6] Kambe, T., Minota, T. & Ikushima, Y. (1985) J. Fluid Mech. **155**, 77-103.

Non-Steady Flow Over a Trailing Edge – An Experimental Study

K. SCHAAF, D. RONNEBERGER, W. PFLÜGER

Drittes Physikalisches Institut, University of Goettingen, Buergerstr. 42-44, D-3400 Goettingen, Federal Republic of Germany

Abstract

The forced non-steady velocity field of the flow over the edge of a back-facing step is measured by means of a Laser-Doppler-Anemometer. The velocity field is obtained as close as 0.2 mm away from the wall and we are able to resolve the Stokes layer of the incoming flow. The perturbation of the shear-layer is discussed and the vorticity in the downstream region is evaluated from the velocity field. Additionally the displacement of the separation streak line was visualized. The wavelength of the shear-layer perturbation as obtained from the streak line was found to be significantly smaller than that obtained from the velocity field.

Introduction

It is well established that steady or weakly non-steady flow over a trailing edge satisfies the Kutta condition at low Strouhal numbers $St = 2\pi f\theta_o/U_\infty$ (θ_o = momentum thickness at the edge). However, for higher Strouhal numbers, where amplification of the instability wave is still possible, both experimental and theoretical work have not yet yielded satisfying results.

Nevertheless, the exact knowledge of the perturbation vorticity field in the vicinity of the edge is important. It is absolutely necessary in order to predict e.g. the sound transmission through discontinuous cross sections of flow ducts, the impedance of orifices exposed to grazing flow, or the lift in unsteady aerodynamics.

Since non-intrusive techniques must be used for getting further insight, we used a Laser-Doppler-Anemometer as the tool to overcome restrictions of other methods such as hot wire anemometry or visualized streak-lines.

Aero- and Hydro-Acoustics
IUTAM Symposium Lyon 1985
Editors: G. Comte-Bellot and J. E. Ffowcs Williams

This paper reports on results of first experiments which were conducted in order to explore the measurement technique.

Experimental Set-Up

Fig. 1 shows a sketch of the water channel we used for our investigation. The flow enters the flow duct through a sintered bronze plate with highly uniform flow resistance. The cross section of the channel is 47 x 28 mm^2 upstream from the edge, and 150 x 28 mm^2 downstream. The mean velocity for the experiments was $U_\infty = 4.8 \times 10^{-2}$ m/s, the momentum thickness at the edge was θ_o = 1.4 mm, giving a Reynolds number $Re = \theta_o \cdot U_\infty / \nu = 63$.

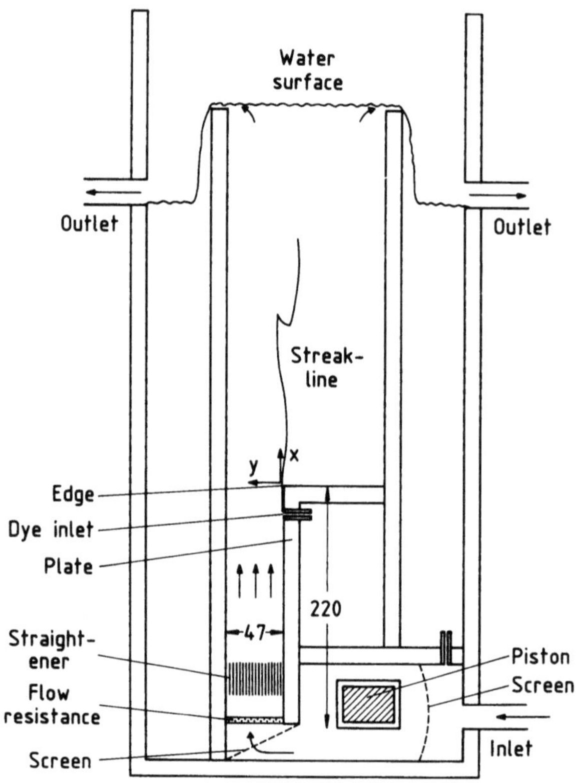

Fig. 1. Flow Channel

The unsteady forcing of the flow is achieved by a piston in one of the side-walls of the plenum. The unsteady flow behind the edge is visualized by means of a streak line emanating from a

1 mm dia dye inlet 20 mm upstream the edge; the amplitude and the frequency of the excitation were adjusted to yield a well behaved instability wave. The Strouhal number corresponding to this frequency turns out to be close to the Strouhal number of the maximum amplification (St = 0.08). The Laser-Doppler-Anemometer was self made and a tracker was used for the velocity measurement. The diameter of the measuring volume was approximately 0.1 mm, so we were able to measure the velocity as close as 0.2 mm away from the wall. By phase averaging we were able to resolve the perturbation velocity down to $4 \cdot 10^{-5}$ m/s, which is 10 % of the excitation velocity and 0.1 % of the mean velocity.

Discussion

Mean Flow

The mean flow velocity profile upstream the edge (Fig. 2), which proved to be independent of the superimposed unsteady velocity, changes from a Blasius-like profile (X = -30 mm) to a more

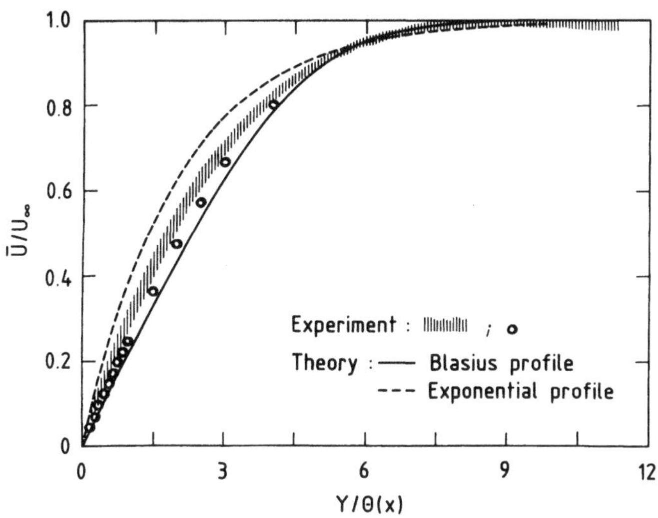

Fig. 2. Wall boundary layer profiles at different locations upstream the trailing edge.
▮▮▮ : -15, -10, -5, 0 mm; o : -30 mm.

exponential one. Behind the edge one observes the transition from the wall boundary layer profile to a symmetrical tanh-like profile (Fig. 3). Differences to the tanh-profile are still visible after 70 mm (i.e. after more than 1 wavelength of the

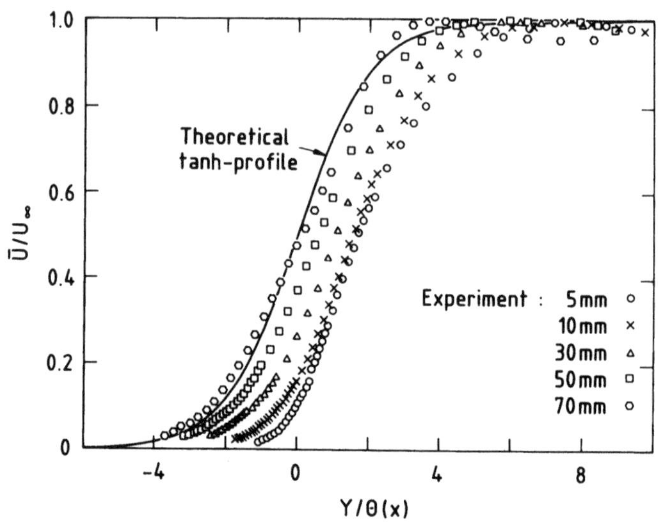

Fig. 3. Shear layer profiles at different locations downstream the trailing edge.

instability wave). Fig. 5a shows the development of the momentum thickness of the shear layer. A linear growth can be stated over a wide range. This measurement was made with the excited flow, so non-linear effects cannot be excluded for the downstream region. The perturbation velocity, which is no longer small compared to the mean velocity could interact with the mean flow and thus alter the result artifactually.

Unsteady Flow

Fig. 4 shows that upstream from the edge, the Stokes-layer can be distinguished. Far upstream (X = -30 mm) and within the boundary-layer, theoretical data are in good agreement with measured data. Approaching the edge, it is obvious that the

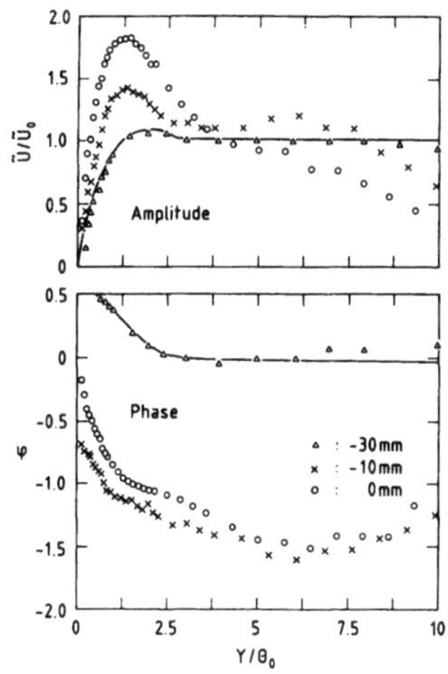

Fig. 4. Amplitude and Phase of the perturbation velocity near the wall at different locations upstream the trailing edge (solid line: Stokes theory).

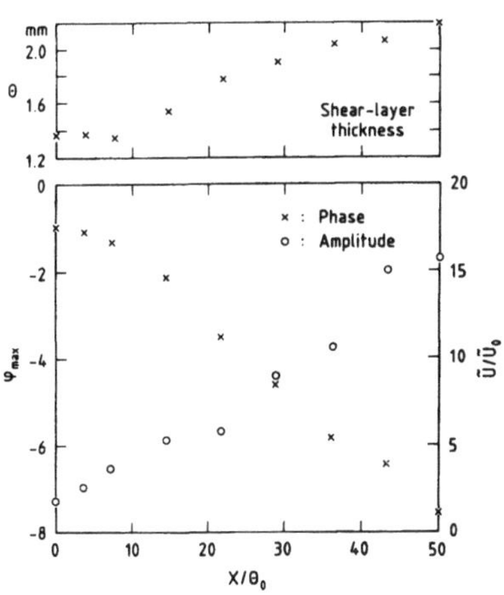

Fig. 5. a) Momentum thickness of shear layer
b) Phase and Amplitude of the maximum perturbation velocity downstream the trailing edge.

maximum of the perturbation velocity is growing. Note, that the phase is at first delayed, but as the flow approaches the edge, the phase delay is partly compensated. The strong influence from the edge probably scales with the wavelength of the instability wave. Downstream from the edge the amplification of the perturbation velocity continues and becomes even larger, as Fig. 5b indicates. The expected exponential growth rate cannot be stated unequivocally. From the corresponding phase data we obtained the real part of the nondimensional wavenumber $k_x \cdot \theta_o = 0.14$. This gives a wavelength of 62 mm and a phase velocity of $2.4 \cdot 10^{-2}$ m/s ($= 1/2\ U_\infty$). Theoretical data predict phase velocities from 0.49 U_o - 0.64 U_o for the different Strouhal numbers, which are due to the growing momentum thickness (St: .073 - 0.115).

Fig. 6 shows the amplitude of the unsteady U-component normalized to $U_{max}(X)$ in the x-y-plane. Two distinct minima develope, one of them in the flow region, the other one in the non-flow region. The latter is predicted in the theory by Michalke [1]. It is connected with a sudden phase reversal. The minimum in the flow region does not appear in the theory but was also found by Freymuth [2]. Two possible reasons are: a) that the shear-layer spreads and b) that the perturbation velocity is no longer small compared to the mean velocity and leads to distortions.

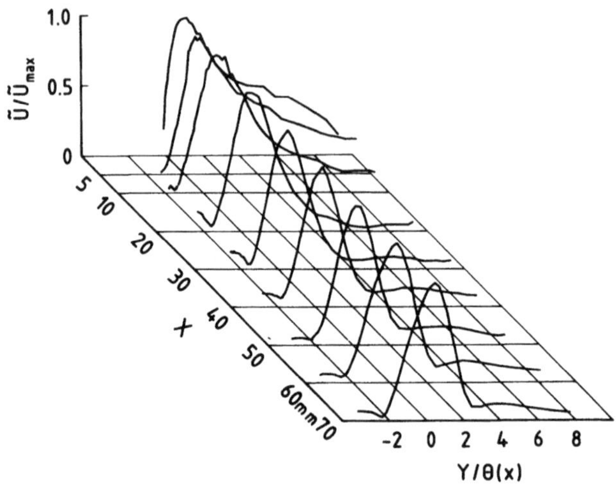

Fig. 6. Normalized amplitudes of the perturbation velocity at different distances from the trailing edge.

Streaklines

To compare our results with streakline experiments, we introduced dye into the flow as described before. Photos of these streaklines were taken at different phases relative to the exciting piston velocity. Thus we obtained the apparent wave speed, which is considerably less than the phase speed of the velocity field. The wavelengths were 47 mm and 62 mm respectively. Another observation (Fig. 7) is that while the average deflection of the separating streakline is shifted towards the resting fluid, the maximum of the perturbation velocity is located constantly at $y/\theta(x) = 1$ ($\theta(x)$: the momentum thickness at the respective x-coordinate). Thus the streakline does not depict the center of the broadening mixing-layer but tends to follow its low-velocity limit, probably caused by secondary flow in the "resting" fluid. From this, as must be expected, one can see that determining characteristic parameters of unsteady flows on the basis of streaklines is problematic.

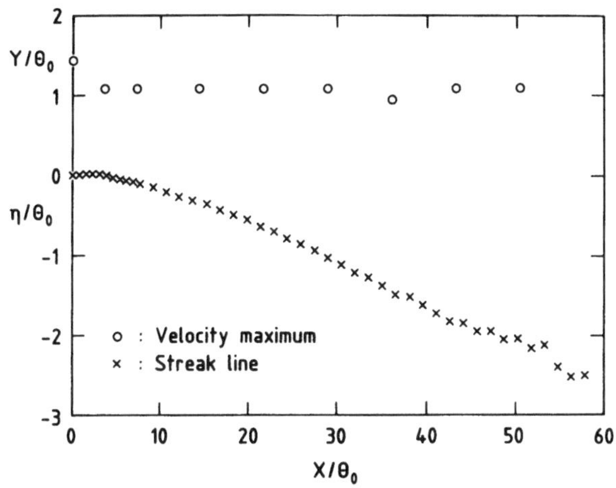

Fig. 7. Location of the maximum perturbation velocity and mean deflection of the streakline as a function of the distances from the trailing edge.

Vorticity

The vorticity of a 2-dimensional flow is given by $\frac{\partial v}{\partial x} - \frac{\partial u}{\partial y}$. The second term of the sum is the derivative with respect to y of the measured velocity field. The first term could not be determined from actual measurements of the velocity component v, but was instead obtained from the measurement of u and by applying the continuity equation which was rewritten yielding $\frac{\partial v}{\partial x} =$ $= k_x^2 \int_\infty^y u(y')dy'$ for the unsteady part. The vorticity 70 mm downstream from the edge is shown in Fig. 8. In accordance with theory [1] the maximum of the vorticity in the non-flow region is higher than the maximum in the flow region. The phase, which is not shown, changes linearly by 2π across the shear layer. Thus we can distinguish which part of the unsteady velocity field actually contains vorticity and where the potential flow region begins.

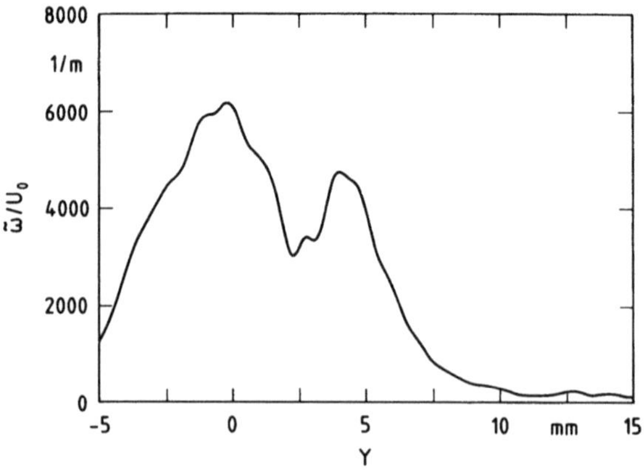

Fig. 8. Amplitude of the perturbation vorticity as a function of cross-stream direction, 70 mm downstream the trailing edge.

Future Perspectives

Our coming investigation will use another water channel to achieve a better 2-dimensional flow. Our aim is to measure, at higher Strouhal numbers, where the actual distribution of vorticity plays an essential role. It depends possibly on the ratio of Stokes layer and mean flow momentum thickness, which is given by $\sqrt{Re \cdot St}$. By means of different excitation mechanisms (e.g. Sound, Tollmien-Schlichting waves) we want to check influences from different incoming vorticity distributions. We want to interpolate and extrapolate the unsteady vorticity field by means of the vorticity transport equation so that the field in the very proximity of the edge can be obtained. We hope to evaluate the amplitude and phase of a transfer function, which relates the external excitation to the shed circulation which describes the effect onto the surrounding potential flow.

References

1. Michalke, A.: On Spatially Growing Disturbances in an Inviscid Shear Layer. JFM, Vol. 23, Part 3, 1965, pp. 521-544.

2. Freymuth, P.: On Transition in a Separated Laminar Boundary Layer. JFM, Vol. 25, Part 4, 1966, pp. 683-704.

An Experimental Study of the Influence of Mean Flow on Acoustic Dissipation by Vorticity Production at Edges.

M.S. VASUDEVAN and P.A. NELSON
Institute of Sound and Vibration Research
M.S. HOWE*

Faculty of Engineering and Applied Science,
University of Southampton,
England.

1. INTRODUCTION

A significant absorption of acoustic energy can result from the interaction of an applied sound field with the separated flow at an edge. The sound induces unsteady vortex shedding at the edge and the vorticity is convected away into regions where it is unable to interact significantly with the sound, and where its energy is ultimately dissipated as heat. The U.K. Science and Engineering Research Council is supporting an experimental study at Southampton University of the use of perforated screens in tangential mean flows to attenuate sound in ducts and cavities. This paper presents the analytical frame-work essential for the experimental study. Although the full experimental programme aims to examine this process in three stages, this paper confines itself to the first - namely the interaction of separated mean flow at a trailing edge with the reciprocating potential flow generated by two up-stream sound sources with a controlled phase relationship.

2. THEORETICAL ESTIMATE OF DISSIPATION CAUSED BY VORTEX SHEDDING

The theory outlined here is due to Howe [1] and is an analysis of the problem of diffraction of a plane sound wave by a splitter plate in a mean flow duct. In performing the analysis, the splitter plate is assumed to be sufficiently thin that the frequency of natural vortex shedding from the plate (which occurs at a Strouhal no. of $\frac{\omega d}{U} \simeq 1.45$ where ω is the radian frequency, d the thickness of the trailing edge, and U is the velocity of the main stream) is very much greater than the

*Also: BBN Laboratories,
10 Moulton Street, Cambridge. MA 02238

Aero- and Hydro-Acoustics
IUTAM Symposium Lyon 1985
Editors: G. Comte-Bellot and J. E. Ffowcs Williams

frequency of the incident sound. In a first approximation, the diffraction problem is then decoupled from complications associated with the natural shedding of vorticity [2].

In the presence of mean flow, linear diffraction theory predicts that both the perturbation pressure and velocity are singular at the edge of the plate. These singularities are removed by application of the unsteady Kutta condition - a procedure which incorporates the influence of viscosity into the idealised boundary conditions of acoustic theory. This results in the production of vorticity at the trailing edge, which is subsequently swept downstream by the flow to form a wake.

Consider first a plane wave to be incident in the upper half section (Fig. 1)

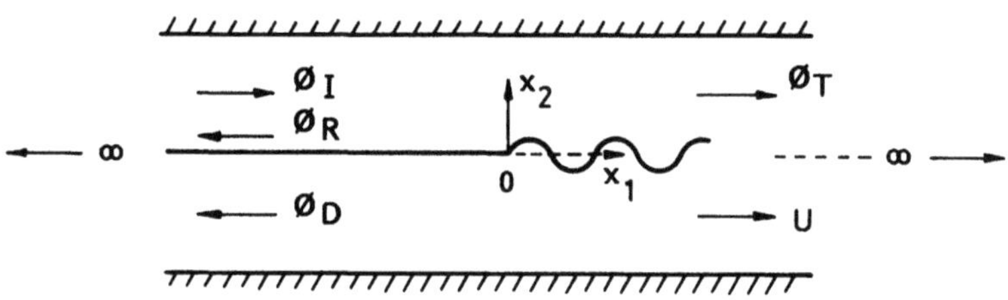

FIGURE 1: Schematic of the Configuration

The power dissipated is related to the strength of the wake vorticity by

$$W_{diss} = \rho_o \int_V \overline{\underline{\omega}\wedge\underline{U}\cdot\underline{u}}\, dV$$

where $\underline{\omega}$ is the strength of the wake vorticity, U the mean flow velocity and u the "compressible velocity component". Let ϕ_I, ϕ_R, ϕ_D and ϕ_T represent the velocity potentials of the incident, reflected, diffracted and transmitted waves respectively (Fig. 1), where ϕ_I is of the form:

$$\phi_I = e^{i\left[K_o x_1/(1+M) - \omega t\right]} \quad , \quad K_o = \frac{\omega}{c_o} \; , \quad M = \frac{U}{c_o} \; .$$

$\underline{\omega}$ is then specified by requiring that: i) the total perturbation velocity potential satisfies the time reduced convected wave equation, (ii) that $\partial\phi/\partial x_2$ be equal to zero on rigid walls (duct and plate) and (iii) that the perturbation pressure be continuous across the wake. Thus, there must be a jump in ϕ across the x_1 axis of the form:

$$\phi(x_1, +0) - \phi(x_1, -0) = \alpha e^{iK_H x_1}$$

where $K_H = \frac{\omega}{U}$ is the hydrodynamic wave number and the constant α is chosen to ensure that the unsteady Kutta condition is satisfied, in which case:

$$\underline{\omega} = \text{Re}\left\{-i\ K_H\alpha\ e^{i(K_H\ x_1-\omega t)}\right\}\delta(x_2).\underline{n} \qquad x_1>0$$

where $\underline{n}$ is a unit vector directed out of the plane of paper (Fig. 1) and $\delta(x_2)$ is the Dirac Delta function.

3. THEORETICAL RESULTS AT FREQUENCIES BELOW THE CUT ON FREQUENCY OF THE FIRST TRANSVERSE DUCT MODE

By reference to Figure 1, the corresponding pressure waves below cut on can be shown to be:

$$P_I=e^{iK_o x_1/(1+M)};\quad P_R= -\frac{1}{2}e^{-iK_o x_1/(1-M)};\quad P_T=\frac{1}{2}e^{iK_o x_1/(1+M)}$$

and

$$P_D=\frac{1}{2}e^{-iK_o x_1/(1-M)}$$

The dissipated sound power can be expressed in the form of an absorption coefficient:

$$\Delta = -\ 10\ \log_{10} \left\{ \frac{W_o - W_{diss}}{W_o} \right\}$$

where W_o is the power flux due to the incident wave alone. The variation of Δ with Mach no. has been computed by Howe and is shown in Figure 2.

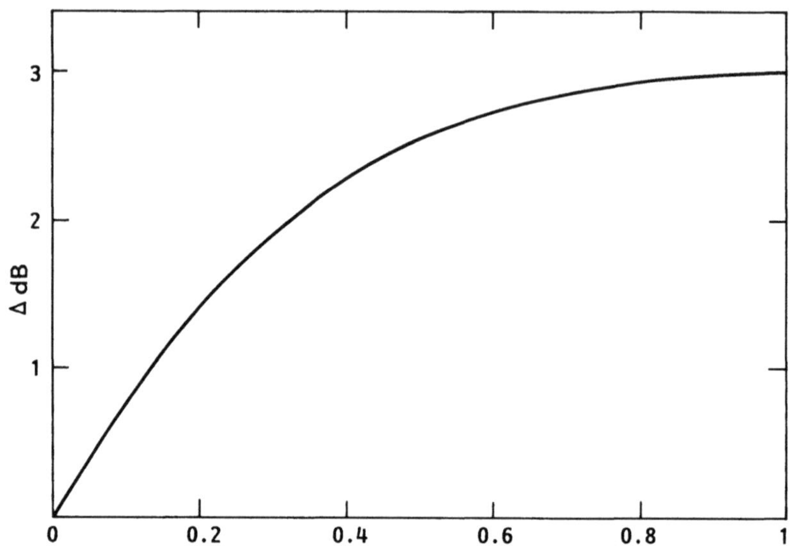

FIGURE 2: Variation of Δ with M

These results, however, are not directly applicable to a realistic experimental situation involving a finite duct, and the predicted magnitude of the dissipation is not easily measurable. Nevertheless, this solution to the basic diffraction problem can be used to predict measurable quantities in a situation which can be produced experimentally with relative ease.

4. SCHEME FOR EXPERIMENTAL PRODUCTION OF THEORETICAL CONDITIONS

The experimental configuration being studied has plane waves incident from both above and below the plate of the form:

$P_{I1}=e^{iK_o x_1/(1+M)}$, and $P_{I2}=e^{(iK_o x_1/(1+M)-i\theta)}$ where θ is the phase difference between incident waves of equal amplitude. The experimental situation is illustrated in Fig. 3:

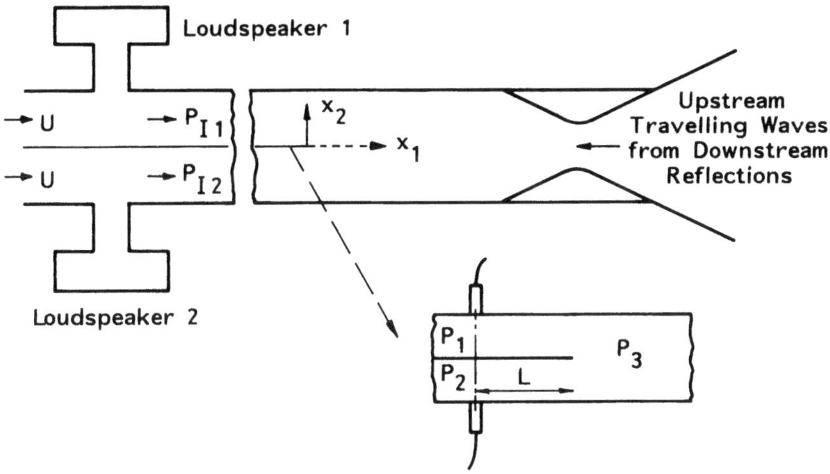

FIGURE 3: The Experimental Configuration

The corresponding net pressure field is indicated in the inset to Figure 3 and it can be shown that

$$P_1=e^{iK_o x_1/(1+M)}+\frac{1}{2}\left\{R(1+e^{-i\theta})-(1-e^{-i\theta})\right\}e^{-iK_o x_1/(1-M)} \quad (1)$$

$$P_2=e^{iK_o x_1/(1+M)-i\theta}+\frac{1}{2}\left\{R(1+e^{-i\theta})+(1-e^{-i\theta})\right\}e^{-iK_o x_1/(1-M)} \quad (2)$$

$$P_3=\frac{1}{2}(1+e^{-i\theta})\left\{e^{iK_o x_1/(1+M)}+R\,e^{-iK_o x_1/(1-M)}\right\} \quad (3)$$

where R is the ratio of complex pressures in the upstream and downstream travelling waves in section 3 at $x_1 = 0$. Experimental production of theoretical conditions therefore specifies, for given values of θ and M, the complex ratio of pressures P_2 to P_1. The required transfer function is:

$$P_2/P_1 = \frac{e^{-iK_oL/(1+M)-i\theta} + \frac{1}{2}\, e^{iK_oL/(1-M)}\left[R(1+e^{-i\theta})+(1-e^{-i\theta})\right]}{e^{-iK_oL/(1+M)} + \frac{1}{2}\, e^{iK_oL/(1-M)}\left[R(1+e^{-i\theta})-(1-e^{-i\theta})\right]}$$

where L is the distance of microphones 1 and 2 from the edge of the plate (see inset Fig. 3). There is a need, therefore, to measure the complex reflection coefficient R. This is done by implementing the Chung & Blaser [3] two microphone transfer function method - illustrated in Figure 4.

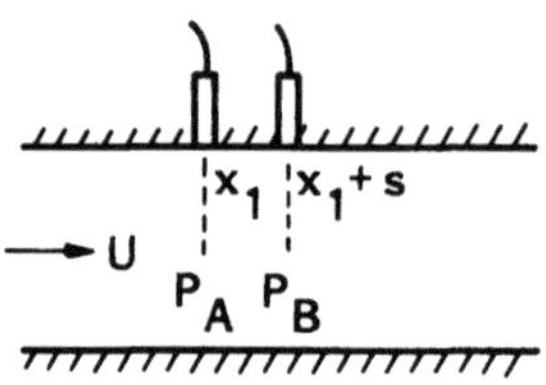

FIGURE 4: Two Microphone Transfer Function Method for Reflection Coefficient

The required reflection coefficient R can be shown to be given by

$$R = \left[\frac{e^{iK_+S} - H}{H - e^{-iK_-S}}\right] e^{2iK_ox_1/(1-M^2)}$$

where $K_+ = K_o/(1+M)$; $K_- = \frac{K_o}{(1-M)}$ and S the microphone separation, and $H = P_B/P_A$ is the transfer function between the pressures at the two microphones.

The procedure to be employed in experimental production of theoretical conditions is therefore to first measure R for the desired frequency and to then physically set the transfer function P_2/P_1 to that required at $x_1 = -L$.

5. CALCULATION OF DISSIPATED SOUND POWER

The dissipated sound power can be calculated from the time averaged acoustic intensity formulation of Morfey [4]. The intensity can be obtained by measuring the two microphone transfer function [5] and at a single frequency is given by

$$\underline{I} = \frac{|P_A|^2}{8\rho_o c_o \sin^2 (KS/1-M^2)} \left[(1+M)^2 \left| e^{-iK_o S/(1-M)} - H \right|^2 - (1-M)^2 \left| H - e^{iK_o S/(1+M)} \right|^2 \right]$$

where $H = {}^{P}B/P_A$ is, as before, the two microphone transfer function. The validity of this equation is restricted to plane wave propagation and this is ensured in practice by driving the sources below the cut on frequency. The dissipated sound power is given by:

$$W_{diss} = \left\{ (I_1 + I_2 - 2I_3) \right\} S$$

where I_1, I_2 and I_3 are the measured time averaged intensities corresponding to power fluxes W_1, W_2 and W_3 respectively and S the cross sectional area of the downstream section (see Fig. 5).

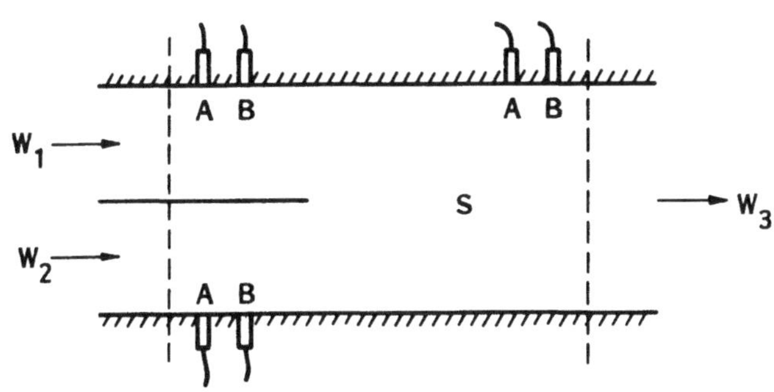

FIGURE 5: Two Microphone Transfer Function Method for Measuring Sound Intensities

The power flux in the upstream and downstream duct sections can also be calculated from the plane wave solutions given by equations 1-3. The resulting transmission factor defined by

$$T = -10 \, \mathrm{Log}_{10} \left(\frac{W_3}{W_1 + W_2} \right)$$

has been computed assuming no downstream reflections [1] and the variation of TdB with θ and M is shown in Figure 6.

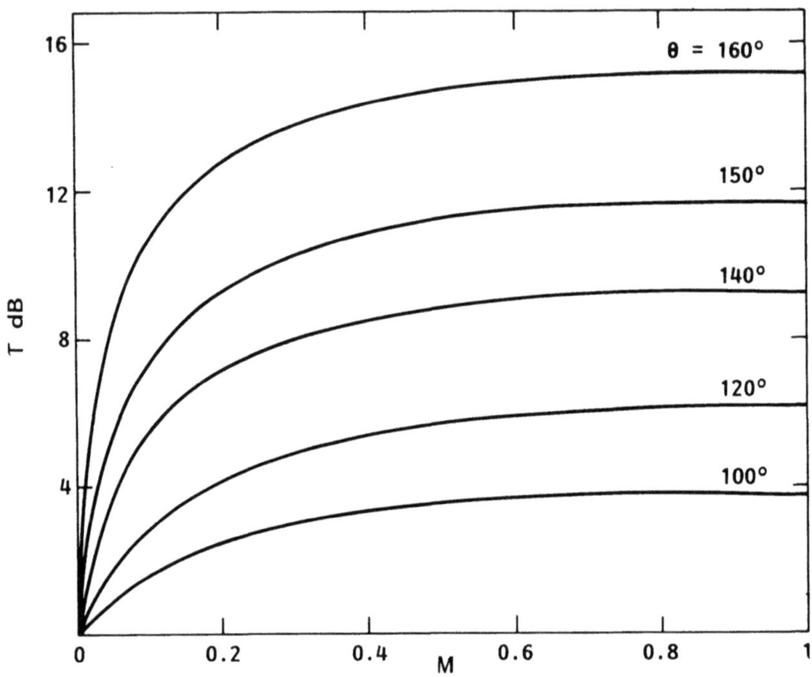

FIGURE 6: Variation of T(dB) with θ and M for R=0

Preliminary experiments in a flow/acoustic tunnel have confirmed the general trend indicated in Fig. 6 for the variation of T with M and θ. An error analysis is essential for both the intensity and reflection coefficient measurements, before definitive experiments to quantify the dissipation (in a situation where downstream reflections are present) can be undertaken. This analysis is currently in progress.

REFERENCES

1. HOWE, M.S.: Attenuation of sound due to vortex shedding from a splitter plate in a mean flow duct. Faculty of Mathematical Studies. Preprint Series No. 144, March '85 To appear in J. Sound & Vibration.

2. WELSH, M.C., STOKES, A.N. and PARKER, R.: Flow resonant sound interaction in a duct containing a plate. Part I Semi-Circular Leading Edge. Journal of Sound and Vibration 95, 305-323, 1984.

3. CHUNG, J.Y. & BLASER, D.A.: Transfer function method of measuring in duct acoustic properties.II .Experiment. J.Acou.Soc.America, 68(3), 914, 1980.

4. MORFEY, C.L.: Acoustic energy in non uniform flows. J.Sound & Vibration, 14, 159-170, 1971.

5. CHUNG, J.Y. & BLASER, D.A.: Transfer function method of measuring acoustic intensity in a duct system with flow. J.Acou.Soc.America, 68, (16), 1570, 1980.

Active Control

Active Noise Control

I. ROEBUCK

Admiralty Research Establishment,
Portland, Dorset, UNITED KINGDOM

Summary

After several years in which the practitioners and advocates of Active Noise Control have had as their major aim to demonstrate to their peers that the concept does not necessarily violate the Laws of Physics, the topic has suddenly become accepted and respectable. Indeed, the possibility now arises of its being regarded as a universal panacea for all noise control problems, with equally dangerous effects on its rational development for the long term. For that reason, the objective of this paper is to indicate the fundamental limitations of Active Noise Control-those of acoustic principle and of controllability - as well as the areas where major advances in technology will be needed to overcome current problems.

Introduction

Except in the final section of this paper, linear sources of sound will alone be considered ; thus our analyses are built on acceptance from the outset that the use of superposition of single-frequency Fourier components to produce solutions appropriate to a broad-band frequency scenario is valid. The fundamental problem is then that of active control of a localized source in unbounded space by a source - or finite distribution of sources - at a different location. We begin by considering this problem in two distinct idealized situations - total A PRIORI knowledge and total initial ignorance - to discover whether or not active control is possible, and what constraints exist on the nature of the controlling ("secondary") sources.

Aero- and Hydro-Acoustics
IUTAM Symposium Lyon 1985
Editors: G. Comte-Bellot and J. E. Ffowcs Williams

Control with complete information

For this case, we begin with the assumption that there is no other noise source in the entire universe and that we know for all time the location, phase and amplitude of the signal to be controlled, a harmonic monopole. We recognize immediately (see, e.g, Jones, ref. 1) that the field outside any surface which contains the source may equally well be represented as being produced by an acoustic dipole distribution on that surface, which can be prescribed by the Helmholtz integral formulation. Then we may reverse this process (as shown in Jessel & Mangiante, ref. 2) to represent the dipole distribution by a multipole source situated at an arbitrary interior point of the volume enclosed by that surface.

Again, a Helmholtz integral transformation is required, but in this case instead of the retarded potential known from electromagnetic theory, an advanced potential is required. The reason for this apparent departure from causality is obvious when the physics of wave propagation is considered. There are volumes of space nearer to the primary source than to the controller ; thus, with finite speed of sound, transmissions from the controller must precede the emissions they are to control if they are to arrive simultaneously at the region of intended silence. We note (c.f. Kempson, ref. 3) that the Helmholtz transforms are global in their operation, and therefore the total time history is necessary, as well as sufficient, to allow specification of the unique multipole which, placed at a predetermined but arbitrary point, will result in a zero total fluctuating pressure field throughout space.

If the prior information this implies is not available, then we cannot in general achieve the representation of one localized source by another. If, indeed, we lack that information and must proceed to determine the signal output to be controlled by measurement, and thence prescribe the control signal, a distributed controller is implied and causality constrains the positions of the measuring and transmit elements for even idealized systems. The following theorem gives a minimum condition for controllability :

> the closed surface of minimum volume, convex everywhere, that contains the measurement sensors must also have the primary source location as an interior point of that volume, and must itself be entirely contained within the hull of the similar volume defined by the position of the secondary transmitters.

Conceptually, this idea of "surrounding" the source with "antinoise" controllers is intuitively much more acceptable than is a differently located point multipole. (As an aside, we note that we can "surround" a location in one-dimensional half-space by a single point, so the control of duct noise by a single controller immediately satisfies both views !). In particular, it would appear that an observer situated between two "point sources" and equipped with a sensor array should, in principle, be able to make directional measurements and distinguish between these sources, counter to our basic premise as proven by Kempson. The implicit paradox in this may be resolved by considering in more detail the concept of a point source of waves of finite wavenumber, which is shewn to be untenable without modification.

The effective size of a "point source"

To illustrate the analysis (which is given in full in ref. 4), and to introduce the concept of MUTUAL IMPEDANCE, which plays a major role in the later sections of this paper in determining performance limits, we consider the far field radiated by two sources, each a single frequency "point monopole". The acoustic potential of each, in geocentred co-ordinates, may be written as :

$$\phi (r, t) = A \exp [i (kr - \omega t)]/r \qquad (1)$$

k, ω , being respectively wavenumber and radian frequency of the source, A its (generally complex) amplitude, and r, t, our chosen co-ordinate variables. By linear superposition, the total potential of the joint system is the sum of the separate potentials, and the total power radiated to the far field is proportional to the square of the modulus of this potential, integrated over the surface of a sphere centred on the points, in the limit as its radius tends to infinity.

From symmetry considerations, we expect the power output to be minimized when the two sources have equal amplitude, but are opposed in phase ; we assume this condition and obtain the result

$$\Pi = 8\pi A^2 \left\{1 - \left(\frac{\sin kd}{kd}\right)\right\} \qquad (2)$$

where d is the separation between the two elementary points. It is apparent from this (cf Nelson, ref. 5) that the symmetry assumption, though correct as d tends to zero, is not valid in an unconstrained sense as for values of d greater than about one-third of a wavelength the total output is greater than that in the absence of a second source. It is, however, easy to show that (2) is the desired

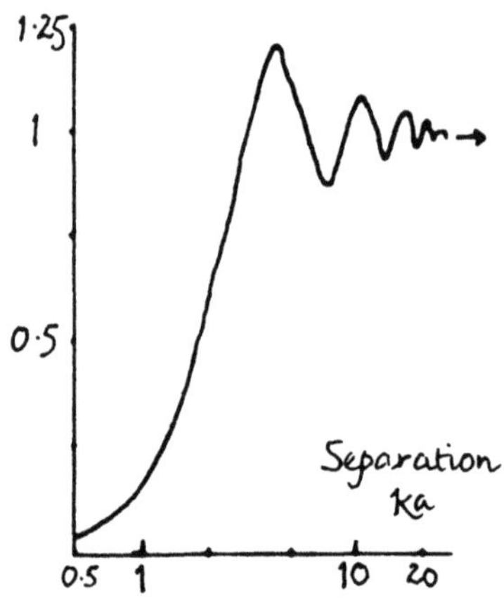

Fig. 1 Normalized total output of source and anti-source as a function of element separation

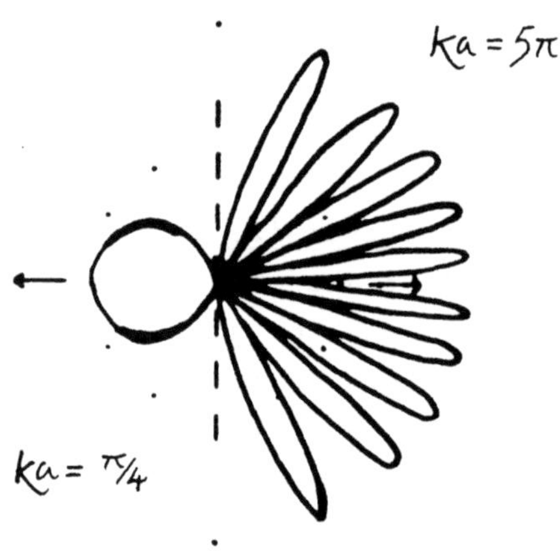

Fig. 2 Directional output of a dipole for representative inter-element separations

minimum value if the sum of the amplitudes of the two potentials is held constant at 2.A, and further that in this case, uniquely the instantaneous total volume flux at infinity becomes identically zero, so we have the possibility of controlling the residual field by dipole and higher order sources alone.

What we have generated is, of course, a dipole moment, as is illustrated by Figures 1 and 2, plotting respectively the total power output versus normalized element separation, from expression (2), and the directional output at small and large separations. Fig. 1 demonstrates that for separations greater than half-wavelength the total output tends rapidly to the sum of the individual source powers. It is emphasized that this is NOT linear superposition applied to a quadratic quantity - the reason for the result is apparent when we consider Fig. 2 as an array beampattern-the narrow main null becomes irrelevant as a contributor in comparison with the many sidelobes of equal mean height.

The obvious interpretation of these results is that a monopole source has an interaction range (or effective size) of radius half the acoustic radiation wavelength - a result analogous to Heisenberg's Uncertainty Principle in quantum theory. This is equally the effective radius of the velocity near field of the monopole, and suggests that a formulation of the power output in terms of mutual impedance concepts might be instructive. We define the mutual impedance coefficient z_{yx} of source X on source Y as the (complex) pressure induced at Y by a unit drive velocity imposed at the source X. Then, suitably normalized, the total power output of a set of sources may be written as

$$\Pi \sim \Sigma_{\text{sources}}\, v_i\, p_i \equiv \Sigma\, v_i\, z_{ij}\, v_j \qquad (3)$$

where we identify the self-induced impedance terms with

our usual driving point impedances.

In this formulation, (2) may be written in matrix form as

$$(A , -A)\begin{pmatrix} 1 & \exp[ikd]/ikd \\ \exp[-ikd]/-ikd & 1 \end{pmatrix}\begin{pmatrix} A \\ -A \end{pmatrix} \quad (4)$$

and this has an immediate generalization to cover sources of higher multipole order. Either by adding to (4) an opposed localized dipole moment situated at our control point, and finding the range of significant total output reduction, or geometrically by packing spheres of influence, we may show that the effective size of a multipole of order 2^n is around n/2 wavelenghts. So we may expect the coefficients in the control expansion to have a maximum value (and maximum effect!) when we reach that order at which the control multipole includes the original source location within its effective size extent.

This is the resolution of the paradox mentioned above - the "decoy" source the observer seeks is NOT separately located at a position distinct from the original ; rather, in this case also we may consider the control source theorem to be satisfied in any practical realization, the decoy source extent including the primary. This statement rests on the recognition (see Fig. 1) that multipoles are inefficient as acoustic radiators except when we separate the individual point monopoles of which they are composed until the mutual impedance contributions are small in comparison with the self impedance.

The large effective diameter of the controlling multipole, and the position of its centre distinct from the location of the original monopole source, together set a further potential limit to the use of active noise control. Again, the mutual impedance formulation makes this clear. Our analysis has shown the identity of observable fields between original and duplicate in free space, but also that the presence of other sources (or boundaries inducing image sources) within the interaction range of a source will interactively change those observable fields. Hence the intrusion of a boundary within the effective size of our control multipole may totally change the form of the control required from that of the free-space configuration, despite having little or no observable effect on the source field. (We note as an aside that similar arguments may be used to estimate, in general acoustic problems, the range from edges to which multipole amplification occurs - cf the paper by Levine, ref. 6).

Causal Control

Since the impedance matrix, the generalization of expression (4), may in principle be obtained by the measurement at each control location of the response to individual unit imputs at the rest, taken one at a time, we see that for a single, time-varying source in an unchanging environment the optimal control problem is reduced to that of minimizing $v^T Z v$ under the constraint of a prescribed value for the first component of v - the source amplitude. This interpretation makes clear that the problem is in two discrete parts - determination of the appropriate eigenvector of Z, and measurement of the "normalizing factor", the time-varying output of the source. In more physical terms, these two operations may be viewed as : determining the desired filter characteristics for the control source feeds, and measuring the time-varying single desired input to them.

It is immediately apparent that for total extinction of the sound field to occur, Z must be a singular matrix ; in all practical situations, the influence of external and measurement "noise", which together add some constant to the diagonal terms, ensures that Z is positive definite and precludes total success. However, selection of control sites will affect the extent to which the sound can be controlled, and examination of the ill-conditioning of mutual impedance matrices composed of subsets of these potential sites gives a useful guide. The clearest (negative) example is the choice of kd = π in (4), when no control can possibly be exercised. An additional aid in deciding on good control locations is that a near-optimal choice will have all the elements of the eigenvector of similar modulus - a small amplitude suggests that there is little influence from that element (which is therefore probably superfluous), whereas an overly large one points to potential stability problems and indicates that an extra control location should be included.

Since v consisting solely of the primary drive defines the uncontrolled condition, the performance of the controller is easily calculated (see, for example, Ref. 7) in analogous fashion to that of an optimal beamformer in antenna design. The result achieved may be expressed as leaving the "noise" unchanged while completely removing the coherent component - this generalizing the one degree of freedom result that the controlled output has the ratio (1 - C) to the uncontrolled, where C is the correlation coefficient between the measured field at the control and far-field monitoring points in the uncontrolled situation.

Two immediate consequences flow from this observation - the first being a direct computation of the effect of sensor errors on controllability, with it becoming very apparent that phase errors are much more likely to limit performance than amplitude mismatch, the second being the definition of an efficient and fail-safe algorithm for filter update if optimal control is not being achieved. Figures 3 and 4 directly translate sensor mismatch into correlation degradation to plot limits of achievable control in a theoretically infinite signal-to-noise ratio case ; the two branches in the case of amplitude mismatch at large errors reflect the result that a correctly phased but too small control signal will always give some gain, whereas too large a control signal can lead to a situation worse than the original, once the mismatch exceeds 6 dB.

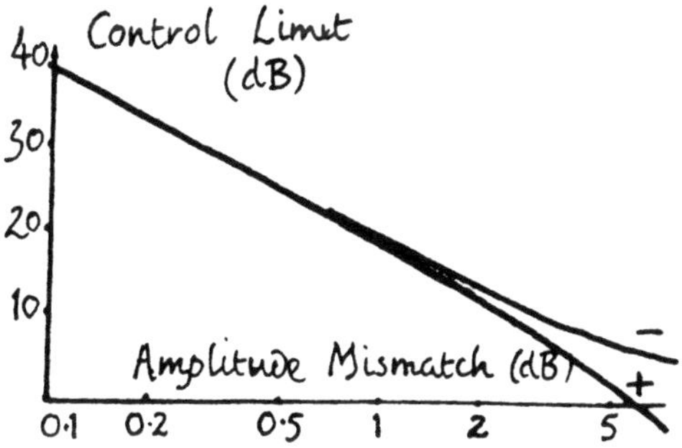

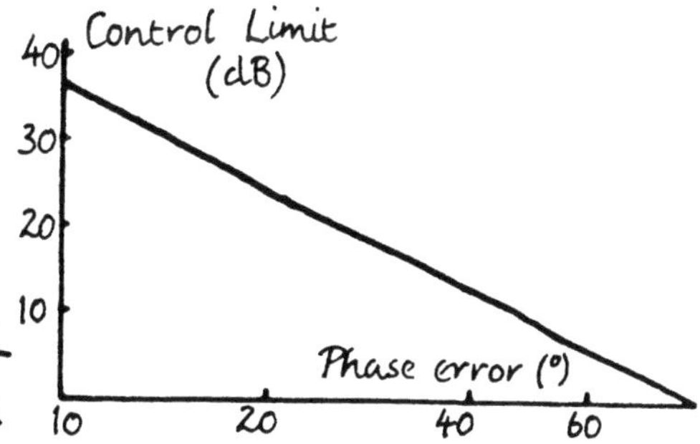

Effect of system or sensor errors on achievable control

Fig. 3 Amplitude mismatch

Fig. 4 Phase mismatch

It is apparent that, in a real situation, where mismatches occur and the environment changes, some form of adaptation by the filter is essential if there is to be a guarantee that Active Control will be beneficial, rather than suffer from stability problems. This implies the use of a feedback algorithm to monitor performance ; the simplest to implement is the introduction of a (small) random change in the filter characteristic, and direct comparison of before and after residual noise power. Unfortunately this is not efficient in convergence, and may well give poor results if the input amplitude varies on the time scale of the filter update.

A more effective process may be deduced from the observation that, if optimal control has been achieved, the residual noise field is uncorrelated both with the measured input to the filters, and with the control signal applied to the secondary sources. Thus, if there is correlation between the control signal and the residual output, our filter is wrong, and the subtraction of a linear function of this correlation from the filter characteristic will improve performance. This procedure is an exact analogue of the Howells-Applebaum algorithm for adaptive interference cancellation in receiver arrays ; here (as set out in detail in reference 8) it must be supplemented by a second identical loop to ensure that the control signal is not being picked up at our measurement points to produce positive feedback - which bedevilled early implementations of Active Noise Control.

Filter Complexity and Broad-Band Control

So far, the discussion has concentrated on a single (known) frequency input, but in many realizations of Active Noise Control the requirement is to quieten the output over a band of frequencies. As noted earlier, in principle the analysis is unchanged, but implementation then demands trade-offs between potential performance and allowable computational load. We note initially that a finite number of frequency bins implies a phase spread across each bin and thus built-in errors, even at infinite signal-to-noise ratios, for frequencies not precisely at the band centre. This error increases with path lenght, so suggesting that the distance from measurement sensor to control injection should be kept small, but countering that is the need to carry out computations within the acoustic travel time-which requires either a long path or a small number of bins, both tending to increased errors. It is this feature which, for the present, renders Active Noise Control essentially a solution only for low frequency control (or for very restricted volumes - c.f. active ear defenders).

But technology is not all ; physical limits arise too. Phase resolution needs a long sample, and real physical systems vary within finite times. Even for linear systems, the ability to achieve perfect control in a varying environment is not available to us. The ambiguity of time and frequency from finite samples cannot be circumvented, and attempts to do so are inherently unstable. It may be that we can use some extra A PRIORI knowledge to force rapid convergence, but such attempts - of which the most popular is the use of Maximum Entropy enhancement with truncated discrete Fourier Transforms - "fail hard" if their (untestable) underlying assumptions are violated. In general, we can only control systems whose rates of change are slow in comparison with the control response time - and so again we are constrained to low frequencies.

Active Control of non-linearly described Sources

It is stressed throughout the foregoing discussion that the stability of active noise control for changing source inputs is one of the most fundamental requirements, and among the most difficult to assure except by some form of direct performance measure on the residual field being applied to close the feedback loop. This is straightforward when we are dealing with the restricted problem of providing finite and delineated "quiet zones", but it is not immediately apparent how to provide the appropriate measure (or even what it is) if we are dealing with the unbounded problem of suppressing the total radiated field from a source. As is the case with the related problems of adaptive beamforming, there is no guarantee of overall suppression with even an infinite number of independent linear point constraints, and any rational, measurable global penalty function is inherently non-linear, so vitiates many of the analyses of this paper.

Nevertheless, progress can be made. Although the algorithm described earlier, with its guiding principle of minimizing correlation would seem, at first sight, to be restricted only to linear systems, it appears that there can be, in certain classes of problem, the possibility of control of noise even when the information to be used is non-linearly related to the radiated field we wish to "cancel". What is required is the sort of functional behaviour for which a Newton-Raphson root-finding procedure would succeed.

For this to be the case, however, we need both

(i) a clear understanding of the way in which convergence can be expected - involving in general the demonstration that a Lyapunov stability criterion can be satisfied at the desired optimum ;

and

(ii) a sufficiently accurate a priori starting estimate of the required control signal to be within the convergence radius of this optimum point.

The danger in this second requirement is similar to that which occurs in such non-linear estimation procedures as the Prony technique, which can "converge" to entirely spurious line spectra if a bad initial hypothesis is chosen.

For the most likely type of active noise control problem when we must use non-linear controls - suppressing a single line whose frequency is slowly time-varying, the difficulty is precisely that of using an initial hypothesis to speed frequency estimation while avoiding catastrophic failure. The key to progress is a technique described by Byrne & Fitzgerald in reference 8 - a short sample of the measured data is used to test (using Bayesian statistics) the hypotheses that the wanted signal is band-limited to each of a set of relatively narrow but overlapping bands.

If success is obtained, then we iteratively apply a filter to suppress contributions from outside that band and test for its subdivisions - thus rapidly converging to the spectral estimate on which we can use our now (quasi-) linear null correlation algorithm. If, however, the Bayesian test fails, we know not to look for a single dominant line, and may either abandon the attempt at control or, given adequate computing power, refine our hypothesis to look for two or more lines. Thus we have achieved a "fail-safe" situation with some potential in circumstances which previously could only have unstable open-loop control. But much remains to be done in this area.

Conclusion

The conclusion of this study is unchanged from that derived earlier - that Active Noise Control is one extra weapon in the acoustician's armoury, but not a blunt instrument. The physics of control demand accurate recognition of the causes of the fields to be controlled, and this understanding may well suggest conventional methods of noise reduction. Both physics and technology imply that low- frequency, highlystructured time varying fields are the most suitable for active control. It is here that the interaction between the source and the controller, which is essentially an impedance changer, will be most sensitive to source changes, and the wider use of active control depends on the technology of implementation of fail-safe adaptive control algorithms.

But we depend on our understanding of the acoustic environment to develop the technology of solutions which are robust in particular non-linear problems, and on improvements in technology to provide the data from which to verify that understanding, adaptation and interaction are as inherent in our study of active noise control as they are in its implentation.

Acknowledgement

This work is published by permission of the Ministry of Defence (Procurement Executive), United Kingdom.

References

1. Jones, D.S., 1964 : Theory of Electromagnetism, chapter 1. (Pergamon Press).

2. Jessel, M. & Mangiante, G.A., 1972 : Active Sound Absorbers, J. Sound Vib., Vol. 23, p. 383.

3. Kempson, A.J., 1975 : PhD Thesis, Dept. of Engineering, University of Cambridge.

4. Roebuck, I., 1985 : On the Effective Size of a Point Multipole, 109th meeting, ASA, Austin.

5. Nelson, P.A., Curtis, A.R.D. & Elliott, S.J., 1985 : Active Minimization of Harmonic Enclosed Sound Fields, IUTAM, Lyon.

6. Levine, H., 1985 : Diffraction and Transition Radiation Phenomena, IUTAM, Lyon.

7. Ffowcs Williams, J.E., Roebuck, I. & Ross, C.F., 1985 : AntiPhase Noise Reduction, Phys. Technol., Vol. 16, p. 19.

8. Byrne, C. & Fitzgerald, R., 1984 : Spectral Estimators that extend the Maximum Entropy and Maximum Likelihood Methods, SIAM J. Appl. Math., Vol. 44, p. 425.

The Constraint of Causality on an Active Control System

G P EATWELL, C F ROSS and J F SCOTT

Topexpress Ltd
13/14 Round Church Street
Cambridge CB5 8AD

This paper discusses a method of ascertaining the optimal performance of a single channel active control system. It looks specifically at the loss of performance introduced by some of the features that are likely to be inherent in real systems. These include the limitations on the control system caused by finite computational power and the need for the controller to be causal. We also consider the effect of noise corruption of the sensor signal and unwanted noise production by the transducer. In order to quantify the effects of these real features on an active control system we first study a perfect system and then observe the effects on the system performance of introducing these features. It is most convenient for clarity in this paper to discuss the simplest active control system; the waveguide.

Below the cut-on frequency for transverse modes of propagation (which for a rectangular waveguide is when the wavelength is equal to twice the cross dimension) only plane waves propagate. If the all the spectral components of the sound propagating in the waveguide are below this frequency then it is appropriate to introduce a single channel control system to stop the downstream propagation of sound waves. A microphone positioned in the waveguide can be used to measure the pressure variation of the downstream propagating wave and thus provide a sensor signal related to this sound wave. This signal can then be fed into the digital control system which will filter it and produce an output to drive the cancelling transducer (or loudspeaker). The characteristic of this filter is chosen so that no sound is heard downstream of the system. The sensor will detect the waves produced by the downstream transducer,

Aero- and Hydro-Acoustics
IUTAM Symposium Lyon 1985
Editors: G. Comte-Bellot and J. E. Ffowcs Williams

consequently the filter will have a compensation loop to subtract the effect of the feedback from the sensor output. In this ideal, noise-free, linear system perfect cancellation is possible. We will introduce some realistic features and observe how the system performance is degraded.

If the spectrum of the sound propagating down the wave guide extends beyond the cut-on frequency then some of acoustic energy will propagate in these higher order modes. Since the sound transmission process will be different for these modes the single channel control system will be unable to stop their propagation. In fact a multi-channel device would be needed, with one channel dedicated to each mode, in order to eliminate all of the propagation. In a single channel the sensor microphone will respond to the pressure fluctuations of all of the modes and since the system will not be controlling the higher order modes the sensor response to these higher modes constitutes unwanted noise contamination of the signal representative of the plane wave mode. The sensor system should be designed to minimise its sensitivity to these higher order modes and it may be more appropriate to use an array of microphones rather than a single one in order to achieve this. In this case it is more usual to call the noise corruption from higher order modes spatial aliases of the sensor system.

There will be a similar problem with the transducer that produces the cancelling sound waves. If it is driven at a frequency above cut-on then it will inevitably excite not only the plane wave mode but also the higher order transverse modes. Since the transducer will be driven to cancel the plane wave mode and the excitation of the higher modes will be related to this drive the higher mode excitation will not cancel the sound propagating in these modes but will augment it. Again, if our transducer system was in the form of an array of loudspeakers then this array could be designed so that the efficiency of excitation of the higher order modes was as low as possible when compared with the plane wave mode excitation efficiency.

It is now clear that in addition to our system cancelling the plane wave mode it will generate extra energy in this mode due to the sensor noise output and also generate energy in the higher modes due to the imperfect transducer array characteristics.

The effect on the system performance may be assessed if we quantify these features (and we will do this initially for one frequency). Let us denote the waveguide modal amplitudes as P for the plane wave mode and N_i for the i-th higher order mode. The sensitivity of the sensor array to these modes is defined as unity for the plane wave and α_i for the higher ones. Thus the sensor output will be

$P + \sum_i N_i \alpha_i$ This sensor output is fed through the filter with gain -A generating a signal for the transducer.

For simplicity we have assumed that the sound generated by the transducer array is not detected by the sensor array. In reality this feedback is accounted for by subtracting, from the sensor output, that part which is predicted (from knowledge of the transducer output) to be due to the feedback. This is done electrically by an additional filter in tandem with filter "A". The output of the filter "A" is

$$-A \; (P + \sum_i N_i \; \alpha_i)$$

If we now write the transmitter modal generating efficiencies as unity for the plane wave mode and β_i for the i-th mode then the resultant amplitudes of the plane wave mode and the k-th higher mode will be

$$P(1 - A) \; -A \sum_i N_i \; \alpha_i$$

and

$$N_k - A(P + \sum_i N_i \alpha_i)\beta_k$$

respectively.

Here, for simplicity, we have made an assumption that the modes have all propagated unchanged.

The mean square amplitude, I, of all the modes is

$$I = P^2|1 - A|^2 + |A|^2Q^2 + \sum_k|N_k|^2$$

where

$$Q^2 = [|\sum_i N_i \alpha_i|^2 + |P + \sum_i N_i \alpha_i|^2 \sum_k|\beta_k|^2]$$

Here we have neglected the cancellation of higher order modes. It can be included at the expense of more algebra but it will only have a significant effect when A is small. In our optimal system we wish to choose A both to cancel the plane wave mode (i.e. for A to be unity) and not to generate a significant level of higher modes (i.e. for A to be zero). It may be desirable to weight these competing aims differently and so we write the problem as a minimisation of I, where I is now defined as

$$I = P^2|1 - A|^2 + \lambda|A|^2Q^2$$

by variations of A. Here we have ignored the constant term in the minimisation. P, Q, and A are functions of frequency but we are not able to do this minimisation independently at each frequency because A must at least be constrained to be causal.

So we minimise

$$\min_{A(\omega)} \int I(\omega)d\omega \qquad \text{subject to } A(t) = 0 \text{ for } t < 0$$

If, additionally $A(t)$ is implemented as a finite impulse response digital filter then $A(t) = 0$ for $t > nT$ where n is the number of taps and T is the sampling period.

Let us suppose then we have a filter which responds to input x_n at $t = nT$ with an output y_n in the linear fashion

$$y_n = \sum_{m=-\infty}^{\infty} a_{n-m}\, x_m,$$

where the impulse response a_n is constrained to be zero outside some range $n_1 \leqslant n \leqslant n_2$.

Defining the discrete Fourier transform (DFT) of x_n as

$$X(\omega) = \sum_{n=-\infty}^{\infty} x_n e^{in\omega T}$$

for ω in the range $|\omega| < \omega_0$ where $\omega_0 = \frac{\pi}{T}$ is the Nyquist frequency. The DFT of a_n and y_n are defined similarly as $A(\omega)$ and $Y(\omega)$ respectively. We have

$$Y(\omega) = A(\omega)\, X(\omega),$$

describing the filter response in the frequency domain. The inverse DFT is

$$x_n = \frac{1}{2\omega_0} \int_{-\omega_0}^{\omega_0} X(\omega)e^{-i\omega nT}\, d\omega.$$

Our problem is to minimise the quantity:

$$I = \int_{-\omega_0}^{\omega_0} (P^2(\omega)|1-A|^2 + \lambda Q^2(\omega)|A|^2)d\omega$$

where P^2 and Q^2 are real, positive functions with $P^2(-\omega) = P^2(\omega)$ and $Q^2(-\omega) = Q^2(\omega)$. λ is also real, $\lambda \geqslant 0$.

Let

$$g_n = \int_{-\omega_0}^{\omega_0} P^2(\omega)\, e^{-i\omega nT} d\omega$$

and

$$h_n = \int_{-\omega_0}^{\omega_0} Q^2(\omega) e^{-i\omega nT} d\omega.$$

We have $g_{-n} = g_n$ and $h_{-n} = h_n$. The g_n and h_n are real numbers.

Now, if x_n is any time series it is not difficult, using the above properties of the DFT, to obtain the results

$$\int_{-\omega_0}^{\omega_0} |X|^2\, P^2\, d\omega = \sum_{n,m=-\infty}^{\infty} x_n x_m g_{n-m}$$

and likewise

$$\int_{-\omega_0}^{\omega_0} |X|^2\, Q^2\, d\omega = \sum_{n,m=-\infty}^{\infty} x_n x_m h_{n-m}.$$

Let

$$J = \int_{-\omega_0}^{\omega_0} |1 - A|^2\, P^2\, d\omega$$

and

$$K = \int_{-\omega_0}^{\omega_0} |A|^2\, Q^2\, d\omega$$

the measures of how well the filter achieves "transparency" ($A = 1$), and yet rejects components which are not controllable ($A = 0$). The combined "cost" function is $I = J + \lambda K$, where λ determines how much weight is placed on each of the two parts of the problem. By putting $x_n = \delta_n - a_n$ ($\delta_n = 1$ for $n = 0$ and $\delta_n = 0$ for $n \neq 0$) into the first of the above results and $x_n = a_n$ into the second we have

$$J[a_n] = \sum_{n,m=n_1}^{n_2} G_{nm}\, a_n a_m - 2 \sum_{n=n_1}^{n_2} g_n a_n + g_0$$

and

$$K[a_n] = \sum_{n,m=n_1}^{n_2} H_{nm}\, a_n a_m,$$

where $G_{nm} = g_{n-m}$ and $H_{nm} = h_{n-m}$ are symmetric, Toeplitz matrices defined for $n_1 \leqslant n,m \leqslant n_2$.

The results also indicate that G and H are positive definite unless P^2 or Q^2 is identically zero in the Nyquist frequency band (in which case the corresponding matrix is identically zero). This can be shown for G as follows. Given x_n, zero outside the range $n_1 \leqslant n \leqslant n_2$, the DFT

$$X(\omega) = \sum_{n=n_1}^{n_2} x_n e^{in\omega T}$$

is an analytic function of complex ω. It can therefore only be identically zero in an i n t e r v a l of the real frequency axis if it is zero everywhere and this would imply that all the x_n are zero. Otherwise, the function $|X|^2 P^2$ must be p o s i t i v e over some interval of the frequency range $|\omega| < \omega_0$ and so using the first result we have

$$\sum_{n,m=n_1}^{n_2} G_{nm}\, x_n x_m > 0$$

so that G is positive definite.

Now

$$I = J + \lambda K = \sum_{n,m=n_1}^{n_2} B_{nm}\, a_n a_m - 2 \sum_{n,m=n_1}^{n_2} g_n a_n + g_0$$

where $B_{nm} = G_{nm} + \lambda H_{nm}$. Setting the variation of I with respect to each a_n $(n_1 \leqslant n \leqslant n_2)$ to zero we obtain the matrix set

$$\sum_{n,m=n_1}^{n_2} B_{nm}\, a_m = g_n \quad (n_1 \leqslant n \leqslant n_2)$$

to solve for the optimal filter a_n. Because of the properties of G and H, the matrix B is real, symmetric, Toeplitz and positive definite.

Referring back to the above expression for I we see that for variations away from the optimal filter we have

$$\delta I = \sum_{n,m=n_1}^{n_2} B_{nm}\, \delta a_n\, \delta a_m$$

so that the positive definite nature of B_{nm} implies that we have a global minimum for I.

Having obtained the equations for the optimal filter, we examine the behaviour of our two measures of optimality, J and K, as the weight factor λ is changed. Thus we look at $J_0(\lambda) = J[a_n]$, $K_0(\lambda) = K[a_n]$, evaluated at the optimal filter $a_n(\lambda)$ for any value of $\lambda \geqslant 0$. We show that $J_0(\lambda)$ is an increasing function of λ, while $K_0(\lambda)$ is a decreasing function of λ.

Let $0 \leqslant \lambda_1 < \lambda_2$ and $a_n(\lambda_1)$, $a_n(\lambda_2)$ be the optimal filter coefficients. Since $a_n(\lambda_1)$ is optimal for $\lambda = \lambda_1$, we have

$$J[a_n(\lambda_1)] + \lambda_1 K[a_n(\lambda_1)] \leqslant J[a_n(\lambda_2)] + \lambda_1 K[a_n(\lambda_2)]$$

or

$$J_0(\lambda_1) - J_0(\lambda_2) \leqslant \lambda_1(K_0(\lambda_2) - K_0(\lambda_1))$$

Likewise optimality of $a_n(\lambda_2)$ implies

$$J_0(\lambda_1) - J_0(\lambda_2) \geqslant \lambda_2(K_0(\lambda_2) - K_0(\lambda_1)).$$

Supppose that $J_0(\lambda_1) > J_0(\lambda_2)$, then the penultimate inequality implies $K_0(\lambda_2) > K_0(\lambda_1)$. We may then divide by $K_0(\lambda_2) - K_0(\lambda_1)$ to show that $\lambda_2 \leqslant \lambda_1$ which contradicts our choice of λ_1 and λ_2. Thus $J_0(\lambda_1) \leqslant J_0(\lambda_2)$. Similarly $K_0(\lambda_1) \geqslant K_0(\lambda_2)$. Thus J_0 is an increasing and K_0 a decreasing function of λ. This is a reasonable result since increasing λ place more weight on K and less on J.

As $\lambda \to \infty$ the matrix equations for $a_n(\lambda)$ imply that $a_n(\lambda) \to 0$ which is not surprising since $\lambda = \infty$ places all the weight on making $|A|^2 = 0$. The functions $K_0(\lambda) \to 0$ and $J_0(\lambda) \to g_0$.

As $\lambda \to 0$, if the range $n_1 \leqslant n \leqslant n_2$ includes $n = 0$, then the matrix equations for $a_n(\lambda)$ show that $a_n(\lambda) \to \delta_n$ and the filter is the identity $A = 1$. This also is expected. If this is the case, then $J_0(\lambda) \to 0$ and $K_0(\lambda) \to h_0$. If the range $n_1 \leqslant n \leqslant n_2$ does not include $n = 0$, then in general we will have $J_0(\lambda) \to j_0$; $K_0(\lambda) \to k_0$; some positive values, as $\lambda \to 0$.

Taking the two inequalities for J_0 and K_0 and dividing by $\lambda_2 - \lambda_1$, then taking $\lambda_1 \to \lambda_2$ shows that

$$\frac{dJ_0}{d\lambda} + \lambda \frac{dK_0}{d\lambda} = 0$$

and so

$$\frac{dJ_0}{dK_0} = -\lambda.$$

Let us now sketch the curve of optimal solutions in the (K_0, J_0) plane as λ varies. At $\lambda = 0$ we have $\frac{dJ_0}{dK_0} = 0$, $J_0 = j_0$,

$K_0 = k_0$. At $\lambda = \infty$ we have $\frac{dJ_0}{dK_0} = -\infty$, $K_0 = 0$, $J_0 = g_0$. The curve must look like the sketch in Figure 1 if $n = 0$ lies in the range $n_1 < n < n_2$ and like the sketch in Figure 2 if it does not.

These curves show the trade-off that can be made between the competing objectives of transparency and rejection of uncontrollable components of the sound field.

In order to illustrate the solution presented here we have calculated the optimal filter for an example problem. In this example the spectral levels of P^2 and Q^2 are shown in Figure 3. Figure 4 shows the curve of I as a function of frequency on a decibel scale for three cases i) with $A = 0$ (i.e. no control), ii) with A determined according to the above analysis (i.e. with optimal control), and iii) with A determined independently for each frequency (i.e. without the constraint that A should be causal). The latter case is that which is normally presented as being the best achievable performance and is often called the coherence limit. It is clear that the constraint that the filter A be causal severely limits the performance and thus the use of the coherence as a measure of achievable performance should be used with care.

FIGURE 1

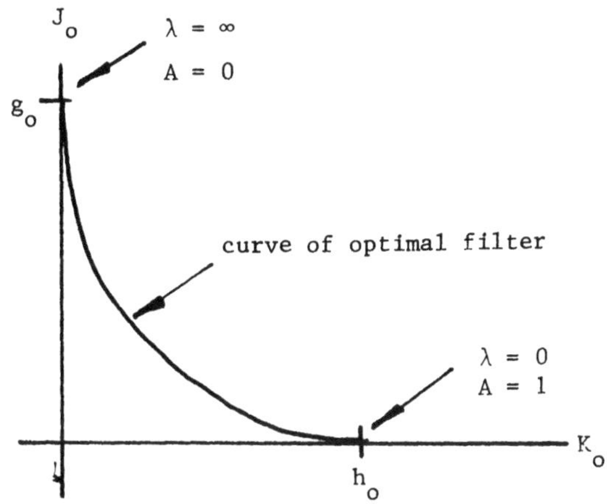

FIGURE 2

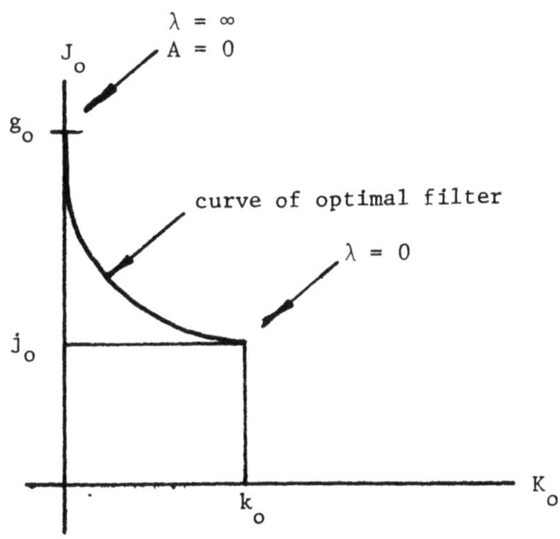

FIGURE 3

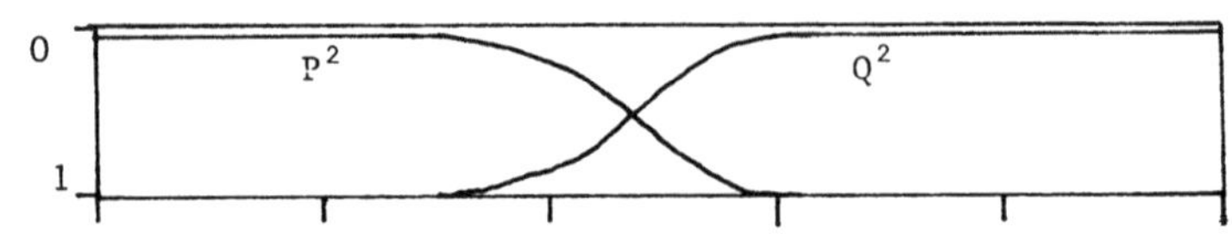

FIGURE 4

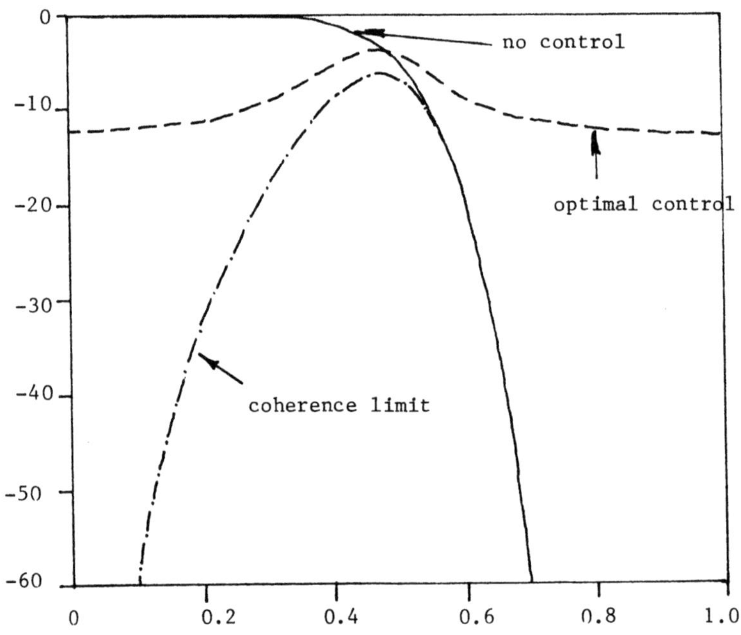

Acoustic Pressure and Intensity Distributions in Finite Length Active Noise Control Systems

M.A. GALLAND and M. SUNYACH

Laboratoire de Mécanique des Fluides et d'Acoustique

UA CNRS 263
Ecole Centrale de Lyon - B.P. 163 - 69131 ECULLY FRANCE

Summary

This paper deals with monopolar active sound absorption in a finite length duct. A time dependent numerical technique for computation of isopressure contours and maps of intensity vectors yields results which are in good accord with experimental findings. It appears that a reduction in the level of reflections in cancelling source location is acheived when a lining covers one wall in the duct. Suitable adjustment of the internal source impedances shorters the range of cancelling distance, wherin absorption takes place.

Introduction

The performance obtained by an active noise control system is intimately related to the characteristics of the duct in which the system is installed (geometry, wall impedance). This is particularly true for the tight-coupled monopole system first described by OLSON and MAY and later successfully used by EGHTESADI and LEVENTHALL. Indeed TRINDER and NELSON demonstrated theoretically that a tight-coupled monopole is especially effective at the longitudinal resonant frequencies of the duct. These authors also pointed out that some acoustic lining of the duct is necessary to minimize the disturbing effect of the transverse modes.
The lining seems also to be important in the system described by LUEG and by CHAPLIN and others at Essex Univ., in which the detection microphone is located at some distance upstream of the cancelling source.
This last case has been investigated in the work of SUNYACH and COMTE-BELLOT, which included a detailed analysis of the acoustic fields computed in an infinite length duct, as well as measurements made with anechoic terminations. Our attention is

Aero- and Hydro-Acoustics
IUTAM Symposium Lyon 1985
Editors: G. Comte-Bellot and J. E. Ffowcs Williams

now mainly focused on more practical cases and the study of finite length or acoustic loading effects.
In the first part we discuss, for the purpose of computation, the choice of a time dependent difference technique, which can handle sound propagation in different cases including variations of cross sectional area, wall liner impedance and gradients in flow velocity. In a second part we give a comparison between computed pressure fields and those experimentally obtained. Such comparisons are made for different cases in a hard-walled or fiberglass-lined rectangular duct at frequencies below cut-off. Finally an analysis of intensity vector fields demonstrates the mutual impedance effect discussed by ROEBUCK and the exact role played by the lining on the acoustic energy balance.

The time dependent difference approach in duct acoustics

1) Choice of a convenient technique

Time dependent numerical procedures for computing pressure distributions offer an advantage, as compared with steady state of finite element procedures, insofar as less computer storage is required ; however, the calculations for a cut-off or evanescent mode in a duct (BAUMEISTER) encountered convergence difficulties or instabilities which were attributed to singularities of the transient solution. For this reason CABELLI developed an implicit procedure which generates a false transient to enhance convergence to steady state solution.
In the present work, the geometrical configuration differs from that of BAUMEISTER in two significant aspects.
i) the sources are located at the wall and not in the initial cross section
ii) the length is finite. Furthermore, as we are concerned mainly with low frequencies the major component at the terminal sections of the duct is a plane wave, whereas evanescent modes are only present in the vicinity of the source. For these reasons a strong asymmetry of the field is expected and it is

not clear that the instability observed in other cases persists. At will be shown later, the solutions seem to be correct after the transient effects have disappeared. The modal decomposition of the field used by CABELLI does not appear to be necessary for our purpose.

2) Computation of acoustic fields in finite length ducts

Consider the rectangular duct sketched in Fig. 1 and the relevant two-dimensional wave equation for the pressure

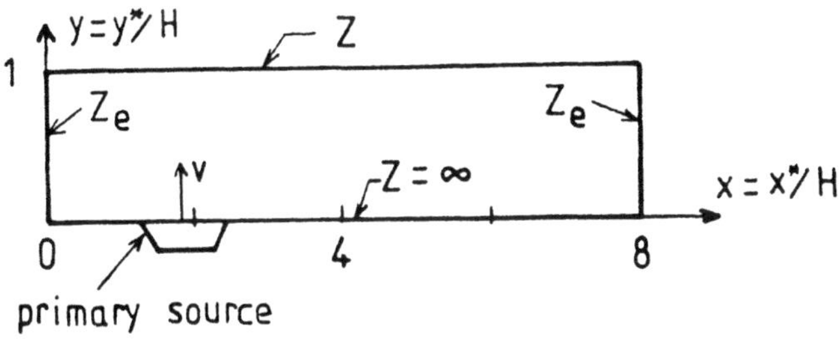

Fig. 1 - The coordinate system and characteristics of the duct.

$$(1) \qquad \eta^2 \frac{\partial^2 p}{\partial t^2} = \frac{\partial^2 p}{\partial x^2} + \frac{\partial^2 p}{\partial y^2}$$

where the dimensionless variables x,y,t and the parameter η, are scaled with reference to the height H^* and the frequency $\omega^*/2\pi$; thus

$$x = \frac{x^*}{H^*} \quad , \quad y = \frac{y^*}{H^*} \quad , \quad t = \frac{\omega^*}{2\pi} \cdot t^* \quad , \quad \eta = (\frac{H^*}{2\pi}) \frac{\omega}{C_0}$$

in term of the dimensional variables x^*, y^*, t^*, and the appropriate ranges are $0 < x < L^*/H^*$

$0 < y < 1$

If the plane terminations of the duct have a common specific acoustic impedance Z_e, the upper (soft-walled) surface an impedance Z and the source, with uniform normal velocity v(x,0,t) occupes a portion S of the lower rigid surface, the boundary conditions which specify the appropriate solution of the wave equation (1) take the form

$$\frac{\partial p}{\partial x} = \pm \frac{\eta}{Z_e} \frac{\partial p}{\partial t} \qquad x = 0, \quad \frac{L^*}{H^*} \quad ; \quad 0 < y < 1$$

$$\frac{\partial p}{\partial y} = - \frac{\eta}{Z} \frac{\partial p}{\partial t} \qquad y = 1 \; ; \; 0 < x < \frac{L^*}{H^*}$$

$$\frac{\partial p}{\partial y} = - 2i\pi\eta \; v(x,0,t) \qquad x \text{ in } S \quad y = 0$$

$$\frac{\partial p}{\partial y} = 0 \qquad y = 0 \quad x \text{ outside } S$$

The dimensionless wave equation is integrated over area and time for each boundary cell, including the source, as reported by Baumeister, and the computation starts from the quiescent state at t less or equal to zero.

An example of computation is given in Fig. 2a for the case of a short duct ($L^*/H^* = 3$), with a source located symmetrically in the center of the wall. The terminations are assumed to be anechoical. It can be seen that the results are in complete agreement whith those obtained by a modal computation in which several evanescent modes are added to the first longitudinal mode (Fig. 2b).

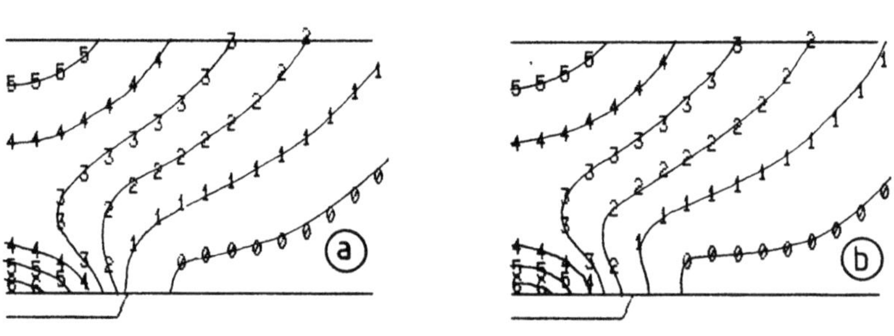

Fig. 2 - Sound pressure contours for a duct with anechoic terminations, scale .5 dB. frequency of 300 Hz.
a. Finite difference technique ;
b. Modal computation.

When reflections are produced at the ends of the duct (Fig. 1), further comparisons have been made between distributions of computed and measured pressure levels on lines parallel to the X axis. Typical examples are displayed in Fig. 3, with and without a 3 CM FIBERGLASS PANEL OPPOSITE THE SOURCES. There is a general agreement in all cases, with the largest discrepancies (less than 1.5 dB) occurring in the vicinity of the sources.

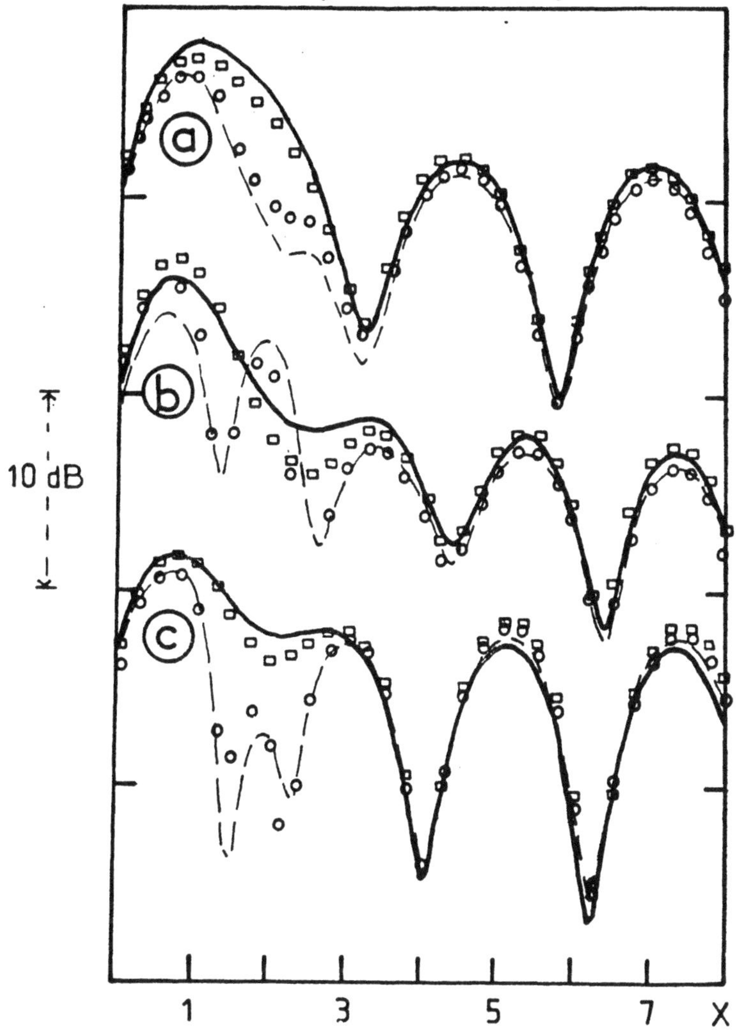

Fig. 3 - Distributions of computed and measured pressure levels.
—— , y = 0.9 ; — — , y = 0.2 (Theory) ;
□ , y = 0.9 ; o , y = 0.2 (Experiment).
a. $f^* = 300$ Hz ; $z = 1.16 - 3.63i$; $Ze = 0.15 + 0.4i$
b. $f^* = 400$ Hz ; $z = 2.49 - 2.79i$; $Ze = 0.2 + 0.6i$
c. $f^* = 400$ Hz ; $z = 4.41 + 7.35i$; $Ze = 0.2 + 0.75i$.

The latter can be attributed to back reflections produced by the loudspeaker baffles, which are neglected in the computations. We notice that in all cases the homogeneity of the measured pressure field is greater than that obtained by computation.

In addition, for one case under investigation, the time pressure profile at a given location is reported in Fig. 4. The convergence of the computation can then be observed at t = 15 and t = 25 for the lined duct and the hard-walled duct respectively. The explicit procedure is therefore expected to be suitable in the present work, despite the contribution of a multiplicity of evanescent modes affecting the acoustic field in the vicinity of the driving source.

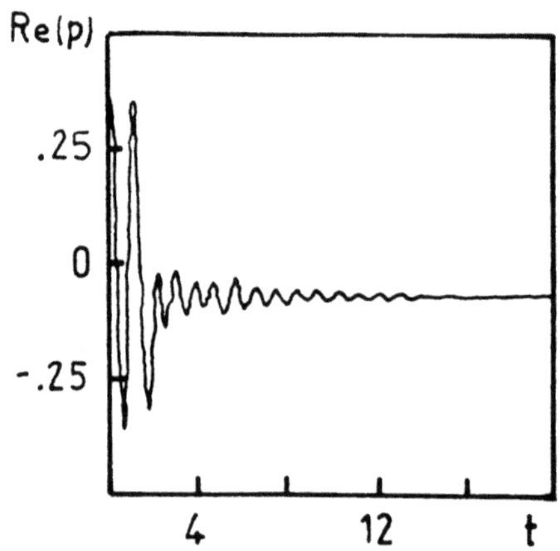

Fig. 4 - A typical time pressure profile, x = 8 ; y = .5 ; $f^* = 400$ Hz ; z = 2.49 - 2.76i ; Ze = 0.2 + 0.6i.

3) Pressure fields in the monopole system in operation

These pressure fields are obtained by superposition of the field produced by the primary source, as computed in the previous section, and a cancelling field due to a secondary source which is symmetrically placed in the duct (Fig. 5).

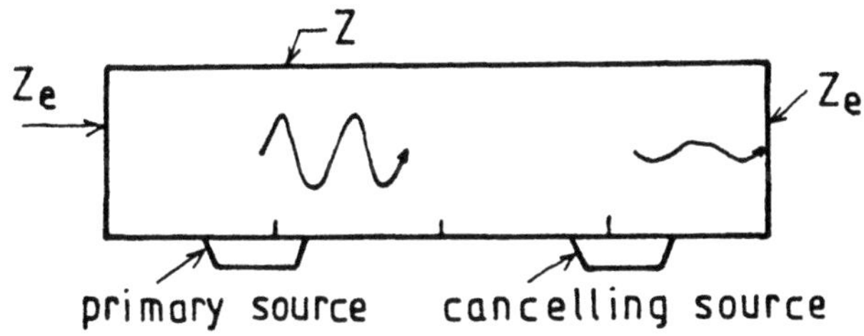

Fig. 5 - Sketch of the monopole system.

The relative amplitudes of these two fields are adjusted to achieve a theoretical attenuation of about 30 dB. Results are presented in Fig. 6 for the different cases already considered. Iso-attenuation contours are drawn, each of them being numbered by a value which represents the attenuation divided by a factor of 4. The Fig. 7 report the corresponding experimental results. The experimental and computed iso-attenuation contours show in all cases a good agreement, except downstream of the cancelling source (X^*/H^* greater than 6.5) where strong attenuation is produced. For example the curves ≠ 0 corresponding to a 28 dB attenuation extend practically up to the cancelling source axis ($X^*/h^* = 6.2$) in the computation (Fig. 6), whereas they are only observed at the extreme end of the duct in the experimental maps (Fig. 7). On the contrary, all the experimental contours numbered by 3 or higher are well predicted. Such a behaviour can be clearly related to the small differences observed between computed and measured pressure fields reported in the previous section. Indeed, from comparison of Fig. 6 and 7, it can be seen that a significant effect of the lining at a given frequency, is the strong reduction of the inhomogeneity in the vicinity of the sources. A similar effect can be produced by the abrupt change in impedance occurring in the rigid wall at the source location. As a consequence the final result is an enhancement of the cancelling mechanism in the experiment.

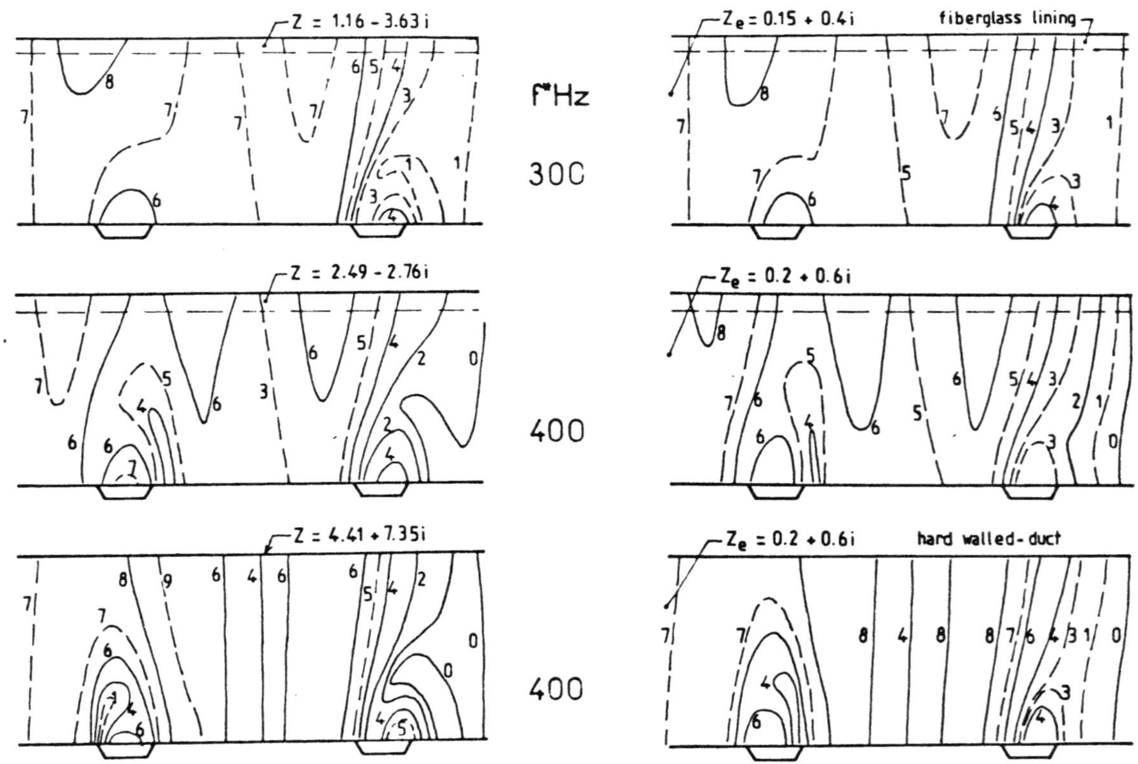

a. Theory.

b. Experimental results.

Fig. 6 - Iso-attenuation contours for on operating monopole system.

Intensity vector fields

1) Computed intensity fields

The intensity vector fields are expressed by the classical formulas :

$$I_x = \frac{1}{2} R_e \ (p_\omega \ v^*_{\omega x}) \quad ; \quad I_y = \frac{1}{2} R_e \ (p_\omega \ v^*_{\omega y})$$

where

$$v_{\omega x} = \frac{1}{i\omega\rho_0} \frac{\partial p_\omega}{\partial x} \quad ; \quad v_{\omega y} = \frac{1}{i\omega\rho_0} \frac{\partial p_\omega}{\partial y}$$

results are obtained readily from the finite difference procedure and are given for a forcing frequency of 400 Hz.
For the primary source alone, Fig. 7a, the intensity vector field permits us to define the scale of the incident field to be suppressed. The attenuation effect of the wall lining is also illustrated. The regular pattern obtained gives us confidence in the computation process.

When the active sound absorber is in operation, Fig. 7b, it can be observed that the acoustic energy involved is radiated by the primary source as well as by the cancelling source. The total flux is then absorbed by the lined wall in the high pressure regions (Cf. Fig. 6a - 6b). The energy flux radiated downstream (depending on the frequency and wall admittance) is in the present case only 1/2 of that which would be radiated if this source were acting alone. For the limiting case where $Z = \infty$, it is possible to show that the monopole system appears to behave as a perfect reflector. In other words, all the pressure field downstream of the primary source would be reactive everywhere. The fiberglass lining will then weaken the reflexion at the cancelling source and consequently reduce the stationary wave ratio in the duct.

2) Experimental fields

The probe and the equipment are similar to those previously used by Sunyach and Comte-Bellot. The results are reported in

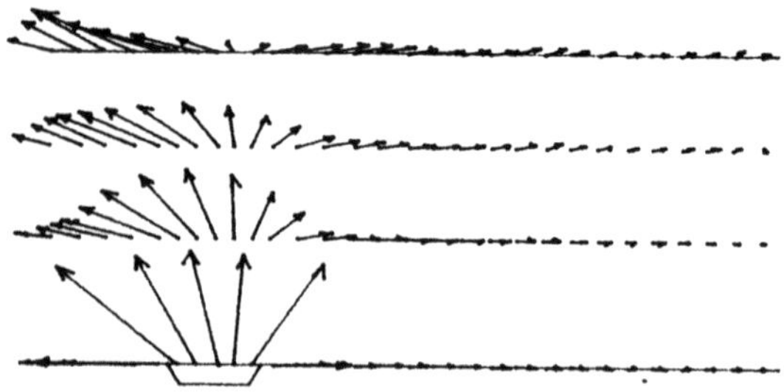

a. Primary source alone (Theory).

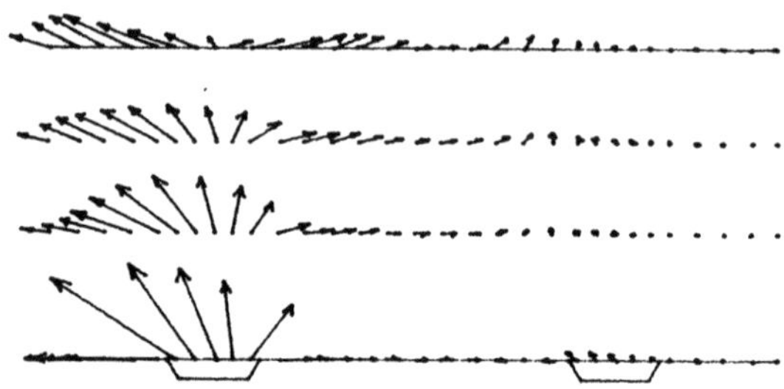

b. Monopole system in operation (Theory).

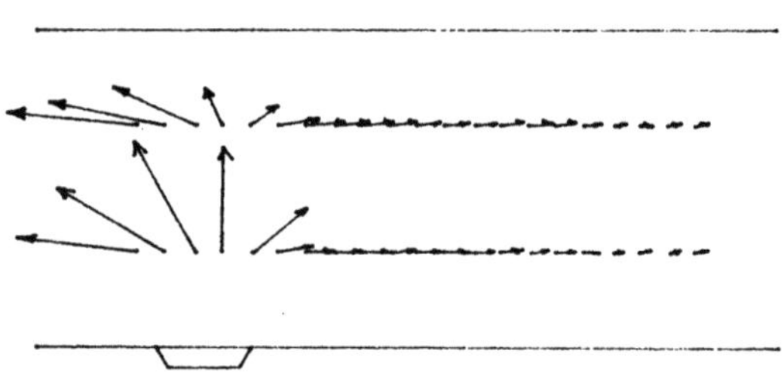

c. Primary source alone (Experiment).

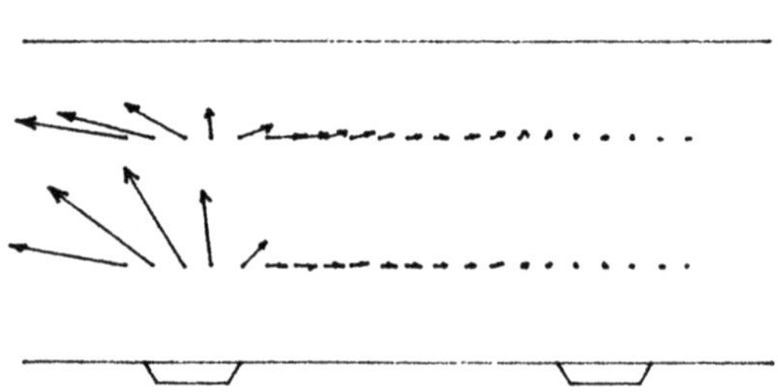

d. Monopole system in operation (Experiment) .

Fig. 7 - Distribution of acoustic intensity vector.
$Z = 2.49-2.76i$; $Ze = 0.2+0.6i$; frequency of 400 Hz.

Fig. 7c and 7d for the lined duct and the frequency of 400 Hz. The measured and the computed energy flux patterns are in general agreement when the fiberglass lining is in place. The fact that energy flux issues from both the primary and the cancelling source, and is directed towards the fiberglass is then verified.

Conclusions

Investigations relative to pressure and intensity distributions within a short duct which contains a monopole active sound absorption system indicate that a lining on the duct wall may be used to improve the wide band performance of the system. An improved capability results from the redistribution of energy flux towards the lined wall and a reduction in the stationary wave ratio, which permit a simpler realisation of transfer function or control.

References

1. Olson, H.F. ; May, E.G. ; "Electronic sound absorber" J. Acoust. Soc. Am. 25 (1953) 1130-1136.
2. Egthesadi, Kh. ; Leventhall, G. ; "Active attenuation of noise. The monopole system" J. Acous. Soc. Am. 71 (1982) 608-610.
3. Trinder, M.C.J. ; Nelson, P.A. "Active noise control in finite length ducts" J. Sound Vib. 89 (1983) 95-105.
4. Lueg, P. "Process of silencing sound oscillations" U.S. Patent n° 2 043 416, 1936.
5. Chaplin, B. ; "Antinoise. The ESSEX breakthrough" Chartered Mechanical Engineer (Jan. 1983).
6. Sunyach, M. ; Comte-Bellot, G. ; "Acoustic pressure and intensity fields in active noise control system" A.I.A.A. Paper 84.2271, Williamsburg, Virginia (1984).
7. Roebuck, I. ; "Active noise control" I.U.T.A.M. Symposium Aero and Hydro-acoustics, Ecole Centrale de Lyon, July 1985.
8. Baumeister, K.J. ; "Time-dependent difference theory for noise propagation in a two-dimensional duct" A.I.A.A. Journal 18 (1980) 1470-1476.
9. Cabelli, A. ; "Duct acoustics-A time dependent difference approach for steady state solutions" J. Sound Vib. 85 (1982) 423-434.
10. Baumeister, K.J. ; "Influence of exit impedance on finite difference solutions of transient acoustic mode propagation in ducts" J. Eng. Ind. Transactions A.S.M.E. 104 (1982) 113-120.

The Active Minimization of Harmonic Enclosed Sound Fields

P.A. NELSON, A.R.D. CURTIS & S.J. ELLIOTT

Institute of Sound and Vibration Research,
University of Southampton,
England.

ABSTRACT

The total time averaged acoustic potential energy in a steady harmonic sound field can be expressed as a positive definite quadratic function of the complex strengths of a number of "secondary sources" of sound introduced into the enclosure. For a given number and location of secondary sources, there is a unique set of complex source strengths which minimises this potential energy. This analysis is applied to the case of a lightly damped enclosure excited by a point primary source at a frequency well above the Schroeder large room frequency. It is demonstrated that in this case, the maximum reduction that can be achieved in the potential energy of the sound field is critically dependent on the relative locations of primary and secondary sources.

1. INTRODUCTION

The work presented here is primarily motivated by the problem of the low frequency sound produced in the cabins of propeller driven aircraft. The sound field produced is of a (relatively) steady periodic nature and consists of a series of tones harmonically related to the blade passage frequency. At the lower end of this spectrum, transmission of sound into the cabin is very difficult to prevent by conventional means. In view of the form of this sound field, and especially since the acoustic wavelength at the frequencies involved is typically of the order of the cabin dimensions, the application of active techniques for the suppression of the field is worthy of consideration. The theoretical techniques outlined below are intended to help establish the extent to which a sound field of this type can be suppressed by the introduction of secondary sources. In particular one seeks to determine the degree to which global reductions in the level of the sound

Aero- and Hydro-Acoustics
IUTAM Symposium Lyon 1985
Editors: G. Comte-Bellot and J. E. Ffowcs Williams

field are possible. This for example, could dictate whether in practice one should seek to produce a widespread reduction in pressure fluctuations or attempt to selectively redistribute the acoustic energy in the field.

2. THEORETICAL BACKGROUND

The analysis presented here uses a similar approach to those presented by Gaudefroy [1] and Piraux and Nayroles [2] in dealing with the active suppression ofthree dimensional sound fields. In this case, attention is restricted to enclosed sound fields (having a harmonic time dependence of the form $e^{j\omega t}$) where it is assumed that the field can be adequately represented by a finite series of "N" modal contributions. This enables the complex pressure amplitude at position $\underline{x}$ to be expressed as

$$p(\underline{x},\omega) = \underline{\psi}^T \underline{a} \tag{1}$$

where $\underline{\psi}$ is the N'th order vector of normalised characteristic functions and $\underline{a}$ is the N'th order vector of complex mode amplitudes. The latter can be expressed as a linear superposition of contributions from the primary source distribution and a finite number "M" of point secondary sources introduced into the enclosure. Thus

$$\underline{a} = \underline{a}_p + \underline{B}\,\underline{q}_s \tag{2}$$

where $\underline{a}_p$ is the N'th order vector of mode amplitudes due to the primary source distribution only and $\underline{q}_s$ is the M'th order vector of complex secondary source strengths. The matrix $\underline{B}$ is of order NxM and consists of an array of complex coefficients which quantify the extent to which each mode is excited by each secondary source. Note that in the (hypothetical) case of a finite number N of contributing modes and a number M=N of secondary sources, then $\underline{a}$ could be set to zero by choosing $\underline{q}_s = \underline{B}^{-1}\underline{a}_p$. This implies that $p(\underline{x},\omega)$ could be made zero throughout the enclosure irrespective of the form of the primary source distribution, provided the secondary sources are placed such that $\underline{B}$ is non-singular. Such a result is, of course, impossible to realise in practice since an infinite number of modes is required to give an exact representation of the sound field. It is more relevant to deal with the case where M<<N, when N is large enough for equation (1) to give a resonable approximation to the sound field considered. In this instance one may seek a least squares solution to the system of complex linear equations represented by (2). In fact, it is logical to seek a solution for $\underline{q}_s$ which minimises the sum of the squared moduli of the complex mode amplitudes. This in turn is directly proportional to the total time averaged acoustic potential energy in the enclosure [3] which can be written as

$$E_p = \frac{V}{4\rho_o c_o^2} \underline{a}^H \underline{a} \tag{3}$$

where V is the enclosure volume, ρ_o and c_o are the density and sound speed of the enclosed medium and the symbol H denotes the Hermitian transpose of a vector or matrix. Substitution of equation (2) into equation (3) shows that the energy E_p is related to the vector of complex secondary source strengths $\underline{q}_s$ by

$$E_p = \frac{V}{4\rho_o c_o^2} \left[\underline{q}_s^H \underline{B}^H \underline{B}\, \underline{q}_s + \underline{q}_s^H \underline{B}^H \underline{a}_p + \underline{a}_p^H \underline{B}\, \underline{q}_s + \underline{a}_p^H \underline{a}_p \right] \tag{4}$$

This is a real quadratic function of the complex vector $\underline{q}_s$ which can be written in its most general form as

$$E_p = \underline{q}_s^H \underline{A}\, \underline{q}_s + \underline{q}_s^H \underline{b} + \underline{b}^H \underline{q}_s + c \tag{5}$$

where the matrix $\underline{A} = (V/4\rho_o c_o^2)\underline{B}^H \underline{B}$ is in general Hermitian (although real symmetric in this case), the vector $\underline{b} = (V/4\rho_o c_o^2)\underline{B}^H \underline{a}_p$ is a complex vector or order M, and c is given by $(V/4\rho_o c_o^2)\underline{a}_p^H \underline{a}_p$. The latter corresponds to the acoustic potential energy E_{pp} in the enclosure due to the primary source distribution only. This particular form of function has the property that, provided the matrix $\underline{A}$ is positive definite, then there is a unique value of the vector $\underline{q}_s$ which defines the minimum value of the function. In the general case, then this minimum is defined by $\underline{q}_{so} = -\underline{A}^{-1}\underline{b}$ with a corresponding minimum value E_{po} of the acoustic potential energy given by $E_{po} = c - \underline{b}^H\underline{A}^{-1}\underline{b}$. (An introductory treatment of the use of quadratic forms in the complex case is given in [4]). Thus for a given number and location of secondary sources, there is a unique set of complex secondary source strengths which minimises E_p. This optimal vector of secondary source strengths is given by

$$\underline{q}_{so} = - [\underline{B}^H\underline{B}]^{-1} \underline{B}^H\underline{a}_p \tag{6}$$

The corresponding minimum value E_{po} of the acoustic potential energy can be expressed relative to the acoustic potential energy E_{pp} due to the primary source distribution only. This ratio is given by

$$\frac{E_{po}}{E_{pp}} = 1 - \frac{\underline{a}_p{}^H\underline{B}\ [\underline{B}^H\underline{B}]^{-1}\ \underline{B}^H\underline{a}_p}{\underline{a}_p{}^H\ \underline{a}_p} \tag{7}$$

It is therefore possible, for an enclosure excited by a known harmonic primary source distribution, to produce an unequivocal determination of the maximum achievable reduction in the acoustic potential energy, given a certain number and location of the secondary sources of sound.

3. RESULTS AT HIGH MODAL DENSITIES

For illustrative purposes, this result can now be applied to the particular case of a lightly damped rectangular enclosure excited by a point primary source located at $\underline{x}_p$. The energy in the enclosure is minimised by adjusting the complex strength of a single point secondary source located at $\underline{x}_s$. Using Morse's

[5] solution for a lightly damped enclosed sound field, it can be shown [6] that equation (7) reduces to

$$\frac{E_{po}}{E_{pp}} = 1 - \frac{\left[\sum_{n=1}^{N} \left|A_n(\omega)\right|^2 \psi_n\left(\underline{x}_p\right)\psi_n\left(\underline{x}_s\right)\right]^2}{\left[\sum_{n=1}^{N} \left|A_n(\omega)\right|^2 \psi_n^2\left(\underline{x}_p\right) \sum_{n=1}^{N} \left|A_n(\omega)\right|^2 \psi_n^2\left(\underline{x}_s\right)\right]} \tag{8}$$

where $\psi_n(\underline{x})$ is the normalised characteristic function associated with the n'th mode, and $A_n(\omega) = \omega/[2\zeta_n\omega_n\omega - j(\omega_n^2-\omega^2)]$ where ω_n and ζ_n are the natural frequency and damping ratio of the n'th mode. If the acoustic wavelength at the frequency of interest is sufficiently small compared to the enclosure dimensions, then the modal summations in (8) can be expressed as integrals [6]. Evaluation of these yields

$$\frac{E_{po}}{E_{pp}} = 1 - \left[\frac{\sin\left(\omega r/c_o\right)}{\omega r/c_o}\right]^2 \tag{9}$$

where $r = |\underline{x}_p - \underline{x}_s|$ is the distance separating the primary and secondary sources. This demonstrates the (intuitively obvious) result that at high frequencies, no appreciable reduction in E_p can be produced by a secondary source that is placed a distance from the primary source that is greater than one half wavelength at the excitation frequency. Also, note that an identical dependence on non-dimensional separation distance can be demonstrated for the maximum achievable reduction in radiated power for a pair of harmonic point sources in a free field [4].

4. CONCLUDING REMARKS

The application of the above theoretical approach consistently demonstrates that global energy reductions are only possible at high frequencies (modal densities) provided primary and secondary sources are not widely separated relative to the wavelength at the frequency of interest. However, results of computer simulations [7] of low modal density sound fields also demonstrate that strongly excited resonances can be suppressed by appropriately placed secondary sources. These may be separated from the primary source by a distance of greater than half a wavelength and still achieve substantial global reductions in acoustic energy.

REFERENCES

1. GAUDEFROY, A: 1980 Examples d'attenuation active de L'energie acoustique du moyen de sources monopolaires voisines d'une source primaire poncteulle. Comptes Rendus Acad. Sc. Paris, Serie B 291 299-301.

2. PIRAUX, J & NAYROLES, B: 1980 A theoretical model for active noise attenuation in three dimensional space. Proc. Internoise 703-706.

3. PIERCE, A.D.: 1981 Acoustics: an introduction to its physical principles and applications. (McGraw-Hill, New York).

4. NELSON, P.A., CURTIS, A.R.D., & ELLIOTT, S.J.: 1985 Quadratic optimisation problems in the active control of free and enclosed sound fields. Proc. IOA 7 45-54.

5. MORSE, P.M.: 1948 Vibration and Sound. McGraw-Hill.

6. NELSON, P.A., CURTIS, A.R.D., ELLIOTT, S.J. & BULLMORE, A.J: 1985 The active minimisation of harmonic enclosed sound fields Part I:Theory. (to be submitted to the Journal of Sound and Vibration).

7. BULLMORE, A.J., NELSON, P.A., ELLIOTT, S.J.: 1985 Active control of harmonic sound fields of low modal density; a computer simulation. Proc. IOA 7 55-64.

Source Mechanisms

Source Identification Techniques – A Critical Evaluation

W. G. RICHARZ

Department of Mechanical and Aeronautical Engineering
Carleton University, Ottawa, Ontario, Canada

Abstract

Experimental methods aimed at source identification continue to evolve. In the last decade, computers, programmable or dedicated, have provided the experimenter with unprecedented data processing capabilities. Highlights of several techniques are presented along with comments regarding limitations and characteristic results.

1. Introduction

Source identification has found application in both fundamental and applied aeroacoustic research. The techniques usually combine acoustic and other relevant data to provide estimates of the nature of the sources of sound and their spatial distribution. Each technique makes explicit or implicit use of an underlying source model. Thus the resulting estimates are not unique [1]. This is especially true for source identification systems that rely solely on acoustic data measured in the far field. It is beyond the scope of this note to examine and comment on the considerable literature devoted to the subject. Thus only a few source identification techniques are examined.

2. Far Field Methods

Source identification is a diagnostic tool for applied as well as basic research. On the applied side the techniques find application in rank ordering the various sources, and permit one to obtain a virtual image of the sources. Far field methods employ instruments such as the acoustic mirror [2], the acoustic telescope [3], and the polar correlation technique [4]. The acoustic mirror (Fig.1) is an elliptical reflector: a microphone located at one focus collects energy emitted by sources of sound at the other. The construction of such a mirror is

Aero- and Hydro-Acoustics
IUTAM Symposium Lyon 1985
Editors: G. Comte-Bellot and J. E. Ffowcs Williams

straightforward. With a certain degree of care it is possible to deduce source distributions and to demonstrate closure, that is accounting for virtually all of the energy. The limitations of the method appear to be well understood, and is seen as a cost-effective diagnostic tool.

The constraint of a fixed focal length has led to alternate systems. The acoustic telescope and the polar correlation technique provide this flexibility at the expense of system complexity. Both reconstruct the source field by fitting, in a least squares sense, certain coherence data to an assumed source model. The investment in hardware (microphones and computer) is considerable. Thus such high performance systems are not likely to be used for 'quick-look' applications. Innovative data processing and (in the case of the acoustic telescope) array configuration has expanded the capabilities of the systems permitting, as shown in figure 2, impressive detail in the reconstruction of the source region [5,6]. Still, the spatial resolution is limited to order $(S/d)(f/c_0)$.

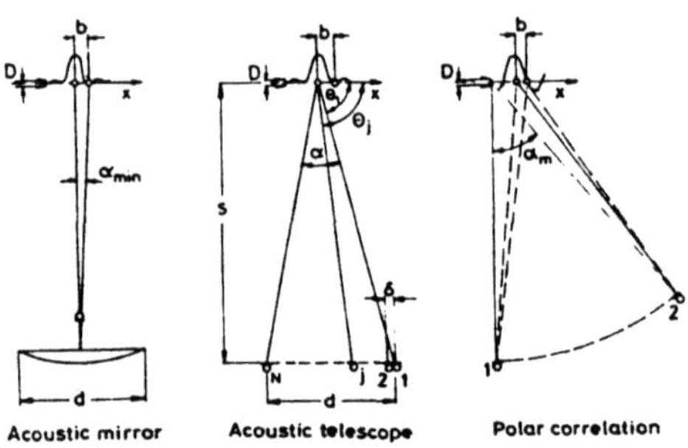

Fig.1 Schematic of Far Field Source Location Systems [8]

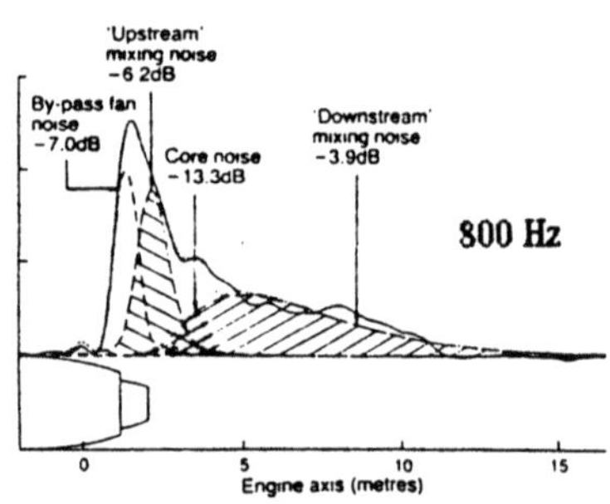

Fig.2 Capabilities of Far Field Source Location Technique [5]

3. Other Correlation Techniques

If one is concerned solely with the identification of the most likely position of a source emitting energy in a narrow frequency band, then a two microphone near-field far-field correlation can be used. The method makes no assumptions regarding the nature of the sources, relying on time delay information to locate the source on a hyperbola [7]. Superposition of results for several near-field positions yields the point (or region) of intersection of the hyperbolae. This then is taken as the most likely source position. Source distributions deduced therefrom compare with the polar correlation technique (Fig.3).

A detailed image of the source region is possible, when the correlation of a quantity related to the sound generated and the far field pressure is measured (causality correlation) [8]. These source identification techniques require highly sophisticated measurement instruments. Thus their application has been limited to basic research projects. In addition, one is usually constrained to the measurement of but a single family of sources. To data, credible measurements are available for flow noise [9,10] combustion noise [11], and airframe noise [12].

Correlations and cross-spectral densities play important roles in the diagnostic processes. Correlations are most useful in the identification of distinct sources. The time delay figuring strongly; it permits one to identify the location of sources. Seiner and his colleagues [13] were able to identify the source location of shock cell noise by means of such a technique. This author has performed diagnostics of an acoustically excited jet [14] and observed a variety of disturbances propagating at various velocities (Fig.4).

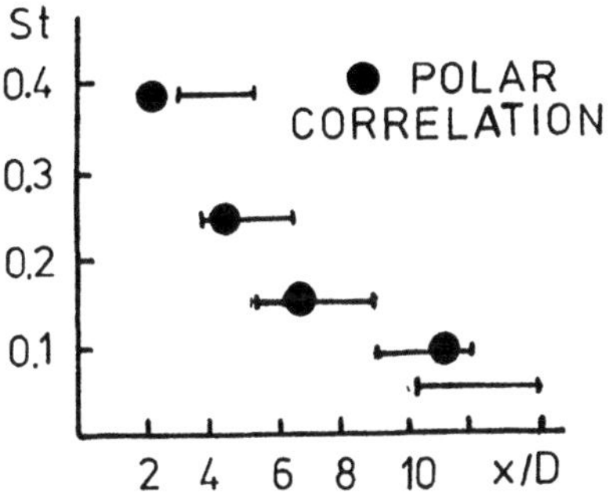

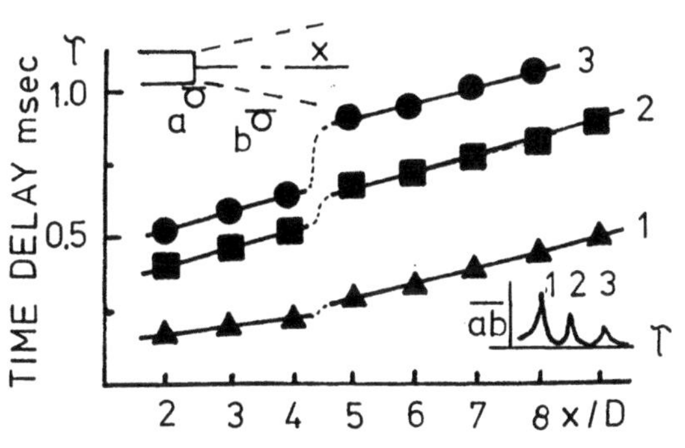

Fig.3 Source Location via Near Field Far Field Correlations ├──┤

Fig.4 X-T Diagram for Pressure Fields near an Excited Subsonic Jet [14]

The cross-spectral density contains the same information, albeit in the frequency domain. At times, interpretation is hampered by the subtle changes in phase, and temporal data is perferable. Cross-spectral densities have the attribute of allowing post-processing of data multiplication by $(2\pi if)^n$, for example, is equivalent to differentiating n times in the time domain, an operation fraught with error. Although cross-correlations and cross-spectral densities are powerful diagnostic tools, one must be mindful of their limitations. In many instances the variables under examination are sensed with dissimilar transducers. The relative difference in their respective phase and amplitude responses may introduce spurious information (Fig.5). Even similar transducers such as microphones may require precise phase matching in critical applications such as the estimation of acoustic intensity.

The examples cited above deal with source identification. The basic technique however is not limited to source identification. Source path identification is another possibility. Ver [15] has illustrated a rather clever means of source path identification using a reciprocity principle. Here vibrations induced by sound generated by a controlled source in an aircraft cabin was monitored at various points of the airframe. The resultant transfer function can be taken as a measure of the effectiveness of the vibration/sound transmission path.

Although correlation techniques are powerful tools, the user must take care not to forget the physics of the sound generation mechanism. Propeller noise, for example, is known to scale with thrust and applied torque. A direct correlation between the local lift of a blade section and the far field sound would yield virtually a null result. The reason, clearly, is the neglect of the kinematics of the alleged source.

4. Signal Averaging

The continued improvements in data handling has provided enhanced signal processing capabilities. Signal (phase) averaging, for example, is an analytical as well as experimental tool used in the diagnosis of orderly structures. The technique requires that the process under observation be a periodic one. This requirement can be relaxed somewhat, in certain instances. A periodic signal can also be identified by means of an autocorrelation as the asymptotic behaviour will be a periodic one. The relative phase, governing the precise time history, is lost however. The signal averaging technique "locks onto" the periodic portion of the signal and rejects spurious "noise". The rejection improves as the number of periods is increased. The result is a time-domain representation of the periodic signal.

Today there exists a large data base dealing with the flow field. Limited data is available for the sound field. The work of Morrison and his colleagues [16] and relevant experiments performed at the University of Toronto [17] illustrate the current capabilities (Fig.6). The results suggest that the sound of typical subsonic jets cannot be direftly attributed to large scale structures.

5. Acoustic Intensity

Without doubt the acoustic intensity meter will have a significant impact on source diagnostics in general. The principal applications in the aeroacoustics field will likely be an the cabin noise [18]. The technique can be applied to flow noise: one potential drawback is the need to survey an extended region near the jet. When the measurement location is moved away from the jet, the information available differs little from a single point measurement. Given the geometric constraints of the sound intensity probe, it is likely that the method should only be applied to source regions whose smallest dimension is considerably larger than that of the transducer. It is not likely that the technique will have a significant impact on flow noise diagnostics. The work of Pao and Maestrello [19] illustrates the capabilities.

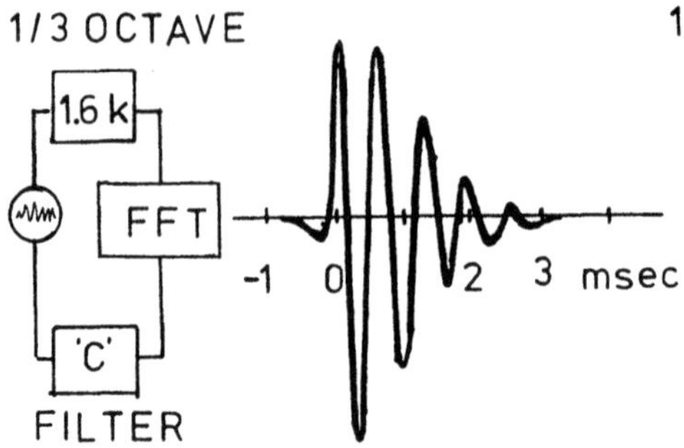

Fig.5 Influence of Phase and Amplitude Mismatch on a Cross-Correlation

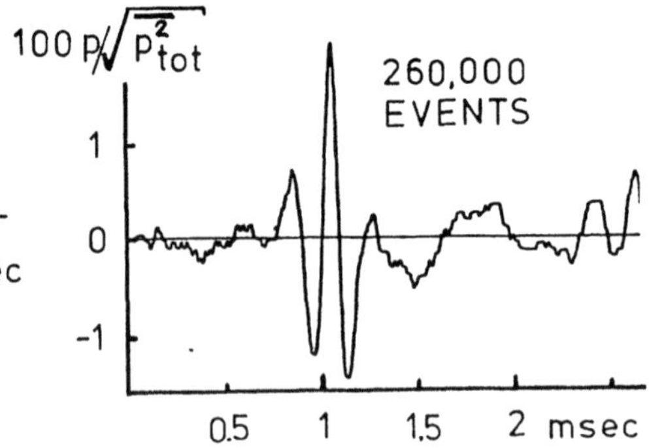

Fig.6 Direct Measurement of Sound of Large Scale Structures [17]

6. Optical Techniques

When the flow velocity approaches or exceeds the local speed of sound, many conventional flow measurement devices no longer operate effectively. The rather severe gradients in density, and pressure can be exploited by optical means. Shadow, schlieren, and interferometric techniques are the most common ones. In many instances the radiation field near the flow is sufficiently energetic to permit visualization. The present techniques lend themselves to phase (signal) averaging, allowing one to study the dynamics and to some extent the acoustics of orderly structures. Care must be taken to survey the flow field from several directions [20]: this permits identification of certain hidden modes.

It is apparent that these optical techniques will find application only in dedicated facilities, as the equipment is not portable. Still, a single picture conveys a wealth of information not available through other means. Advances in high speed photography and computer analysis and enhancement will provide considerable detail and make quantitative analysis a relatively straightforward task. In addition, there is considerable scope for development of systems with three dimensional, even holographic, capabilities.

7. Concluding Remarks

There is a continued need for reliable as well as innovative source location techniques. They aid the theorist in identification of principal physical processes, thereby permitting the introduction of reasonable approximations to the often bewildering complexity of theory of sound generation. In addition, they are a means of testing various theoretical models, and may give guidance to the noise control engineer. The virtually exponential growth of our data handling capabilities should foster development of even more sophisticated systems.

Acknowledgement

The work was supported by an operating grant of the Natural Sciences and Engineering Research Council of Canada.

References

1. Kempton, A.J.: The ambiguity of acoustic sources - A possibility for active control? J. Sound Vib. Vol. 48 (1976), 475-483.

2. Grosche, F.-R.: Distribution of sound source intensities in subsonic and supersonic jets. AGARD CP-131, Brussels 1973.

3. Billingsley, J.; Kinns, R.: The acoustic telescope. J. Sound Vib. Vol. 48 (1976) 485-510.

4. Fisher, M.J.; Harper-Bourne, M.; Glegg, S.A.L.: Jet engine noise source location: The polar correlation technique. J. Sound Vib. Vol. 51 (1977) 23-54.

5. Strange, P.J.R.; Podmore, G.; Fisher, M.J.; Tester, B.J.: Coaxial jet noise source distributions. AIAA paper 84-2361, 1984.

6. Boone, M.M.; Berkhout, A.J.: Theroy and applications of a high-resolution synthetic acoustic antenna for industrial noise measurement. Noise Control Eng. J. Vol. 23 (1984) 60-68.

7. McLean, I.R.; Richarz, W.G.: Jet noise source location via near-field far-field correlations. J. Acoust. Soc. Am., Vol. 67 S1 (1980) S4.

8. Grosche, F.-R.: How powerful are the experimental techniques of sound source location in aeroacoustics; What prospects are there for an improvement? Mechanics of Sound Generation in Flows E.-A. Mueller ed. Springer 1979.

9. Schaffar, M.: Direct measurement of the correlation between axial in-jet velocity fluctuations and far field noise near the axis of a cold jet. J. Sound Vib. 64 (1979) 73-83.

10. Richarz, W.G.: Direct Correlation of noise and flow of a jet using laser doppler. AIAA J. Vol. 18 (1980) 759-765.

11. Shivashankara, B.N.; Strahle, W.C.; Handley, J.C.; Muthukrishnan, M.: Combustion generated noise. Proc. 2nd. Interagency Symp. on University Research in Transportation, Raleigh N.C. (1974) 708-713.

12. Miller, W.R.; Meecham, W.C.; Ahtye, W.F.: Large scale model measurements of airframe noise using cross-correlation techniques. J. Acoust. Soc. Am., Vol 71 (1982) 591-599.

13. Seiner, J.M.; Yu, J.C.: Acoustic near-field properties associated with broadband shock noise. AIAA J. , Vol. 22 (1984) 1207-1215.

14. Richarz, W.G.: Diagnostics of acoustically excited jets. AIAA paper 84-2315, 1984.

15. Ver. I.L.; Mayers, W.H.; McGary, M.C.: Use of reciprocity in predicting the sound radiation of comples structures. J. Acoust. Soc. Am., Vol. 70 S1 (1981) s63.

16. Emani, S.; Morrison, G.L.: Phase averaged acoustic measurements for a Mach number 0.6 jet. AIAA paper 84-1657, 1984.

17. Richarz, W.G.; Keith, S.E.: Direct measurement of sound from large scale structures in jet flows. AIAA paper 83-0662, 1983.

18. Wang, Y.S.; Crocker, M.J.: Direct measurement of transmission loss of aircraft structures using the acoustic intensity approach. Noise Control Eng. J., Vol. 19 (1982) 80-86.

19. Pao, S.P.; Maestrello, L.: Evidence of the beam pattern concept of subsonic noise emission. NASA TN D-8104, 1976.

20. Westley, R.; Woolley, J.H.: The near field sound pressures of a chocked jet when oscillating in the spinning mode. AIAA paper 75-479, 1975.

Sound Generation in the Ocean Under a Turbulent Airflow

Y.P.GUO

Cambridge University, Engineering Department
Trumpington Street, Cambridge CB2 1PZ, England

ABSTRACT

A general theory is developed, based on the extension of the Lighthill-Curle theory of aerodynamic sound by Ffowcs Williams and Hawkings, to account for various features in the problem of sound generation in the ocean. Some particular situations are examined in detail. It is shown that, whenever the ocean surface is non-linearly deformed, any quadrupole radiations become negligible; the most significant sound arises from surface motions. This surface source sound is dominant over the aerial quadrupole radiation by the inverse of the characteristic Mach number in the air and bigger than any quadrupole field due to non-linear motions in the water body by the ratio of a typical surface wavelength to a typical acoustic wavelength. A simple relation is obtained to relate the frequency spectrum of the sound pressure to the surface wave spectrum, which predicts the underwater ambient sound in good agreement with meassurements. The theory is also applied to examine the sound generation mechanism advocated by Brekhovskikh. It is shown that, theoretically, the sound calculated according to Brekhovskikh's theory overwhelms the aerial turbulence radiation when the gravity surface waves have wavelengths much smaller than the height of the turbulence source layer. However, in practical applications, this theory underestimates the sound level and numerical calculations show that this mechanism is less important in the production of underwater sound than that of fully non-linear surface motions. Finally, an expression is derived to include the effects of wave foamings and rainstorms which introduce a powerful dipole distribution on the ocean surface.

1.GENERAL THEORY

Lighthill's (1952) acoustic analogy deals with sound generation by turbulence in a finite region embedded in an unbounded acoustic medium of uniform mean density. It can be applied to the problem of sound production by airflow over the ocean surface in that a local airflow over the wavy ocean surface is inevitably turbulent and chaotic, exactly the kind of sources described by this theory. However, great care must

Aero- and Hydro-Acoustics
IUTAM Symposium Lyon 1985
Editors: G. Comte-Bellot and J. E. Ffowcs Williams

be taken in this application because of the unique features of the particular problem.

The first complexity comes from the fact that the turbulent airflow is ajacent to an infinite density interface which is nearly of 'pressure release' type because of the big difference between the mean density of water and that of air. Pressure variations in the vicinity of the interface may result in non-linear surface motions which can become important sources of the underwater sound. Another significant feature that makes the problem distinctive is the existence of gravity waves on the air-water interface due to gravitational effects and static stratifications of the fluids. These surface waves may alter the properties of sound radiation by non-linear interactions in various manners. Besides, the occurrance of wave foamings (whitecaps) on the ocean surface and rain showers in storms, which certainly have their own effects on the underwater sound production, make the problem even more full of character.

To account for the features outlined above and to understand the important mechanisms in the ocean sound generation, a comprehensive theory is needed. That is the subject underlying this paper.

Our theory is developed by the method due to Ffowcs Williams and Hawkings (1969), which can efficiently tackle the situation of arbitrary surface motions. We incorporate gravitational effects and ignore viscosity in the problem so that the F-H equations can be written as

$$\{\frac{\partial^2}{\partial\tau^2} - c_n^2\nabla^2 - g\frac{\partial}{\partial y_3}\}H_n\rho_n' = \frac{\partial^2 H_n T_{ij}}{\partial y_i \partial y_j} + \frac{\partial}{\partial\tau}\{\bar{\rho}_n u_i \frac{\partial H_n}{\partial y_i}\} - \frac{\partial}{\partial y_i}\{p_n' \frac{\partial H_n}{\partial y_i}\} \tag{1.1}$$

where $T_{ij} = \rho_n u_i u_j + \delta_{ij}(p_n' - c_n^2\rho_n')$ (1.2)

is the turbulence stress tensor and n=1 and 2 standing respectively for 'air' and 'water'; H_1 is a Heaviside function defined to be unity in the air and zero in the water and H_2 is equal to $1-H_1$; g is the value of the gravitational acceleration and the other symbols have their conventional meanings.

The equations can be sovled by defining two Green functions; the solution can be expressed in terms of the quadrupole fields and a surface source field which is represented by an integral on the moving surface. From (1.1), we notice that the surface source terms contain only the pressure perturbation and the surface displacement, both of which are continuous across the density interface. The surface integral can be expanded from the moving surface to its mean position if we introduce some assumptions. First, if the situation of wavebreaking is excluded, the ocean surface can be regarded as continuous and single-valued. Secondly though the surface displacement may be non-linear, it can still be supposed small in comparision with the dimension of the turbulence source layer. Then, the physical boundary conditions and those imposed on the Green functions can be used to eliminate the difficult-to-interpret linear source terms. That procedure gives rise to a solution in terms of the quadrupole fields and the non-linear surface source field. Small terms in the surface integral can be identified as negligible by careful dimensional analyses, allowing us to write the leading terms in the solution as

$$\rho_2^1(\underline{x},t) = \rho_{q.air} + \rho_{q.water} + \rho_{surf.} \tag{1.3}$$

where the first two terms are quadrupole field due respectively to the turbulent flow and the non-linear motions in the water body and the last term is the surface source field which is a surface integral on the mean position of the ocean surface and involves only the square of the surface elevation. The neglect of heiger order terms is the approximation we have already indicated to be justified. Therefore, once the turbulence sources are given, the quadrupole field can easily be found and, once the surface displacement is known, the dipole field can also be calculated.

2.SOUND PRODUCTION BY NONLINEAR SURFACE MOTIONS

Because of the smallness of the ratio between the air mean density and that of water, the ocean surface can hardly subject to any pressure variations; the turbulent airflow and the natrual wave motions in the water can violently distort and deform the air-water interface, causing non-linear surface motions. Indeed, such nonlinear surface motions are inevitable provided that the turbulent airflow exists in a large area for a sufficient long period (Phillips 1958). Studies on ocean waves also show that, in a well-developed sea, surface motions are ramdon both in space and in time, showing the same character as turbulent flow. This suggests that the mechanism of sound production by non-linear surface motions should be similar to that of turbulence sound generation; ramdon sources generate a relatively well ordered response.

According to (1.3), the sound field in the water comprises three parts, two quadrupole fields and a surface source field. We first compare the relative importance of the three by dimensional analysis and then estimate the sound level from the dominant term.

Since it has heen assumed that the airflow can be clearly distinguished from the water by a continuous surface and since the turbulence Mach number is very low, we may approximate the turbulence stress tensors by the Reynolds stress tensors. And from the kinematic boundary condition, it can be shown that the local velocity in the turbulent airflow and that in the non-linear water motions can all be scaled as the product of the surface displacement and a typical frequency. Hence it follows that

$$\frac{\rho_{q.air}}{\rho_{surf.}} \sim \frac{\rho_{01}}{\rho_{02}} \frac{U_0}{c_1} \ll 1, \tag{2.1}$$

and $$\frac{\rho_{q.water}}{\rho_{surf.}} \sim \frac{\Delta_2 \omega}{c_2} \ll 1 \tag{2.2}$$

where U_0 is the free stream velocity beyond the turbulent flow; ρ_{01} and ρ_{02} are the mean densities of air and water and c_1 and c_2 are corresponding sound speeds; Δ_2 is the vertical dimension of the source layer in the water and ω a typical frequency. The first inequality (2.1) is straightforward while the second one (2.2) becomes obvious once it is recognized that Δ_2 is at most of the same order as a typical surface wavelength so that $\Delta_2\omega/c_2$ is at most equal to the ratio of a typical surface wavelength to a typical acoustic wavelength, which is very much less than unity.

Therefore it can be concluded that, whenever the ocean surface is non-linearly disturbed, the surface motions contribute the most significant part to the sound field; the aerial turbulence radiation is smaller than this sound by a factor equal to the typical wind Mach number times the ratio of the air mean density to that of water and the quadrupole field due to non-linear motions in the water body is also smaller than it by the ratio of a typical surface wavelength to a typical sound wavelength.

In a particular simple model where gravity is ignored and the sound speeds of the two fluids are assumed identical, Ffowcs Williams (1983) has obtained similar conclusions. In fact, it can be shown that our results reduce precisely to his if we let g, the value of gravitational acceleration, vanish and $c_1 = c_2$.

Having decided that the dominant sound arises from the ramdon surface motions, we seek to determine its magnitude. As mentioned early, waves found on the surface of the sea are almost always ramdon and chaotic so that only the statistical meassurements can be regarded as significant. One of the well-studied quantities concerning ocean surface waves is the wave spectrum, to which we find that the frequnency spectrum of sound pressure can be related by

$$P(\omega) = \frac{\rho_{0\,2}^{2}\omega^{4}}{4\pi}\,\Phi(\omega)*\Phi(\omega) \qquad (2.3)$$

where $\Phi(\omega)$ is the surface wave spectrum and the star means the convolution operator.

3.SOUND FIELD BY BREKHOVSKIKH'S THEORY

In this section, we examine the situation where only linear motions are involved on the ocean surface.There are some subtleties in this case because of the admittance of both weak nonlinearity and gravity.

On the one hand, in conventional gravity-free sound problems, the effects of non-linear interaction of linear surface waves are usually identified as negligible. The role of the linearly disturbed, infinite homogeneous surface is merely a perfect acoustic reflector; it does not fundamentally change the character of quadrupole radiation (Ffowcs Williams 1965). Ffowcs Williams (1983) also proved that the sound field in this linear problem is uniquely determined by quadrupole sources regardless of the choice of control surfaces. On the other hand, when gravity is admitted, subsonic surface waves can still not generate sound in strictly linear theory; sound and surface gravity waves are energetically uncoupled and the effects of gravity waves on sound can be characterised by a modified reflection coefficient on the surface (Guo 1985). However, the consideration of both weak non-linearity and gravity changes all that. Subsonic surface waves with nearly equal frequencies travelling in opposite directions can non-linearly interact with each other to form surface motions of supersonic phase velocity. This kind of surface motion is certainly capable of generating sound, releasing energy from surface waves to sound. And since surface gravity waves have a different length scale from that of surface motions in the gravity-free problem, the once negligible surface source term, though still proportional to the product of two linear quantities, is now no longer neccssarily of smaller order.

This mechanism of sound generation by second order interaction of linear surface waves was first advocated by Brekhovskikh (1966). Since then, much attention has been paid, by many others (e.g.Hughes 1976), to the application of this theory to estimate the ambient noise level in the ocean. However, surprisingly, little has been done to examine the relative importance of the sound field predicted by this second order interaction theory and those due to other mechanisms such as turbulence radiation. It seems that this lack of theoretical investigation is mainly due to the fact that, in Brekhovskikh's original theory and all its later applications, the effects of airflow are ignored based on the argument that the ratio of the mean density of air to that of water, of the order 10^{-3}, is negligible. However this practically reasonable approximation also neglects a quadrupole sound field due to the aerial turbulence that generated the surface waves and therefore makes it impossible to examine this problem on the theoretical ground.

The incorporations of airflow and gravity in our theory enable us to fill the void. Under linear assumption, our theory can not only predict the sound radiation (Lighthill's quadrupole field) but also describe the surface gravity waves, from which the sound by Brekhovskikh's interaction mechanism can be calculated. Hence both quadrupole and surface source fields can be expressed in terms of the aerial turbulent flow. The result turns to be

$$\frac{p_{surf.}}{p_{qua.}} = \frac{A}{2\pi} \frac{\rho_{01}}{\rho_{02}} \frac{c_2}{U_0} \left(\frac{\Delta}{\lambda_s}\right)^6 \tag{3.1}$$

where Δ is the height of the turbulence source layer and λ_s is a typical surface wavelength; A is a function of the observation angle and is of order unity. Since U_0/c_2 is of the order 10^{-2} or higher, (3.1) can be written as

$$\frac{p_{surf.}}{p_{qua.}} \sim \left(\frac{\Delta}{\lambda_s}\right)^6 \tag{3.2}$$

which states that the relative importance of the two fields depends on the surface wavelength and the height of the turbulence source layer. Theoretically, the interaction of short surface gravity waves with wavelengths much smaller than the height of the source layer can generate a surface field dominant over the quadrupole radiation, but this mechanism of second order interaction is not important for long surface waves. In particular, when surface wavelengths are comparable to that of acoustic waves, which is the limiting case of no gravity, the ratio of surface source sound to the quadrupole sound is proportional to the sixth power of the typical turbulence Mach number; the surface field is then negligible.

Though our theoretical analysis shows that sound field due to Brekhovskikh's theory may be bigger than the quadrupole radiaton, we still argue that this mechanism may not be important in the prediction of the underwater sound. This is because the basic assumption of linear surface deformation in Brekhovskikh's theory, characterised by the classical linear dispersion relation, is obviously a poor representation of the real ocean surface conditions. From (3.2), it can be shown that the inequality $p_{surf.} \gg p_{qua.}$ is true only if

$$\omega \gg g/U_0. \tag{3.2}$$

However this is exactly the condition for which non-linear surface motions occur (Phillips 1966). Therefore, under this condition, there should be another sound field predicted by the theory described in the previous section and we are left to the task of deciding which of the two surface source fields is more important. We answer this question by numerical calculation. It is shown that the random and fully non-linear surface motions indeed contribute the main part of the low frequency underwater sound; the sound level predicted by this theory is higher and more consistent with the measured data than that by Brekhovskikh's theory.

4.SOUND GENERATION BY MIXED TURBULENT FLOW

So far, we have been assuming that the ocean surface is continuous and single-valued so that Taylor's expansion can be applied to the surface integrals. This assumption actually requires the turbulent airflow to be clearly distinquished from the water. However this is not always the case. Even casual observation shows that a characteristic property of the ocean waves under a storm is the occurrence of intermittent patches of wave foaming; breaking waves make the ocean surface neither continuous nor single-valued, not to mention the situations of rain showers where the turbulent airflow is completely mixed with water drops. We investigate those situations in this section.

In order to utilize the theory developed in the first section, we need to move the control surface, which was chosen as the air-water interface in the previous sections, down to the linear acoustic medium in the water so that the integrals to be performed on this surface can still be expanded to its mean position. Hence the theory can formally apply; the solution (1.3) still specifies the sound field in the water.However it is important to notice that the terms in this solution now have different meanings from what they had before.

It should be pointed out that the generalised function H_1 and H_2 have changed their definitions; the region $H_1=1$ is no longer identical to the space occupied by air, instead it simply denotes the region above the chosen control surface. Since there are no non-linear motions below the control surface (in the region where $H_2=1$), the second term in (1.3) must vanish. The surface source term, the third term in (1.3), is also negligible for only linear acoustic motions are involved on this surface. In this case, the sound field is given only by the first term. The crucial point in interpreting this term is that the turbulence stress tensor cannot be approximated by the Reynolds stress tensor in this case because of the mixing of airflow with water. In fact, for very low Mach number and compact flow (compact in comparision with the typical sound wavelength), the dominant source term arises from the density element in the turbulence stress tensor; the Reynolds stress terms and the sources associated with pressure fluctuations are all negligible by comparision. Therefore, after some mathematics, the sound field is found to be

$$\rho_2'(\underline{x},t) = \frac{C_t}{2\pi c_2^3|\underline{x}|}\frac{-x_3}{|\underline{x}|}\frac{\partial}{\partial t}\int_{V_0}[\rho\frac{Du_3}{D\tau}]d^3y \tag{4.1}$$

where v_0 is a fixed volume containing the mixed turbulent flow; the square bracket means the conventional retarded time implication and C_t is the acoustic transmission coefficint defined by

$$C_t = \frac{\rho_{02}\sqrt{c_2^2/c_1^2 - x_\alpha^2/|\underline{x}|^2}}{\rho_{02}\sqrt{c_2^2/c_1^2 - x_\alpha^2/|\underline{x}|^2} - \rho_{01}x_3/|\underline{x}|}. \tag{4.2}$$

The sound determined by (4.1) is due to the nonlinear density variations in the source region and is bigger than that from the Reynolds stress sources by the inverse of the typical turbulence Mach number. It is clear that this sound field, characterised by the factor $-x_3/|\underline{x}|$, is of dipole type. This is expected considering the fact that the most important mechanism of sound generation in this situation is the impulsive forces exerted on the ocean surface by breaking waves and raindrops, which are equavelent to acoustic dipole distributions. It can also be seen that the effects of the difference in sound speeds of the two fluids can be completely accounted for by the transmission coefficient C_t; abrupt change in sound speeds across the interface does not introduce any important sources.

At first glance, it seams to be a surprising result for monopole sources are usually expected whenever change of sound speeds is involved. However, our result is perfectly reasionable under the assumption of very low Mach number turbulent flow which can be approximately regarded as incompressible. The source region, and hence the sound generation process, is not influenced by the sound speeds.

5.REFERENCES

1. Brekhovskikh,L.M. 1966 Underwater sound waves generated by surface waves in the ocean. Izv. Atmos. Ocean Phys. 2, 582-587

2. Ffowcs Williams, J.E. 1965 Sound radiation from turbulent boundary layers formed on compliant surfaces. J.Fluid Mech. 22(2) 374-385

3. Ffowcs Williams,J.E. and Hawkings,D.L. 1969 Sound generation by turbulence and surfaces in arbitrary motion. Trans. Roy. Soc. A 264 321-342

4. Ffowcs Williams,J.E. 1983 Waves in turbulent mixing layer. Proc. the international symposium on recent advances in aerodynamics and aeroacoustics. Stanford

5. Guo,Y.P. 1985 Sound radiation from sources near the ocean surface. (submitted to Proc. Roy. Soc. Lon.)

6. Hughes,B. 1976 Estimates of underwater sound produced by non-linearly interacting ocean waves. J. Acoust. Soc. Am. 60 1032-1039

7. Lighthill,M.J. 1952 On sound generated aerodynamically. 1. General thoery. Proc. Roy. Soc. A 221 564-587

8. Phillips,O.M. 1958 The equilibrium range in the spectrum of wind-generated waves. J. Fluid Mech. 4 426-436

9. Phillips,O.M. 1966 The dynamics of the upper ocean. CUP.

Acoustic Radiation from Vortex Solitons

by

L.M.LYAMSHEV, A.T.SKVORTSOV

N.N.Andreev Acoustics Institute,
Academy of Science of the USSR,
Shvernik St., 4
Moscow 117036, USSR

A theory of acoustic radiation from large-scale vortex disturbances in axisymmetric shear flow at low Mach number is proposed.

The problem of acoustic radiation from large-scale vortex disturbances in shear flow arises in connection with aerodynamically generated sound and detection of coherent structures in turbulence [1].

Let us consider an axisymmetric flow of incompressible inviscid fluid in which the velocity vector is everywhere parallel to z-axis and $u_z = U(r)$, where r is the distance from this axis. Profile $U(r) \to U_0$ at $r \to 0$ $(U'(0)=0)$ and $U(r) \to 0$ (exponentially) at $r > a$ (a is characteristic width of the flow). Under these conditions the evolution of large-scale fluctuations of u_z-velocity is governed by the Korteweg-de Vries equation [2]

$$\frac{\partial u}{\partial t} + U_0 \frac{\partial u}{\partial z} + u \frac{\partial u}{\partial z} - \alpha \frac{\partial^3 u}{\partial z^3} = 0 \tag{1}$$

where u is u_z-fluctuation at $r=0$: $u_z = U_0 + u$; $\alpha = U_0 a^2/2$.

According to the solutions of the equation (1) [3] the disturbances of the u_z-velocity and the induced disturbances of the vorticity propagate in the considered flow in the form of solitons. (Note that in the presence of dissipation a number of

Professor LIAMSHEV did not appear at the Meeting but afterwards communicated this paper we include in the proceedings.

Aero- and Hydro-Acoustics
IUTAM Symposium Lyon 1985
Editors: G. Comte-Bellot and J. E. Ffowcs Williams

solutions self-organized in a coherent state [4]).

In slightly compressible fluid (i.e. when the Mach number $M=U_0/c\ll 1$) propagation of the solitons generates acoustic waves which far from the flow region ($|\vec{R}|\gg\lambda$) are discribed by the equation

$$\frac{1}{c^2}\frac{\partial^2 \varphi}{\partial t^2} - \Delta \varphi = 0 , \tag{2}$$

where C is the speed of sound; $R^2 = z^2 + r^2$; λ is the wave length; Δ is three dimentional Laplas operator; φ is the velocity potential of sound field.

At $a\ll|\vec{R}|\ll\lambda$ solution (2) have to be "fitted" to discribe the potential motion corresponding to disturbances (1) (at $a\ll|\vec{R}|\ll\lambda$ the equation $\Delta\varphi=0$ is valid and hence the expression for the potential φ is known [2]). This aim is achived by applying the method of matched asymptotic expansion with Mach number as a small parameter; the final formula may be wrote as follows [5]

$$\varphi_{\underline{IV}} = \frac{a^2\cos^3\theta}{c^3 R}\dddot{\mathcal{P}}(t-R/c). \tag{3}$$

Here θ is the polar angle of the observation point $\vec{R}$ from the z -axis; $\mathcal{P}(t)=\int_{-\infty}^{+\infty} z^2 u(z,t)\,dz$; $u(z,t)$ is any solutions of the equation (1); dot denotes differentiation with respect to time .

The expression (3) represents an octupole term in the multi pole expantion. The quadrupole term which corresponds to Lighthill theory is given by the expression

$$\varphi_{\underline{III}} = \frac{a^2\cos^2\theta}{c^2 R}\ddot{Q}(t-R/c). \tag{4}$$

This term vanishes because

$$\ddot{Q}(t) = \frac{d^2}{dt^2}\int_{-\infty}^{+\infty} z\, u(z,t)\,dz = \frac{d}{dt}\int_{-\infty}^{+\infty} u^2(z,t) = 0$$

for Korteweg-de Vries equation [3] .

Thus, the acoustic field associated with the vortex soliton-soliton interaction has an octupole nature, i.e. $W\sim M^{10}$ where M denotes the sound intensity.

Furthermore, it is of interest to know the specific features of sound waves emitted by the vortex soliton's being affected by the fine-scale turbulence. This phenomenon is discribed by (4), but now $u(z,t)$ is the solution of the equation

$$\frac{\partial u}{\partial t} + U_0 \frac{\partial u}{\partial z} + u \frac{\partial u}{\partial z} - \alpha \frac{\partial^3 u}{\partial z^3} = \nu_t \frac{\partial^2 u}{\partial z^2} \qquad (5)$$

and so $\ddot{Q}(t) \neq 0$. Parameter $\nu_t \approx v_t \ell$ is the turbulent viscousity; v_t and ℓ are characteristic fluctuating velocity and scale of turbulent flow. The equation (5) may be found by averaging (1) over fast turbulent oscillation (see, for ex. [6]).

The approximate solution of equation (5) may be derived by using well-known one-soliton solution (1) [3] , which is

$$u = u_0 / ch^2[(\eta + \frac{u_0}{3} t)/\Delta] ; \; \Delta^2 u_0 = \alpha^2 U_0 / 2 = const \qquad (6)$$

Here $\eta = z - U_0 t$; Δ , $u_0 = const$ denote the width and amplitude of the solitons. Being affected by the turbulent viscousity, parameters Δ and u_0 in (6) become slight functions of time. These functions may be obtained by multiplying equation (5) on u , substituting (6) and integrating with respect to z .
Thus

$$\Delta^2(t) = \Delta^2(0) + \nu_t t \; , \quad \frac{1}{u_0(t)} = \frac{1}{u_*} + \frac{\nu_t}{2\alpha^2 U_0} t$$

where u_* denotes $u_0(t)$ at $t=0$.
The total sound intensity in this case is

$$W(t) = W(0)(1 + t/\tau)^{-7} ; \; 1/\tau \sim \frac{\nu_t u_*}{\alpha^2 U_0} ; \; W(0) = \frac{\rho_0}{c^5 \alpha^2} \frac{\nu_t^4 u_*^7}{U_0^3}$$

where ρ_0 is the density of understurbed field. It is important that the characteristics sound intensity in such mechanism is proportional to the turbulent viscousity in the fourth power and to the amplitude of large-scale vortex disturbances in the seventh power.

The equations (3),(4) are very convenient for studing of acoustic radiation in different processes with partisipation of vortex solitons (fusion, coalescence, collision ect.) because corresponding solutions of equation (1),(5) are available [3]. Once this solutions are known the time evolution $\mathcal{P}(t)$ and $Q(t)$ and hence the intensity of acoustic radiation can easily be obtained.

---

1. A.K.Hussian: Phys.Fluids, 26 (1983), 2816.
2. V.N.Trigube: Izv. AN SSSR. Mech. Jid. Gas. (transl. as Fluid Dynamics), 4 (1983), 35.
3. G.B.Whitham. Linear and Nonlinear Waves (J.Willey and sons, New York, 1974).
4. A.Hasegawa, Y.Kodawa, K.Watanable: Phys. Review Letters, 47 (1981), 21, 1525.
5. L.M.Lyamshev, A.T.Skvortsov: Pis'ma v Zh. Tech.Fiz.(Soviet Technical Physics Letters), 10,(1984), 988.
6. S.S.Moiseev, R.Z.Sagdeev, A.V.Ture, G.A.Khomenko: Zh. Exp. Teor. Fiz. (Soviet Physics - JETF), 85 (1984), 1979.

Rotor Noise

Noise of High Speed Surfaces

A. R. GEORGE

Sibley School of Mechanical and Aerospace Engineering
Cornell University, 105 Upson Hall
Ithaca, New York 14853, U. S. A.

Summary

Aerodynamic noise can be radiated by 1) internal fluid motions (e.g., turbulence), 2) small motions of surfaces which are essentially fixed with respect to the fluid (e.g., loudspeakers), or 3) by surfaces moving with respect to the fluid. This paper discusses the last case. The noise radiated can be either due to the motion of the volume of the body directly, due to the steady or unsteady forces on its surfaces from their motion relative to the fluid, or due to nonlinear internal fluid motion effects. These noise sources arise frequently in conjunction with rotors and with supersonic flight. This paper reviews the analysis of sonic boom due to supersonic motion; rotor noise due to the acceleration of rotor volumes, steady forces and nonlinear effects; and the noise due to various unsteady force effects on rotors. The unsteady forces include effects from harmonically-varying inflow, from impulsive blade-vortex interaction loading (including transonic effects), and broadband noise due to loading from inflow turbulence and self generated turbulence. The state-of-the-art of predicting noise from practical configurations is assessed.

Introduction

Sound can be looked at quite generally as the propagating part of the full solution of some fluid flow problem. In a few cases the radiated sound can be found this way. However, in most cases of practical interest one cannot find a full solution and instead looks for an approximate near field solution. Thus it is often advantageous, or even necessary to develop ways of finding the far field noise from near field solutions.

The various approaches used to find the far field noise radiated by high speed surfaces can be classified as follows:

Aero- and Hydro-Acoustics
IUTAM Symposium Lyon 1985
Editors: G. Comte-Bellot and J. E. Ffowcs Williams

FULL FLOW FIELD - Calculation of the full nonlinear flow field including far field waves

NONLINEAR PLUS KIRCHHOFF - Calculation of the nonlinear near and mid field with the far field solutions found from a linear Kirchhoff formulation evaluated on a surface surrounding the nonlinear field.

NONLINEAR PLUS FFOWCS WILLIAMS AND HAWKINGS (ACOUSTIC ANALOGY) - Calculation of the nonlinear near and mid field with the far field found from a linear Green's function formulation evaluated in terms of volume and surface integrals over retarded-time transformations of the near and mid field flow and body surfaces.

LINEAR VOLUMES AND FORCES - For linear cases the appropriate equations can be manipulated to express the far field in terms only of surface geometry and surface forces on retarded-time transformations of the moving bodies.

FAR FIELD CORRECTIONS - Correction for far field nonlinearities and nonuniform medium effects using ray acoustics plus weak nonlinear corrections.

The parallel successive stages of the development of analyses and experiments are shown in fig. 1. Initially, in stage one, oversimplified, qualitative, but mathematically tractable model problems are analyzed in conjunction with exploratory experiments. As shown in the figure, the second stage involves research type analyses for particular phenomena and experiments specialized to emphasize these phenomena. The third stage includes quantitative analysis and more details of effects found to be important and approximations of effects found to be less important. The pilot codes can be run for design studies. Also in the third stage the experiments are producing specific data for code input and overall cases for code verification. Finally, by stage four, production codes exist for use in design, while development tests of full configurations are used to check the analyses and codes and to indicate needed incremental improvements to them.

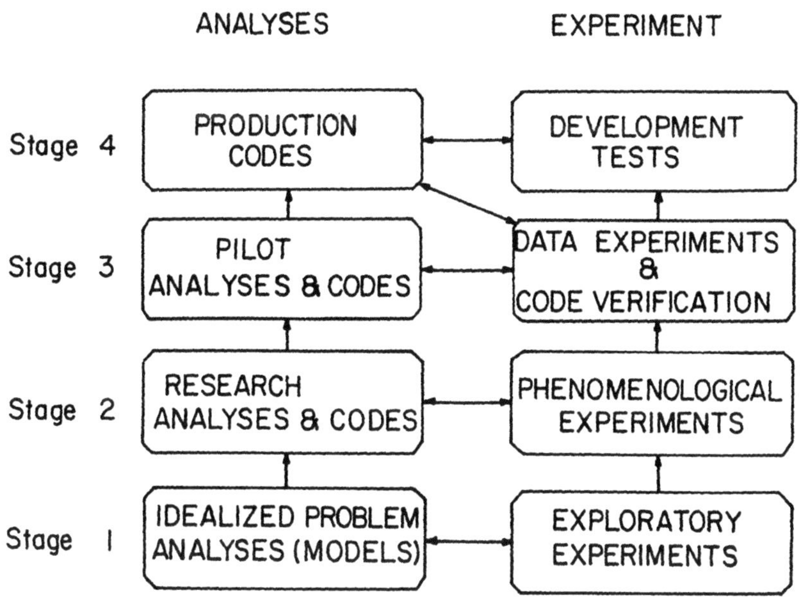

Fig. 1. Stages of analyses and experiments.

General Considerations

It is informative to first consider the classical "acoustic analogy" method of determining the sound field radiated by a flow. As developed by Lighthill [1] and Ffowcs Williams and Hawkings [2] the exact equations of motion can be put into the familiar acoustic analogy form for the perturbation density in terms of the surface's velocities v_i and stress tensor P_{ij} and the Lighthill stress tensor in the fluid T_{ij}:

$$c^2 \Box^2 \rho = \frac{\partial}{\partial t}[\rho_0 v_i \frac{\partial f}{\partial x_i} \delta(f)] - \frac{\partial}{\partial x_i}[P_{ij} \frac{\partial f}{\partial x_j} \delta(f)] + \frac{\partial^2 T_{ij}}{\partial x_i \partial x_j} \equiv Q(x_i,t) , \quad (1)$$

where the body surface is defined by $f(x_i,t) = 0$. Using a Green's function approach, the far field solution to equation 1 is straightforward in principle if the right hand (source) terms

$Q(x_i,t)$ are known, a priori, from some near and mid field aerodynamic results. Equivalently the right hand side can be written: $Q(x_i,t) = \partial q/\partial t - \partial F_i/\partial xi + \partial^2 T_{ij}/\partial x_i \partial x_j$, where q and F_i a the source and force strengths per unit volume applied to the fluid Although it is less general, the second interpretation is often use to model aeroacoustic phenomena in a simple way.

Equation 1 can be used to show that sound is only radiated by disturbances in subsonic motion if their source terms vary in time or if they accelerate. This is analogous to the radiation from charges in electromagnetic theory. Equivalently, some components o the disturbance must have a phase velocity which is supersonic relative to the undisturbed fluid. If a disturbance component move at a subsonic phase velocity relative to the fluid then an "evanescent wave" is generated (see e.g. [3]. Evanescent waves do not propagate to the far field but generally decay exponentially with distance. The transonic rotating blade case is complicated by the fact that in a rotating coordinate system the governing equations change type from elliptic to hyperbolic at a sonic surfac near the blade tip. Thus, as will be discussed below, the evanescent waves make a transition to propagating waves in this transonic region [4].

An important parameter in an aeroacoustic problem is compactness, i.e., whether the acoustic signals from a certain frequency radiation will arrive in phase from different parts of a body. For non-compact cases and significant Mach numbers the approach of equation 1 is most needed. For some applications, such as broadband rotor noise, the non-compact directionality is less important as it tends to be averaged out as the blade rotates.

Sonic Boom

Sonic boom is the far field sound generated by objects moving at supersonic velocities (as opposed to the subsonic tip velocities of most rotors). The sonic boom of supersonic aircraft is the simplest case to discuss as the near field is essentially steady an the rotation of body-fixed coordinates is not important. Sonic boom is usually analyzed by solving the complete supersonic linearized flow field for straight line motion, solving for the asymptotic linear far field, and then correcting it for weak nonlinear propagation and for ray curvature effects [5,6]. In sonic boom theory the near field flow nonlinearities are fairly easily separated from the propagation nonlinearities, in contrast to the transonic rotor case. Farassat [7] has shown that some classical linearized supersonic flow results can be derived alternatively from the acoustic analogy of equation 1.

The equations for the analyses can be succinctly summarized as follows:

Pressure in terms of volume (A=area) and force (L'=force/length) components (see fig. 2):

$$\frac{\delta p}{q} = \frac{2^{1/2}}{\beta r} \frac{1}{2\pi} \int_0^{x-\beta r} \frac{[A''(\xi,\theta) + (2q)^{-1} \beta L'(\xi,\theta)]}{(x-\beta r-\xi)^{1/2}} d\xi$$

Propagation by geometrical acoustics (S=ray tube area):

$$\Delta p (S/\rho a)^{1/2} = \text{Constant}$$

Nonlinear advance: $\dfrac{dt}{ds} = \dfrac{1}{a} + \dfrac{\gamma+1}{2\gamma a} \dfrac{\Delta p}{p}$

Shock by Landau equal area rule: $\int p \, dt = \text{Constant}.$

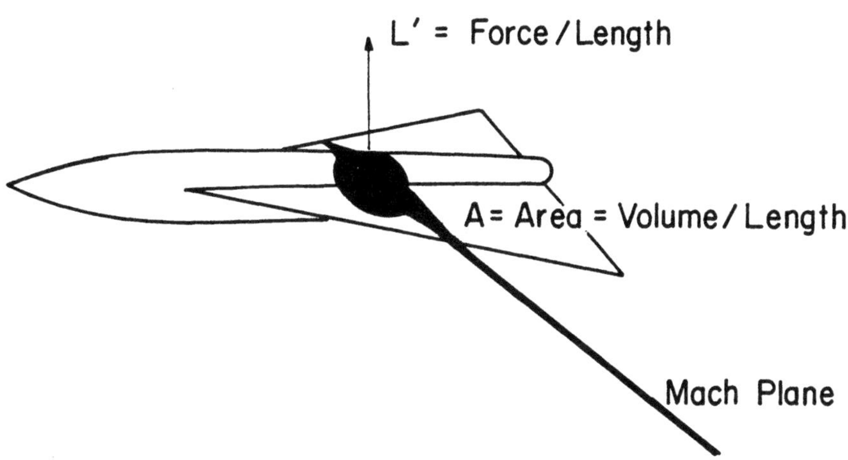

Fig. 2. Linearized pressure depends upon force and volume along Mach plane sections.

As shown in fig. 3, these predictive procedures are quite accurate, including the effects of maneuvers and a layered atmosphere. Thus the combined "linear volumes and forces" and "far field corrections" analyses have reached stage 4 (production codes), while experiments remain at stage 3, due to the lack of development of low sonic boom aircraft. (The case of sonic boom from laterally accelerated subsonic aircraft, which is analogous to subsonic rotor noise, has not been analyzed to date.)

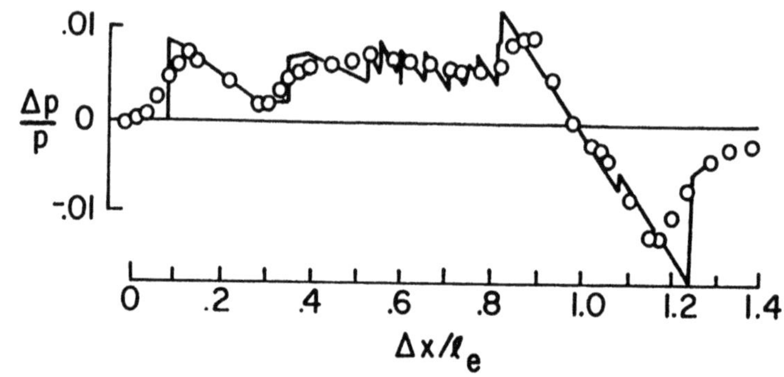

Fig. 3. Comparison of measured and theoretical boom signatures for M = 1.5 arrow wing model. From [6].

Rotational Noise

Rotational noise of rotors has been studied primarily by the "linear volumes and forces" approach, with some attention to nonlinear near field effects. The initial studies of rotor noise due to blade force and volume by Gutin [8] and Deming [9] were based on the interpretation of moving sources as a distribution of fixed sources which unsteadily wax and wane to define the path of the motion. The alternative moving source interpretation was studied in the compact case by Lowson [10] who explicitly showed the separate unsteady and acceleration effects. The general case of equation 1 was extensively treated by Farassat, Hanson, and others [11,12], as reviewed by Farassat elsewhere in this volume. These very successful analyses include source volume and force effects which vary periodically with the motion of the rotor. Analyses based on equation 1 are generally quite complex and difficult to program for computation. Thus the less demanding applications are often treated by the compact analyses similar to those of Lowson and Ollerhead and Wright [13,14]. For all of these "linear volumes and forces" approaches the analyses and experiments have reached stages three and four, and are in good agreement, as shown in fig. 4 [15].

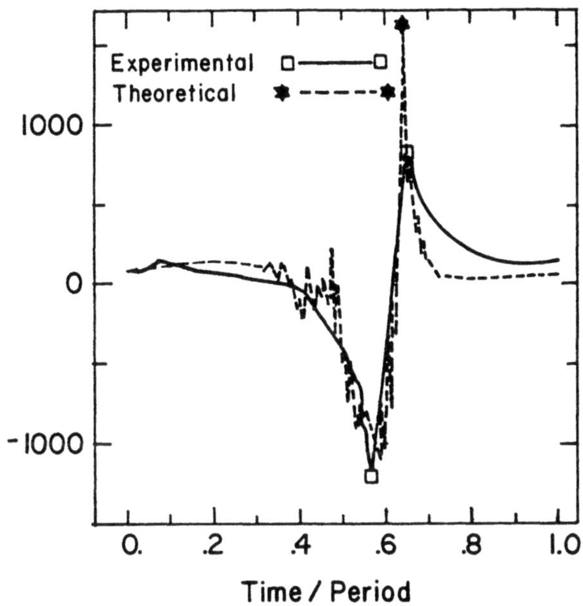

Fig. 4. Comparison of experiment and calculations. From [15].

The nonlinear T_{ij} (quadrupole) terms have been treated for rotors, but with much less generality and detail [16-19]. The evaluation of these terms depends on full three-dimensional transonic flow solutions and also adds considerable practical, if not fundamental, complication to the numerical evaluation of results. In fact, the nonlinear quadrupole terms are extremely important to transonic flow noise; they are essentially accounting for the local Mach number varying about $M = 1$ and the resulting fundamental changes in the local propagation of disturbances. Part of the difficulty with the quadrupole terms is also that the local nonlinearities associated with transonic flow are very difficult to separate from the simpler propagation effects in the mid field. The local flow equations in blade-fixed coordinates near a transonic rotor tip change type from elliptic to hyperbolic because of both the local flow velocity changes and because of the increase of roto: velocity with radius. This leads to a "delocalization" (direct propagation to the far field) of waves generated in the transonic flow on the rotor tips [4,20]. The extent of the hyperbolic region in the rotating coordinate frame is shown shaded in fig. 5, adapted from ref. 21.

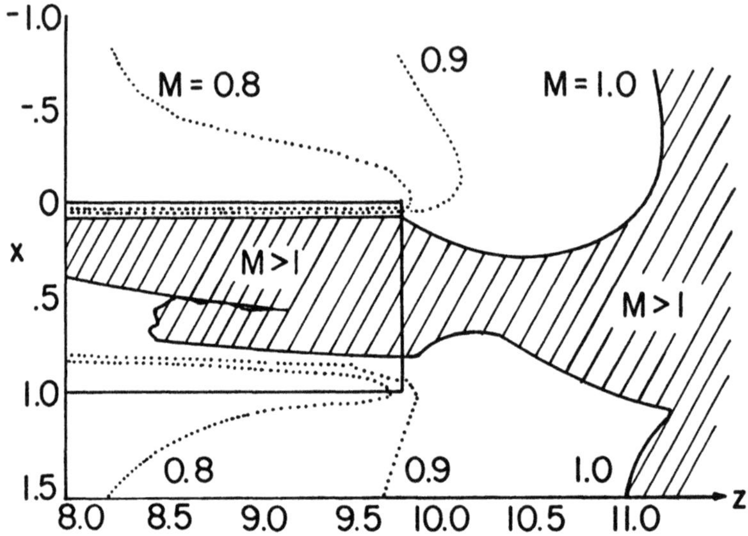

Fig. 5. Hyperbolic regions (M > 1) near rotating blade tip allow waves to propagate from blade to far field. From [21].

An alternative approach to deal with these nonlinear effects is a numerical solution for the near field of the blades and a Kirchhoff theorem method to extend the flow to the far field [22]. All of th(approaches to nonlinear rotor noise have only reached stage two (research analyses and phenomenological experiments).

Blade-Vortex Interactions

An important helicopter rotor noise source is blade-vortex interaction, where a rotor blade interacts with discrete tip vortices shed by the same or some other blade. In this case the blade forces are highly unsteady and the radiation is primarily determined by the local interaction rather than the rotational

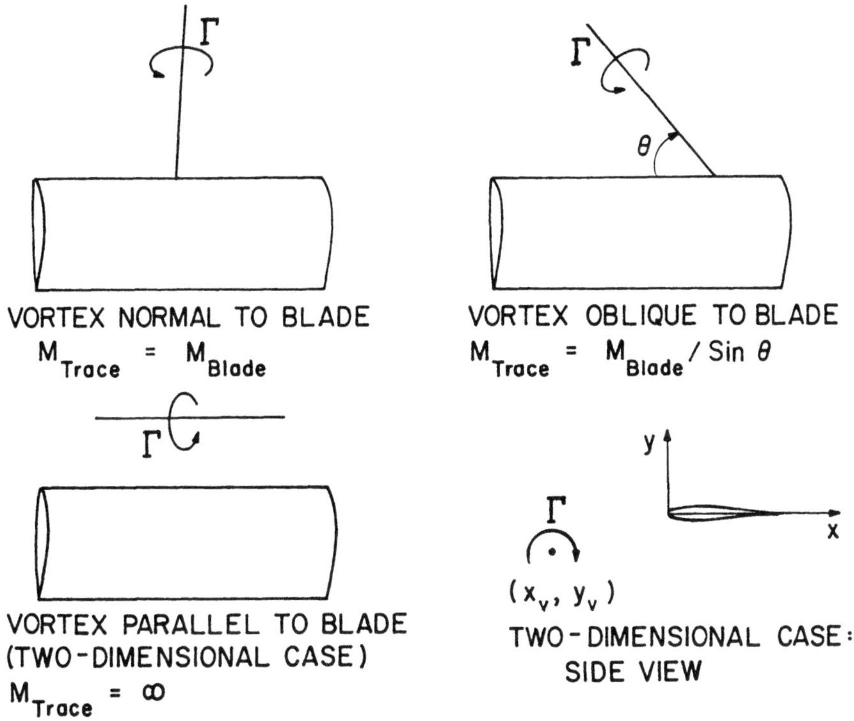

Fig. 6. Types of blade-vortex interaction geometry. From [23].

motion. The fundamental parameter is the "trace Mach number" of the interaction with respect to the undisturbed fluid [23]. As shown in fig. 6, the geometry of the interaction can range from a basically steady normal interaction with trace Mach number equal to blade Mach number to a fully unsteady parallel interaction where the trace Mach number is infinite. Graham [24] has shown for the linear, infinite-span case that general interactions can be transformed into either the normal or parallel cases for subsonic or supersonic trace Mach numbers respectively. If the trace Mach number is supersonic then strong radiation will result, and the local flow is analogous to two-dimensional unsteady flow. The geometry and structure of the incident vortex, as well as the aerodynamic and acoustic modeling are important to the results.

Widnall and others [25-28] have analyzed many aspects of linear blade-vortex interactions using both the acoustic analogy and complete flow field analyses. Nakamura [29] predicted the far field sound using the acoustic analogy with measured surface pressures but had difficulties due to the low resolution of the available measurements. George and Chang [23] have shown that in many practical cases blade-vortex interaction phenomena are intrinsically transonic. George and Chang [23] and McCroskey and Goorjian [30] analyzed unsteady, transonic, two-dimensional interaction flow fields using a numerical, small-disturbance approach, including the introduction of finite core-size vortices which are convected by the local disturbed flow. Srinivason et al. [31] also studied this flow using a numerical thin-layer Navier-Stokes approach.

George and Chang identified some of the parameters which affect transonic blade-vortex interaction noise, and showed that strong shock-like disturbances could be propagated off the airfoil surface into the mid field. Fig. 7 shows the timewise progression of $C_p(x)$ on an airfoil surface as a vortex is convected past it [23]. A compression wave can be seen to move forward in the process of radiating off the airfoil It was found to be difficult to maintain accuracy in following the numerically-determined radiated waves into the far field. Here an acoustic analogy or Kirchhoff theorem approach might be helpful. None of the present approaches can accurately predict general blade-vortex interactions yet, but recent progress has been rapid. The transonic analyses are at stage one

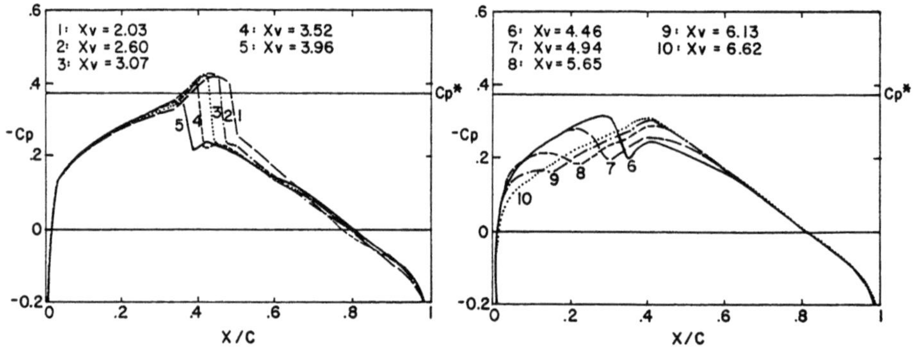

Fig. 7. Lower surface $C_p(x)$ for various vortex positions x_v, showing progression of shock off blade. (NACA64A006, $M_\infty = 0.822$, $C_{lv} = 0.4$). From [23].

(idealized models), while experiments have progressed to stage two (phenomenological experiments).

Broadband Noise

Broadband noise from high speed surfaces is due to random aerodynamic loadings on the surfaces. These random loadings are usually induced by turbulence. Fig. 8 indicates some of the mechanisms causing these loadings. In one approach to finding roto: noise, Homicz and George [32] used the acoustic analogy to determin the radiated sound from an arbitrary distribution of loads on the plane of a rotor. After making various aerodynamic and acoustic approximations they were able to predict broadband rotor noise including the low frequency peak/valley structure of noise due to large scale turbulent downwash velocity components being convected through the rotor plane. In another approach to the problem, Amiet [33] analyzed the radiation from an airfoil in straight line motion and then used that result to numerically sum the sound from rotor blades in successive straight line motion segments approximating th actual accelerated motion. The neglect of explicit acceleration effects is justified for high frequency noise.

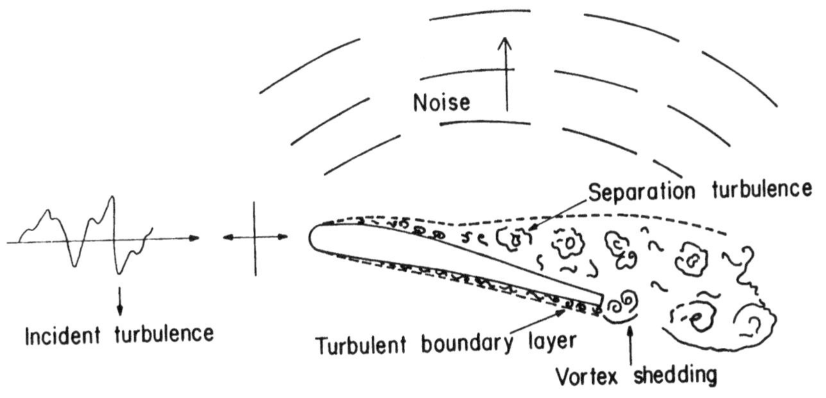

Fig. 8. Broadband noise sources of an airfoil.

For the higher frequency noise associated with smaller scale load fluctuations, the computationally more economical rotating dipole approach of Ffowcs Williams and Hawkings [34] has been successfully applied by George and others to analyze the noise due to: 1) turbulent inflow, 2) the turbulent boundary layer passing the trailing edge, 3) tip vortex formation turbulence, and 4) trailing edge bluntness [35-38]. The crucial parts of such analyses were in modeling the turbulence associated with the various mechanisms and the resulting fluctuating loadings on the blades' surfaces. For example, in modeling tip vortex noise the turbulent surface pressure fluctuations were found by appropriate scaling of those due to leading edge vortices on delta wings [37]. Each of the various broadband mechanisms can be important for various rotor and inflow parameters, but existing codes satisfactorily predict this and compare well to experiment, as shown in fig. 9. Broadband noise predictions are becoming satisfactory, with codes and experiments at stage three (pilot codes and data experiments). Remaining problems

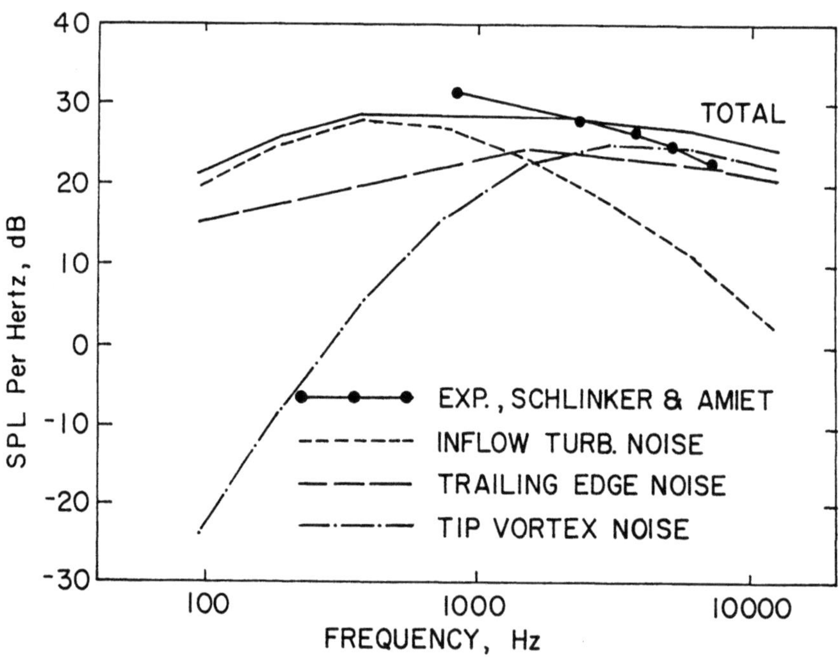

Fig. 9. Broadband noise predictions and experiment for small subsonic rotor. From [37].

are in determining inflow and loading parameters in particular cases (such as tail rotors [39]) and in predicting noise near the rotor plane where fluctuating forces in the drag (torque) direction are important.

Conclusions

The prediction of noise from high speed surfaces is beginning to approach a satisfactory state of understanding and application. Progress is needed in accounting for local and propagation nonlinearities (quadrupole effects), in problems involving transonic flows, and in the fluctuating load predictions for blade-vortex and broadband rotor noise.

References

1. Lighthill, M.J.: On sound generated aerodynamically - I: general theory. Proc. Roy. Soc., London A211 (1954) 564-587.

2. Ffowcs Williams, J.E.; Hawkings, D.L.: Sound generation by turbulence and surfaces in arbitrary motion. Phil. Trans. Royal Soc. of London A264 (1969) 321-342.

3. Dowling, A.P.; Ffowcs Williams, J.E.: Sound and sources of sound. Chichester, Ellis Horwood Ltd. (1983).

4. Isom, M.P.: Acoustic shock waves generated by a transonic helicopter blade. Preprint No. 80-63, Amer. Helicopter Soc. 36th Annual Forum Proceedings, Washington, D. C. (1980).

5. Seebass, A.R.; George, A.R.: Design and operation of aircraft to minimize their sonic boom. J. Aircraft 11 (1974) 509-517.

6. Mack, R.J.; Darden, C.M.: Some effects of applying sonic boom minimization to supersonic cruise aircraft design. J. Aircraft 17 (1980) 182-186.

7. Farassat, F.: The acoustic far-field of rigid bodies in arbitrary motion. J. Sound and Vibr. 32 (1974) 387-405.

8 Gutin, L.: On the sound field of a rotating propeller. NACA TM-1195 (1948).

9. Deming, A.F.: Noise from propellers with symmetrical sections at zero blade angle, II. NACA TM-679 (1938).

10. Lowson, M.V.: The sound field for singularities in motion. Proc. Roy. Soc., London A286 (1965) 559-572.

11. Farassat, F.: Linear acoustic formulations for calculation of rotating blade noise. AIAA J. 19 (1981) 1122-1130.

12. Hanson, D.B.: Compressible helicoidal surface theory for propeller aerodynamics and noise. AIAA J. 21 (1983) 881-889.

13. Lowson, M.V.; Ollerhead, J.B.: A theoretical study of helicopter rotor noise. J. Sound and Vibr. 9 (1969) 197-222.

14. Wright, S.E.: Sound radiation from a lifting rotor generated by asymmetric disc loading. J. Sound and Vibr. 9 (1969) 223-240.

15. Farassat, F.; Succi, G.P.: The prediction of helicopter rotor discrete frequency noise. Vertica 7 (1983) 309-320.

16. Hanson, D.B.; Fink, M.R.: The importance of quadrupole sources in prediction of transonic tip speed propeller noise. J. Sound and Vibr. 62 (1979) 19-38.

17. Schmitz, F.H.; Yu, Y.H.: Theoretical modeling of high-speed helicopter impulsive noise. J. Amer. Helicopter Soc. 24 (1979) 10-19.

18. Ffowcs Williams, J.E.: On the role of quadrupole source terms generated by moving blades. AIAA Paper 79-0576, AIAA 5th Aeroacoustics Conf., Seattle, Washington (1979).

19. Tam, C.K.W.; Salikuddin, M.: Weakly nonlinear theory of supersonic turbopropeller noise. AIAA Paper 84-2302, AIAA 9th Aeroacoustics Conf., Williamsburg, Virginia (1984).

20. Schmitz, F.H.; Yu, Y.H.: Helicopter impulsive noise: theoretical and experimental status. Intl. Symp. on Recent Advances in Aerodynamics and Acoustics, Stanford University (1983).

21. Tauber, M.E.: Computerized aerodynamic design of a transonically quiet blade. Paper A-84-40-46-0000, 40th Annual Forum of the Amer. Helicopter Soc., Arlington, Virginia (1984)

22. Hawkings, D.L.: Noise generation by transonic open rotors. in Mechanics of Sound Generation in Flows, E. A. Müller (ed.), Springer Verlag 1979, Berlin, 294-300.

23. George, A.R.; Chang, S.-B.: Flow field and acoustics of two-dimensional transonic blade-vortex interactions. AIAA Paper 84-2309, AIAA 9th Aeroacoustics Conf., Williamsburg, Virginia (1984).

24. Graham, J.M.R.: Similarity rules for thin aerofoils in non-stationary subsonic flows. J. Fluid Mech. 43 (1970) 753-766.

25. Widnall, S.E.: Helicopter noise due to blade-vortex interaction. J. Acoustical Soc. Amer. 50 (1971) 354-365.

26. Martinez, R.; Widnall, S.E.: An aeroacoustic model for high-speed, unsteady blade-vortex interaction. AIAA J. 21 (1983) 1225-1231.

27. Widnall, S.E.; Wolf, T.L.: Effect of tip vortex structure on helicopter noise due to blade-vortex interaction. J. Aircraft 17 (1980) 705-711.

28. Amiet, R.K.: Airfoil gust response and the sound produced by airfoil-vortex interaction. AIAA Paper 84-2268, AIAA/NASA 9th Aeroacoustics Conf., Williamsburg, Virginia (1984).

29. Nakamura, Y.: Prediction of blade-vortex interaction noise from measured blade pressure. Paper 32, 7th European Rotorcraft and Powered Lift Conf., Garmisch-Partenkirchen, W. Germany (1981).

30. McCroskey, W.J.; Goorjian, P.M.: Interactions of airfoils with gusts and concentrated vortices in unsteady transonic flow. AIAA Paper 83-1691, AIAA 16th Fluid and Plasma Dynamics Conf., Danvers, Massachusetts (1983).

31. Srinivasan, G.R.; McCroskey, W.J.; Baeder, J.D.: Aerodynamics of two-dimensional blade-vortex interaction. AIAA Paper 85-1560, AIAA 18th Fluid Dynamics and Plasmadynamics and Lasers Conference, Cincinatti, Ohio (1985)

32. Homicz, G.F.; George, A.R.: Broadband and discrete frequency radiation from subsonic rotors. J. Sound and Vibr. 36 (1974) 151-177.

33. Amiet, R.K.: Noise produced by turbulent flow into a propeller or helicopter rotor. AIAA J. 15 (1977) 307-308.

34. Ffowcs Williams, J.E.; Hawkings, D.L.: Theory relating to the noise of rotating machinery. J. Sound and Vibr. 10 (1969) 10-21.

35. George, A.R.; Kim, Y.N.: High frequency broadband rotor noise. AIAA J. 15 (1977) 538-545.

36. Kim, Y.N.; George, A.R.: Trailing-edge noise from hovering rotors. AIAA J. 20 (1982) 1167-1174.

37. George, A.R.; Chou, S.-T.: Broadband rotor noise analyses. NASA CR-3797 (1984).

38. Chou, S.-T.; George, A.R.: Effect of blunt trailing edges on rotor broadband noise. Cornell Univ., Sibley School of Mech. & Aero. Engr. Rep. FDA-85-01 (1985).

39. George, A.R.; Chou, S.-T.: A comparative study of tail rotor noise mechanisms. Proceedings, Amer. Helicopter Soc. 41st Annual Forum, Fort Worth, Texas (1985).

Comparison of Two Propeller Source Models for Aircraft Interior Noise Studies

J. R. MAHAN and C. R. FULLER

Department of Mechanical Engineering
Virginia Polytechnic Institute and State University
Blacksburg, Virginia

Summary

The advanced turboprop (ATP) engines currently under development for transport aircraft are expected to be significantly noisier than existing turboprop engines. If this proves to be the case, public acceptance and thus the ultimate success of the ATP will depend on the degree to which the passenger cabin can be isolated from propeller noise. This raises the possibility that much of the gain in fuel efficiency anticipated for the ATP-powered transport could be lost because of the weight penalty associated with passive interior noise control techniques. This concern has led to a concerted effort to develop noise control methods that do not significantly increase take-off weight. One such technique being considered is synchrophasing, in which the phase angle between propellers on opposite sides of the fuselage is controlled to a value which minimizes the sound pressure level at typical cabin interior locations. The present article investigates the sensitivity of the predicted synchrophasing effectiveness trends to the propeller source model used. The influence on synchrophasing effectiveness of the relative direction of rotation of the two propellers is also investigated.

Introduction

A new class of transport aircraft engine is currently under development as an alternative to the turbofan engine. This new engine, sometimes called the advanced turboprop, or ATP, is attractive because it promises twenty to thirty per cent increases in fuel efficiency without reduction in flight speed. However, preliminary tests using scale models of the ATP [1,2] indicate that their sound source strength will greatly exceed that associated with current technology turboprop engines. It is therefore conceivable that the additional weight associated with the mass and absorptive materials needed to limit penetration of propeller noise into the cabin could effectively wipe out much of the gain in fuel efficiency afforded by the ATP. This possibility has led to renewed efforts to develop alternative aircraft interior noise control methods that do not significantly increase take-off weight.

Aero- and Hydro-Acoustics
IUTAM Symposium Lyon 1985
Editors: G. Comte-Bellot and J. E. Ffowcs Williams

One of several promising methods of aircraft interior noise control which does not increase take-off weight is synchrophasing [3,4]. The idea behind synchrophasing is that two propellers producing essentially the same noise signature but with a specified relative phase relationship will force the fuselage in such a way that the concomitant sound pressure field at typical interior locations is significantly reduced. Until recently the development of synchrophasing and other aircraft interior noise reduction strategies has been hampered by a lack of physical understanding of the transmission mechanism. However, Fuller [5] has recently formulated an analytical model for sound transmission into aircraft cabins which apparently describes all of the important mechanisms. The model, which predicts the interior sound pressure field due to an arbitrary distribution of elemental sources exterior to the aircraft cabin, is developed in detail in Ref. 5, and its use with simple dipole and virtually rotating dipole propeller source models is described in Refs. 5 and 6, respectively.

In the original application of his aircraft interior noise model to the synchrophasing problem, Fuller modeled the propellers on either side of a twin engine aircraft as acoustic dipoles whose axes were oriented toward the fuselage. Variation of the synchrophase angle, or the phase angle between propeller blades moving past the fuselage on either side of the aircraft, was simulated by varying the phase relationship between the two dipoles. Fuller found that there is an optimum synchrophase angle that minimizes the sound pressure level at each interior location. He was able to obtain reductions on the order of 10 to 15 dB at typical locations in the cabin. Jones and Fuller [7] have subsequently verified these analytical results experimentally using a long aluminum cylindrical shell driven by two exterior monopole sources whose relative phase could be controlled.

The synchrophasing effectiveness is defined as the degree to which the interior sound pressure field is reduced by the synchrophasing mechanism. Mahan and Fuller [6] have recently shown that the synchrophasing effectiveness predicted using Fuller's model is sensitive not only to the magnitude of the acoustic pressure distribution on the fuselage surface but also to its phase. In the cited analytical study the authors replaced the dipole sources used in Fuller's original work with virtually rotating dipole sources. This latter type of elemental source, described in detail in Ref. 6, produces fuselage surface phase distributions similar to those reported in Ref. 8 for a current technology turboprop. This work has subsequently led to further studies of the relationship between propeller source models and their corresponding synchrophasing effectiveness. In particular, evolution of the synchrophasing effectiveness as the source structure varies continuously from that of a simple dipole to that of a virtually rotating dipole producing essentially the same fuselage surface pressure magnitude distribution but a different phase distribution has now been studied. Results are also reported for the comparison of the synchrophasing effectiveness for counter- and co-rotation of the propellers on either side of the fuselage, assuming that the difference in phase distribution for the two cases is indeed important.

The Propeller Model

As mentioned above, previous synchrophasing studies using Fuller's aircraft interior noise model have made use of one of two types of propeller source model: the simple dipole model, shown in Fig. 1(a), and the virtually rotating dipole model, shown in Fig. 1(b). The details of these two propeller source models and descriptions of their application to the synchrophasing problem can be found in Refs. 5 and 6, respectively.

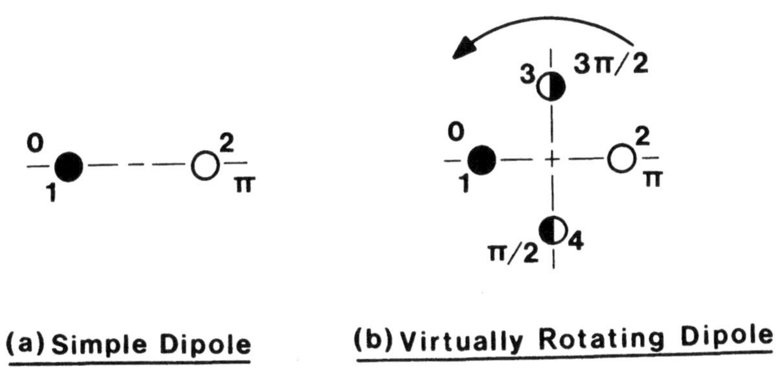

Fig. 1. Schematic representation of the simple dipole (a) and the virtually rotating dipole (b) sources used in previous synchrophasing studies

The propeller source model used in the present study is a hybrid of the simple dipole and the virtually rotating dipole models from the cited studies. As in the case of the virtually rotating dipole source model of Fig. 1(b), the new model consists of two mutually orthogonal dipole elements per engine with the vertically oriented dipole 90 deg out of phase with the horizontally oriented dipole. Then if the propeller being modeled is to rotate counter-clockwise, monopole number 1 in Fig. 1(b) would lead monopole number 3 by 90 deg, whereas if the propeller is to rotate clockwise, monopole number 1 would lead monopole number 4 by 90 deg. Of course, the two monopoles comprising a given dipole are always 180 deg out of phase with each other. In fact, the only characteristic which distinguishes this new propeller source model from the virtually rotating dipole model of Ref. 6 is the fact that now the strength of the vertical element is continuously varied as a parameter. This makes it possible to study the effect on synchrophasing effectiveness of gradually transforming the source model from a simple dipole (vertical dipole source strength equal to zero) to a virtually rotating dipole (equal vertical and horizontal dipole source strengths). Also, by changing the direction of rotation of the propeller source model on one side of the fuselage, it is possible, assuming that the fuselage surface phase angle distribution is indeed important, to compare the relative synchrophasing effectiveness trends of counter-rotation versus co-rotation.

Results and Discussion

Figure 2 compares the predicted synchrophasing effectiveness for three types of source distribution: simple dipole sources, virtually co-rotating dipole sources, and virtually counter-rotating dipole sources. In Fig. 2(a) the source distribution is centered at three shell radii on opposite sides of the shell, and in Fig. 2(b) it is centered at 4.3 shell radii. The synchrophasing effectiveness is represented here in terms of the relative attenuation, in decibels, of the acoustic pressure at a given interior location and synchrophase angle. The curves in Fig. 2 represent the synchrophasing effectiveness in the plane of the sources at the interior wall of the shell and at an angle of 45 deg from the horizontal. The nondimensional frequency is defined as $\omega a/c_L$, where ω is the angular frequency of the harmonic of interest, a is the radius of the cylindrical shell, and c_L is the extensional phase speed of the shell material. The results in Fig. 2 correspond to a nondimensional frequency of 0.2, which is representative of an aluminum aircraft fuselage.

While the difference between the synchrophasing effectiveness curves for the simple dipole and the virtually counter-rotating dipole in Fig. 2 is small, there is a very significant difference between these and the curve corresponding to co-rotation. In the former two cases, the optimum synchrophase angle is around 310 deg, whereas in the latter case the optimum is about 80 deg. Also, the maximum attenuation is much greater, more than 20 dB, in the two former cases than in the latter case, where it is only about 7 dB.

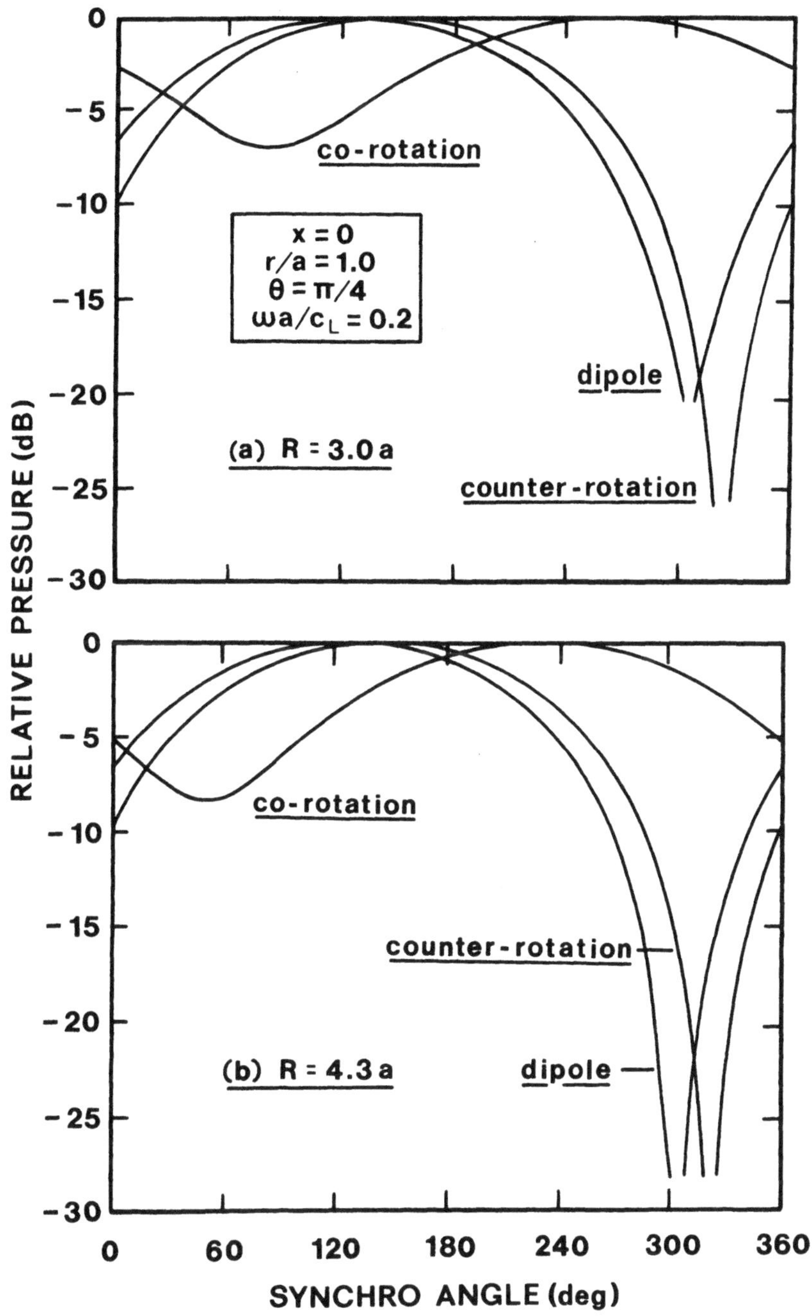

Fig. 2. Comparison of synchrophasing effectiveness for two radial locations of co-rotating, counter-rotating, and simple dipole sources for $\omega a/c_L = 0.2$

The phase distribution on the fuselage exterior surface depends on the distance from the fuselage to the source distribution. This is true because of geometrical considerations, of course, but also because the acoustic waves radiating from the source have both a radial and a circumferential component. While a virtually rotating dipole source does produce a fuselage surface phase distribution which is in qualitative agreement with observed distributions, quantitative agreement is difficult to obtain over all of the fuselage. However, comparison of Figs. 2(a) and 2(b) reveals that the influence of radial location of the source distribution within reasonable limits is relatively minor in this case. This implies that while it is important to include the rotational component of the incident fuselage surface pressure field, the results obtained are relative insensitive to its exact distribution.

Figure 3 compares synchrophasing effectiveness for the same cases as Fig. 2, except now the nondimensional frequency is lower, 0.14. At this frequency the difference between all three curves is small except near the optimum synchrophase angle ($\sim$ 350 deg). The principal physical difference between the cases of Fig. 2 and those of Fig. 3 is that the dominant interior acoustic mode being excited in the former case is even ($n = 2$), whereas in the latter case the dominant mode is odd ($n = 1$). An even acoustic mode is evidently more sensitive to the directions of the exterior phase velocities on either side of the fuselage than is an odd acoustic mode. The effect of radial translation of the source distribution, as revealed by comparing corresponding curves in Figs. 3(a) and 3(b), is once again relatively minor.

Figure 4 shows the evolution of the synchrophasing effectiveness as the sources are gradually varied from simple dipoles ($Q = 0$) to virtually rotating dipoles ($Q = 1$). In this figure, the nondimensional frequency is 0.2 and the source and interior locations are as in Figs 2(a) and 3(a). The quantity Q in Fig. 4 is the relative strength of the vertical element of the virtually rotating dipole in Fig. 1(b) to the horizontal element. As the rotational component of the source field is imposed in the case of corotation, Fig. 4(a), the optimum synchrophase angle shifts "to the right" from about 305 deg to about 85 deg, and the maximum effectiveness decreases from a value greater than 20 dB to a value of about 7 dB of relative attenuation. However, in the case of counter-rotation, Fig. 4(b), imposition of the rotating component has a relatively small effect, especially on the optimum synchrophase angle. Then to the degree that variation of the rotating component of the source field can be related to propeller speed and location,

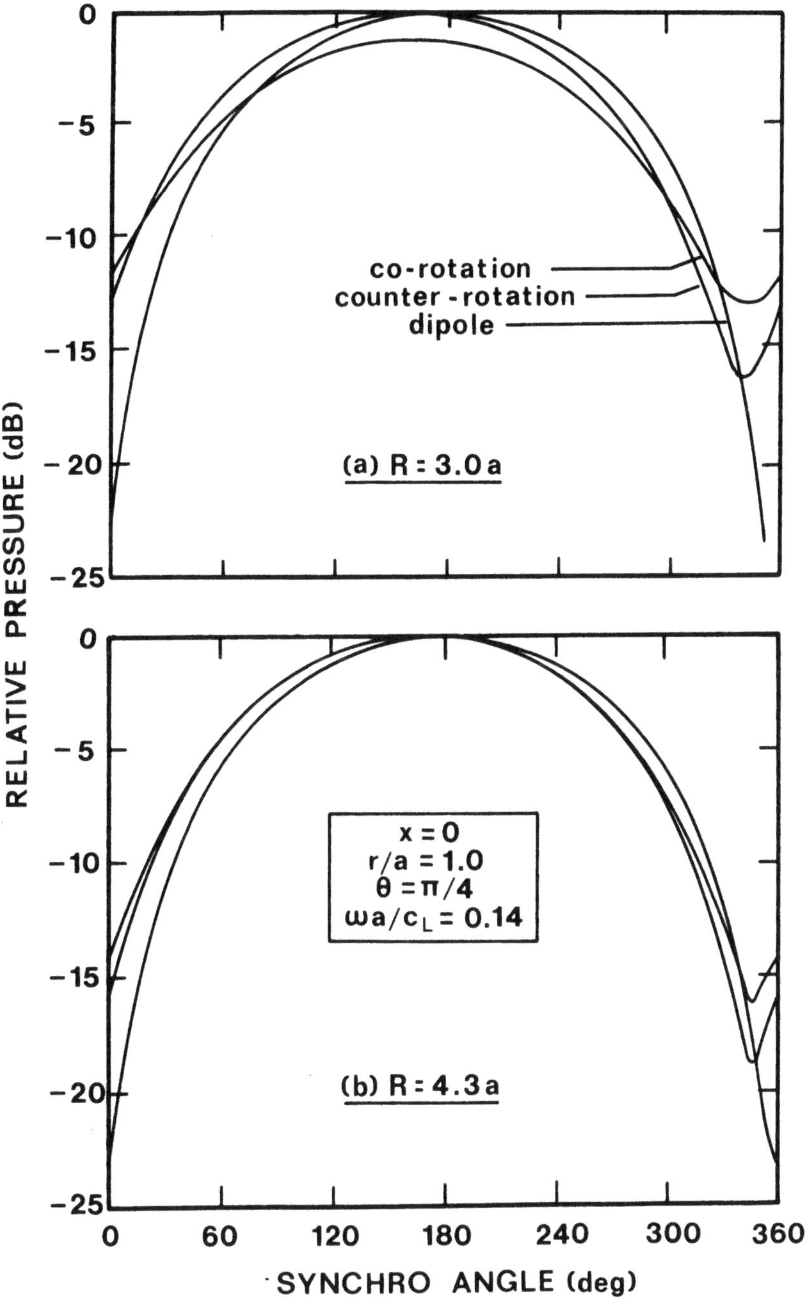

Fig. 3. Comparison of synchrophasing effectiveness for two radial locations of co-rotating, counter-rotating, and simple dipole sources for $\omega a/c_L = 0.14$

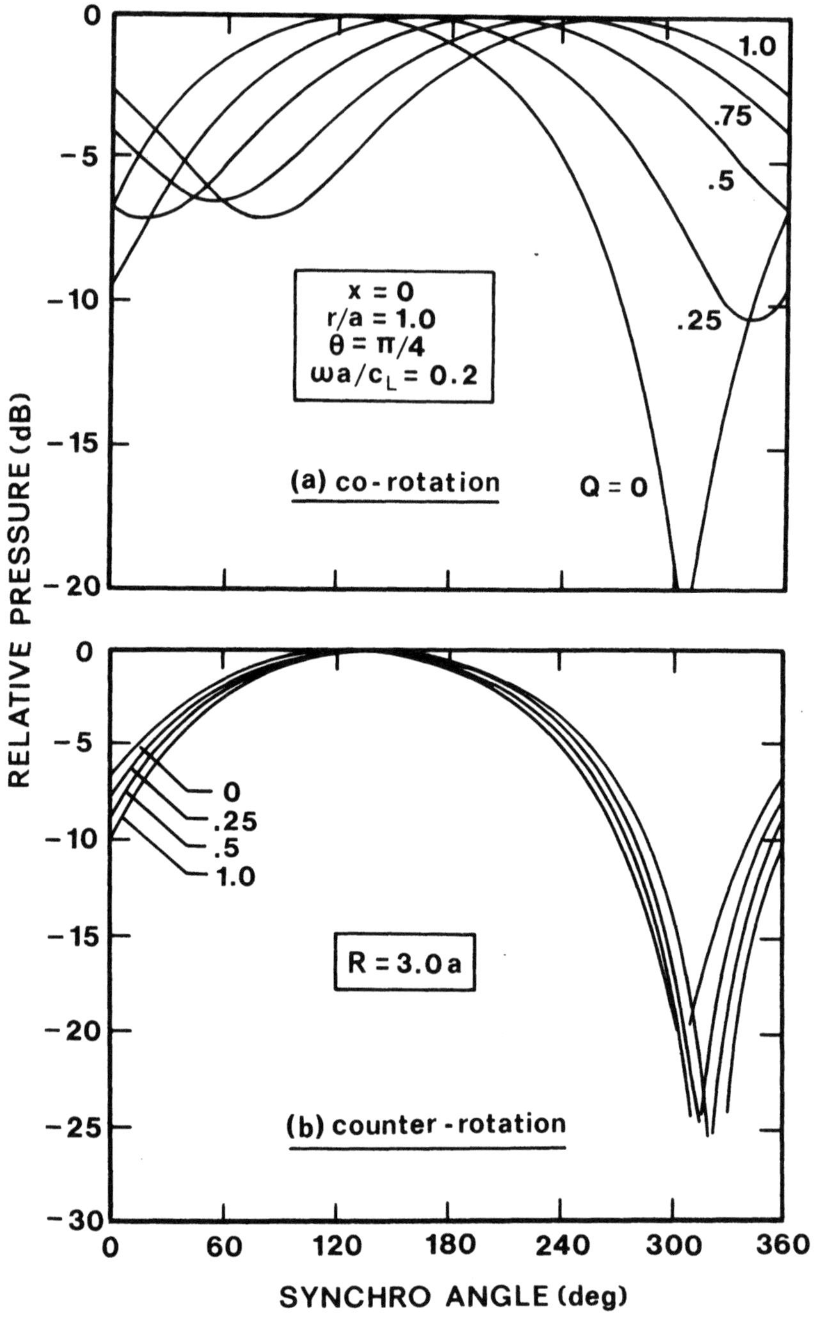

Fig. 4. Evolution of the synchrophasing effectiveness as the source varies continuously from a simple dipole (Q = 0) to a virtually rotating dipole (Q = 1) with (a) co-rotation and (b) counter-rotation for $\omega a/c_L = 0.2$

the results of Fig. 4 suggest that for co-rotating propellers these factors are important to the success of the synchrophasing approach to aircraft interior noise control, whereas for counter-rotating propellers, these factors are relatively unimportant.

Conclusions

The results presented above lead to the following conclusions:

(1) the synchrophasing effectiveness is sensitive to the type of source model used

(2) for the virtually rotating dipole source model, the synchrophasing effectiveness is sensitive to the direction of rotation at some frequencies but not at others

(3) the synchrophasing effectiveness obtained from the virtually rotating dipole model is not very sensitive to the radial location of the source distribution within reasonable limits

(4) the predicted synchrophasing effectiveness is more sensitive to the details of the source model used for the case of co-rotation than for the case of counter-rotation.

Acknowledgment

The research reported in this article was supported by NASA Grants NAG1-390 and NAG1-493.

References

1. Zorumski, W. E.: Propeller noise prediction, NASA TM 85636, May 1983.
2. Magliozzi, B.: Acoustic pressures on a propfan aircraft fuselage surface, AIAA paper 80-1002, June 1980.
3. Johnston, J. F.; Donham, R. E.; Guinn, W. A.: Propeller signatures and their use, AIAA paper 80-1035, June 1980.
4. Magliozzi, B.: Synchrophasing for cabin noise reduction of propeller-driven aircraft, AIAA paper 83-0717, April 1983.
5. Fuller, C. R.: Noise control characteristics of synchrophasing--an analytical investigation, AIAA paper 84-2369, October 1984.
6. Mahan, J. R.; Fuller, C. R.: An improved source model for aircraft interior noise studies, AIAA paper 85-0787-CP, April 1985.
7. Jones, J. D.; Fuller, C. R.: Noise control characteristics of synchrophasing--an experimental investigation, AIAA paper 84-2370, October 1984.
8. Piersol, A. G.; Wilby E. F.; Wilby J. F.: Evaluation of Aero Commander propeller acoustic data: static operations, NASA CR 158919, May 1978.

Modelling of Trailing Edge Flow Tones in Elastic Structures

W.K. BLAKE, J.L. GERSHFELD, L.J. MAGA

David Taylor Naval Ship R&D Center, Bethesda, MD

ABSTRACT

The hydroelastic coupling of turbine blade vibration modes and trailing edge vortex shedding from the blade is a well known phenomenon that leads to intense audible tones and shortened fatigue life. Historically, most researches on such turbo-blade "singing" have focussed on the prediction of the vortex shedding frequency and on the reduction of offending vibration levels by reshaping the trailing edges. The associated control measures effectively permit the "detuning" of the vortex shedding from the blade modes, and the effectiveness of such measures is more or less qualitatively known. This paper builds on and extends a different line of research that is aimed at merging the controling fluid dynamic and structural dynamic factors of these flow-induced tones into a quantitative semiempirical model of the non linear flow-induced vibration of turbo machine blading. The elements of the model development involve the use of measured nonlinear "wake impedances" by the forced motion of trailing edges, the development of a mathematical model for the response of blade structures that are coupled to the wake dynamics, and finally the testing of the modeling against measured flow-induced vibration of simple hydrofoils. The parameters used in both the modeling and the experimental verification include structure mass density, hydrofoil geometry, damping, and trailing edge geometry.

1. INTRODUCTION

The self-excitation of hydrofoils and cylinders by a coupling of vibration and vortex street development is just one of the many nonlinear fluid-structure interactions of hydroacoustics. The flow-elastic coupling involves a phase locking between the vortex formation in the near-wake and the motion of the trailing edge. In the scenario of a cylinder or hydrofoil in a cross flow with a slowly increasing fluid velocity, the natural vortex shedding frequency approaches and then exceeds a natural frequency of vibration

Aero- and Hydro-Acoustics
IUTAM Symposium Lyon 1985
Editors: G. Comte-Bellot and J. E. Ffowcs Williams

of the structure. As the vortex shedding frequecy ω_s approaches the resonance frequency, ω_{mn} of the mn mode, the elastic structural dynamics and the wake vortex dynamics combine as coupled ocillators and a self excited state is reached at $\omega_s/\omega_{mn} \approx 1$. In this state, energy is transferred to and from elastic and fluid oscillators at the points on the body where the flow separates. In the case of the circular cylinder our knowledge of this process is quite detailed. The development of the instabilities in the near wake and of the vibration-enhanced lift forces, have all received considerable attention. Though a formal theoretical analysis of the process has not been put forth, the physics of the fluid mechanical processes of vortex entrainment by a transversely oscillating cylinder is now well characterised experimentally and is reasonably well understood. The theoretical treatment of the self-excited circular cylinder as an elastic wake-coupled oscillator, though developed using somewhat heuristic analytical modeling, has also attained enough empirical support that engineering predictions are now quite common place. It is now possible to predict the limiting cycle behavior of elastic circular cylinders in terms of both the damping and the inertance of the cylinder under certain circumstances.

In the case of the hydrofoil, vortex-edge coupling processes have only recently been examined in enough detail to allow a comparable level of understanding. The vortex formation in the near wake depends on the geometry of the edge, the upstream boundary layer characteristics, and the degree of vibration of the edge transversely to the wake. Thus, both geometry and Reynolds number determine the vortex shedding from lifting surfaces, while only Reynolds number must be considered in the case of circular cylinders. The self-excited behavior of hydrofoils has been given only limited attention. In work reported by the author, the aspect self excitation has been examined as a problem of structural response. In separate research, however, the features of wake production downstream of a trailing edge flap that is subject to forced oscillation have also been examined in some detail. For a given geometry of trailing edge those measurements provide the input impedance of the near wake as a function both the fluid velocity and the amplitude of oscillation. Thus, this paper will show how it is now possible to model mathematically the interaction of the modal vibration of a lifting surface with the forcing features of the vortex street. The analytical methods developed here are similar to those which have found utility in modeling the self-excitation of cylinders. Specifically, the pressure at the edge is modelled as a nonlinear Rayleigh oscillator to generalize measured pressures as functions of vibration amplitude, velocity of flow, and thickness of the trailing edge. The wake induced surface pressure is then used to derive a modal lift co-efficient which couples to the mode resonances of the hydrofoil. This step has been made possible by measurements which confirmed that the chordwise variation of the enhanced wake-induced pressures under the influence of vibration is identical to that of the naturally imposed pressures on the rigid (or fixed) edge. In this paper, the problem of self-excited vibration of blunt-edged hydrofoils will be examined for the first time in a model format that combines aerodynamic measurements of the nonlinear "input impedance" of the wake with observed hydroelastic response of singing hydrofoils. The new feature of the work is a quantitative relationship between details in the features of hydoelastic response as well as an extension of observed hydroelastic behavior to ratios of added mass to dry mass which greatly exceed unity.

2.0 SCOPE OF THE MEASUREMENTS

The empirical support for the analytical developments in this paper originate from a variety of measurement programs. Thus this paper draws heavily on previous work. This section gives an overview of those measurement programs, the details are to be found in the cited references.

2.1 Hydroelastic Observations

The hydrofoils are all of a simple flat plate section with a circular leading edge and a tapered trailing after section. This simple geometry was selected to facilitate the application of damping materials during various phases of the work. These hydrofoils had spans on the order of 20 inches and were clamped at one end in a block of 200 lbs. mass. The arrangement has remained essentially unchanged since it was initially reported by Blake and Maga (1975a). Table I gives a list of the pertinent geometric and structural features of the hydrofoils. With the exception of the 0.375 inch thick 5 inch chord hydrofoil, the hydrofoils all had a 12.5° taper. Various trailing edge thicknesses were obtained on these foils by removing material from the trailing edge as indicated by the dashed lines at the bottom of the table. Two families of hydrofoils were thus obtained with base line chords W_o of 2.75 inches and 5 inches with initially sharp trailing edges. Because of the clamping system used, the vibration modes of the hydrofoils were quite similar to those expected on theoretical grounds, Blake and Maga (1975). A leading difference between the modal characters of the 2.75 inch and 5-inch chord families is that the 2.75 inch hydrofoil had only beam (m,o) modes in the frequency range studied, while the 5-inch family had plate modes (m,n). This has provided an opportunity to examine tendices of modes with different degrees of chord-wise flexure to couple with the vortex street.

Measurements on these hydrofoils under flow excitaiton has included damping (loss factor) following methods outlined by Blake, Gershfeld, and Maga (1986). The variable $a^2(x)$ will be used to represent the time average of the surface aceleration a(x) squared, and $\langle a^2 \rangle$ will represent an area average of the time mean square values. Measurements of of resonance frequencies in air and water give the modal added mass, $(m_a)_{mn}$.

2.2 Aerodynamic Observations

Aerodynamic observations have been made on a rigid 8 x wooden model of the W_0xh = 2.75 in. x 0.25 in. hydrofoil with either the blunt 0.072 in. edge or the basic sharp edge. Results of these measurements have been reported previously by Blake, Maga, and Finkelstein (1977) and by Blake (1984). Measurements of the aerodynamic pressures and near-wake properties of the 5x 0.375 inch hydrofoil using a large rigid 5.33x scale model have also been made at various wind velocities with some results reported by Blake and Maga (1979) in which a movable flap was installed at the trailing edge. The basic aerodynamics of the rigid version of this airfoil had been reported by Blake (1975) but have been recently extended and re-analyzed in (1985). Figure (1) is a sketch of the arrangement used for the measurement of the "wake impedance". The main airfoil was wooden, but the flap was wood-reinforced styrofoam with the surface smoothed with a coating of epoxy resin. The flap was oscillated with a yoke driven by two matched shakers. Oscillation amplitude was determined with an accelerometer. Surface pressures were measured with small taps fixed to flexible tubes which transducted pressures into the microphones located in the rigid portion of the airfoil. Motion of the flap was rigid-body and electrical power requirements of the shakers contributed to high frequency limitations of the experiment. The scope of measurements included chordwise distribution of pressure, spanwise correlation, and disturbance growth in the wake. (See Blake and Maga (1979) for a selection of the results of the measurements.) Velocity profiles in the wake were also obtained to determine the wake thickness y_f. This thickness is determined at the point downstream of the edge at which the shear-layer disturbance reaches its maximum growth in accordance with a definition by Bearman (1965); y_f differs from the base thickness t_0 by up to 20% depending on the thickness of the boundary layer on the foil upstream of the edge and on the Reynolds number.

3 DESCRIPTION OF THE OSCILLATING LIFT

3.1 Review of Measurements

Characteristics of the distribution of unsteady pressure on both the fixed and moving trailing edges have been reported from time to time (e.g. Blake and Maga (1979), Blake (1984)). On the fixed edge, the pressure is tonal, and may be described as

$$p(\vec{x},t) = P_o(t) g(x_1)\, h(x_3)$$

$$p(\vec{x},t) = P_o\, g(x_1)\, h(x_3)\, e^{-i\omega_s t} \tag{1}$$

where P_o is an amplitude that is determined by the shed vortex strength κ or by

$$P_o = \rho_o\, \kappa\, U_s/(2\pi y_f) \tag{2}$$

The functions of $g(x_1)$ and $h(x_3)$ describe the chord and spanwise variations of pressure. The velocity $U_s = U_\infty \sqrt{(1 - C_{pb})}$ is the effective mean velocity past the edge with a base pressure coefficient $C_{pb} \approx 0.56$ as shown first by Bearman (1965) and later corroborated in the studies by Blake (1974). The function $g(x_1)$ is the chordwise distribution which is maximum at the trailing edge and decreases monotonically with increasing distance x_1 from the edge. Theoretically (Blake 1984b), in the immediate vicinity of the edge

$$g(x_1) \sim 1/\sqrt{x_1/y_f} \tag{3}$$

where y_f is the wake thickness at the end of the formation zone. However, when $x_1 > \sim (5 - 10)\, t_o$ the decrease in $p(\vec{x},t)$ with x_1 is more pronounced. Accordingly a match with experimental results gives for a blunt trailing edge (Blake 1984a)

$$g(x_1) = 0.35\ (x_1/y_f)^{-1/2} \quad \text{for } x_1/y_f < \sim 6 \tag{4}$$

and substantially reduced levels for $x_1/y_f > 6$. Accordingly, $x_1/y_f = 6$ is viewed simply as an extremum of applicability for equation (4) that is implemented in determining the lift fluctuation.

The function $h(x_3)$ is random and describes the spanwise correlation of both the surface pressure and the shed vorticity. For spanwise uniform flow, the correlation of $h(x_3)$ is simply the spatial correlation function

$$R\,(r_3) = \langle h(x_3)\, h(x_3 + r_3)\rangle \qquad (5)$$

which is symmetric in r_3 and yields the integral scale

$$\Lambda_3 = \int_0^\infty R\,(r_3)\, dr_3 \qquad (6)$$

There have been measurements of Λ_3 (or of $R(r_3)$) for the pressures on blunt edges (Blake (1984a,b)), for the velocity fluctuations in the wakes of blunt edges, Graham (1969), of flat plates, and for the vortex pressures on a blunted NACA 0012 airfoil (Dr. T. Brooks of NASA, Langley - see Blake (1984a)). Collectively, these measurements disclose a range of $1.5 < \Lambda_3/y_f < 3.5$ for $10^4 < {}_{y_f} < 2x\ 10^5$ for rigid airfoils.

The amplitude factor P_o has been examined at length in previous work and is linearly dependent on the magnitude of the velocity fluctuation at the end of the formation zone, u_s, and the apparent vortex core size (Blake (1984b) and Blake and Maga (1979)). When the edge is oscillated at a frequency ω_v equal to the vortex shedding frequency ω_s, the pressure amplitude P_o increases in direct proportion to u_s and u_s increases with the amplitude of vibration (Blake and Maga (1979)). The wake thickness and apparent vortex core size are not measurably influenced by small edge vibration. Thus, the edge vibration increases the strengths κ of the shed vorticity which accounts for a direct increase in P_o and $g(x_1)$ appears to be un-influenced, at least for $x_1/y_f < \sim 5$. Unfortunately, for $x_1/y_f > 5$ the mechanically induced vibrations of the apparatus were great enough to obscure further aerodynamic behavior. Accordingly, Equation (4), with its extremum on x_1, has been adopted for the moving edge as well until future work extends its useful range. Also, when $\omega_v = \omega_s$, the spanwise correlation increases. In recent work (Blake (1984a)) the increase in correlation has been viewed as due to a vibration-induced component, say p_v, that is fully correlated with the vibration along the span, and a "naturally" - induced component, p_s, which has spanwise stochastic qualities much like those of the pressure exerted by the wake of the rigid edge. In the scenario of increasing the frequency of vibration of the edge while maintaining constant amplitude y_v and constant wind speed, pressures at both frequencies ω_v and ω_s are generated on the edge. When ω_v/ω_s is outside a range that may be roughly defined by $0.95 < \omega_v/\omega_s < 1.11$ (according to the current measurements for which $0 < y_v/t_o < 27 \times 10^{-3}$) the pressure at ω_s is the same as that which is naturally induced and the pressure at ω_v has reactive and dissipative properties that are uniquely determined by ω_v/ω_s and y_v/t_o (or y_v/y_f). Figure 2 shows typical examples of $p(\omega_v, y_v)$ and the phase angle

$$\phi = \underline{\angle p(\omega_v, y_v) \quad y_v}$$

for 3 amplitudes of oscillation at various velocities ($\mathcal{R}_{y_f} = 2.4 \times 10^4$). In these examples the separate pressures at ω_v and ω_s are no longer distinguishable when ω_v/ω_s is near unity. The bandwidth of entrainment $\Delta\omega_e/\omega_s$ that relates to sustained oscillations increases with y_v/y_f. If a comparison is made between pressures at different speeds, but at equal values of y_v, then the patterns of $p(y_v, \omega_v)$ at the two wind velocities are qualitatively similar. For example the $\Delta\omega_e$ for 100 f/s and 50 f/s and $(y_v)_{rms/to} = 4 - 5 \times 10^{-3}$ are comparable as they are roughly so for $(y_v)rms/t_o = 9 \times 10^{-3}$ at 50 f/s and 25 f/s, $(y_v)_{rms} = y_v/\sqrt{2}$.

3.2 The Rayleigh Oscillator Model

Hartlen and Currie (1970) proposed that the unsteady lift coefficient may be modeled by the non-linear Rayleigh equation for the purpose of modeling the self-sustained oscillations of circular cylinders. The pressure may be described by the same nonlinear equation. The measurements in Figures 2 and 3 show that a similar relationship

$$\frac{\ddot{p}}{q} - \alpha' \left(\frac{\dot{p}}{q}\right) + \gamma' \left(\frac{\dot{p}}{q}\right)^3 - \omega_s^2 \left(\frac{p}{q}\right) = \beta\left(\frac{\dot{y}_v}{y_f}\right) \qquad (7)$$

where p is the pressure at the edge q = the dynamic pressure $= \rho_o U_\infty^2/2$, and y_v is the vibration velocity of edge, appears to describe the pressures in the entrainment region. The notation $\dot{p}$ and $\ddot{p}$ denote derivatives with respect to time. The coefficients α' and γ' are dimensionless constants and β is a dimensional constant coupling factor. That the coupling is through the velocity y_v has been suggested by the comparison of the $p(\omega_v, y_v)/q$ at different speeds. Note that the ratio $p(\omega_v, y_v)/p(\omega_s, 0)$, where $p(\omega_s, 0)$ is the pressure induced at zero vibration amplitude, and both $p(\omega_v, y_v)$ and $p(\omega_s, 0)$ are measured at the same location $\vec{x}$, is not dependent on proximity to the edge, x_1.

By definition, in the entrained region the function $h(x_3)$ in equation (1) may be replaced by unity. Thus we define a lift coefficient

$$C_L = \frac{dL/dx_3}{q\ y_f} = \frac{2P_o \int g(x_1)dx_1}{q\ y_f} = 2K\left(\frac{P_o}{q}\right) \qquad (8)$$

since the lift is generated by a differential pressure Δp which is 2p and where according to equation (4)

$$K = 0.7\sqrt{m} \qquad (9)$$

Now scaling the frequency on the oscillating frequency ω_v, then we find the dimensionless and scaled variables

$$\ddot{C}_L = \omega_v^2\, C_L'' \quad , \dot{C}_L = \omega_v\, C_L'$$
$$T = \omega_v t, \qquad \Omega = \omega/\omega_v, \qquad (10)$$
$$\Omega_s = \omega_s/\omega_v, \text{ and } \Omega_o = U_\infty/\ (y_f\ \omega_v)$$

so that the equation (7) yields one of similar form for the lift coefficient appears in the dimensionless canonical form of Hartlen and Currie (1970)

$$C_L'' - \alpha\, C_L' + \gamma(C_L')^3 - \Omega_S^2\, C_L = b\ y_v/y_f \qquad (11)$$

where

$$\alpha = \left(\frac{\alpha}{\omega_s}\right)\cdot\left(\frac{\omega_s}{\omega_v}\right), \quad \gamma = \left(\frac{\gamma\omega_s}{K^2}\right)\left(\frac{\omega_v}{\omega_s}\right), \quad b = \frac{\beta K}{\omega_v} \tag{12}$$

Various solutions to equation (11) have been examined by Nayfeh and Mook, (1979) and we shall be interested only in steady state solutions very near $\Omega = \Omega_s$. Assumed solutions of equation (11) are thus of the form

$$C_L(t) = C_L \sin(\Omega T + \phi) \tag{13}$$

with a driving oscillation of the form

$$Y_v(t) = Y_v \sin(\Omega T) \text{ and } Y_v = y_v/y_f \tag{14}$$

We are interested in steady state solutions of equation (11) under the assumption that contributions from harmonics $\omega = 3\omega_s$ and subharmonics $\omega = \omega_s/3$ are negligable compared with the fundamental $\omega = \omega_s$. The measurements support these assumptions in indicating no significant contributions from these harmonics. Thus, in substitutions of equations (13) and (14) into equation (10) the approximation

$$\cos^3(\Omega T + \phi) \approx 3 \cos(\Omega T + \phi)$$

follows from neglect of the second term of $4\cos^3\theta = 3\cos\theta + \cos 3\theta$. Then as shown by Nayfeh and Mook (1979) the solution to the homogeneous form of equation (11), i.e. $y_v = 0$ is at $\omega_v = \omega_s$ and

$$C_L(t) = C_{Lo} \sin \omega_s t \tag{15}$$

where

$$C_{Lo} = \left(\frac{4}{3}\frac{\alpha}{\gamma}\right)^{1/2} \tag{16}$$

The forced solution very near $\omega_v = \omega_s$ is

$$\left(\frac{C_L}{C_{Lo}}\right)\left[1 - \left(\frac{C_L}{C_{Lo}}\right)\right]^2 = \left(\frac{b}{\alpha}\right)^2\left(\frac{\omega_v}{\omega_s}\right)^2\left[\frac{(y_v/y_f)}{C_{Lo}}\right]^2 + \frac{12}{\alpha^2\,\omega_s^2}(\omega_s - \omega_v)^2 \tag{17}$$

where, in terms of the pressure on the edge, C_L/C_{Lo} = $p(\omega_v, y_v) / p(\omega_s, 0)$. As ω_v departs from ω_s, C_L approaches C_{Lo} so that an approximate band of frequencies $\Delta\omega_e$ that straddles $\omega_v \approx \omega_s$ for which entrainment may occur is given by

$$\frac{\Delta\omega_e}{\omega_s} \approx \frac{2b}{\sqrt{12}} \left[\frac{y_v/y_f}{C_{Lo}}\right] \qquad (18)$$

Figure 3 shows that the measured amplification and phase conforms reasonably well to the solutions of equations (11): particularly in the behavior of the entainment zone, $\Delta\omega_e$ with vibration amplitude y_v. Comparison of equation (18) with the measurements in Figure 2 suggests that the coupling factor β in equation (7) scales on ω_s^2 if b is to remain a wind velocity and amplitude - independent parameter.

The rise of (C_L/C_{Lo}) with vibration amplitude is shown in Figure 4; a fit to equation (17) with $\omega_s = \omega_v$ gives a factor (b/α) of order 47.5. The measurements, however, do not give large values of either C_L/C_{Lo} or $|y_v|$ and future work is necessary to extend the ranges. Thus, the measurements offer both qualitative and quantitative support to the use of the Rayleigh oscillator model to describe the influence of trailing edge vibration to the lift induced by vortex shedding.

4 FLOW INDUCED VIBRATION

4.1 Review of Measurements

The general behavior of the vibration of the hydrofoils as velocity is increased has been discussed rather completely in previous papers (Blake (1984a), Blake, Maga, Finkelsten (1977)) so here we examine some of the detailed features of flow excitation. As previously described, the normalized time and area-averaged mean square vibration amplitude $\overline{\langle y^2\rangle}/y_f^2$ is a function of a damping parameter,

$$\frac{\overline{\langle y^2\rangle}}{y_f^2} = F(D_a)$$

where

$$D_a = \frac{2\,\eta_s\, mW}{\rho_o y_f} \qquad (19)$$

The linear response of the hydrofoil, i.e. the vibration level which results from an imposed vortex street force that is not influenced by the edge displacement it induces, depends on D_a as (Blake (1984a))

$$\frac{\langle Y_{mn}^2 \rangle}{y_f^2} = \frac{C_{Lo}^2}{2} \frac{2\Lambda_3}{L_3} \frac{\langle \Psi_{mn}^2 (\vec{x}_{edge}) \rangle}{D_a^2} \tag{20}$$

where

$$C_{Lo}^2 = 2 \left(\frac{\rho_o U_s \kappa}{2 \pi y_f q} \right)^2 \cdot (0.49\ m) \tag{21}$$

is the amplitude -squared of the unsteady section lift coefficient and where $\langle \Psi_m^2 (\vec{x}_{edge}) \rangle$ represents the magnitude of the mode-shape squared evaluated at the trailing edge and averaged along the span. It is assumed in this equation that the frequency bandwidth of the wake-induced pressures is smaller than that of the resonance, $\eta_s \omega_m$. The factor $C_{Lo}^2/2$ is the temporal mean square of the harmonic lift coefficient of amplitude C_{Lo}. Equation (20) holds for the response of heavily damped modes, or those for which the ratio η_s / y_f^2 is large requiring moderate damping and a relatively thin edge.

The nonlinear response involves lightly damped modes of all classifications. Figure 5 shows the mean square displacement as a function of velocity for two beam modes of hydrofoils of two chords. The abcissa is a normalized velocity which also represents the ratio of natural vortex shedding frequency to resonance frequency of the mode. The hydrofoil with the lower value of D_a has the larger modal displacement. The orders of modes m,n in this paper follow the convention that m equals the number of chord-wise node lines and n equals the number of span-wise lines. Examples are diagrammed in Figures 5 and 6. Figure 6 shows the displacement for two plate modes of a brass and an aluminum hydrofoil of W = 4.66 inches. The values of D_a are, again, comparable to each other and comparable to the beam modes featured in Figure 5. Although the magnitudes of displacement are comparable for both the plate modes and the beam modes, the velocity range for self excitation appears somewhat larger for the plate modes. The self excited range of velocities is indicated by the breadth of the $\langle y^2 \rangle$ v.s. Ω_o "bump" which departs from the linear response of the sharp-edged beam. This self excited range is also a range of entrainment in which the vortex shedding and modal vibration frequencies are not distinguishable. Its breadth may be defined by the $\Delta\Omega_o$ at some amplitude below the maximum response displacement. Thus, at a point 20dB down from the maximum amplitude, the normalized speed range $\Delta\Omega_o$ appears to be roughly 0.15 for the beam modes and 0.20 for these plate modes. Figure 6 brings into the discussion the issue of the relative added masses of the structures. As summarized in Table 1, the ratio of added mass to dry mass for the aluminum hydrofoil is 3 to 4 times that for the brass and steel beams. This increased mass ratio of the lower density hydrofoils raises the possibility of modification of self-excited response by any altered character of the fluid-elastic coupling as occurs, for example, an analogous problem of flutter response of low mass-ratio hydrofoils.

Figure 7 shows the frequencies of vortex shedding and modal resonances for the aluminum W = 4.66 inch hydrofoil. The behavior with speed through the various coincidence regions is of the classical form. From this illustration it may be seen that the entrainment regions, here indicated by the velocity range ΔU_e of the various f, U_∞ plateaus, do not appear to be systematically different for plate and beam modes. This figure also shows the method generally used for determining the wake thickness y_f when a velocity profile is unavailable. The open points show vortex shedding frequencies that are distinct from the modal resonances and these determine a continuous f_s, U_∞ line which gives the indicated value of y_f. That $y_f \approx 1.12 t_o$ is a typically encountered result and that y_f is greater than the thickness t_o is due to the upstream boundary layer on the hydrofoil thickening the wake.

4.2 The Coupled Oscillator Model - The Non-linear Lift Oscillator

The non-linear lift oscillator will now be coupled to the linear modal amplitude in the fashion used by Hartlen and Currie (1970). Defining the response of the hydrofoils as a modal sum of time-dependent modal amplitudes

$$y(\vec{x},t) = \sum_{mn} y_{mn}(t)\, \psi_{mn}(\vec{x}) \tag{22}$$

with normalized in-vacuo mode shape

$$\int_0^{L_3}\int_0^{W} \psi_{mn}^{\,2}(\vec{x})\; d^2\vec{x} = L_3 W \tag{23}$$

the equation for $Y_{mn}(t)$ is (Blake (1984c), Ch 9)

$$[\rho_p h+m_a]\, A_o[\ddot{y}_{mn}(t)+\omega_{mn}\eta_s\, \dot{y}_{mn}(t)+\omega_{mn}^2\, y_{mn}(t)]$$

$$= \int_0^{L_3}\int_0^{W} \Delta p(\vec{x},t)\, \psi_{mn}(\vec{x})\, d^2\vec{x}$$

$$= \int_0^{L_3}\int_0^{W} 2(P_o(t))\, g(x_1)\, [\mathrm{Sgn}\, \psi_{mn}(\vec{x})]\, \psi_{mn}(\vec{x})\, dx_1 dx_3$$

$$\approx 2P_o(t)[\mathrm{Sgn}(x_1)dx_1]\quad \psi_{mn}(\vec{x}_{edge})\, L_3$$

$$\approx C_L(t)\, q y_f \quad \psi_{mn}(\vec{x}_{edge})\quad L_3 \tag{24}$$

where $C_L(t) = C_L e^{-i\omega t}$ is the time varying lift coefficient. We let

$$|\psi_{mn}(\vec{x}_{edge})| = \psi_e \tag{25}$$

be the absolute value of the mode shape function along the edge, and $2\pi\, S = \omega_s y_f/U$ is the natural Strouhal number ($2\pi\, S \approx 1$, Blake (1984 a,b)).

Following the normalization on ω_{mn} that led to equation (15) and normalizing the displacement on the wake thickness y_f we find

$$Y''_{mn}(T) + \eta_s\, Y'_{mn}(T) + Y_{mn}(T) = \Omega_o^2 a C_L(T) \qquad (26)$$

where

$$\Omega_o = (U_\infty / y_f\, \omega_{mn})(2\pi\, S)$$

$$Y_{mn} = y_{mn}/y_f$$

$$a = \rho_o y_f^2/[2(\rho_p h + m_a)W]$$

$$D_a = \eta_s/a \qquad (27)$$

$$C_{Le} = C_L\, \psi_e$$

$$2\pi\, S \approx 1$$

and the remainder of parameters are given in equation (10) but with ω_{mn} replacing ω_v. Note that ω_s in the case of free vortex shedding. Since

$$[Y_{mn}(T)]_e = Y_{mn}(T)\, \psi_e$$

equation (11) can be rewritten in terms of C_{Le} as

$$C''_{Le} - \propto C'_{Le} + \frac{\gamma}{\psi_e^2}\left(C'_{Le}\right)^3 + \Omega_s^2 C_{Le} = b\psi_e{}^2 Y'_{mn} \qquad (28)$$

Equations (26) and (27) are now in the convienent forms used by Hartlen and Currie (1970), differing only in the appearance of the mode shape factors ψ_e which arise because the hydrofoil and the wake are coupled only at the free trailing edge. It is to be noted that considering the definition of C_{Le} in equation (27), equations (11) and (28) give identical solutions for the rigid-edge lift coefficient, equation (16). Exact solutions of the coupled equations (26) and (28) may be obtained numerically or on analog computers, but it is instructive to derive certain asymptotic forms. We do this following the lead of Hartlen and Currie (1970) who sought steady state solutions of the form

$$Y_{mn}(T) = Y_{mn} \sin \Omega T \tag{29}$$

$$C_{L_e}(T) = C_{L_e} \sin(\Omega t + \phi) \tag{30}$$

by neglecting the third harmonic as in Section 3.2. The $\Omega = \omega/\omega_{mn}$ is the flow-elastic "system" resonance given by

$$(1 - \Omega^2)^2 = \eta_s^2\Omega^2 \left\{ \frac{b\psi_e^2 \ \Omega_o^2}{D_a(\Omega_o^2 - \Omega^2)} - 1 \right\} \tag{31}$$

which yields solutions for Ω close enough to unity that typically

$$(\Omega_o - 1) \gg (1 - \Omega) \tag{31A}$$

except when $\Omega_o = 1$. As shown by Hartlen and Currie (1970), equation (31) gives a requirement that the two solutions exist only when $D_a < b\,\Psi_e^2$ and that $\Omega < \Omega_o$ meaning that the flow-elastic "system" resonance given by this model is slightly less than the natural resonance frequency of the mode in still fluid. The modal amplitude is rewritten in terms of the subject variables, and in light of equation (31A), we obtain the approximation for the self-excited modal amplitude

$$Y_{mn}^2 = \frac{4}{3} \frac{a \left(\frac{\Omega_o}{\Omega}\right)^3}{\gamma \eta_s^2 \ b \ \Omega} \left\{ \frac{b \ \Psi_e^2 \ \Omega_o^2}{D_a} - (\Omega_o^2 - \Omega^2) \right\}^{1/2} \tag{32}$$

where, in the cases of the lightly-damped hydrofoils considered in this study,

$$\propto \eta_s < (\Omega_o^2 - \Omega^2)(1 - \Omega^2) \tag{33}$$

Important limiting forms of equation (32) may be determined. The first is the upper limit of Ω_o for self excitation. Recognizing that $\Omega \approx 1$, since the bracketed term in equation (32) must always be positive, then the approximate upper limit is

$$\Omega_\ell \approx [D_a/(D_a + \psi_e b)]^2 \tag{34}$$

The second limiting form is the upper limit on Y_{mn}^2 which is obtained by taking the derivative of equation (33) and equating the result to zero. This gives the frequency Ω_m of the maximum in Y_{mn}^2 at the value of Ω given by

$$\Omega_m \approx (1/2) \quad (1 + 3\ \Omega_\lambda^{\ 2})^{1/2} \qquad (35)$$

and $1 < \Omega_m < \Omega_e$, where a factor $2\ \omega_o^2 - 1$ has been approximated by unity. The maximum value of Y_{mn}^2 which occurs at $\Omega_o = \Omega_m$ is

$$[Y_{mn}^{\ 2}]\ max \simeq \frac{3}{4} \frac{(\Psi_e^{\ 2} b)}{D_a^{\ 3}(\gamma/\ \Psi_e^2)} \left(1 - \frac{\psi_e^2 b}{D_a}\right) \qquad (36)$$

Figure (8) illustrates the form of y_{mn}^2 v.s. Ω_o and indicates the locations of the limiting values of parameters. Equation (36) shows that $[Y_{mn}^{\ 2}]_{max}$ is much more strongly dependent on D_a than is the linear response, equation (20). Also, the mode shape at the edge is very important. Assuming equal values of b and γ for all modes, equation (36) shows that selected modes with extra-ordinarily large values of Ψ_e are likely to experience larger limiting levels of vibration, than other modes. The time and area averaged mean square vibration level $\langle y^2\rangle_{mn}$ may be determined form y_{mn} by equations (22) and (23) to be

$$\langle y^2\rangle_{mn} = \frac{1}{2}\ y_{mn}^{\ 2} \qquad (37)$$

and recall that $Y_{mn} = y_{mn}/y_f$. Figure 9 shows the asymptotic dependences of equations (20) and (36) as well as the behavior of the more complete equation (32) for $\langle y^2\rangle_{mn}$ as a function of D_a for some representative values of $\propto$, b, and γ . The heavy solid line is an evaluation of equation (32) for the example parameters and it shows a general approach to a linear response to a hypothetically spanwise-correlated vortex pressure, but a blocked lift coefficient. In this example, as D_a exceeds 20, the amplitude decreases to less than $(3 \times 10^{-4})y_f$, which yields a value of $y_{mn}/(y_f C_{Lo})$ of an order found necessary to just correlate the vortex shedding in the aerodynamic experiment; i.e. the value $100\ y_{mn} / (C_{Lo}\ y_f) \approx .03/.017 \simeq 1.5$ should be compared to the abcissa in Figure 4. Thus, it is expected that spanwise decorrelation of the wake will occur at vibration amplitudes of this order and less for hydrofoils having parameters like the ones used in this example. Accordingly, further linear response for $D_a >> 20$ will follow the lower asymptote in the Figure 9 which is determined both by the blocked lift coefficient C_{Lo} and the spanwise integral scale, Λ_3 rather than the upper line for equation (32).

5. Comparison Of Hydroelastic and Aerodynamic Experiments

Equations (16-18) for the forced aerodynamic wake (see Figures 3 and 4) and equations (32), (34) and (36) for the hydroelastic response (see Figures 8 and 9) may be used with the measurements to give competing values of the parameters. This was done using the hydroelastic response of the brass beam and the aerodynamic data. The results are shown in Table II for an assignment of $\psi_e = 1.4$ which represents an approximate median between the trailing edge mode shape functions for beam and torsion modes. Thus, no attempt has yet been made to distinguish between possible differences between these modes as none was demonstrated by the measurements on the brass beams (Blake, Gershfeld and Maga (1986)). To arrive at parameters for the hydrofoils from the aerodynamic experiments conducted at a somewhat larger Reynolds number, it was assumed that values of (b/α), Figure 4, and of b or α, Figure 3, determined from measurements in air are Reynolds number - independent, but C_{Lo} is dependent on Reynolds number as determined by observations of $\rho_o U_s \kappa / (2\pi y_f)$ (see eq. (21) and Blake (1984b). For example, in the hydroelastic experiments, $\mathcal{R}_{y_f} = 1600$ to 14000 and (Blake (1984a)) $20 \log[\rho_o U_s \kappa / (2\pi q y_f)$ is roughly -45 to -40, while at the higher Reynolds numbers of the moving-edge experiment $\mathcal{R}_{y_f} \sim 26000 - 100{,}000$ for which $20 \log[\rho_o U_s \kappa / (2\pi q y_f)] \sim -22$ to -18.

Figure 10 shows measured and computed values of $\langle y^2 \rangle$ using both sets of parameters in Table 2 for the non-linear response. For the linear response, the upper and lower bounds were determined by the limits in 20 log $\{\rho_0 U_s \kappa / (2 \pi q y_f)\}$ of -45 to -40 and by $\Lambda_3 = 1.5\ y_f$ to $3.5\ y_f$. The computed curves compare with the measured vibration levels of the brass beams quite well over the range of mode types, chords, and damping. The results also emphasize the factor D_a (equation (19)) as the dominant controlling parameter. This factor, recall, includes the in-water structural damping (measured at $U_\infty = 0$) and the total (entrained plus dry) mass per unit length $(m_a + \rho_p h)W$ of the hydrofoil. Figure 11 shows a comparsion of $y_v^2\ \Omega_o$ as obtained from measurements and theory. The point denoted by "linear response" was calculated with equations (16) and (20).

Figure 12 shows, however, that for the low density hydrofoils D_a remains an essential parameter, but an important dependence on mode order appears to emerge. All of the darkened points, circles and squares, and the open square points are for the W_o = 5 inch hydrofoils with base thicknesses of t_o = 0.04, 0.072, or 0.11 inch for various specimens. Of two other samples, the one from Heskestad and Olberts (1960) falls in line with both the calculation and the other measured brass beams. The closed triangles are for a blunt edged flat plate, $h = t_o \approx 1.2\ y_f$. This hydrofoil had a moderately large y_f that was in the range of the aeroelastic measurements. However, the measured point for the (0,1) mode falls beneath the prediction.

6.0 CONCLUSIONS

A simple non-linear Rayleigh oscillator model was found to describe the hydroelastic vortex-street-driven response of hydrofoils just as such modeling is a well known descriptor of the forcing of circular cylinders. In this study the modeling brings into correspondence both aerodynamic and hydroelastic measurements. The aerodynamic measurements gave the non-linear input impedance of the wake as determined by the wake-induced surface presures resulting from an imposed amplitude and frequency of an oscillated trailing edge.

To first order, both the analysis and the measurements show that the parameter D_a (equation (19)) determines the wake-induced response of most mode types of both low and high density hydrofoils. However, the analysis also gives an indication that variations in the blade admittance distribution over its surface from mode-to-mode, or from geometry-to-geometry, as determined by the ratio

$$\Psi_e^{\,2}/\langle \Psi^2 \rangle = \Psi^{\,2}(\vec{x}_{edge})\;/\langle \Psi^2\rangle$$

may also influence the hydrofoil response. This is suggested by the appearance of two crucial parameters emerging in the analysis to determine response: the damping parameter D_a and the coupling parameter $b\,\Psi_e^2$. Although the parameter b determines the frequency range (or velocity range) for which vortex shedding may be entrained by the vibrating edge, the weighted parameter $b\,\Psi_e^2$ determines hydroelastic response. dominant parameter in determining the responses of a given class of hydrofoil section geometries of high density structures, it apparently cannot uniquely determine response for certain modes of low-density hydrofoils. A more general analysis which would include the fluid reaction on the structural modes would possibly give better agreement to the measurements. It is to be noted that the trailing edge displacement for the plate modes of the tapered hydrofoils may in fact substantially exceed the representative value of $\psi_e^2/\langle \psi^2\rangle$ assumed in this paper (viZ. 2). Future structural measurements on the hydrofoils may disclose this behavior. The results of this paper also give further support to the suggestion given previously (Blake (1984b)) that the limiting maximum singing response of propeller blades is also a function of a $(D_a)_p$ that is defined as

$$(D_a)_p = 2\,(\rho_p/\rho_s)\,\eta_s\,h\,C/y_f^2$$

where h and C are the radially-averaged thickness and chord of each blade.

References

1. Bearman, P.W. (1965) "Investigation of the Flow Behind a Two-Dimensional Model With a Blunt Trailing Edge and Fitted with Splitter Plates" J. Fluid Mech. 21, 241-255, 1965.

2. Blake, W.K. (1984a) "Excitation of Plates and Hydrofoils by Trailing Edge Flows" Trans. A.S.M.E. Journal of Vibration, Acoustics, Stress, and Reliability in Design 106, 351-363, 1984.

3. Blake, W.K. (1984b) "Trailing Edge Flow and Aerodynamic Sound, Part I. Tonal Pressure and Velocity Fluctuations, Part II. Random Pressure and Velocity Fluctuations" DTNSRDC Report 83/113 Dec. 1984.

4. Blake, W.K. (1984c) "Aero-Hydroacoustics For Ships" Vol. I and Vol. II DTNSRDC Report 84/010 June 1984.

5. Blake, W.K., Gershfeld, J. and Maga, L.J. (1986) "Self-Excited Vortex Driven Vibration of Hydrofoils of Varying Densities and Geometries" DTNSRDC Report in Preparation

6. Blake, W.K. and Maga, L.J. (1975) "On the Flow-excited Vibrations of Cantilever Struts in Water. 1. Flow-induced Vibration and Damping" J. Acous. Soc. Am. 57, 610-625, 1975.

7. Blake, W.K. and Maga, L.J. (1979) "Near-Wake Stucture and Unsteady Pressures at Trailing Edges of Airfoils" IUTAM/ICA/AIAA Proc. Symp. Mechanics of Sound Generation in Flows, Gottungen, Pg. 1979 Springer-Verlag, Pub.

8. Blake, W.K., Maga, L.J., and Finkelstein G. (1977) "Hydroelastic Variables Influencing Propeller and Hydrofoil Singing" ASME Symposium on Flow Induced Noise and Vibration. Atlanta, Ga.

9. Graham, J.M.R. (1967) "The Effect of End Plates on the Two-Dimensionality of a Vortex Wake" Aero. Quarterly. 20, 237-247, 1969.

10. Heskestad, G. and Olberts, D.R. (1960) "Influence of Trailing Edge Geometry on Hydraulic-Turbine Blade Vibration" Trans. A.S.M.E., J. Eng. for Power 82, 103-110, 1960.

11. Hartlen, R.T. and Currie, I.G. (1970) "Lift-Oscillator Model of Vortex-Induced Vibration" Proc. J. Eng. Mech. Div., Am. Soc. Civ. Eng. 96, 577-591,

12. Nayfeh, A.H. and Mook, D.T. (1979) "Non-linear Oscillations" J. Wiley and Sons.

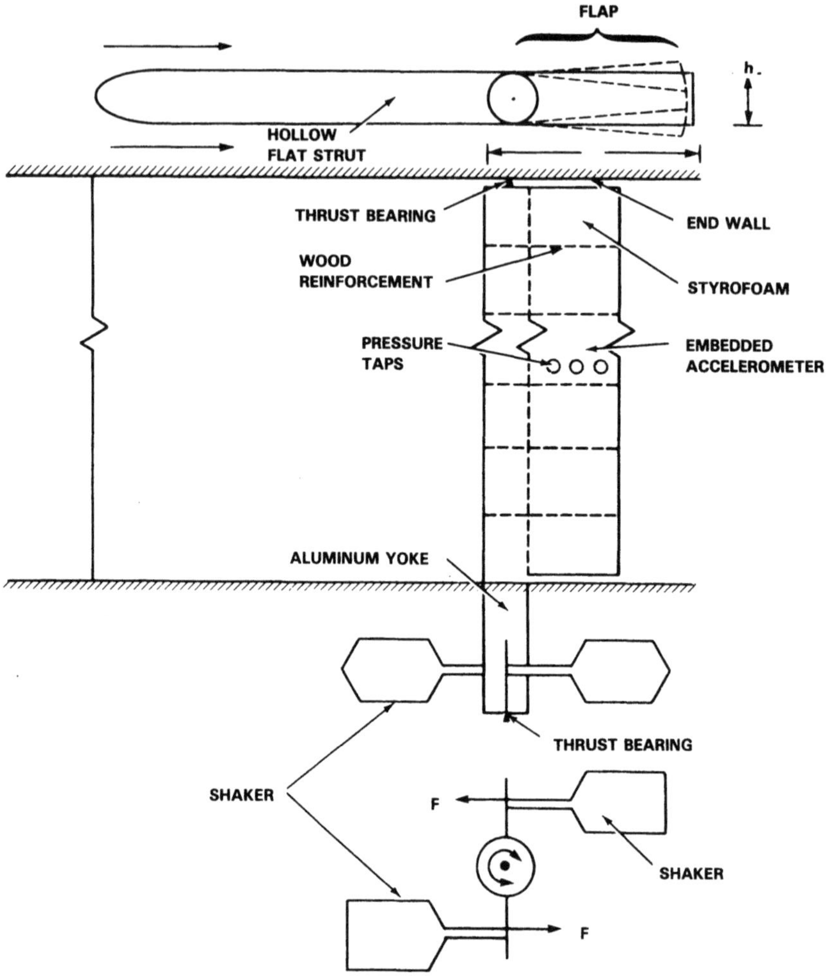

Figure 1. Arrangement of the Moving Trailing Edge Flap for Aerodynamic Measurements.

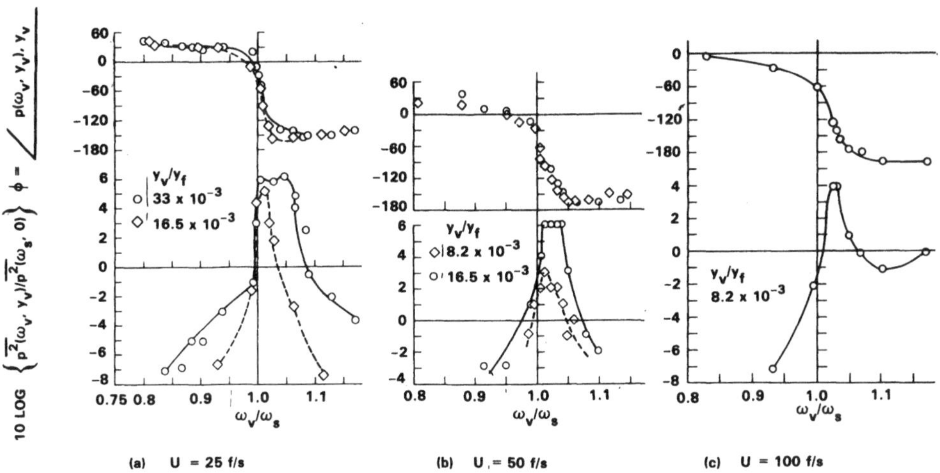

Figure 2. Surface Pressure Amplitude and Phase for the Forced Wake and $x_1/y_f = 0.15$.

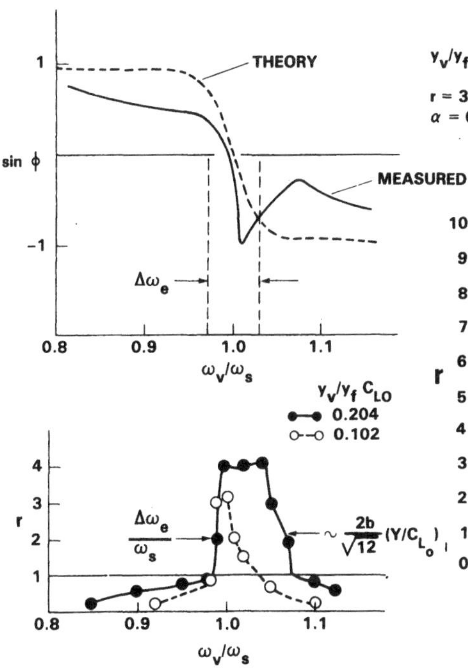

Figure 3. Behavior of $r = (C_L/C_{LO})^2$ and sin ϕ Consistent with Rayleigh Oscillator

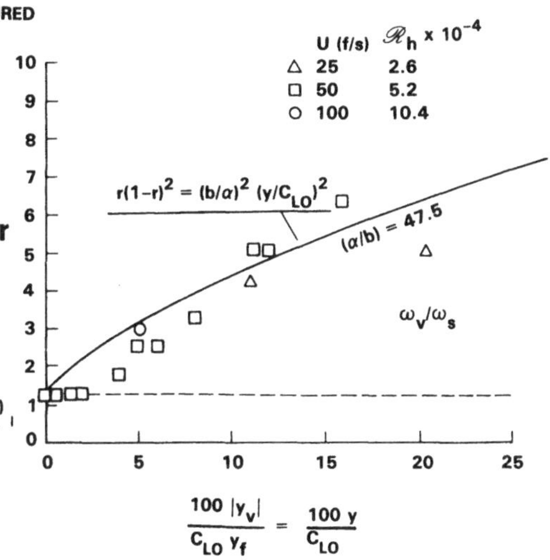

Figure 4. Behavior of $r = (C_L/C_{LO})^2$ at Coincidence as a Function of Normalized Edge Displacement

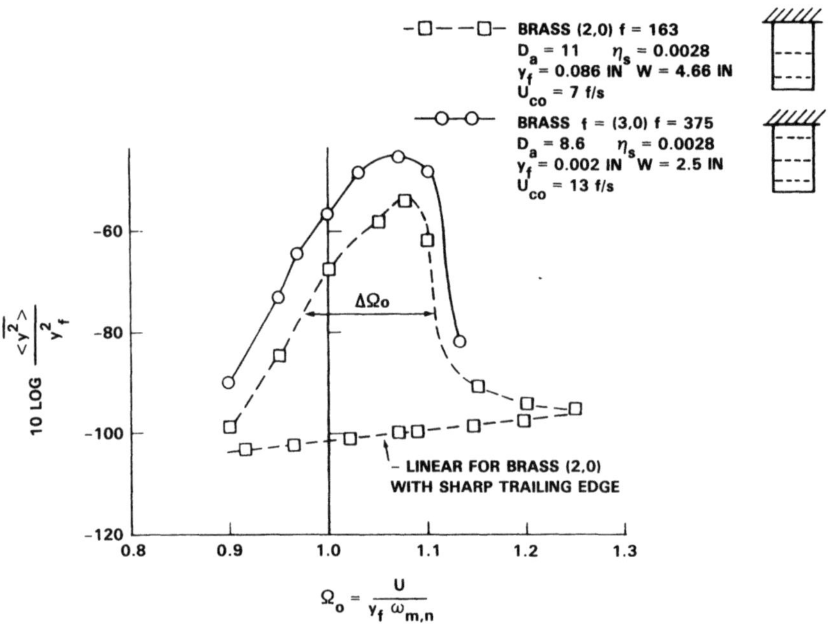

Figure 5. Flow Induced Vibration of Beam Modes for Hydrofoils of Different Chord

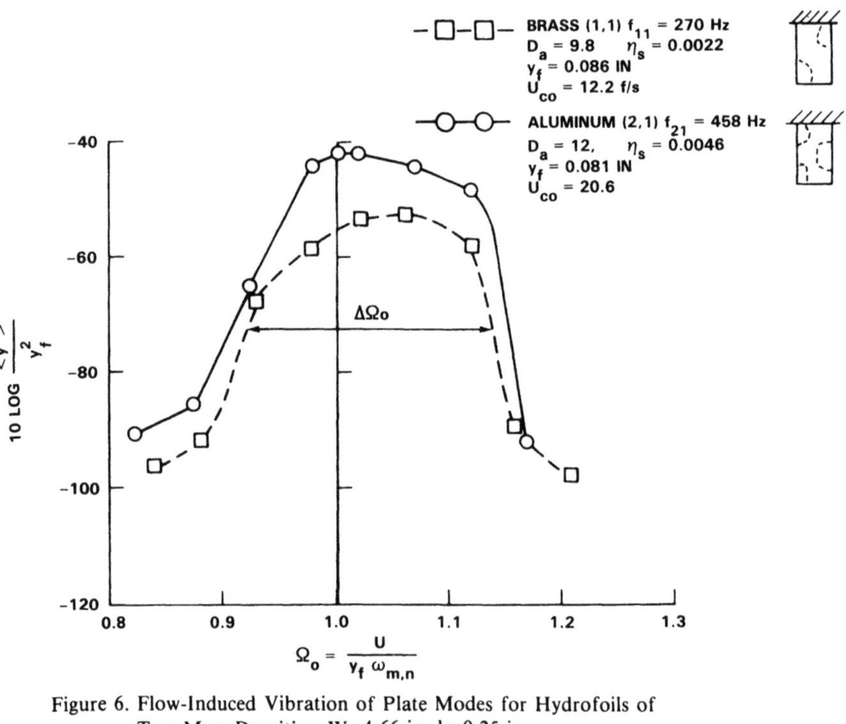

Figure 6. Flow-Induced Vibration of Plate Modes for Hydrofoils of Two Mass Densities, W=4.66 in, h=0.25 in

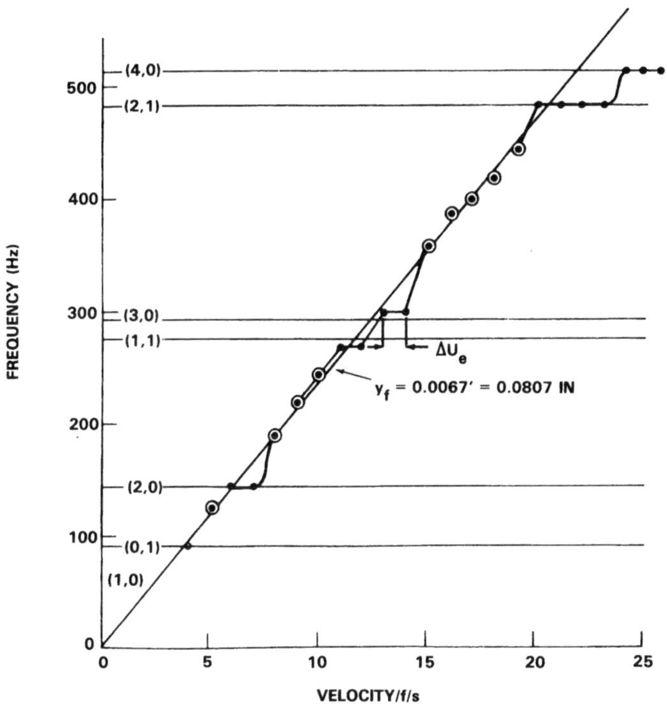

Figure 7. Vortex-Shedding Frequencies of an Aluminum Hydrofoil W = 4.66 in

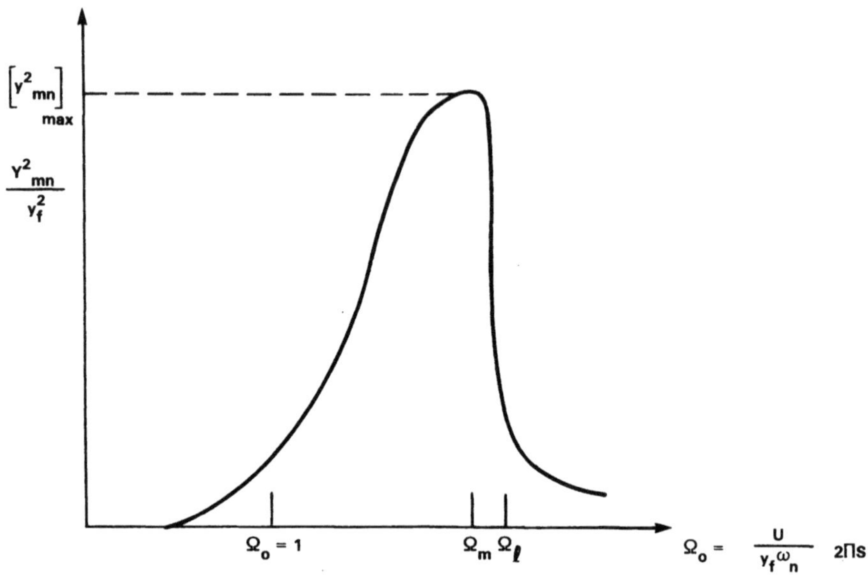

Figure 8. Parameters in Lift Oscillator Divulged by Vibration Behavior

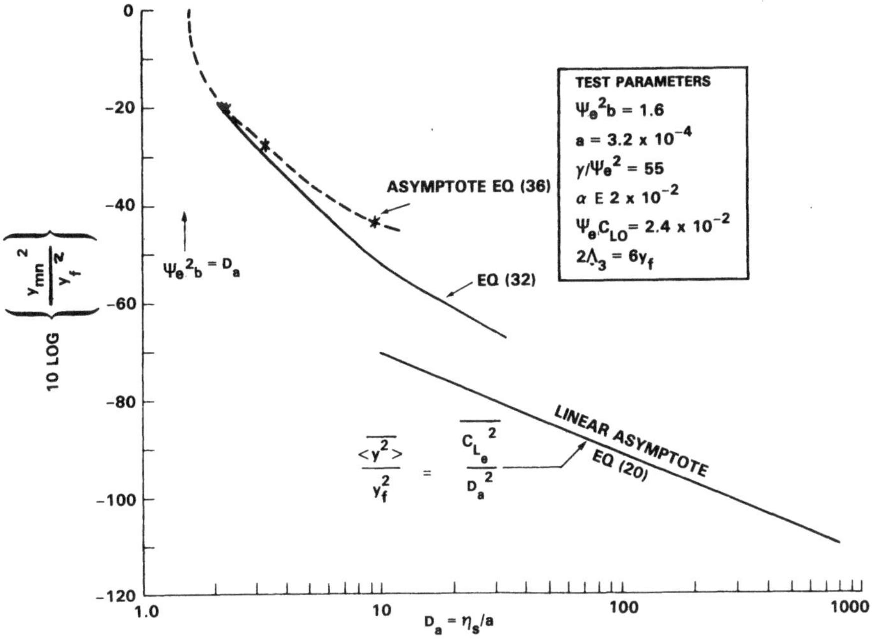

Figure 9. Examples of the Asymptotic Functions for Nonlinear and Linear Response

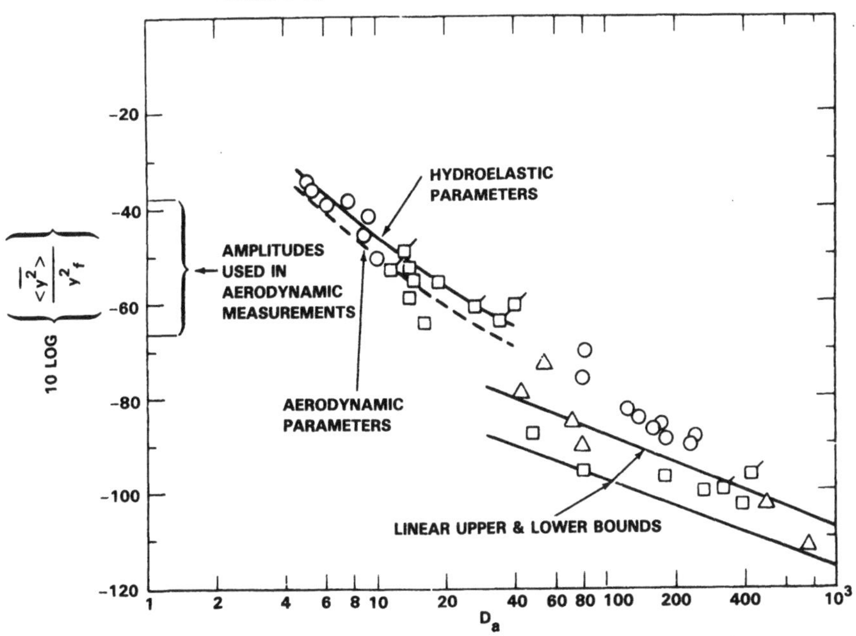

Figure 10. Comparison of Predicted and Measured Responses of Brass Hydrofoils of Various W, t, and η_s. Nonlinear Response Estimates Are Based on Alternative Parameters Determined from Hydroelastic and Aerodynamic Measurements.

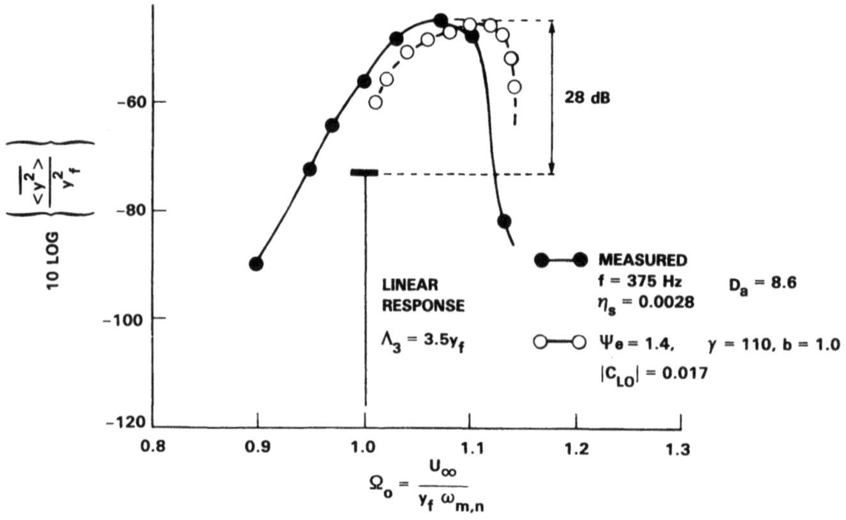

Figure 11. Example of Calculated Modal Response Using Indicated Oscillator Parameters for the (3,0) Beam Mode of a Brass Hydrofoil

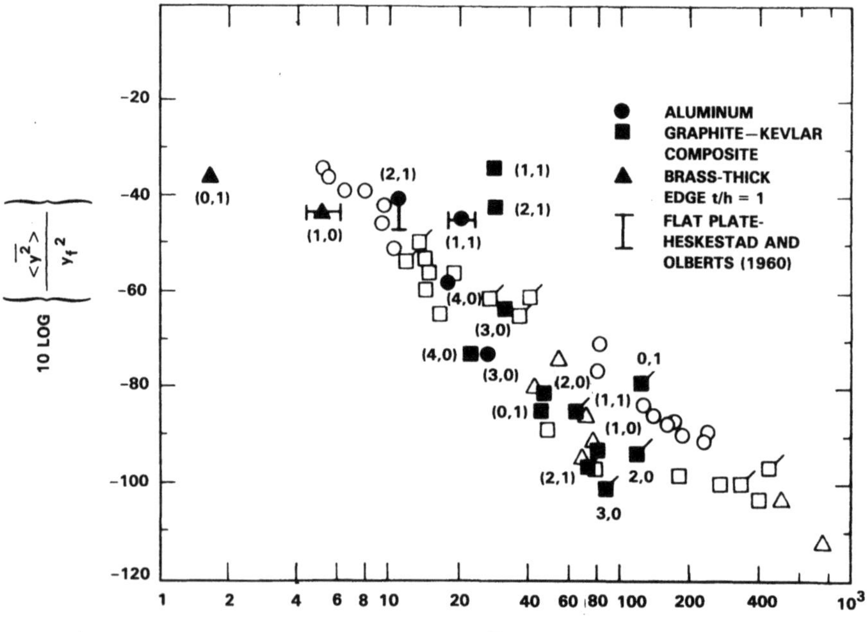

Figure 12. Flow-Induced Responses of Various Other Hydrofoils — See Table 1 for Parameters

Table 1.

HYDROFOIL CONFIGURATIONS FOR HYDROELASTIC OBSERVATIONS							
W_o	W_{in}	h_{in}	$t_{o\,in}$	M_a/M_s	$\eta_s \times 10^3$	HYDROFOIL MATERIALS	$_t$ RANGE
2.75	2.5	0.25	0.036	0.7 - 1.0	2 - 70	BRASS, STEEL	820 - 6,000
2.75	2.5	0.25	0.072	0.7 - 1.0	2 - 70	BRASS, STEEL	1,640 - 12,000
5	4.7	0.025	0.072	0.6 - 1.0	2 - 60	BRASS	1,640 - 12,000
5	4.6	0.025	0.072	2.7 - 4.2	3 - 5	ALUMINUM	2,000 - 14,000
5	4.5	0.025	0.11	4.4 - 6.5	12 - 15	FIBER COMPOSITE	2,600 - 18,000
5	4.8	0.025	0.04	4.4 - 6.5	12 -15	FIBER COMPOSITE	945 - 6,500
	5	0.375	0.375	0.6	2.8	BRASS	30,000 - 39,000

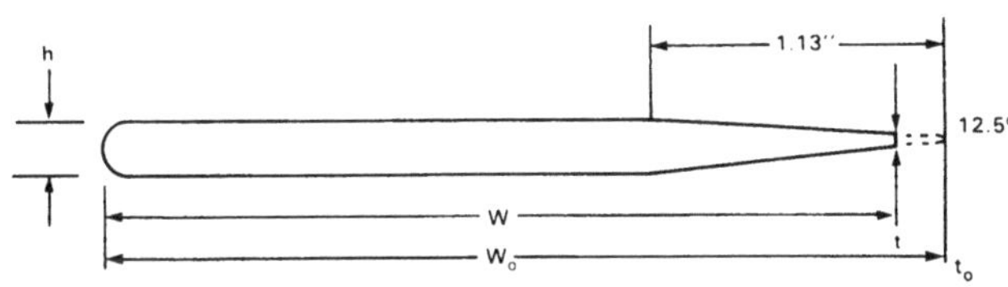

Table 2.

$2.7 \times 10^3 < \mathscr{R}_{y_f} < 1.6 \times 10^4$

	HYDROELASTIC	FORCED WAKE
Ψ_e	1.4	–
$\lvert C_{LO} \rvert$	0.017	0.017
b	1.0	0.52 to 0.81
γ	110	45 to 81
α	0.023	0.01 to 0.022

(UNDERLINED VALUES ARE THOSE USED FOR COMPUTING $\overline{y^2}/y_f^2$)

$\gamma_o = \gamma/(\Psi e)^2$ $\qquad$ $C_{Le} = \Psi e\, C_{Lo}$

The Ffowcs Williams-Hawkings Equation – Fifteen Years of Research

F. Farassat

NASA Langley Research Center
Hampton, Virginia 23665

ABSTRACT

A brief review of Ffowcs Williams-Hawkings equation and the research on applying this equation to problems of aeroacoustics is presented.

INTRODUCTION

In 1969 J. E. Ffowcs Williams and D. L. Hawkings published a paper in the Philosophical Transactions of the Royal Society in which they derived a fundamental equation now known as the Ffowcs Williams-Hawkings (FW-H) equation[1]. This equation governs the generation of sound in fluids in the presence of solid boundaries in motion--a situation very common in engineering practice. The FW-H equation was based on the acoustic analogy and, in fact, is a generalization of Lighthill's jet noise equation[2]. As is well known by the aeroacoustic community, the effects of fluid motion and solid boundaries in generation of sound are accounted for in acoustic analogy by fictitious sources. This approach has both merits and shortcomings which were ably described by Ffowcs Williams in 1984[3]. Acoustic analogy should only be viewed as a tool which can be useful depending on the level of approximation expected from the solution and the amount of effort spent on obtaining the results.

In general, many of the boundary value problems of acoustics are very difficult to solve analytically without some level of idealization and simplification. In particular, the problem of noise prediction of rotating blade machinery is governed by nonlinear partial differential equations in regions with moving boundaries. These equations also describe the blade aerodynamics. Acoustic analogy which separates the problem of aerodynamics and acoustics appears to introduce a major simplification in solving noise prediction problems. Of course, the simplification is achieved by shifting some of the burden of solving the problem to the aerodynamicist. One can

Aero- and Hydro-Acoustics
IUTAM Symposium Lyon 1985
Editors: G. Comte-Bellot and J. E. Ffowcs Williams

then utilize the powerful methods of linear analysis in solving the purely linear problem of acoustics thus obtained and do substantial amounts of analytic work. However, the acoustic analogy approach should never be viewed as a substitute for solving the nonlinear boundary value problems of flow acoustics. For this reason, this review is limited to situations where linearization of the governing equations is allowed.

THE FFOWCS WILLIAMS-HAWKINGS EQUATION

The FW-H equation is[1]

$$c^2\Box^2\rho = \frac{\partial}{\partial t}\left[\rho_o v_n\left|\nabla f\right|\delta(f)\right] - \frac{\partial}{\partial x_i}\left[\ell_i\left|\nabla f\right|\delta(f)\right] + \frac{\partial^2}{\partial x_i \partial x_j}\left[H(f)T_{ij}\right] \qquad (1)$$

where ρ is the perturbation density, v_n and ℓ_i are the local surface normal velocity and force intensity, respectively, and T_{ij} is the Lighthill stress tensor. The equation of the surface in motion is $f=0$ ($f>0$ outside body in the flow region) and ρ_o and c are the density and speed of sound in the undisturbed medium. We will use the notation p' for $c^2\rho$ which stands for the acoustic pressure away from the immediate vicinity of the surface. When the original paper of Ffowcs Williams and Hawkings was published in 1969, it was immediately recognized by some researchers in acoustics as a major contribution to the field. However, it took almost five years until frequent referencing to this article became common[4,5].

Before discussing the works on the FW-H equation it is necessary to highlight the main achievements of Ffowcs Williams and Hawkings. First the sources on the surface in motion were treated as inhomogeneous terms of the differential equation (1). The mathematical procedure for doing this was known[6] but was first used for problems of acoustics by Ffowcs Williams and Hawkings. This rigorously explained the many ad hoc methods used by earlier researchers to include the effect of solid boundaries[7]. Second, by casting the problem in terms of generalized functions, the powerful machinery of this mathematical theory was made available to the users. This fact itself, although an important aspect of FW-H theory, is also responsible for avoidance of the use of the FW-H equation by many researchers. Some feel that whatever can be done by using generalized function theory can be done just as easily by classical mathematics. This feeling is totally erroneous. The operational properties of generalized functions allows solutions of linear differential equations in a more direct and simpler manner than classical methods. For this reason one should make an effort to learn the basics of generalized function theory[8-10]. Simple derivations of FW-H equation are now available[9-11].

RESEARCH ON THE FW-H EQUATION

In surveying the acoustic literature refering to the FW-H equation, it soon becomes obvious that many refer to this equation because it is fashionable or use it with classical mathematics and thus defeat the purpose intended. Leaving this group of workers aside, two groups emerge which have successfully utilized the FW-H equation: (i) those who use the equation for qualitative or approximate results, and (ii) those who use the equation without the quadrupole term for exact numerical noise calculations. The first group of people are represented mainly by Ffowcs Williams and his students[12-16]. The second group includes Hawkings and Lowson[5], the author[4,17-22], and Blackburn[16].

The first group has tackled the more difficult problems of flow noise such as turbulent flow near an edge. These works are characterized by the use of a good deal of physics and a fair amount of mathematics. They are also highly intuitive and ingenious. It appears that, guided by intuition and physics, an almost endless number of methods of approximation are available for qualitative results. For this reason, a fair review of the works of this group is not possible. One should refer to periodic reviews of Ffowcs Williams[3,15,23].

The second group has tried to find an exact solution of FW-H equation using the surface terms (thickness and loading) only. Although the inclusion of the quadrupole term in the solution is almost trivial[16,17,19], this is rarely used in calculations because of the lack of information on T_{ij}. It should not be misinterpreted that the quadrupole term in FW-H equation is unimportant[24,25]. It is now generally believed that to include the effect of nonlinearities, one should use methods other than the acoustic analogy. For this reason, the results reported below only apply to linear problems where flow perturbations near the body surface are small.

A brief review of Hawkings and Lowson's work was given by the author[19]. Their result is valid for static propellers or hovering rotor and is in the frequency domain. It is interesting to note that there are many exact and equivalent forms of the solution of equation (1)[19-22]. Some of these are more suitable for numerical work. A search for deriving useful forms of the solution and computer coding for prediction of the noise of rotating blades started more than a decade ago at Langley Research Center. Basically four time domain formulations emerged which are designated Formulation 1, 1-A, 2 and 3. At present there are several codes at Langley using these results. All these include thickness and loading noise and are valid in the near and far fields.

Formulation 1 is the simplest solution of FW-H equation known to the author. The solution involves two surface integrals, one of which has an observer time derivative before the integral. This result is valid for both subsonic and supersonic sources[18-20]. Several codes for prediction of helicopter rotor and propellers have been developed by Mall, Brown, Nystrom and the author. The observer time differentiation necessitated change of Formulation to 1-A where the time derivative is taken inside the integral and the integrand is differentiated explicitly[21]. This formulation has integrands with terms in the denominator which have the Doppler factor $1-M_r$ to the second or third power. Therefore, the result is used only for subsonic sources on blades.

Formulation 2 was developed for supersonic sources moving in uniform propeller-like motion. It was restricted to infinitely thin blades[20]. The idea behind this result was borrowed from aerodynamics. It utilized the fact that the pressure field of the propeller rotated with the blades. Thus, the observer time differentiation was equivalent to minus the product of the helical speed of the observer and the directional derivative in the direction of helical velocity at the observer in the rotating frame. After proving the correctness of the result, this formulation was abandoned in favor of Formulation 3 due to restrictions in deriving the result.

To derive Formulation 3, FW-H equation is written in the form

$$\Box^2 p' = \nabla_4 \cdot [\vec{Q}|\nabla f|\delta(f)] \tag{2}$$

where $\vec{Q} = (-p\vec{n}, M_n)$, $\nabla_4 = (\nabla, 1/c\ \partial/\partial t)$ and $M_n = v_n/c$. Here both p and p' are nondimensionalized with respect to $\rho_o c^2$ and $\vec{n}$ is the unit outward normal. To include discontinuities in blade slope (e.g., at TE) and in p (e.g., at a shock trace) equation (2) is solved for an open surface[22]. The source $\vec{Q}$ is decomposed into two components normal and tangent to the surface f=0 in four dimensions. The normal component needs considerable manipulation. The use of special curvilinear coordinates and general tensor analysis were essential to obtaining an expression which was useful for numerical work. The exact expression is very long and complicated and the details of blade geometry such as surface curvature enter this result.

There are two codes at Langley Research Center which use a combination of Formulation 1-A and 3 for subsonic and supersonic sources, respectively. The results are highly encouraging. Some very interesting results for subsonic propellers have been published by Block[26]. It must be mentioned that Blackburn's solution of the identical problem looks very different and even more complicated than the author's because of differences in details and objectives[16]. It has not been successfully coded for numerical work.

A recent development is the use of Formulations 1-A by Long[27] and 3 by the author[28] to derive linear integral equations for aerodynamic calculations in terms of the unknown surface pressure. These are obtained by bringing the observer onto the surface and then regularizing the resultant divergent integrals. The integral equations must then be solved numerically. Such an approach can also supply the needed surface pressure data for acoustic calculation.

In conclusion, many previously known acoustic results such as those by Gutin, Lowson[29], Garrick and Watkins[30] and Succi[21], and aerodynamic results such as the supersonic area rule[4] can be derived by using FW-H equation. Judged by its success in acoustics and the proven potential for linear aerodynamic calculations, the FW-H equation must be viewed as a fundamental result in aeroacoustics.

REFERENCES

1. J. E. Ffowcs Williams and D. L. Hawkings, Phil. Trans. Roy. Soc. London, A264, 1969, 321-342.
2. M. J. Lighthill, Proc. Roy. Soc. London, A211, 564-587.
3. J. E. Ffowcs Williams, IMA J. Appl. Math., 32, 1984, 113-124.
4. F. Farassat, J. Sound and Vib., 32, 1974, 387-405.
5. D. L. Hawkings and M. V. Lowson, J. Sound and Vib., 36, 1974, 1-20.
6. B. Friedman, Principles and Techniques of Applied Mathematics, John Wiley and Sons, Ltd., 1956.
7. E. J. Richards and D. J. Mead, Noise and Acoustic Fatigue in Aeronautics, John Wiley and Sons, Ltd., 1968.
8. I. M. Gel'fand and G. E. Shilov, Generalized Functions, Academic Press, Vol. 1, 1964.
9. R. P. Kanwal, Generalized Functions-Theory and Technique, Academic Press, 1983.
10. F. Farassat, J. Sound and Vib., 55, 1977, 165-193.

11. M. E. Goldstein, Aeroacoustics, McGraw-Hill International Book Co., 1976.

12. J. E. Ffowcs Williams, J. Fluid Mech., 66, 1974, 791-816.

13. J. E. Ffowcs Williams and D. J. Lovely, J. Sound and Vib., 50, 1977, 333-343.

14. J. E. Ffowcs Williams, AIAA Paper 79-0576, AIAA 5th Aeroacoustics Conf., Seattle Washington, 1979.

15. J. E. Ffowcs Williams, AIAA Jour., 20, 1983, 307-315.

16. H. W. Blackburn, Sound Sources on High Speed Surfaces, Ph.D. Dissertation, Clare College, University of Cambridge, 1983.

17. F. Farassat, The Sound from Rigid Bodies in Arbitrary Motion, Ph.D. Thesis, Cornell University, 1973.

18. F. Farassat, NASA TR R-451, 1975.

19. F. Farassat, AIAA J., 19, 1981, 1122-1130.

20. F. Farassat, Advanced Theoretical Treatment of Propeller Noise, von Karman Inst. Lec. Ser. 1982-08, 1982.

21. F. Farassat and J. P. Succi, Vertica, 7, 1983, 309-320.

22. F. Farassat, AIAA Paper 84-2303, AIAA/NASA 9th Aeroacoustics Conf., Williamsburg, Virginia, 1984.

23. J. E. Ffowcs Williams, Aeroacoustics, Ann. Rev. Fl. Mech., 9, 1977, 447-468.

24. D. B. Hanson and M. R. Fink, J. Sound and Vib., 62, 1979, 19-38.

25. F. H. Schmitz and Y. H. Yu, J. Am. Helicopter Soc., 24, 1979.

26. P. J. W. Block, AIAA Paper 84-2263, AIAA/NASA 9th Aeroacoustics Conf., Williamsburg, Virginia, 1984.

27. Lyle N. Long, NASA TP-2197, 1983.

28. F. Farassat, AIAA J., 22, 1984, 1337-1340.

29. M. V. Lowson, Proc. Roy. Soc. Lond., A286, 1965, 559--572.

30. I. E. Garrick and C. E. Watkins, NACA Rep. 1198, 1954.

Thermo Acoustics

Thermoacoustic Sources

A. P. DOWLING

The University Engineering Department
Trumpington Street, Cambridge, ENGLAND

1. Introduction

Thermoacoustics deals with the acoustics of flows in which the variation of entropy plays a significant role. We derive an integral representation for the sound field which enables us to review the literature on thermoacoustic sources within a unified framework. The noise generated by a selection of thermoacoustic sources, including unsteady combustion, heat and mass transfer, and turbulent two-phase flows, is discussed.

2. An integral representation for the sound field

Unsteady heating can be an efficient acoustic source and the Lighthill theory [1] provides a convenient description of the sound generation. The Navier-Stokes equation and the equation of mass conservation may be combined to give an inhomogeneous wave equation, which we write in the form

$$\frac{1}{c_o^2}\frac{\partial^2 p}{\partial t^2} - \nabla^2 p = \frac{\partial^2}{\partial x_i \partial x_j}(\rho u_i u_j - \tau_{ij}) - \frac{\partial^2 \rho_e}{\partial t^2}, \qquad (1)$$

where p is the pressure, ρ the density, u the particle velocity, c the sound speed and τ_{ij} the viscous stress tensor. ρ_e is the 'excess' density

$$\rho_e = \rho - \rho_o - (p - p_o)/c_o^2 \quad . \qquad (2)$$

The suffix o denotes a mean value in the distant acoustic field. ρ_e therefore vanishes in the far-field but is non-zero in regions where the entropy is significantly different from ambient.

Aero- and Hydro-Acoustics
IUTAM Symposium Lyon 1985
Editors: G. Comte-Bellot and J. E. Ffowcs Williams

While the quadrupole source term $\rho u_i u_j$ is the subject of most aeroacoustic studies, the remaining source terms $-\partial^2\rho_e/\partial t^2 - \partial^2\tau_{ij}/\partial x_i \partial x_j$ are the topic of this paper. They are 'thermo-acoustic' sources and are non-zero in regions of appreciable irreversible processes. To determine the strength of these sources we need to investigate the thermodynamics of the source region.

It is shown in the Appendix that the energy equation for a gas made up of N (possibly reacting) species leads to a relationship of the form

$$\frac{D\rho}{Dt} = \frac{1}{c^2}\frac{Dp}{Dt} + \frac{\alpha}{c_p}\left\{ \sum_{n=1}^{N} \rho \left.\frac{\partial h}{\partial Y_n}\right|_{\rho,p,Y_m} \frac{DY_n}{Dt} + \nabla\cdot\underline{q} - \frac{\partial u_i}{\partial x_j}\tau_{ij} \right\}. \tag{3}$$

Y_n is the mass fraction of the nth species, h the enthalpy, $\underline{q}$ the heat flux. α is the volumetric expansion coefficient and for an ideal gas is equal to T^{-1}. Conservation of species can be expressed as

$$\rho\frac{DY_n}{Dt} = w_n - \nabla\cdot\underline{J}_n\ , \tag{4}$$

where w_n is the production rate per unit volume of species n by reaction, and $\underline{J}_n$ is the flux of species n by diffusion.

We will now manipulate the form of the Lighthill equation (1) so that the thermodynamic relationship (3) can be used to substitute for ρ_e. In integral form (1) is

$$4\pi(p-p_o)(\underline{x},t) = \frac{\partial^2}{\partial x_i \partial x_j}\int\left[\frac{\rho u_i u_j - \tau_{ij}}{r}\right] d^3\underline{y} - \frac{\partial^2}{\partial t^2}\int\left[\frac{\rho_e(\underline{y},\tau)}{r}\right] d^3\underline{y} \tag{5}$$

$r = |\underline{x}-\underline{y}|$ and the square brackets denote that the function they enclose is to be evaluated at a retarded time $\tau = t-r/c_o$. The procedure described by Ffowcs Williams [2] enables one of the partial time derivatives of ρ_e to be expressed as a material derivative. The steps are as follows. First the retarded time integral is displayed explicitly, and the spatial integration variable changed from $\underline{y}$ to material co-ordinates $\underline{\eta}$, to show that

$$\frac{\partial^2}{\partial t^2}\int\left[\frac{\rho_e}{r}\right] d^3\underline{y} = \frac{\partial}{\partial t}\int\frac{\rho_e(\underline{y},\tau)}{r} \frac{\partial}{\partial t} \delta\left(t-\tau-\frac{|\underline{x}-\underline{y}|}{c_o}\right)\frac{\rho^*}{\rho} d^3\underline{\eta}\, d\tau \tag{6}$$

$\underline{y} = \underline{\eta} + \int_o^\tau \underline{v}\, d\tau'$ and ρ^*/ρ is the Jacobian of the transformation from $\underline{y}$ to $\underline{\eta}$, where ρ^* is a reference density at the time when the two co-ordinate systems coincide. Now, since $\underline{u} = \partial\underline{y}/\partial\tau|_{\underline{\eta}}$,

$$\frac{\partial}{\partial\tau}\left(\tau - \tau - \frac{|\underline{x}-\underline{y}|}{c_o}\right)\Bigg|_{\underline{\eta}} = -1 + M_r , \tag{7}$$

with $M_r = \underline{u}\cdot(\underline{x}-\underline{y})/c_o|\underline{x}-\underline{y}|$. The right hand side of equation (6) is therefore equivalent to

$$-\frac{\partial}{\partial t}\int\frac{\rho_e}{r(1-M_r)} \frac{D}{D\tau} \delta\left(t-\tau-\frac{|\underline{x}-\underline{y}|}{c_o}\right) \frac{\rho^*}{\rho} d^3\underline{\eta} d\tau .$$

After integration by parts and a return to the $\underline{y}$ coordinate system this gives

$$\frac{\partial^2}{\partial t^2}\int\left[\frac{\rho_e}{r}\right] d^3\underline{y} = \frac{\partial}{\partial t}\int\left[\rho\frac{D}{Dt}\left(\frac{\rho_e}{\rho(1-M_r)r}\right)\right]d^3\underline{y} . \tag{8}$$

When this is substituted into (5) it shows that the pressure perturbation at a far-field position $\underline{x}$ is given by

$$4\pi|\underline{x}|(p-p_o)(\underline{x},t) = \frac{\partial^2}{\partial x_i \partial x_j}\int\left[\rho u_i u_j - \tau_{ij}\right] d^3\underline{y}$$

$$-\frac{\partial}{\partial t}\int\left[\frac{\rho}{1-M_r}\frac{D}{Dt}\left(\frac{\rho_e}{\rho}\right) + \frac{\rho_e}{(1-M_r)^2}\frac{DM_r}{Dt}\right] d^3\underline{y} \tag{9}$$

We see from (9) that there is a thermoacoustic source if ρ_e/ρ varies following a material particle but, even a constant but non-zero ρ_e can generate sound if the material particles accelerate.

Using the relationship (3) to substitute for $D\rho/Dt$ in (9) gives

$$4\pi|\underline{x}|(p-p_o)(\underline{x},t) = -\frac{\partial}{\partial t}\int\left[\frac{\rho_o\alpha}{\rho c_p(1-M_r)}\left\{\sum_{n=1}^{N}\frac{\partial h}{\partial Y_n}\bigg|_{\rho,p,Y_m}(w_n-\nabla\cdot\underline{J}_n)+\right.\right.$$

$$\left.\left.+\ \nabla\cdot\underline{q} - \frac{\partial u_i}{\partial x_j}\tau_{ij}\right\}\right]d^3\underline{y}$$

$$+\ \frac{\partial^2}{\partial x_i\partial x_j}\int\left[\rho u_i u_j - \tau_{ij}\right]d^3\underline{y}$$

$$-\ \frac{\partial}{\partial t}\int\left[\frac{1}{(1-M_r)c_o^2}\left\{\left(\frac{\rho_o c_o{}^2}{\rho c^2}-1\right)\frac{Dp}{Dt}+\frac{p-p_o}{\rho}\frac{D\rho}{Dt}\right\}\right]d^3\underline{y}$$

$$-\ \frac{\partial}{\partial t}\int\left[\frac{\rho_e}{(1-M_r)^2}\frac{DM_r}{Dt}\right]d^3\underline{y} \qquad (10)$$

No approximation (other than $|\underline{x}|$ large) has been made in deriving this expression. It conveniently displays the effect of all the thermoacoustic sources, and enables us to review the literature on thermoacoustic source mechanisms within a unified framework. The first term on the right-hand side of equation (10) describes the sound generated by irreversible flow processes and is of monopole type. The second term involves the familiar quadrupole source of Lighthill's jet noise theory. It is well known that it leads to an acoustic intensity that scales on Mach number to the eighth power. The third term is appreciable if there are regions of unsteady flow with different mean density and sound speed from the ambient fluid. We will see that this term also leads to an M^8 scaling, but with a different coefficient from the jet noise term. The last term in (10) is dipole in nature and describes the effect of accelerating density inhomogeneities. In the remainder of this paper we will summarise the effects of these various sources.

3. Combustion noise

In a reacting gas w_n is non-zero and the first term on the right-hand side of equation (10) describes the strong monopole sound of combustion. When combustion occurs unsteadily in a low Mach number flow, this term is far larger than the sound due to the other source mechanisms demonstrated in equation (10) and we can write the distant sound field as

$$4\pi|\underline{x}|(p-p_o)(\underline{x},t) = -\frac{\partial}{\partial t}\int\left[\frac{\rho_o\alpha}{\rho c_p}\sum_{n=1}^{N}\frac{\partial h}{\partial Y_n}\bigg|_{\rho,p,Y_m} w_n\right]d^3\underline{y} \tag{11}$$

For an ideal gas $\alpha/c_p = (\gamma-1)/c^2$. ρ and c denote the actual density and sound speed in the combustion zone and, since the gas there is hot, will be different from the far-field values ρ_o and c_o. It is shown in the Appendix that if the gas has constant average molecular wieght, a good approximation for the combustion of hydrocarbons in air,

$$-\sum_{n=1}^{N}\frac{\partial h}{\partial Y_n}\bigg|_{\rho,p,Y_m} w_n = Q , \tag{12}$$

where Q is the heat release rate per unit volume. The combustion noise may therefore be rewritten as

$$4\pi|\underline{x}|(p-p_o)(\underline{x},t) = \frac{\partial}{\partial t}\int\left[\frac{\rho_o(\gamma-1)Q}{\rho c^2}\right]d^3\underline{y} \tag{13}$$

When γ is assumed to be independent of temperature and the combustion takes place at ambient pressure, $\rho c^2 = \gamma p_o = \rho_o c_o{}^2$ and this expression for the acoustic pressure perturbation simplifies to

$$4\pi|\underline{x}|(p-p_o)(\underline{x},t) = \frac{\gamma-1}{c_o{}^2}\int[\dot{Q}]\, d^3\underline{y} \tag{14}$$

Equation (14) predicts that for a compact flame the far-field acoustic pressure perturbation should be proportional to the time derivative of the total rate of heat release evaluated at retarded time. Hurle et al [3] confirmed this relationship experimentally by using the light emission from short-lived CH or

C_2 free radicals to monitor the rate of combustion. Since their pioneering work the relationship has been verified using more advanced signal processing [4].

Equation (13) states that the far-field acoustic pressure perturbation is proportional to ρ_o the density of the surroundings. Let us compare this prediction from the acoustic analogy with the exact solutions of two simple problems.

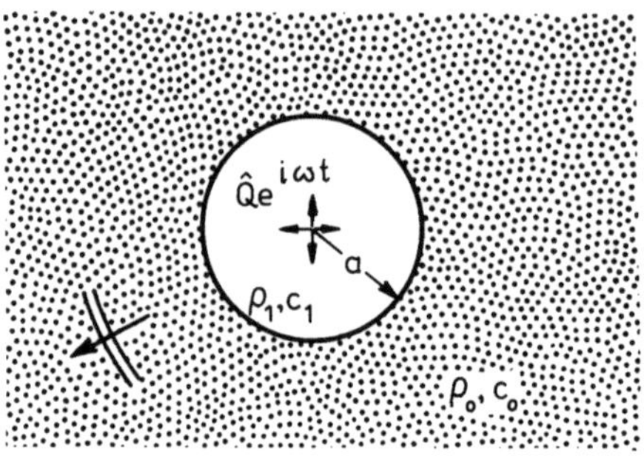

Fig. 1 Geometry of the simple model problem

As a first example, consider combustion occurring with a total rate of heat release $\hat{Q}e^{i\omega t}$, concentrated at the origin. The gas within a sphere of radius a is hot and has mean density ρ_1 and sound speed c_1, while the surrounding colder gas has density ρ_o and sound speed c_o as shown in Figure 1. The pressure perturbation is therefore to satisfy the equations

$$\frac{1}{c_1^2}\frac{\partial^2 p}{\partial t^2} - \nabla^2 p = \frac{i\omega(\gamma-1)}{c_1^2}\hat{Q}\,e^{i\omega t}\,\delta(\underline{x}) \quad \text{in } r < a \qquad (15a)$$

and

$$\frac{1}{c_o^2}\frac{\partial^2 p}{\partial t^2} - \nabla^2 p = 0 \quad \text{in } r > a \qquad (15b)$$

The general solution to (15a) has the form

$$(p-p_o)(r,t) = \left(I\,e^{-i\omega r/c_1} + R\,e^{i\omega r/c_1}\right)\frac{e^{i\omega t}}{r} \quad \text{in } r \le a,$$

and integrating (15a) over a small sphere centred on the origin shows that

$$I + R = \frac{i\omega(\gamma-1)\,\hat{Q}}{4\pi c_1^2}\,. \qquad (16)$$

In $r \ge a$ the pressure perturbation can be expressed as

$$(p-p_o)(r,t) = \frac{T}{r} e^{i\omega(t-r/c_o)} \tag{17}$$

It is a matter of straightforward algebra to use equation (16) and conditions of continuity of pressure and particle velocity across $r = a$ to determine the coefficients I, R and T in terms of $\hat{Q}$. The results simplify when the hot region is compact. Then the solution reduces to

$$(p-p_o)(r,t) = \frac{\rho_o i\omega(\gamma-1)}{4\pi r\rho_1 c_1{}^2} \hat{Q} e^{i\omega(t-r/c_o)} \quad \text{in } r \geq a \tag{18a}$$

$$= \frac{i\omega(\gamma-1)}{4\pi r c_1{}^2} \hat{Q} e^{i\omega t} \left\{1 + \frac{r}{a}\left(\frac{\rho_o}{\rho_1} - 1\right)\right\} \quad \text{in } r \leq a \tag{18b}$$

for $\omega a/c_o$ and $\omega a/c_1$ both small in comparison with unity. We see that throughout $r \geq a$ the solution of the exact problem is identical to the far-field pressure predicted from the representation theorem (equation 13). The distant acoustic pressure perturbation is proportional to $\rho_o\omega(\gamma-1)\hat{Q}/\rho_1 c_1{}^2$. A heavier fluid surrounding the combustion zone increases the radiated acoustic energy by a factor $\rho_o c_1/\rho_1 c_o$ from the value it would have if the combustion occurred in a space entirely occupied by ρ_1, c_1 fluid. It is interesting to note how large this factor is when the combustion takes place in an air pocket surrounded by steel or water!

We will now discuss a second but similar problem with spherical symmetry, which again has a simple exact solution. Unsteady combustion with a total rate of heat release $\hat{Q} e^{i\omega t}$ is distributed uniformly throughout the sphere $r \leq a$. Again the mean density and sound speed have values ρ_1, c_1 within $r < a$ and ρ_o, c_o in $r > a$. The wave equation in $r < a$ is then

$$\frac{1}{c_1{}^2}\frac{\partial^2 p}{\partial t^2} - \nabla^2 p = \frac{i\omega(\gamma-1)3}{4\pi a^3 c_1{}^2} \hat{Q} e^{i\omega t} . \tag{19}$$

This is to be solved together with (15b) and conditions of continuity of pressure and particle velocity across $r = a$. When the combustion zone is compact the solution in $r \geq a$ again has the form expressed in (18a), thereby verifying again the results obtained from the representation theorem.

We have obtained an expression for the sound generation by combustion noise,

$$(p-p_o)(\underline{x},t) = \frac{1}{4\pi|\underline{x}|}\frac{\partial}{\partial t}\int\left[\frac{\rho_o(\gamma-1)Q}{\rho c^2}\right]d^3\underline{y} \quad , \tag{13}$$

and checked it by comparison with two exact solutions. We now go on to compare it with existing theories of combustion noise. Chiu and Summerfield [5] also find the distant acoustic pressure perturbation to be proportional to $\rho_o/\rho_f c^2$. (In their notation ρ_f is the density at the flame.) But since they have expressed the time derivative as a material derivative their source term appears more complicated. In [6] Ffowcs Williams discusses linear source processes and equation (13) is in agreement with his expression for sound generation by the unsteady addition of heat in a fluid with a different density and sound speed from the far-field.

Strahle has written many papers on combustion noise. In [7] he relates the sound field to the reaction rate and obtains

$$(p-p_o)(\underline{x},t) = \frac{\gamma-1}{4\pi|\underline{x}|c_1{}^2}\frac{\partial}{\partial t}\int[Q]\ d^3\underline{y} \tag{20}$$

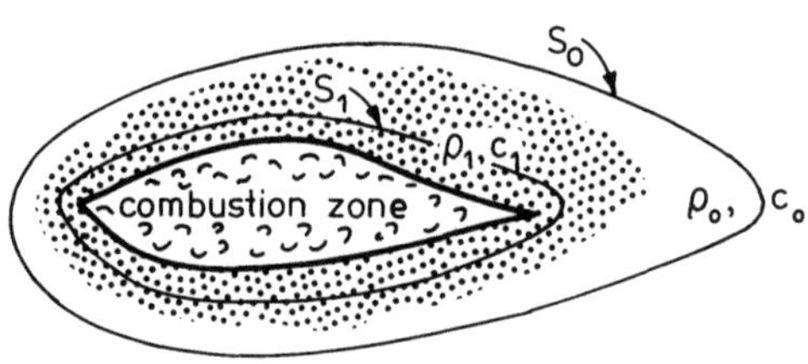

Fig. 2 The geometry of the flame

ρ_1 and c_1 are the mean density and sound speed in the hot gas after the combustion, as shown in Figure 2. The combustion is considered to be in a perfect gas at constant pressure so that $\rho c^2 = \rho_1 c_1{}^2 = \rho_o c_o{}^2$. A comparison of (13) and (20) shows that (20) is a factor $c_o{}^2/c_1{}^2 = \rho_1/\rho_o$ smaller than (13). This factor is about ⅙ for the combustion of hydrocarbons in air. The difference does not affect the scaling laws derived in reference [7]. But, since both (13) and (20) have been derived from exact representations of the sound field and should be identical, this discrepancy must be resolved.

In [7] Strahle has a monopole source whose strength is the time derivative of

$$\int_V \frac{\partial \rho}{\partial t}\, d^3\underline{y} \quad .$$

The integral is to be evaluated over the source region V in which $\rho-\rho_o$ is appreciably different from $(p-p_o)/c_o{}^2$. From the equation of mass conservation

$$\int_V \frac{\partial \rho}{\partial t}\, d^3\underline{y} = -\int_S \rho\ \underline{u}\cdot d\underline{S} \ , \tag{21}$$

where S is the surface enclosing the source region V. In reference [7] this bounding surface S is taken to lie in the hot gas, i.e. the surface S_1 in Figure 2. On this surface ρ is nearly equal to ρ_1 and so

$$\int_S \rho\ \underline{u}\cdot d\underline{S} \simeq \rho_1 \int_{S_1} \underline{u}\cdot d\underline{S} \tag{22}$$

Strahle goes on to show how $\int \underline{u}\cdot d\underline{S}$ over any near-field surface enclosing the combustion zone can be related to the heat release rate to obtain the result in (20). However S_1 does not enclose all the acoustic sources. Outside it ρ has strong gradients as its mean value changes from ρ_1 to ρ_o. Acoustic waves travelling through this region move the interface and produce large values of $\partial\rho/\partial t$. To include these sources the bounding surface S should should be taken to lie in the ρ_o fluid; i.e. to coincide with surface S_o in Figure 2. For this position of S

$$\int_S \rho \underline{u}\cdot d\underline{S} \simeq \rho_o \int_{S_o} \underline{u}\cdot d\underline{S} \qquad (23)$$

A comparison of equations (22) and (23) shows that if all the acoustic sources are to be included Strahle's result (20) should be multiplied by a factor ρ_o/ρ_1. It is then in agreement with equation (13).

Hassan [8] also obtains an expression for the far-field density perturbation in terms of the heat release rate. His result is

$$(p-p_o)(\underline{x},t) = \frac{\gamma-1}{4\pi|\underline{x}|}\frac{\partial}{\partial t}\int\left[\frac{Q}{c^2}\right] d^3\underline{y} \quad . \qquad (24)$$

Comparison with (13) shows that it differs from our result by a multiplicative factor ρ/ρ_o. This discrepancy can be quickly resolved because a source term, which is not small, has been neglected in reference [8]. The author retains $-(\gamma-1)Q/c^2$ as his main source term, but neglects $\rho_e\ \nabla\cdot\underline{u}$. In the combustion zone however $\nabla\cdot\underline{u}$ is related to the heat release rate per unit volume by equation (3) and is not negligible;

$$\rho_e\ \nabla\cdot\underline{u} \simeq -\frac{\rho-\rho_o}{\rho}\frac{D\rho}{Dt} \simeq \frac{\rho-\rho_o}{\rho c^2}(\gamma-1)\ Q\ .$$

Adding this to the term, $-(\gamma-1)\ Q/c^2$, retained by Hassan gives a total combustion source $-(\gamma-1)\ \rho_o\ Q/\rho c^2$, a factor ρ_o/ρ larger than Hassan's source and in agreement with our expression.

In equation (13), then, we have an expression for the sound field radiated by unsteady combustion. This has been checked by comparison with the exact solutions of two problems with simple geometries, and we have been able to reconcile apparent differences in the literature of combustion noise.

Our form for the radiated pressure can be used to derive scaling laws for the acoustic power. Let us assume that combustion occurs in a perfect gas at constant pressure. Then $\rho c^2 = \rho_o c_o{}^2$ and equation (14) is the appropriate form for the distant pressure. The acoustic power can be found by integrating the intensity over a large sphere. For a compact source region this gives

$$\text{acoustic power} = \frac{\overline{(p-p_o)^2}\ 4\pi|\underline{x}|^2}{\rho_o c_o}$$

$$= \frac{(\gamma-1)^2}{4\pi\rho_o c_o{}^5} \int \overline{\frac{\partial Q(\underline{y},t)}{\partial t} \frac{\partial Q(\underline{y}+\underline{\Delta},t)}{\partial t}}\, d^3\underline{\Delta}\, d^3\underline{y} \tag{25}$$

The overbar denotes a time average. The acoustic power can be converted into an efficiency of sound generation, η, by dividing by the heat release rate.

If we consider an open, premixed flame, in which the premixed gas emerges through a circular orifice of diameter D with velocity U and density ρ_o,

$$\int \overline{Q(\underline{y},t)\ d^3\underline{y}} = \rho_o \frac{U\pi D^2}{4} \quad F\Delta H \tag{26}$$

The combustion has been assumed to be totally efficient. F is the ratio of the mass of fuel to premixed gas, and ΔH is the calorific value of the fuel. Before equation (25) can be used to determine the combustion noise two questions need to be answered: how should the time derivatives $\partial/\partial t$, and the correlation volume be scaled? There are two main protagonists of scaling, Putnam and Strahle.

Putnam and his colleagues use an extension of Bragg's expanding monopole theory [9] to determine the combustion noise. But since our expression is exact we can recover their scaling law by making the same physical assumptions. Diammar and Putnam [10] view the combustion as being correlated within eddies of size d_L, the burning within each eddy being controlled by the laminar flame speed S_L. Hence

$$\int \overline{Q(\underline{y},t)\ d^3\underline{\Delta}} = \rho_o\ S_L\ d_L{}^2\ F\Delta H \tag{27}$$

and by comparison with (26)

$$d^3\underline{\Delta} \sim \frac{4S_L d_L^{\,2}}{\pi U D^2}\, d^3\underline{y} \tag{28}$$

They take

$$\frac{\partial}{\partial t} \sim \frac{u'}{d_L} \tag{29}$$

where u' is a typical turbulent velocity scale. These assumptions lead to

$$\eta = \frac{(\gamma-1)^2}{4\pi}\left(\frac{u'}{c_o}\right)^2 \frac{S_L}{c_o}\frac{F\Delta H}{c_o^{\,2}} \tag{30}$$

This is seen to be identical to Diammar and Putnam's result once $(\gamma-1)F\Delta H/c_o^{\,2}$ is rewritten as E-1, where E is the expansion ratio. Putnam and Faulkner [11] show that this efficiency compares well with experiment. However it has been derived by assuming that the frequency of combustion noise scales as u'/d_L, and experiments show the frequency to be virtually independent of U. Now u' is proportional to U, and Diammar and Putnam try to reconcile their assumed frequency scaling with experiment by arguing, not very convincingly, that d_L is also proportional to U. So although equation (30) gives a reasonable form for the acoustic efficiency, it is based on assumptions of the frequency which are not compatible with observation.

Strahle has developed several scaling laws for combustion noise. Probably the most successful for open, premixed flames is based on an empirical frequency [12]. Strahle takes

$$d^3\underline{\Delta} = d^3\underline{y} \qquad \text{and} \qquad \frac{\partial}{\partial t} = St\,\frac{U}{D}\;, \tag{31}$$

where from curve-fitting to a set of experimental results St is taken to be proportional to $M^{-0.40}Re^{0.02}Da^{0.39}F^{-1.1}$, with $Re = \rho_o UD/\mu$, $Da = \rho_o S_L^{\,2} Dc_p/U\lambda$ and λ is the thermal conductivity. With these assumptions it follows directly from (25) that

$$\eta \propto M^{2.6}\; Re^{0.02}\; Da^{0.78}\; F^{-1.1}\; \frac{\Delta H}{c_p T_o}\;. \tag{32}$$

This is in close agreement with measurements of efficiency from the same set of experiments, which gave

$$\eta \propto M^{2\cdot 68}\, Re^{-0\cdot 09}\, Da^{0\cdot 92}\, F^{-1\cdot 26} . \tag{33}$$

4. The diffusion of mass and heat

We have seen that combustion is a strong monopole sound source. The representation theorem (10) is in a convenient form to investigate whether there is a monopole source, in the absence of combustion, due to the diffusion of heat and/or species. Monopole sources may lead to an acoustic power output proportional to M^4 and so be important in a low Mach number flow. The effects of heat diffusion have been investigated by Kempton [13]. Morfey [14] considers the diffusion of both heat and species.

From equation (10) the contribution to the far-field pressure perturbation due to diffusion, $p_d'(\underline{x},t)$, is given by

$$4\pi|\underline{x}|p_d'(x,t) = -\rho_o \frac{\partial}{\partial t}\int\left[\frac{\alpha}{\rho c_p(1-M_r)}\left\{\nabla\cdot\underline{q} - \sum_{n=1}^{N}\left.\frac{\partial h}{\partial Y_n}\right|_{\rho,p,Y_n}\nabla\cdot\underline{J}_n\right\}\right] d^3\underline{y} . \tag{34}$$

The source involves $\nabla\cdot\underline{q}$ and $\nabla\cdot\underline{J}_n$, but their integral over the source region at fixed time is zero. We can demonstrate this near cancellation by rewriting equation (34) in the exactly equivalent form;

$$4\pi|\underline{x}|p_d'(\underline{x},t) =$$

$$\rho_o \frac{\partial}{\partial t}\int\left[\underline{q}\cdot\nabla\left(\frac{\alpha}{\rho c_p(1-M_r)}\right) - \sum_{n=1}^{N}\underline{J}_n\cdot\nabla\left(\frac{\alpha}{\rho c_p(1-M_r)}\left.\frac{\partial h}{\partial Y_n}\right|_{\rho,p,Y_m}\right)\right] d^3\underline{y}$$

$$-\rho_o \frac{\partial^2}{\partial t\,\partial x_i}\int\left[\frac{\alpha}{\rho c_p(1-M_r)}\left\{q_i - \sum_{n=1}^{N}\left.\frac{\partial h}{\partial Y_n}\right|_{\rho,p,Y_m} J_{n_i}\right\}\right] d^3\underline{y} \tag{35}$$

If the very small Dufour effect is neglected, the flux of heat in a gas made up of N species is given by (Williams [15], page 427)

$$\underline{q} = -\lambda\nabla T + \sum_{n=1}^{N} h_n \underline{J}_n \ . \tag{36}$$

where we will use the notation h_n and V_n to denote the partial derivatives

$$\left.\frac{\partial h}{\partial Y_n}\right|_{T,p,Y_m} \qquad \text{and} \qquad \left.\frac{\partial \rho^{-1}}{\partial Y_n}\right|_{T,p,Y_m}$$

respectively. Once $\underline{q}$ has been substituted into (35), the acoustic pressure perturbation can be simplified by using the chain rule,

$$h_n = \left.\frac{\partial h}{\partial Y_n}\right|_{\rho,p,Y_m} + \left.\frac{\partial h}{\partial \rho}\right|_{p,Y_n} \rho_n \ , \tag{37}$$

and the relationship $\partial h/\partial\rho\big|_{p,Y_n} = -\,c_p/\rho\alpha$. This leads to

$$4\pi|\underline{x}|p_d'(x,t)$$

$$= \rho_o \frac{\partial}{\partial t}\int\!\!\left[-\lambda\nabla T.\nabla\left(\frac{\alpha}{\rho c_p(1-M_r)}\right) + \sum_{n=1}^{N} \underline{J}_n \cdot \left\{\nabla\left(\frac{V_n}{1-M_r}\right) - \frac{\alpha\nabla h_n}{\rho c_p(1-M_r)}\right\}\right] d^3\underline{y}$$

$$+ \rho_o \frac{\partial^2}{\partial t\partial x_i}\int\!\!\left[\frac{1}{1-M_r}\left\{\frac{\alpha\lambda}{\rho c_p}\frac{\partial T}{\partial x_i} - \sum_{n=1}^{N} J_{n_i} V_n\right\}\right] d^3\underline{y} \tag{38}$$

Equation (38) shows that there is only a monopole sound field due to heat diffusion in a low Mach number flow if $\alpha/\rho c_p$ is a function of position. For an ideal gas $\alpha/\rho c_p = (\gamma-1)/\gamma p$. Since fractional changes in p are of order M^2, the existence of a monopole source due to heat diffusion relies on variations in γ. As pointed out by Kempton and Morfey, this is a very weak effect.

A low Mach number monopole source due to species diffusion depends on variations of V_n or h_n. Morfey discusses how in an ideal gas V_n and h_n are independent of species concentration and can only vary as a function of position due to temperature differences.

In general then, diffusion does not lead to a significant monopole source and its dominant effect is to generate dipole sound. This is described by terms involving M_r in the first integral on the right-hand side of equation (38) and by the second integral. When diffusion is significant these terms lead to a radiated sound power proportional to M^6. Kempton [13] has calculated this sound power exactly in a number of model problems which are amenable to analytical solution.

Equation (10) also describes the role of viscosity as a sound source. But, since this has recently been treated comprehensively by Obermeier [16], it will not be discussed here.

5. Acceleration of density inhomogeneities

We have seen that diffusion can lead to a dipole-type sound field. A similar field is produced by the acceleration of a compact density inhomogeneity. The acoustic pressure generated by this sound source is described by the last term in equation (10) and we will denote it by $p_a'(\underline{x},t)$:

$$4\pi|\underline{x}|p_a'(\underline{x},t) = -\frac{\partial}{\partial t}\int\left[\frac{\rho_e}{(1-M_r)^2}\frac{DM_r}{Dt}\right]d^3\underline{y} \tag{39}$$

Equation (39) shows that a compact fluid element, even with constant density, radiates sound when it accelerates provided its density is different from ambient - a form of 'acoustic bremsstrahlung'.

For a far-field observer $M_r = \underline{u}\cdot\underline{x}/c_o|\underline{x}|$, and the momentum equation shows that $D\underline{u}/Dt$ can be rewritten as $-\nabla p/\rho$ if viscous effects are neglected. Equation (39) is therefore equivalent to

$$4\pi|\underline{x}|p_a'(\underline{x},t) = \frac{x_i}{|\underline{x}|c_o}\frac{\partial}{\partial t}\int\left[\frac{\rho_e}{\rho(1-M_r)^2}\frac{\partial p}{\partial y_i}\right]d^3\underline{y} \tag{40}$$

showing clearly the dipole nature of the sound field radiated by the density inhomogeneity as it convects through a region of pressure gradients. Such a density inhomogeneity might be due to convected temperature variations downstream of a zone of unsteady combustion for example. The sound field produced by 'hot spots' convected through pressure gradients in a low Mach number nozzle has been investigated by Ffowcs Williams and Howe [17]. There the situation is more complicated than our free-space result because of unsteady forces on the nozzle walls. Ffowcs Williams and Howe overcome this difficulty by using an exact Green function. Cumpsty and Marble [18] consider the convection of entropy inhomogeneities through the intense pressure changes across a turbine, and find it can generate considerable sound energy. From equation (40) the pressure perturbation produced by the interaction of a compact density inhomogeneity with a mean flow pressure gradient is proportional to $(\rho-\rho_o)MU^2$. The radiated acoustic power is therefore proportional to M^6.

Integration by parts shows that

$$\frac{x_i}{|\underline{x}|c_o}\frac{\partial}{\partial t}\int\left[\frac{\rho_e}{\rho(1-M_r)^2}\frac{\partial p}{\partial y_i}\right]d^3\underline{y} =$$

$$-\frac{x_i}{|\underline{x}|c_o}\frac{\partial}{\partial t}\int\left[(p-p_o)\frac{\partial}{\partial y_i}\left(\frac{\rho_e}{\rho(1-M_r)^2}\right)\right]d^3\underline{y}$$

$$+\frac{x_i}{|\underline{x}|}\frac{\partial^2}{\partial t\partial x_i}\int\left[\frac{(p-p_o)\rho_e}{\rho(1-M_r)^2}\right]d^3\underline{y}\,. \tag{41}$$

The first term on the right-hand side of (41) shows that turbulent fluctuations in a region of appreciable density gradients can generate dipole sound. This is one of the source processes we will consider in the next section when we investigate the sound generated by turbulence within two-phase flow.

6. Turbulent two-phase flow

The sound field radiated by turbulence can be changed significantly if that turbulence is within a region of fluid of different density and sound speed from ambient. Variations in ρ and c are particularly dramatic in the case of two-phase flows where, for example, 1% air bubbles by volume in water can cause the sound speed to drop to about 30 m/s, $1/50$ of the sound speed in water. These effects have been investigated by Crighton and Ffowcs Williams [19].

We begin by solving a simple problem with an exact solution. Consider an isotropic quadrupole of strength $\rho_1 U^2 e^{i\omega t}$ concentrated at the origin. The fluid within a sphere of radius a has density ρ_1 and sound speed c_1, while the surrounding fluid has density ρ_o and sound speed c_o. The pressure perturbation satisfies

$$\frac{1}{c_1^2}\frac{\partial^2 p}{\partial t^2} - \nabla^2 p = \nabla^2 \left(\rho_1 U^2 e^{i\omega t}\delta(\underline{x})\right) \qquad \text{in } r < a \tag{42a}$$

and

$$\frac{1}{c_o^2}\frac{\partial^2 p}{\partial t^2} - \nabla^2 p = 0 \qquad \text{in } r > a\,, \tag{42b}$$

with conditions of continuity of pressure and particle velocity across $r = a$. A rearrangement of (42a) shows that it can be written as

$$\left(\frac{1}{c_1^2}\frac{\partial^2}{\partial t^2} - \nabla^2\right)\left(p + \rho_1 U^2 e^{i\omega t}\delta(\underline{x})\right) = -\frac{\omega^2}{c_1^2}\rho_1 U^2 e^{i\omega t}\delta(\underline{x})\,. \tag{43}$$

We have already discussed the solution to an identical set of equations in §2. By comparison with (15a) and (18a), the solution for small $\omega a/c_o$ and $\omega a/c_1$ is

$$(p-p_o)(r,t) = -\frac{\omega^2\rho_o U^2}{4\pi r c_1^2} e^{i\omega(t-r/c_o)} \qquad \text{in } r \geq a. \tag{44}$$

The acoustic power may be found by integrating the intensity over a large sphere and is equal to $\omega^4\rho_o U^4/\sqrt{2}\ 4\pi c_1^4 c_o$ For comparison with Crighton and Ffowcs Williams' work we convert this into an efficiency by dividing by $\rho_1 U^3 L^2$ where L has dimensions of length. This shows that the efficiency is increased by a multiplicative factor

$$\frac{c_o^4\ \rho_o}{c_1^4\ \rho_1} \tag{45}$$

over the value it would have if a quadrupole with the same turbulent velocity fluctuations were entirely in ρ_o, c_o fluid.

For the case of air bubbles in water ρ_o is nearly equal to ρ_1 but c_1 can be very much less than c_o. The factor c_o^4/c_1^4 may be as large as 10^7 and demonstrates how a small quantity of bubbles can dramatically increase the radiation efficiency of turbulence in water. This predicted increase is entirely in agreement with the work of Crighton and Ffowcs Williams. The other example they consider is that of a dusty gas, for which ρ_1 is much larger than ρ_o while $\rho_1 c_1^2 = \rho_o c_o^2$. Then from (45) the efficiency is increased by a factor c_o^2/c_1^2. This is far less augmentation than that predicted in reference [19], where the corresponding result is c_o^6/c_1^6. We will explain this discrepancy later.

The problem considered here had such a simple geometry that an exact solution was possible. However, in general for an arbitrarily shaped source region we need to rely on an acoustic analogy of the sort described by equation (10) to determine the radiated sound. Let us consider turbulent flow in a compact region of fluid with properties different from ambient. In the absence of reactions, diffusion and viscosity the representation (10) reduces to

$$4\pi|\underline{x}|(p-p_o)(\underline{x},t) = -\frac{\partial}{\partial t}\int\left[\frac{1}{1-M_r}\left\{\left(\frac{\rho_o}{\rho c^2}-\frac{1}{c_o^2}\right)\frac{Dp}{Dt}+\frac{p-p_o}{\rho c_o^2}\frac{D\rho}{Dt}\right\}\right]d^3\underline{y}$$

$$+\frac{\partial^2}{\partial x_i\,\partial x_j}\int[\rho u_i u_j]\,d^3\underline{y} + \frac{x_i}{|\underline{x}|c_o}\frac{\partial^2}{\partial x_i\,\partial t}\int\left[\frac{\rho_e(p-p_o)}{\rho(1-M_r)^2}\right]d^3\underline{y}$$

$$-\frac{x_i}{|\underline{x}|c_o}\frac{\partial}{\partial t}\int\left[\frac{\partial}{\partial y_i}\left(\frac{\rho_e}{\rho(1-M_r)^2}\right)(p-p_o)\right]d^3\underline{y} \qquad (46)$$

Equation (41) has been used to expand the last term on the right-hand side of (10). The first three sources in (46) are volume terms, while the last involves gradients of density which are only large near the boundary of the two-phase region.

For air bubbles in water

$$\rho = (1-\beta)\rho_o + \beta\,\rho_a \quad . \qquad (47)$$

β is the volume fraction of bubbles, ρ_o the density of water and ρ_a the density of air. ρ is nearly equal to ρ_o, but in the bubbly region c is very much less than c_o. The expression (46) appears complicated, but only one volume term is important. That is

$$\frac{\partial}{\partial t}\int\left[\frac{1}{c^2}\frac{Dp}{Dt}\right] d^3\underline{y} \ . \tag{48}$$

This term involves division by c^2, whereas all the other volume terms are divided by the very much larger $c_o^{\ 2}$.

Following Crighton and Ffowcs Williams we consider a flow with mean velocity U, turbulent velocity fluctuations of order u' and typical eddy size ℓ, within a bubbly region of linear dimension L. The pressure perturbation is of order $\rho u'U$ and $\partial/\partial t \sim u'/\ell$. The radiated acoustic power can be converted into an efficiency, η, by dividing it by $\rho u'U^2L^2$, the average rate of working of the fluctuating pressure against the mean flow over the source region The term in (48) leads to

$$\eta = \frac{1}{4\pi}\left(\frac{u'}{U}\right)^5 M^5 \left(\frac{c_o}{\bar{c}}\right)^4 \frac{L}{\ell} \tag{49}$$

The M^5 scaling of efficiency, which is equivalent to M^8 in power, demonstrates that this source is of quadrupole type. The presence of bubbles increases the radiation efficiency by a factor $c_o^{\ 4}/\bar{c}^4$ over the value it would have in pure water.

The radiation efficiency of the density gradient term,

$$\frac{-x_i}{|\underline{x}|c_o}\frac{\partial}{\partial t}\int\left[\frac{\partial}{\partial y_i}\left(\frac{\rho_e}{\rho}\right)(p-p_o)\right] d^3\underline{y} \ , \tag{50}$$

depends on how abruptly the density varies at the edge of the bubbly region. If β varies over the turbulent length scale ℓ, so that $|\nabla\beta| \sim \bar{\beta}/\ell$, the term in (50) radiates sound with an efficiency

$$\eta = \frac{\bar{\beta}^2}{4\pi}\left(\frac{u'}{U}\right)^3 M^3 \frac{L}{\ell} \ . \tag{51}$$

The Mach number dependence illustrates (as we predicted in §5) that this is a dipole-type source. This gradient term has been discussed in reference [19]. It happens to be identically zero in our exact example (42) because of the spherical symmetry of that problem. One of the drawbacks of the simple geometry. Whether, for an arbitrarily shaped bubbly region, the dominant source mechanism is described by (49) or (51) depends on the relative magnitudes of $u'^2 c_o^2/\bar{c}^4$ and $\bar{\beta}^2$. But in either case the bubbles lead to a considerable enhancement of the radiation efficiency.

For dusty gases

$$\rho = \rho_o \,(1 + f) \text{ and } c^2 = \frac{c_o^2}{1+f}\,, \tag{52}$$

where f, the mass concentration of dust, is assumed to be large in comparison with unity. Since $\rho c^2 = \rho_o c_o^2$, two terms cancel in the first integrand in equation (46). For a compact dusty region the remaining volume terms generate sound with efficiency

$$\eta = \frac{1}{4\pi}\left(\frac{u'}{U}\right)^5 M^5 \left(\frac{c_o}{\bar{c}}\right)^2 \frac{L}{\ell}\,, \tag{53}$$

while the boundary term has an efficiency

$$\eta = \frac{1}{4\pi}\left(\frac{u'}{U}\right)^3 M^3 \left(\frac{\bar{c}}{c_o}\right)^2 \frac{L}{\ell} \tag{54}$$

if f varies over the turbulent length scale ℓ at the edge of the dust cloud.

The form given in (53) is consistent with our exact solution, and means that the dust produces little increase in the radiation efficiency. This augmentation is less than that predicted in reference [19] by a factor $\bar{c}^4/c_o^4$. There the partial time derivative of $\partial\rho/\partial t - c_o^{-2}\,\partial p/\partial t$ appears as a volume source term and $\partial\rho/\partial t$ is estimated by writing

$$\frac{\partial\rho}{\partial t} = \frac{1}{c^2}\frac{\partial p}{\partial t}\,, \tag{55}$$

so that

$$\frac{\partial \rho}{\partial t} - \frac{1}{c_o^2}\frac{\partial p}{\partial t} = \left(\frac{1}{c^2} - \frac{1}{c_o^2}\right)\frac{\partial p}{\partial t} = \frac{f}{c_o^2}\frac{\partial p}{\partial t} \quad . \tag{56}$$

This leads to the quadrupole radiation efficiency being increased by a factor $c_o^6/\bar{c}^6$ due to dust. But since there are considerable mean density gradients, the material density derivative should have been used in (55);

$$\frac{\partial \rho}{\partial t} = \frac{1}{c^2}\frac{\partial p}{\partial t} - \underline{u}\cdot\nabla\rho \tag{57}$$

The equation of mass conservation enables $\underline{u}\cdot\nabla\rho$ to be rearranged as follows;

$$\underline{u}\cdot\nabla\rho = \nabla\cdot\left\{\underline{u}(\rho-\rho_o)\right\} + \frac{\rho-\rho_o}{\rho}\frac{D\rho}{Dt}$$

$$= \nabla\cdot\left\{\underline{u}(\rho-\rho_o)\right\} + \frac{\rho-\rho_o}{\rho c^2}\frac{Dp}{Dt} \tag{58}$$

When (57) and (58) are used to substitute for $\partial\rho/\partial t$ in the source term all the terms involving pressure cancel giving

$$\frac{\partial \rho}{\partial t} - \frac{1}{c_o^2}\frac{\partial p}{\partial t} = -\nabla\cdot\left\{\underline{u}(\rho-\rho_o)\right\} \quad . \tag{59}$$

After some algebra our scaling law (53) for the efficiency of the sound generated by the volume sources can be recovered from this description of the source strength. The effect of dust was exaggerated then in reference [19] because the poor approximation $\partial\rho/\partial t = c^{-2}\,\partial p/\partial t$ made the source seem larger than when the correct relationship $D\rho/Dt = c^{-2}\,\partial p/\partial t$ is used.

The acoustic analogy as described by (10) can be used to estimate the sound from a compact source region. But when the source is not compact, retarded time variations across the source mean the integrals in (10) are hard to estimate. We will now use the model problem with spherical symmetry to investigate the effect of a non-compact two-phase region.

When both $\omega a/c_o$ and $\omega a/c_1$ are large in comparison with unity, the solution to equation (42) is

$$(p-p_o)(r,t) = -\frac{\omega^2\rho_1 U^2 T}{4\pi r c_1^2}\frac{e^{i\omega(t-(r-a)/c_o-a/c_1)}}{1+Re^{-2i\omega a/c_1}}, \quad (60)$$

in $r \geq a$, where $R = \dfrac{\rho_o c_o - \rho_1 c_1}{\rho_o c_o + \rho_1 c_1}$ and $T = \dfrac{2\rho_o c_o}{\rho_o c_o + \rho_1 c_1}$.

R and T are the pressure reflection and transmission coefficients for sound at normal incidence onto a plane interface separating $\rho_1 c_1$ fluid from $\rho_o c_o$. For large $\omega a/c_1$, p varies rapidly as a function of frequency, particularly for extreme differences between the two fluids when $|R|$ is approximately unity. So, rather than consider a single frequency, we will investigate the sound field produced by a source with time dependence f(t).

$$\frac{1}{c_1^2}\frac{\partial^2 p}{\partial t^2} - \nabla^2 p = \nabla^2(f(t)\delta(\underline{x})) \quad \text{in } r < a \quad (61a)$$

$$\frac{1}{c_o^2}\frac{\partial^2 p}{\partial t^2} - \nabla^2 p = 0 \quad \text{in } r > a. \quad (61b)$$

Provided f(t) only contains frequencies that are high in comparison with both c_o/a and c_1/a, the solution to (61) can be calculated by inverting the solution of the Fourier transformed problem (60). This gives

$$(p-p_o)(r,t) = \frac{T}{4\pi r c_1^2}\int\Big\{-\omega^2\tilde{f}(\omega)e^{i\omega(t-(r-a)/c_o-a/c_1)}$$

$$\times\,(1 - Re^{-2i\omega a/c_1} + R^2e^{-4i\omega a/c_1} - \ldots)\Big\}d\omega \quad (62)$$

Here $\tilde{f}(\omega)$ is the Fourier transform of f(t), and the binomial theorem has been used to expand the demoninator in (60). After integrating term by term we obtain

$$(p-p_o)(r,t) = \frac{T}{4\pi r c_1^2}\left[\frac{\partial^2}{\partial t^2} f\left(t - \frac{r-a}{c_o} - \frac{a}{c_1}\right) - R\frac{\partial^2}{\partial t^2} f\left(t - \frac{r-a}{c_o} - \frac{3a}{c_1}\right) + R^2 \frac{\partial^2}{\partial t^2} f\left(t - \frac{r-a}{c_o} - \frac{5a}{c_1}\right) - \cdots\right] \tag{63}$$

The first term in square brackets in (63) describes sound that has travelled a distance a in the c_1 region, r-a in c_o fluid, and undergone an amplitude of change T when crossing the interface. The sound in the next term has been reflected back once at the r=a interface, travelled an extra distance 2a in c_1 fluid and changed phase by π as it passed through the origin. The higher order terms represent sound waves that have been multiply reflected within r=a.

The total acoustic energy, E, may be found by integrating the power output and is equal to $4\pi r^2 \int (p-p_o)^2 dt/\rho_o c_o$. Substitution for $p-p_o$ from (63) shows that

$$E = \frac{T^2}{4\pi\rho_o c_o c_1^4}(1 + R^2 + R^4 + \ldots.)\int_{-\infty}^{\infty}\left(\frac{\partial^2 f}{\partial t^2}\right)^2 dt\ , \tag{64}$$

if the source f emits for a time less than $2a/c_1$. The series can be summed and substituting for R and T gives

$$E = \frac{1}{4\pi\rho_1 c_1^5}\int_{-\infty}^{\infty}\left(\frac{\partial^2 f}{\partial t^2}\right)^2 dt\ . \tag{65}$$

In this case the turbulent quadrupoles radiate energy as if all the fluid had density ρ_1 and sound speed c_1; the ρ_1, c_1 fluid is so extensive that information that it is bounded does not arrive back at the source during its lifetime. The radiation efficiency is increased by a factor $(c_o/c_1)^5$ from the value it would have if the same velocity fluctuations occurred in ρ_o, c_o fluid. This could represent a considerable enhancement due to both air bubbles in water and dust particles in air.

7. Conclusions

A representation theorem describing sound generation has been derived and used to investigate various effects associated with an unsteady 'excess density' source term. Combustion noise has been discussed and illustrated by a simple problem with an exact solution. Some discrepancies in the literature have been resolved. The diffusion of heat or species and the acceleration of density inhomogeneities have been demonstrated to be dipole sound sources. Finally the effects of turbulence within a region with different mean properties from ambient have been discussed and again illustrated by reference to an exact solution.

Acknowledgement

I would like to thank Professor J. E. Ffowcs Williams and Dr T. P. Hynes for their helpful suggestions and illuminating comments during the preparation of this lecture.

References

1. Lighthill, M.J.: On sound generated aerodynamically. I. General theory. Proc. Roy. Soc. A, 211 (1952) 564-587.

2. Ffowcs Williams, J.E.: Sound production at the edge of a steady flow. J. Fluid Mech., 66 (1974) 791-816.

3. Hurle, I.R.; Price, R.B.; Sugden, T.M.; Thomas, A.: Sound emission from open turbulent premixed flames. Proc. Roy. Soc. Lond. A, 303 (1968) 409-427.

4. Shivashankara, B.N.; Strahle, W.C.; Handley, J.C.: Evaluation of combustion noise scaling laws by an optical technique. AIAA Journal, 13 (1975) 623-627.

5. Chiu, H.H.; Summerfield, M.: Theory of combustion noise. Acta Astro., 1 (1974) 967-984.

6. Ffowcs Williams, J.E.: Sound sources in aerodynamics - fact and fiction. AIAA Journal, 20 (1982) 307-315.

7. Strahle, W.C.: Some results in combustion generated noise. J. Sound Vib., 23 (1972) 113-125.

8. Hassan, H.A.: Scaling of combustion generated noise. J. Fluid Mech., 66 (1974) 445-453.

9. Bragg, S.L.: Combustion noise. J. Instit. of Fuel, 36 (1963) 12-16.

10. Diammar, R.D.; Putnam, A.A.: Combustion roar of premix burners, singly and in pairs. Combustion and Flame, 18 (1972) 435-438.

11. Putnam, A.; Faulkner, L.: An overview of combustion noise. J. Energy, 7 (1983) 458-469.

12. Strahle, W.C.: Combustion noise. Prog. Energy and Combust. Sci., 4 (1978) 157-176.

13. Kempton, A.J.: Heat diffusion as a source of aerodynamic sound. J. Fluid Mech., 78 (1976) 1-31.

14. Morfey, C.L.: Sound radiation due to unsteady dissipation in turbulent flows. J. Sound Vib., 48 (1976) 95-111.

15. Williams, F.A.: Combustion Theory. Addison-Wesley (1965).

16. Obermeier, F.: Aerodynamic sound generation caused by viscous processes. J. Sound Vib., 99 (1985) 111-120.

17. Ffowcs Williams, J.E.; Howe, M.S.: The generation of sound by density inhomogeneities in low Mach number nozzle flows. J. Fluid Mech., 70 (1975) 605-622.

18. Cumpsty, N.A.; Marble, F.E.: The interaction of entropy fluctuations with turbine blade rows; a mechanism of turbojet engine noise. Proc. Roy. Soc. A, 357 (1977) 323-344.

19. Crighton, D.G.; Ffowcs Williams, J.E.: Sound generation by turbulent two-phase flow. J. Fluid Mech., 36 (1969) 585-603.

APPENDIX

The energy equation is

$$\rho \frac{De}{Dt} = - \nabla\cdot\underline{q} - p\,\nabla\cdot\underline{u} + \frac{\partial u_i}{\partial x_j}\tau_{ij} \tag{A1}$$

where e is the internal energy. For a gas made up of N species Williams [15] shows that

$$de = T\,dS + \frac{p}{\rho^2}\,d\rho + \sum_{n=1}^{N} \frac{\mu_n}{W_n}\,dY_n \quad , \tag{A2}$$

with S entropy, μ_n the chemical potential of species n and W_n its molecular weight. Substituting for e from (A2) into (A1) and using the mass equation leads to an expression for entropy changes;

$$\rho T \frac{DS}{Dt} = - \nabla\cdot \underline{q} + \frac{\partial u_i}{\partial x_j} \tau_{ij} - \sum_{n=1}^{N} \rho \frac{\mu_n}{W_n} \frac{DY_n}{Dt} \ . \tag{A3}$$

Now from the chain rule

$$\frac{Dp}{Dt} = c^2 \frac{D\rho}{Dt} + \left.\frac{\partial p}{\partial S}\right|_{\rho,Y_n} \frac{DS}{Dt} + \sum_{n=1}^{N} \left.\frac{\partial p}{\partial Y_n}\right|_{S,\rho,Y_m} \frac{DY_n}{Dt} \tag{A4}$$

When (A4) is rearranged and (A3) used to replace DS/Dt it leads to

$$\frac{D\rho}{Dt} = \frac{1}{c^2}\frac{Dp}{Dt} + \left(\nabla\cdot\underline{q} - \frac{\partial u_i}{\partial x_j}\tau_{ij}\right) \frac{1}{\rho T c^2} \left.\frac{\partial p}{\partial S}\right|_{\rho,Y_n}$$

$$+ \sum_{n=1}^{N} \frac{1}{c^2} \left\{ \frac{1}{T}\frac{\mu_n}{W_n} \left.\frac{\partial p}{\partial S}\right|_{\rho,Y_n} - \left.\frac{\partial p}{\partial Y_n}\right|_{S,\rho,Y_m} \right\} \frac{DY_n}{Dt} \tag{A5}$$

It can be shown by repeated application of the chain rule and using the enthalpy relationship

$$dh = T\,dS + \frac{1}{\rho}\,dp + \sum_{n=1}^{N} \frac{\mu_n}{W_n}\,dY_n \tag{A6}$$

that

$$\frac{1}{\rho T c^2} \left.\frac{\partial p}{\partial S}\right|_{\rho,Y_n} = \frac{\alpha}{c_p} \tag{A7}$$

and

$$\frac{1}{c^2}\left\{ \frac{1}{T}\frac{\mu_n}{W_n} \left.\frac{\partial p}{\partial S}\right|_{\rho,Y_n} - \left.\frac{\partial p}{\partial Y_n}\right|_{S,\rho,Y_m} \right\} = \frac{\rho\alpha}{c_p} \left.\frac{\partial h}{\partial Y_n}\right|_{\rho,p,Y_m} . \tag{A8}$$

Equation (A5) therefore simplifies to

$$\frac{D\rho}{Dt} = \frac{1}{c^2}\frac{Dp}{Dt} + \frac{\alpha}{c_p}\left\{ \sum_{n=1}^{N} \rho \left.\frac{\partial h}{\partial Y_n}\right|_{\rho,p,Y_m} \frac{DY_n}{Dt} + \nabla\cdot\underline{q} - \frac{\partial u_i}{\partial x_j}\tau_{ij} \right\} . \tag{A9}$$

This is Williams' equation (4-40) extended to include transport effects and is used in §2.

From the chain rule

$$\left.\frac{\partial h}{\partial Y_n}\right|_{T,p,Y_m} = \left.\frac{\partial h}{\partial Y_n}\right|_{\rho,p,Y_m} + \left.\frac{\partial h}{\partial \rho}\right|_{p,Y_n} \left.\frac{\partial \rho}{\partial Y_n}\right|_{T,p,Y_m} \tag{A10}$$

A rearrangement of this expression shows that

$$\sum_{n=1}^{N} \rho \left.\frac{\partial h}{\partial Y_n}\right|_{\rho,p,Y_m} \frac{DY_n}{Dt} = \sum_{n=1}^{N} \rho \left.\frac{\partial h}{\partial Y_n}\right|_{T,p,Y_m} \frac{DY_n}{Dt}$$

$$-\rho \left.\frac{\partial h}{\partial \rho}\right|_{p,Y_n} \left(\sum_{n=1}^{N} \left.\frac{\partial \rho}{\partial Y_n}\right|_{T,p,Y_m} \frac{DY_n}{Dt} \right) \tag{A11}$$

The sum on the left-hand side of (A11) appears in (A9). The first term on the right-hand side is $-Q$, where Q is the rate heat release per unit volume due to changes in species concentration. The term in brackets is the change in density at constant p and T due to changes in species concentration. For an ideal gas $p = R\, T\rho/W_o$, where R is the universal gas constant and W_o the average molecular weight. If the average molecular weight is constant

$$\sum_{n=1}^{N} \left.\frac{\partial \rho}{\partial Y_n}\right|_{T,p,Y_m} \frac{DY_n}{Dt} = 0$$

and (A11) reduces to

$$-\sum_{n=1}^{N} \rho \left.\frac{\partial h}{\partial Y_n}\right|_{\rho,p,Y_m} \frac{DY_n}{Dt} = Q \quad . \tag{A12}$$

Active Control of the Noise from a Rijke Tube

Maria Anna HECKL

Department of Engineering
University of Cambridge

Abstract

Instabilities caused by flames or other heat sources in resonators can be suppressed with a feedback system. It consists of a microphone inside the resonator, a phase-shifter, an amplifier and a loudspeaker at one end of the resonator. The pressure field produced by the loudspeaker changes the boundary conditions at the end of the resonator. This annuls the unstable oscillations in a phase range of up to 150 degrees if the amplification is chosen to be sufficiently high. A theory explaining the effect of the feedback system on the stability behaviour will be developed. Experiments using a Rijke tube driven by a hot gauze have been performed which confirm the predictions of the theory. In a practical situation with background noise, the feedback system achieves a noise reduction of over 40 dB.

1. Introduction

The Rijke tube is an 'organ-pipe resonator' with a heat source, e.g. a flame or a hot wire gauze, inside it. The transfer of heat energy from the heat source is coupled to the acoustic velocity of the air in the tube and under certain conditions a thermo-acoustic feedback can occur which gives rise to intense vibrations. This effect is termed thermo-acoustic instability. The stability behaviour can be explained by a balance equation for the acoustic energy in the tube. Acoustic energy is gained at the heat source and lost at the ends of the tube; there is stability if the energy loss is larger than the energy gain and instability if the energy loss is smaller than the energy gain.

The instability in a Rijke tube can be cancelled by the feedback system as shown in fig. 1. A microphone at one end of the tube picks up the sound which consists mainly of the

Aero- and Hydro-Acoustics
IUTAM Symposium Lyon 1985
Editors: G. Comte-Bellot and J. E. Ffowcs Williams

fundamental frequency of the tube. The microphone signal is filtered (so that only the fundamental frequency is let through), phase-shifted, amplified and fed into a loudspeaker at the other end of the tube.

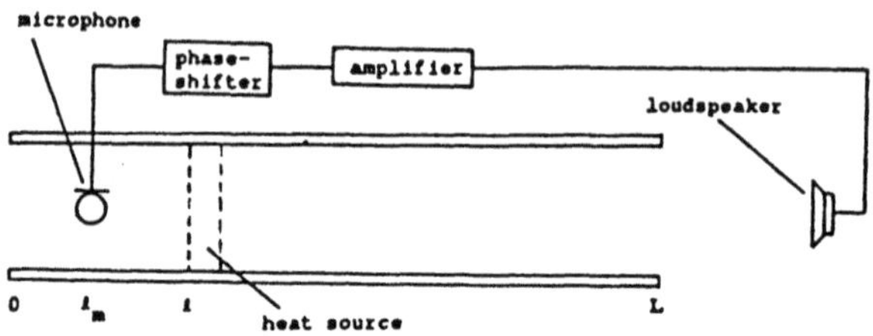

Fig. 1 Rijke tube with the feedback system

The loudspeaker at x=L superposes a pressure field on the original field. This changes the boundary condition at x=L and thus the energy loss at this end; if the phase-shift and amplification are chosen suitably, the energy loss at x=L can be increased to such an extent that the loss of acoustic energy at both ends surpasses the gain of acoustic energy so that stability occurs. The stability behaviour of the Rijke tube with the feedback system can be analysed from the eigenvalue problem.

2. Analysis of the Rijke tube with the feedback system

The heat source at $x=\ell$ and the loudspeaker near the end x=L produce a standing wave field in the tube (fig. 2) with the pressure

$$p(x)=\begin{cases} A\, e^{i\frac{\Omega}{c}(x-\ell)} + B\, e^{-i\frac{\Omega}{c}(x-\ell)} & \text{in } [0,\ell] \qquad (1a) \\ C\, e^{i\frac{\Omega}{c}(x-\ell)} + D\, e^{-i\frac{\Omega}{c}(x-\ell)} & \text{in } [\ell+\Delta,L] \qquad (1b) \end{cases}$$

and the velocity

$$u(x)=\begin{cases} \frac{1}{\rho c}\ (A\ e^{i\frac{\Omega}{c}(x-\ell)} - B\ e^{-i\frac{\Omega}{c}(x-\ell)}) & \text{in } [0,\ell] \qquad (2a) \\ \frac{1}{\rho c}\ (C\ e^{i\frac{\Omega}{c}(x-\ell)} - D\ e^{i\frac{\Omega}{c}(x-\ell)}) & \text{in } [\ell+\Delta,L] \qquad (2b) \end{cases}$$

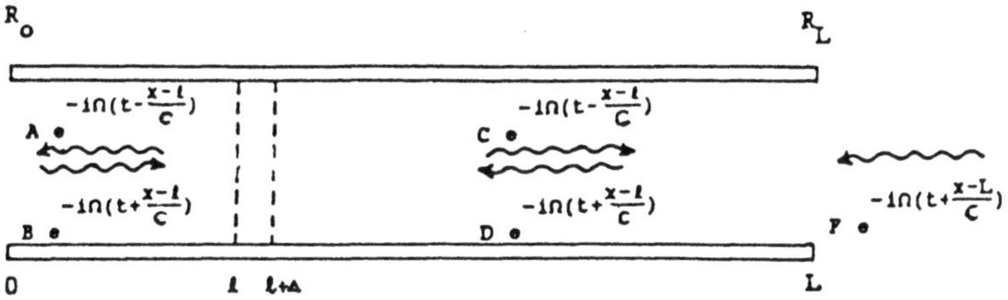

Fig. 2 Acoustic field in the Rijke tube

A complex notation is used with time dependence $e^{-i\Omega t}$; Ω is the complex frequency, c is the speed of sound and ρ is the mean density of the fluid. A, B, C and D are the complex amplitudes of the backward and forward travelling pressure waves in the two regions separated by the heat source, and F is the complex amplitude of the wave that comes from the loudspeaker. These amplitudes are related by the following conditions:

(1) at the tube ends -

The reflection coefficients R_o and R_L are the ratio of reflected and incident pressure wave at x=0 and x=L respectively if control is absent. The reflection at x=0 is described by

$$A\ e^{-i\frac{\Omega}{c}\ell} = R_o\ B\ e^{i\frac{\Omega}{c}\ell}\ . \qquad (3)$$

The internal wave approaching the end x=L is reflected with reflection coefficient R_L and superposed by the wave from the loudspeaker,

$$D\,e^{-i\frac{\Omega}{c}(L-\ell)} = R_L\,C\,e^{i\frac{\Omega}{c}(L-\ell)} + F\,. \qquad (4)$$

(2) across the control volume -

The heat source acts like a monopole with a volume outflow $\frac{\gamma-1}{\rho c^2}\,Q$ which causes a velocity jump

$$\frac{C-D}{\rho c} - \frac{A-B}{\rho c} = \frac{\gamma-1}{\rho c^2}\,\frac{Q}{S}\,; \qquad (5)$$

the pressure is the same on either side of the source,

$$A + B = C + D\,. \qquad (6)$$

Q is the rate of heat transfer, γ the ratio of specific heats and S the cross-sectional area of the duct.

(3) coupling of the heat transfer and the loudspeaker wave to the acoustic field -

The heat transfer is coupled to the acoustic field at the heat source by

$$Q = \beta\,e^{i\Omega\tau}\,\frac{A-B}{\rho c}\,, \qquad (7)$$

where β describes the strength of the heat source (e.g. for a hot wire gauze β depends mainly on the temperature of the wire) and τ is the time-lag between the heat transfer and the velocity. The wave F from the loudspeaker is coupled to the pressure at the microphone by

$$F = \alpha\,e^{i\phi}\,\left(Ae^{i\frac{\Omega}{c}(\ell_m-\ell)} + Be^{-i\frac{\Omega}{c}(\ell_m-\ell)}\right)\,, \qquad (8)$$

where α is the control gain and ϕ the phase difference, which both can be varied experimentally; ℓ_m is the position of the control microphone in the tube.

(3) to (8) form a set of 6 homogeneous equations for the unknowns A, B, C, D, F and Q which has solutions only if its

determinant is zero. The eigenfrequencies Ω can be calculated from this condition. The imaginary part of Ω is the growth rate, its sign gives the stability behaviour,

$$\begin{matrix}\text{stability} \\ \text{instability}\end{matrix} \quad \text{if} \quad \text{Imag}\Omega \lessgtr 0 \ .$$

3. Experimental and predicted results

The stability behaviour of a Rijke tube driven by a hot wire gauze was calculated for different α and ϕ; the results are given in the form of a stability map in fig. 3 which is a plot of the regions of stability in the $\alpha\phi$-plane. The optimum phase difference is ϕ=90 degrees achieving stability if α>0.034. The feedback system is not very sensitive to errors in phase-shift or amplification; stability occurs with any ϕ between 10 and 180 degrees so long as α is sufficiently large.

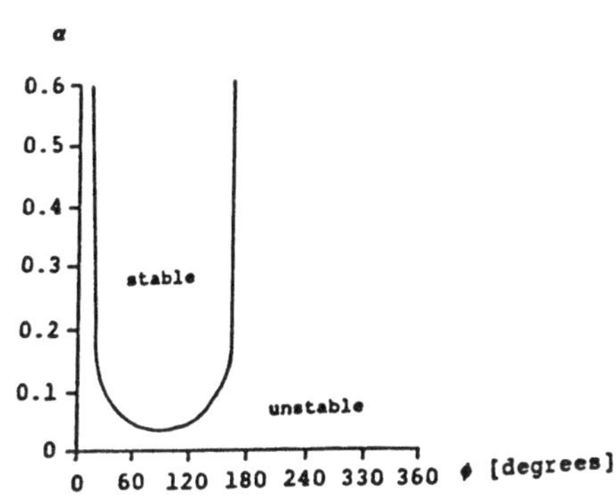

Fig. 3 Stability map of a Rijke tube with a hot wire gauze at ℓ=0.3 L

The theoretical and experimental results are listed in table 1. The sound pressure level was measured inside the tube. It did not become smaller than 96 dB because there was considerable background noise during the experiment.

ϕ [degrees]	$\delta[s^{-1}]$ (predicted, α=0.03)	SPL[dB] (measured, α=0.03)	$\delta[s^{-1}]$ (predicted, α=0.05)		SPL[dB] (measured, α=0.05)
0	3.22	140	3.34		142
30	1.75	136	0.80		117
60	0.63	128	-1.14	↑	96
90	0.15	116	-1.98	stability	96
120	0.45	131	-1.47	↓	96
150	1.45	136	0.27		128
180	2.89	140	2.76		140
210	4.37	143	5.34		143
240	5.49	144	7.30		146
270	5.96	145	8.10		147
300	5.65	145	7.55		147
330	4.65	144	5.81		146
---	3.06	140			

Table 1 Predicted growth rates and corresponding measured sound pressure levels; last line: results for α=0

Coherence Aspects of Acoustic Wave Transmission Through a Medium with Temperature Fluctuations

Ph. BLANC-BENON, S. CHAIZE, D. JUVE

Lab. Mécanique des Fluides et d'Acoustique, CNRS UA 263
Ecole Centrale de Lyon, BP 163, 69131 ECULLY CEDEX - France.

Abstract

The statistical properties of an ultrasonic wave that traverses a medium exhibiting temperature fluctuations are investigated. A heated grid in air is utilized to simulate the random thermal pattern typical of atmospheric and marine environments. The effect of temperature fluctuations on the attenuation of the transmitted pressure field and its mutual coherence is examined.

Experimental results are compared with theoretical estimates based on a solution of the stochastic Helmholtz equation in the parabolic or paraxial approximation. These comparisons suggest that the entire spectrum of the turbulence and the particular form of the incident wave (spherical or collimated) play significant roles.

Introduction

The propagation of sound waves through temperature fluctuations has been investigated under laboratory conditions by several authors e.g. Stone & Mintzer (1), Chotiros & Smith (2), Sederovitz & Favret (3), with a principal emphasis on the statistics of the wave amplitude fluctuations and the related dependence on frequency, range of propagation and turbulence characteristics.

The laboratory experiments employed a water tank in which the random temperature field was produced by a heater array mounted at the bottom. In this arrangement discrete impurities appear, such as minute air bubbles due to thermal degassing, and present an additional and not well controlled parameter, as pointed out by Neubert & Lumley (4). An installation with a grid in air offers advantages and has been chosen ; since anechoïc conditions may be easily realized around the set up continuous wave

Aero- and Hydro-Acoustics
IUTAM Symposium Lyon 1985
Editors: G. Comte-Bellot and J. E. Ffowcs Williams

signals may be used as well as the sound pulses which are exclusively generated in a finite tank.

Measurements in the atmosphere and ocean have also been effected by Daigle, Piercy & Embleton (5) and by Urick (6) ; in such circumstances, however, the uncertainties with regard to relevant environmental parameters, namely the velocity and temperature variations, make for difficulty in assessing their individual influences on wave attenuation or mutual coherence. This investigation was undertaken to clarify the latter aspects, based on a well controlled experimental facility. To facilitate the analysis and achieve practical features in the model, it is assumed that the acoustic wave length λ remains small compared with the integral scale L_T of the temperature field which, in turn, is smaller than the range of propagation X i.e. $X >> L_T >> \lambda$.

Theoretical estimates can then be based on the parabolic approximation which is well suited for the present conceptions and experimental conditions ; and the main objective is to contrast theory and experiment with a view to clarifying the influence of the exact shape of the acoustic beam as well as the entire spectrum of the thermal turbulence field.

Theoretical considerations

The propagation of a time harmonic acoustic wave in a random medium characterized by small temperature fluctuations ($T'/T_0 << 1$) and an integral turbulent scale L_T large in comparison with the acoustic wave length λ, is governed by the stochastic Helmholtz equation (Tatarski (7)),

$$(1)\qquad \begin{aligned} &(\Delta + k_o^2\,(1 + \varepsilon(\vec{r})))p(\vec{r}) = 0 \\ &\varepsilon(\vec{r}) = -\,T'(\vec{r})/T_o \end{aligned}$$

where $p(\vec{r})$ denotes the space dependence of the pressure, k_0 the wave number, $\varepsilon(\vec{r})$ the fluctuation in the index of refraction and $\vec{r} = (x,y,z)$. For paraxial transmission along the x-direction, it follows that $p(\vec{r})$ can be expressed as $p(\vec{r}) = U(\vec{r})\exp(ik_0x)$; after substitution into Eq. 1 and omission of the second derivative $\partial^2 U/\partial x^2$ one obtains,

$$2ik_o \frac{\partial}{\partial \mathbf{x}} U(x,\vec{\rho}) + \Delta_{\vec{\rho}} U(x,\vec{\rho}) + k_o \varepsilon(x,\vec{\rho})\, U(x,\vec{\rho}) = 0 \qquad x \geq 0$$

(2)

$$\Delta_{\vec{\rho}} = \partial^2/\partial y^2 + \partial^2/\partial z^2 \quad ; \quad \vec{\rho} = (y,z)$$

and the appropriate initial condition takes the form $U(0,\vec{\rho}) = U_0(\vec{\rho})$ at $x = 0$. The equation (2), of parabolic type, has been extensively studied (Tatarski (7), Ishimaru (8)) and can be solved for the statistical characteristics of the field if one assumes that the fluctuation $\varepsilon(\vec{r})$ is gaussian and delta-correlated in the x direction, i.e. :

$$(3) \qquad < \varepsilon\,(x,\vec{\rho})\ \varepsilon(\varsigma',\vec{\rho}\,') > = \delta(x - x')\ A\ (\vec{\rho} - \vec{\rho}\,')$$

where $A(\vec{\rho}_d = \vec{\rho} - \vec{\rho}\,')$ is related to the spectrum $\phi_\varepsilon(K)$ by the following two-dimensional Fourier transform

$$(4) \qquad A(\vec{\rho}_d) = 2\pi \int \phi_\varepsilon(K)\ \exp(i\vec{K}.\vec{\rho}_d)\ d\vec{K}$$

For an isotropic turbulence $A(\rho_d)$ can be expressed as :

$$(5) \qquad A(\rho_d) = 4\ \pi^2 \int_0^\infty K\ J_o(K\rho_d)\ \phi_\varepsilon\ (K)\ dK$$

In many practical cases $\phi_\varepsilon(K)$ may be approximated by the modified von Karman spectrum :

$$\phi_\varepsilon(K) = 0.033\ C_\varepsilon^2\ (K^2 + 1/L_o{}^2)^{-11/6}\ \exp\ (-K^2/Km^2)$$

(6)

$$C_\varepsilon^2 = 1.91\ (T'/2T_o)^2\ L_o^{-2/3} \quad ; \quad Km = 5.92/\ell_o$$

C_ε^2 is the structure constant, L_0 is the outer scale of turbulence related to the integral scale $L_T (L_0 = 1.339\ L_T)$ and l_0 is the inner scale.

The ensemble average of Eq. 2 provides an equation governing the coherent field $<U(x,\vec{\rho})>$:

$$(7) \qquad 2ik_o \frac{\partial}{\partial x} < U(x,\vec{\rho}) > + \Delta_{\vec{\rho}} < U(x,\vec{\rho})> + k_o^2 < \varepsilon(x,\vec{\rho}) U(x,\vec{\rho}) > = 0$$

Using the Furutsu-Novikov formula (9) to evaluate the term$<\varepsilon U>$ Tatarski transforms Eq. 7 into :

$$(8) \qquad (2ik_o \frac{\partial}{\partial x} + \Delta_{\vec{\rho}} + i\ \frac{k_o^3}{4} A(o))\ < U(x,\vec{\rho}) > = 0$$

and the appropriate solution is :

$$< U(x,\vec{\rho}) > = U_o(\vec{\rho})\ \exp\ (-\ \alpha x)$$

(9)

$$\alpha = k_o^2\ \frac{A(o)}{8} = 0.391\ c_\varepsilon^2\ k_o^2\ L_o^{5/3} = k_o^2\ (\frac{T'}{2T_o})^2\ L_T$$

We note that the attenuation of the coherent field depends explicitly on turbulent parameters L_T and T'/T_0 and that the transmitted field U becomes almost incoherent when x is large. In a similar way it can be shown that the mutual coherence function $\Gamma(x, \vec{\rho}_s, \vec{\rho}_d) = \langle U(x, \vec{\rho}_1)\, U^*(x, \vec{\rho}_2) \rangle$ ($\vec{\rho}_d = \vec{\rho}_1 - \vec{\rho}_2$, $\vec{\rho}_s = \frac{1}{2}(\vec{\rho}_1 + \vec{\rho}_2)$) obeys the equation :

$$(10) \quad \left(2ik_o \frac{\partial}{\partial x} + \nabla_{\vec{\rho}_s} \nabla_{\vec{\rho}_d} + i\,\frac{k_o^3}{2}\,(A(o) - A(\rho_d))\right) \Gamma(x, \vec{\rho}_s, \vec{\rho}_d) = 0$$

with an initial condition $\Gamma(0, \vec{\rho}_s, \vec{\rho}_d) = \Gamma_0(\vec{\rho}_s, \vec{\rho}_d)$ at $x = 0$. An exact solution of Eq. 10 is available for both plane and spherical waves ; in the case of a beam the solution assumes a particular simple form at points of observation, in any transverse plane, which are equidistant from the axis (Fante (10)). The useful expressions of the transverse coherence function $\Gamma(x, \rho_d)$ for the present investigation are therefore :

$$(11) \quad \Gamma(x,\rho_d) = \frac{1}{x^2} \exp\left(- 4\pi^2 k_o^2 \int_o^x du \int_o^\infty \left(1 - J_o\left(K\frac{u}{x}\rho_d\right)\right) \phi_\varepsilon(K)\,KdK\right)$$

for a spherical wave and,

$$(12) \quad \begin{aligned} \Gamma(x,\vec{\rho}_d) &= \frac{w_o^2}{8\pi} \iint_{-\infty}^{+\infty} \exp(-A\rho_d^2 - BK_d^2 + C\vec{K}_d \cdot \vec{\rho}_d - H)\, d\vec{K}_d \\ H &= \frac{k_o^2}{4} \int_o^x du \int_o^\infty \left(1 - J_o\left(K\left|\vec{\rho}_d - \frac{u}{k_o}\vec{K}_d\right|\right)\right)\phi_\varepsilon(K)\,KdK \end{aligned}$$

$A = 1/2w_o^2$, $B = w_o^2\,(1 + \alpha^2 x^2)/8$, $C = \alpha x/2$, $\alpha = \lambda/\pi w_o^2$

for a collimated gaussian beam $U_0(\rho) = \exp(-\rho^2/w_0^2)$.

Experimental arrangement

The heated grid as well as the locations of the acoustic transmitter and receiver are sketched in Fig. 1. The grid consists of a biplane arrangement of conductors with a square mesh of 9 cm scale. The overall dimensions of the grid are 1.1 m x 2.2 m and the maximum power consumption equals 32 kW. The grid was placed horizontally in a large anechoïc room (11) and the mixing of the free convection plumes above it generates the turbulent thermal field. The acoustic propagation measurements were made at the height H = 1.75 m which corresponds to 20 times the scale of mesh. The mean temperature rise above ambiant was 27°C and the temperature fluctuations had a rms value of $1.7\ 10^{-2}$. The homogeneity of the thermal field was achieved within 0.5°C excluding 1.5 meshes near the edges. The one dimensional spectrum of T' was measured with a FFT analyser Nicolet 660A, in the range .5-200 Hz with a constant bandwidth of .5 Hz. Frequencies were converted into wave numbers K1 by a Taylor hypothesis based on the mean upward velocity (1.25 m/s measured with a hot wire). The results are given in Fig. 2 and compared with the one dimensional spectrum deduced from the modified von Karman form (Eq. 6) by $F_{T'}(K_1) = \int_{K_1}^{\infty} K\ \phi_\varepsilon(K)\ dK$. The integral scale L_T deduced from the spectrum or by integration of spatial correlation equals 7.6 cm. The inner scale l_0 related to the high frequency cutoff of the spectrum was estimated to be 0.1 cm. The index changes which may be induced by the velocity fluctuations in the x direction prove to be negligible ; indeed an upper limit is given by u'_z/c (u'_z component in the upward direction) and measurements indicate that $u'_z/c = 6.10-4$.

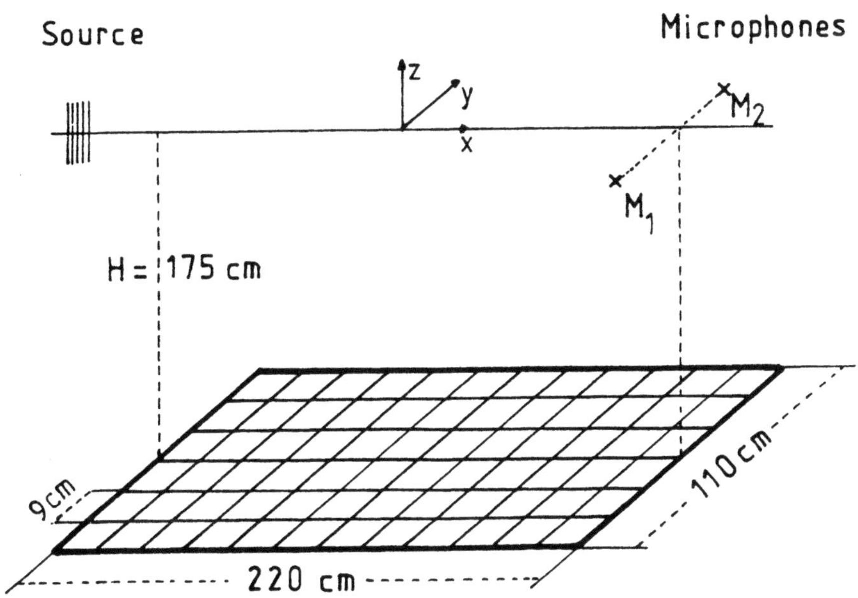

Fig. 1 - Experimental set up.

The spherical acoustic waves were generated by TDK ultrasonic sources (f = 23.4 kHz, 39 kHz) and the collimated beam by a Sell home made transducer whose area is 100 cm^2 (Blanc-Benon

(12)). Frequency is adjustable in the range 20-100 kHz. The transmitted signals were received on 1/4" microphones (Bruel & Kjaer 4135) located in a normal plane to the x-axis. The coherent part was obtained by cross correlating the incident wave with the transmitted wave. The transverse coherence functions were measured over a large range of $\rho_d (0 < \rho_d < 3\ L_T)$. A Nicolet 660A FFT analyser as well as a HP 3721A correlator were used in these measurements. The various distances of propagation through turbulence in the range 0.5 m < x < 2.2 m were obtained by a longitudinal displacement of the microphones, the whole area of the grid being kept heated to insure the homogeneity of the thermal field.

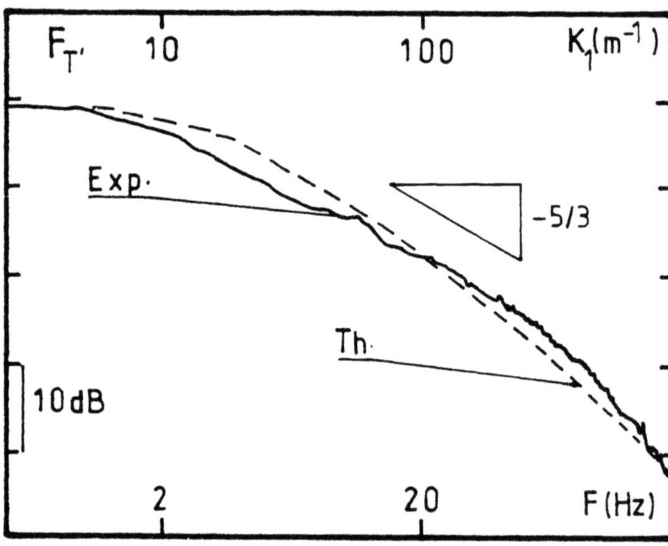

Fig. 2 - Comparison between the experimental (one dimensionnal) spectrum of temperature fluctuations, $F_{T'}$, and the theoretical spectrum deduced from

$$F_{T'}(K_1) = \int_{K_1}^{\infty} K\ \phi_\varepsilon(K)\ dK$$

with the von Karman function $\phi_\varepsilon(K)$.

Results

1. Attenuation of the coherent wave

The experimental results concerning <U> are displayed in Fig.3, which correlates the values of log $(\langle U\rangle/U_0)$ with αx. An exponential rise in attenuation is closely realized on the range

$0 < \alpha x < 4$; the significant and nearly uniform attenuation on the range $\alpha\pi > 4$ cannot be accounted for with the parabolic approximation.

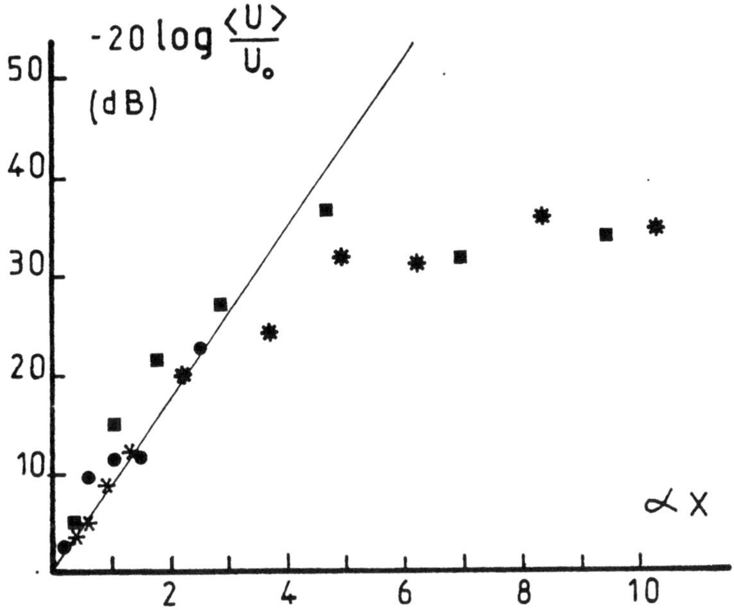

Fig. 3 - Attenuation of the coherent field <U> : a collimated beam (20 kHz $< f <$ 100 kHz) for three ranges of propagation (● x = 54 cm ; ■ x = 103 cm ; ✱ x = 220 cm) and a spherical wave ✱ f = 23.5 kHz for x = 32 cm ; x = 56 cm ; x = 81 cm ; x = 131 cm.

2. Transverse coherence functions

In Fig. 4 the measured values of the coherence function for the spherical wave are plotted in terms of the receiver separation ρ_d for two ranges of propagation (X = 110 cm and X = 220 cm). The solid curves are computed from Eq. 11, employing a modified von Karman spectrum (Eq. 6), and the dotted curves are obtained from Eq. 11 but with a Kolmogorov spectrum $\phi_\varepsilon = 0.033\ C_\varepsilon^2\ K^{-11/3}$. We note that the coherence functions decrease rapidly as the propagation length and frequency increase. The transverse coherence function is accurately described by a von Karman spectrum, which takes into account the outer scale of turbulence.

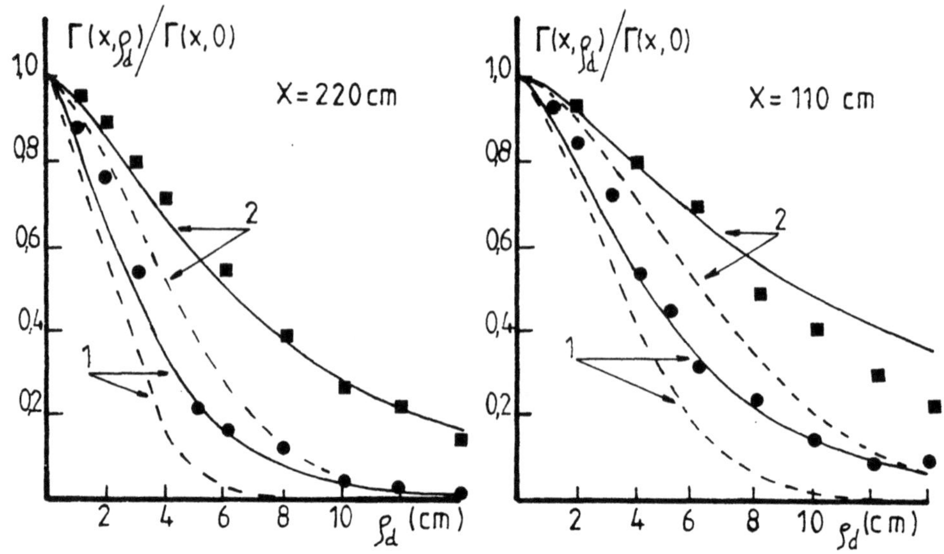

<u>Fig. 4</u> - Transverse coherence functions for a spherical wave (● f = 39 kHz ; ■ f = 23.5 kHz) Numerical estimates with a modified von Karman spectrum (———) and a Kolmogorov spectrum (----------).

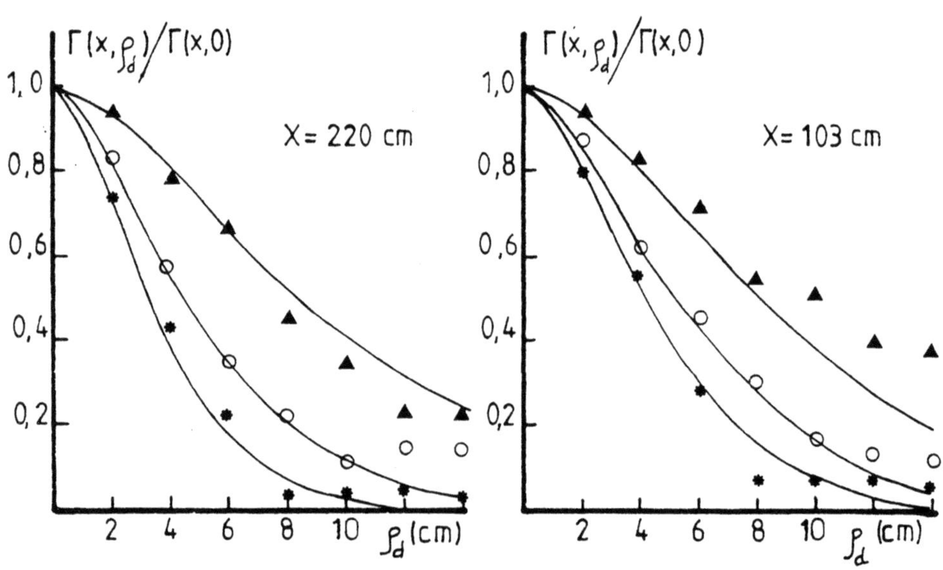

<u>Fig. 5</u> - Transverse coherence functions for a collimated beam (✱ f = 50 kHz, w_0 = 3.3 cm / ○ f = 39 kHz, w_0 = 4.1 cm / ▲ f = 23.5 kHz, w_0 = 6.6 cm). Numerical estimates with a modified von Karman spectrum (———).

The experimental results for a collimated beam are given in Fig. 5 ; here the solid curves, obtained from Eq. 12 with a modified von Karman spectrum, reveal a good accord with measurements on the range $0 < \rho_d < 1.5\,L_T$. For large values of ρ_d the measured transverse coherence increases slighthly, owing to the fact that the actual beam, with side lobes, differs from the gaussian model which is adapted in the analysis.

Conclusion

The effect of temperature fluctuations on the propagation of acoustic waves is revealed with carefully controlled experiments involving a heated air grid. The relevance of the entire spectrum of turbulence has been demonstrated and, in particular, for the transverse coherence function. It appears that the reduction in coherence cannot be estimated by a single turbulence scale, and that a satisfactory estimate is achieved with a modified von Karman spectrum. The differences which arise in connection with different forms of the incident wave necessitate separate attention.

Acknowledgments

This work was supported by the Direction des Recherches Etudes et Techniques under contract n° 83/226.

References

1. Stone, R.G.; Mintzer, D. ; "Range dependence of acoustic fluctuations in a randomly inhomogeneous medium" J. Acous. Soc. Am. 34, pp. 647-653 (1962).
2. Chotiros, N.P. ; Smith, B.V. ; "Sound amplitude fluctuations due to a temperature microstructure" J. Sound Vib. 64 (1979) 349-369.
3. Sederowitz, W.J. ; Favret, A.G. ; "Covariance between acoustic signals traveling diverging paths in a random medium" J. Acous. Soc. Am. 45 (1969) 386-391.
4. Neubert, J.A. ; Lumley, J.L. ; "The nature of sound fluctuation in the upper ocean II" J. Acous. Soc. Am. 64 (1978) 1148-1158.
5. Daigle, G.A. ; Piercy, J.E. ; Embleton, T.F.W. ; "Line-of-sight propagation through atmospheric turbulence near the ground" J. Acoust. Soc. Am. 74 (1983) 1505-1513.

6. Urick, R.J. ; "Sound propagation in the sea" Defense Advanced Research Projects Agency. US Government Printing Office Washington, 1979.

7. Tatarski, V.I. ; "The effects of turbulent atmosphere on wave propagation" IPST Keter Press Jerusalem 1971.

8. Ishimaru, A. ; "Wave propagation and scattering in random media" Academic Press New-York Vol. 2, 1978.

9. Novikov, E.A. ; "Functionals and the random force method in turbulence theory" Soviets Physics JETP 20-5 (1965) 1290-1294.

10. Fante, R.L. "Mutual coherence function and frequency spectrum of laser beam propagating trough atmospheric turbulence" J. Opt. Soc. Am. 64 (1974), 592-598.

11. Sunyach, M. ; Brunel, B. ; Comte-Bellot, G. ; Performances de la soufflerie anéchoïque à grandes vitesses de l'Ecole Centrale de Lyon" Revue d'Acoustique 73 (1985) 316-330.

12. Blanc-Benon, Ph. ; "Effet d'une turbulence cinématique sur la propagation des ondes acoustiques" Thèse de Docteur Ingénieur, ECL 81-02, 1981.

Computational Acoustics

Numerical Methods in Acoustics

Sébastien M. Candel

Office National d'Etudes et de Recherches Aérospatiales,
BP 72, 92322 Châtillon Cedex, FRANCE
and ER 288-CNRS, Ecole Centrale des Arts et Manufactures,
92290 Châtenay Malabry

Abstract

This paper presents a survey of some computational techniques applicable to acoustic wave problems. Recent advances in wave extrapolation methods, spectral methods and boundary integral methods are discussed and illustrated by specific calculations.

1 - INTRODUCTION

This paper presents a survey of some computational techniques which may be used to solve acoustic wave problems. The review is not exhaustive and mainly concerns some recent advances in the following three areas:

(1) Wave extrapolation methods (WEM)

(2) Spectral methods (with emphasis on integral transform techniques)

(3) Boundary integral methods (BIM)

Other methods like those based on the parabolic and geometrical approximations as well as finite difference schemes are considered in a previous review of Candel, 1983. The present report may be considered as a complement to this reference. A different point of view and additional informations are contained in an interesting review paper written by Chin, Hedstrom and Thigpen, 1984, dealing with numerical methods in seismology. We begin by discussing some wave extrapolation methods (section 2). An illustration of WEM is given by treating the problem of radial extrapolation of a near field to the far field. We next consider some advances in spectral methods and concentrate on integral transform methods (section 3). Many studies in this area concern axisymmetric wave fields. The Fourier-Bessel (Hankel) transform and the H_0^1 transform play a central role in this case. Computational techniques for these transforms are re-

Aero- and Hydro-Acoustics
IUTAM Symposium Lyon 1985
Editors: G. Comte-Bellot and J. E. Ffowcs Williams

viewed. A test problem taken from underwater sound is used to underline the possibilities of this type of method. The final section concerns boundary integral methods. These methods are well suited to radiation and diffraction problems and they may be used in combination with finite element methods to solve coupled acousto-elastic problems. We restrict our discussion to some diffraction problems and illustrate one boundary integral method with calculations of the acoustic field scattered by a sphere and by the structure of the Ariane IV launcher.

2 - WAVE EXTRAPOLATION METHODS

Wave extrapolation methods have been developed in many areas. They are useful in seismic reflectivity migration, sound source reconstruction by back projection, acoustic imaging, transducer radiation pattern measurement, scattering cross section determination... Wave extrapolation may be used to deduce the far field radiation of an ensemble of sources from near field data or to obtain the distribution of normal acceleration of the surface of a transducer from field measurements. We shall here describe the last two applications. In the first example sketched in Fig. 1, a circular radiator is located in the z = 0 plane. We assume for simplicity that the normal acceleration at the radiator surface is axisymmetric and monochromatic

$$\ddot{w}(r,\theta,0) = \ddot{w}(r,0) \tag{1}$$

The pressure field induced by this radiator in the half-space $z>0$ satisfies the Helmholtz equation

$$\nabla^2 p + k^2 p = 0 \qquad\qquad k = \omega/c \tag{2}$$

On the plane z = 0 the radiator imposes its normal acceleration and hence

$$\rho_0 \ddot{w}(r,0) = -(\partial p/\partial z)_{z=0} \tag{3}$$

Now let $\tilde{\ddot{w}}$ and $\tilde{p}$ designate the Fourier-Bessel (Hankel) transforms of $\ddot{w}$ and p

$$\tilde{p}(\varsigma,z) = \int_0^\infty p(r,z) J_0(\varsigma r)\, r\, dr \tag{4}$$

$$\tilde{\ddot{w}}(\varsigma,0) = \int_0^\infty \ddot{w}(r,0) J_0(\varsigma r)\, r\, dr \tag{5}$$

It is easily shown that the transformed pressure and acceleration are related by

$$\tilde{p}(\varsigma,z) = i\rho_0 \tilde{\ddot{w}}(\varsigma,0)\, e^{i\gamma z}/\gamma \tag{6}$$

where $\gamma = (k^2-\varsigma^2)^{1/2}$ for $|\varsigma| \leq k$ and $\gamma = i|\varsigma^2-k^2|^{1/2}$ for $|\varsigma| > k$.

Expression (6) may be used to deduce the pressure field corresponding to a given normal acceleration. It also provides a method for the determination of the normal acceleration on the radiator. A formal expression for $\ddot{w}$ is obtained by inverting (6) :

$$\ddot{w}(r,0) = H\{(\gamma/i\rho_o)\, e^{-i\gamma z}\, H[p(r,z)]\} \qquad (7)$$

where H{ } designates the Fourier-Bessel transform. According to this expression the pressure data measured at a distance z from the radiator is first Fourier-Bessel transformed. The result is then multiplied by the back projection operator $(\gamma/i\rho_o)\, e^{-i\gamma z}$. A second Fourier-Bessel transform then provides the surface acceleration distribution. In practice the back projection operator takes exponentially large values for $|\zeta| > k$. These values correspond to the propagation of evanescent (exponentially decaying) components of the spectrum. It is usually not possible to back project this part of the spatial spectrum. Hence the back projection operator must be limited to the wavenumbers ζ belonging to [-k , +k]. The method just described is exploited by Guyomar, 1982, in a theoretical and experimental analysis of ultrasonic transducers. It is used in a more recent study of Medeiros and Stepanishen, 1984, to determine the velocity distribution on a circular vibrator. These authors indicate that expression (7) may be replaced by an approximate band-limited relation if the pressure is measured in the far field. In this situation one makes use of

$$p(R,\theta) = (e^{ikR}/R)\rho_o \tilde{\ddot{w}}(\zeta,0)\big|_{\zeta = k\sin\theta} \qquad (8)$$

where $R = (r^2 + z^2)^{\frac{1}{2}}$ and $\theta = \sin^{-1}(r/R)$

It follows that

$$\tilde{\ddot{w}}(\zeta,0) = (1/\rho_o)\, R e^{-ikR}\, p[R, \sin^{-1}(\zeta/k)] \qquad (9)$$

for $-k \leqslant \zeta \leqslant k$

and hence

$$\ddot{w}(r,0) = \int_0^k (1/\rho_o)\, R e^{-ikR}\, p[R, \sin^{-1}(\zeta/k)] J_o(\zeta r)\zeta\, d\zeta \qquad (10)$$

Evaluation of the Fourier-Bessel transforms arising in expression (7) or (10) may be performed by applying classical integration rules or more specific algorithms reviewed in the next section. Results presented by Guyomar and by Medeiros and Stepanishen clearly indicate that the method is of practical interest.

As a second problem we now consider radial wave extrapolation between the near field and far field of a volume of acoustic sources. Extrapolation of three-dimensional fields usually requires initial data on a two-dimensional aperture. Arrays of 16 x 16 or 64 x 64 transducers have been used in certain cases. Such large arrays are expensive and pose difficult technical problems such as signal conditionning, data multiplexing, transducer calibration. If however the wave field is axisymmetric or nearly so, extrapolation may be performed with great economy from data recorded on a single sideline. This possibility is explored by Candel and Chassaignon, 1984, and Blacodon and Candel, 1984. These authors show that the method remains applicable even when the field is not quite axisymmetric and contains a certain amount of low order azimuthal components. Recent (unpublished) work of Blacodon and Candel indicates that it is possible to replace the transducer array by a synthetic aperture obtained by scanning a sideline with a single probe. This last procedure, when it is applicable, brings a tremendous simplification of the experimental set-up and data collection equipment.

Consider the configuration represented in Fig. 2. A compact volume of acoustic sources radiates a monochromatic wave field. We assume for the moment that the field $p(x,r_o,\theta)$ on a cylinder of radius r_o is known for all x and θ and we wish to extrapolate from this data, the field at a distance r, $p(x,r,\theta)$. Now, let $\tilde{p}(k_x,r,\theta)$ designate the Fourier transform of $p(x,r,\theta)$ and let $\tilde{p}_m(k_x,r)$ represent the coefficients of the Fourier series decomposition of p:

$$p(x,r,\theta) = \frac{1}{2\pi}\int \tilde{p}(k_x,r,\theta)\, e^{ik_x x}\, dk_x \tag{11}$$

and

$$\tilde{p}(k_x,r,\theta) = \sum_{m=-\infty}^{+\infty} \tilde{p}_m(k_x,r)\, e^{im\theta} \tag{12}$$

Each Fourier component $\tilde{p}_m$ satisfies a transformed Helmholtz equation

$$\frac{d^2\tilde{p}_m}{dr^2} + \frac{1}{r}\frac{d\tilde{p}_m}{dr} - \frac{m^2}{r^2}\tilde{p}_m + (k^2 - k_x^2)\tilde{p}_m = 0 \tag{13}$$

outside the source volume.

It is then easily shown that the Fourier components $\tilde{p}_m$ at two distances r_o and r from the x-axis are related by

$$\tilde{p}_m(k_x,r) = \frac{H_m^1(\mu r)}{H_m^1(\mu r_o)}\, \tilde{p}_m(k_x,r_o) \tag{14}$$

where $H_m^{\ 1}$ is the Hankel function of order m and first kind and

$\mu = |k^2 - k_x^2|^{1/2}$ for $k_x \leq k$, $\mu = i|k_x^2 - k^2|^{1/2}$ for $k_x > k$

The previous expression defines a wave extrapolator for the m-th azimuthal component

$$G_m(k_x, r_0, r) = \frac{H_m^1(\mu r)}{H_m^1(\mu r_0)} \tag{15}$$

In the most general three-dimensional situation extrapolation consists of (1) decomposing the data recorded on the cylinder $r = r_0$ in Fourier components $\tilde{p}_m$ (2) multiplying each component by the suitable extrapolator G_m and (3) synthesizing the wave field at distance r from the results obtained in step 2. Mathematically this procedure may be represented by the following set of expressions

$$p(x, r, \theta) = \frac{1}{2\pi} \sum_{m=-\infty}^{+\infty} e^{im\theta} \int e^{ik_x x} G_m(k_x, r_0, r) \tilde{p}_m(k_x, r_0) dk_x \tag{16}$$

where

$$\tilde{p}_m(k_x, r_0) = \frac{1}{2\pi} \int_0^{2\pi} \int p(x, r_0, \theta) e^{-ik_x x - im\theta} dx\, d\theta \tag{17}$$

Now if the radiated field is axisymmetric the procedure becomes much simpler. The only non zero Fourier component is in that case $m = 0$ and it is possible to write

$$\tilde{p}(k_x, r) = G_0(k_x, r_0, r) \tilde{p}(k_x, r_0) \tag{18}$$

The initial data is extrapolated by taking its Fourier transform multiplying the transformed data by the zero-th order extrapolator

$$G_0(k_x, r_0, r) = H_0^1(\mu r) / H_0^1(\mu r_0) \tag{19}$$

and taking the inverse transform of the result. This procedure is represented in compact form by

$$p(x, r) = F^{-1}\{ G_0(k_x, r_0, r) F[p(x, r_0)] \} \tag{20}$$

where F and F^{-1} respectively designate direct and inverse Fourier transforms. A further simplification is obtained by replacing the exact extrapolator G_0 by the asymptotic operator obtained by expanding the Hankel functions $H^1{}_0(\mu r)$ and $H^1{}_0(\mu r_0)$:

$$G_0 \simeq N_0 = (r_0/r)^{1/2} \exp[i\mu(r - r_0)] \tag{21}$$

It is shown by Candel and Chassaignon, 1984, that this approximation is justified if $kr_0 \gtrsim 2$. It is also shown that the asymptotic kernel N_0 constitutes a good approximation for the m-th extrapolator G_m provided that $kr_0 \gtrsim 4m + 2$ or equivalently if $(r_0/\lambda) \gtrsim (2m + 1)/\pi$.

Now if the wave field only contains low order azimuthal components i.e. if $\tilde{p}_m \simeq 0$ for $m > Q$, then the field may be extrapolated with the simplified procedure

$$p(x, r, \theta) \simeq F^{-1}\{ N_o(k_x, r_o, r)\, F[p(x, r_o, \theta)] \} \qquad (22)$$

provided that $kr_o \gtrsim 4(Q + \frac{1}{2})$

This procedure is now illustrated with two examples taken from Blacodon and Candel, 1984. The first case shown in Fig. 3 corresponds to a numerical simulation. A simulated acoustic dipole is placed at a distance $r_o = \lambda$ from an array of 27 transducers. The length of this array is $d = 9\lambda$. The initial data is first augmented by inserting three additional virtual transducers between each pair of "real" transducers. This is achieved with a nonlinear interpolation scheme. This augmented initial data is then extrapolated and the result is plotted in Fig. 3. The extrapolated field p(x,r) closely follows the exact field p(x,r) corresponding to the dipole source. Windowing and aliasing errors are in this case and for the range considered, quite small.

The second example corresponds to an experiment sketched in Figure 4. The recording array extends over a length d = 1.30 m. The source is a driver unit JBL 2420 placed at r_o = 0.70 m and radiating a 1/3 oct band centered at 2 kHz. The sampling array comprises 27 Bruel and Kjaer 4135 (1/4 inch) microphones. The signals are digitized at a sampling rate of 13 kHz and Fourier transformed by blocks of 256 points (Δf = 51 Hz). For each time slice the initial data is formed of 27 complex samples corresponding to a particular frequency in the spectrum. Extrapolation is performed in each time slice and the results are averaged coherently to obtain the wave fronts and incoherently to obtain the field amplitude. The results displayed in Figure 5 are averages of a set of 50 extrapolated wave fields. The wave fronts are correctly centered on the source but the field amplitude exhibits oscillations related to an incomplete sampling of the initial field. In fact this field is not vanishingly small outside the recording array and hence truncation effects are not negligible. Nevertheless the results indicate that the method is indeed applicable to the processing of nearfield data.

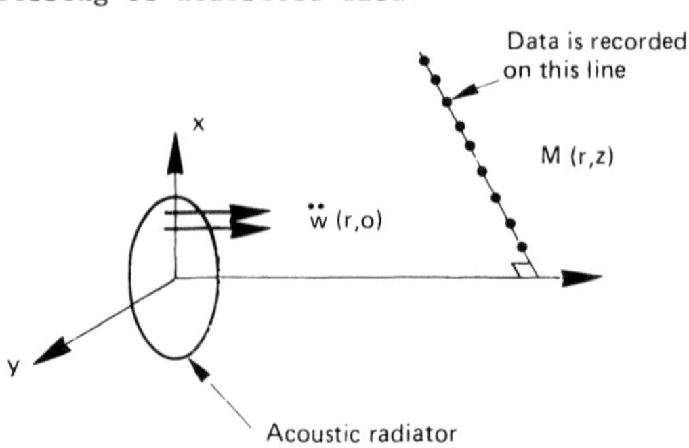

Fig. 1 - Geometry of the back-projection problem. A circular radiator is located in the z = 0 plane. The normal acceleration on the surface of this radiator is axisymmetric $\ddot{w}(r,\theta,o) = \ddot{w}(r,o)$.

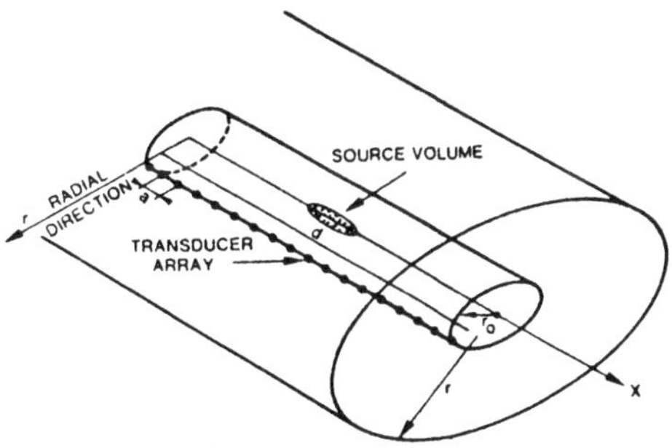

Fig. 2 - Geometry of the wave extrapolation problem. A sampling array is located at a distance r_o from the x-axis. The wave field is extrapolated in the radial direction.

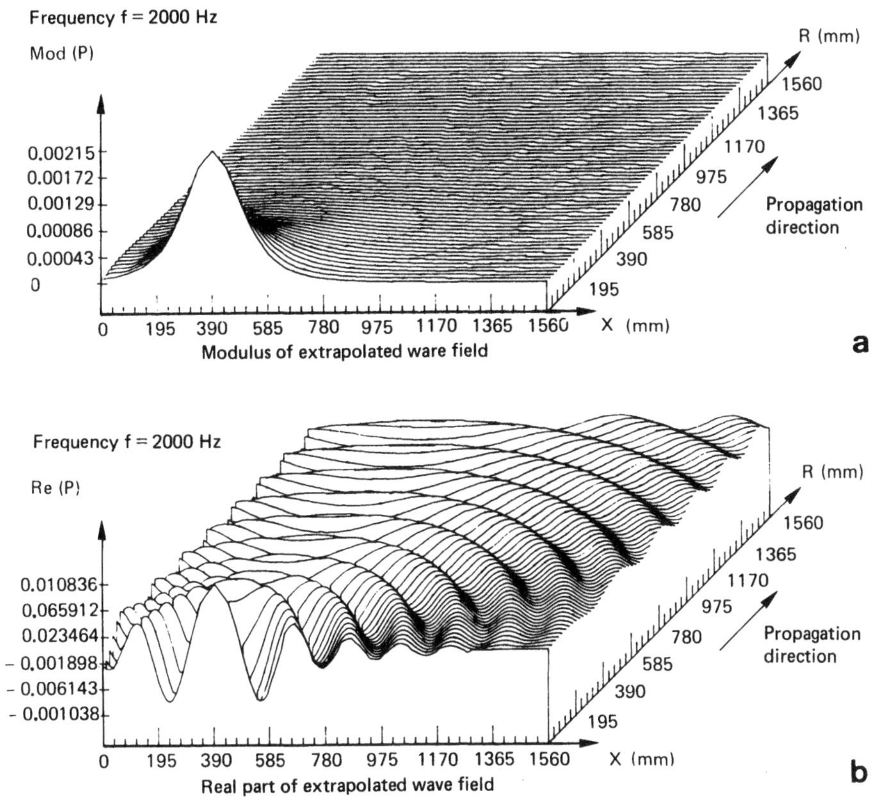

Fig. 3 - Extrapolated wave field (numerical simulation). A dipole source is placed in the vicinity of an array comprising 27 uniformly distributed transducers. The source is at $r_o = \lambda$, the array extends over $d = 9\lambda$. (a) Modulus of the field. (b) Real part.

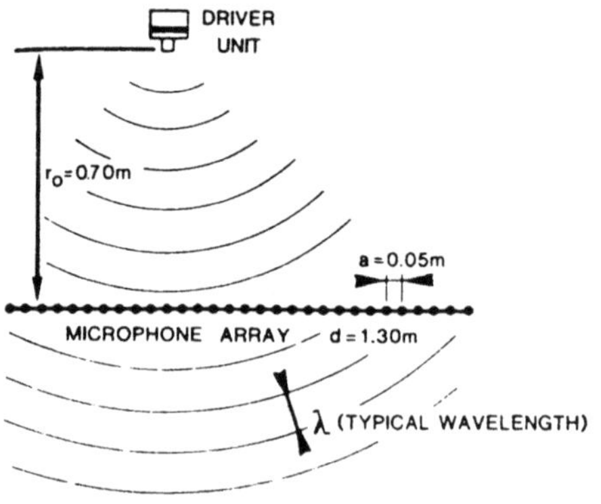

Fig. 4 - Experimental configuration.

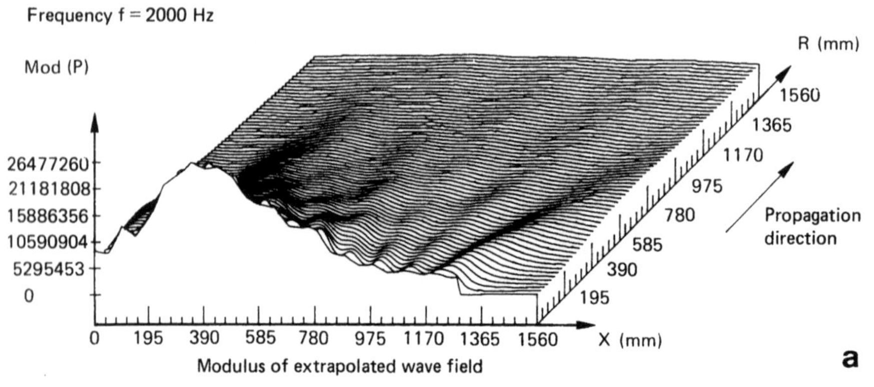

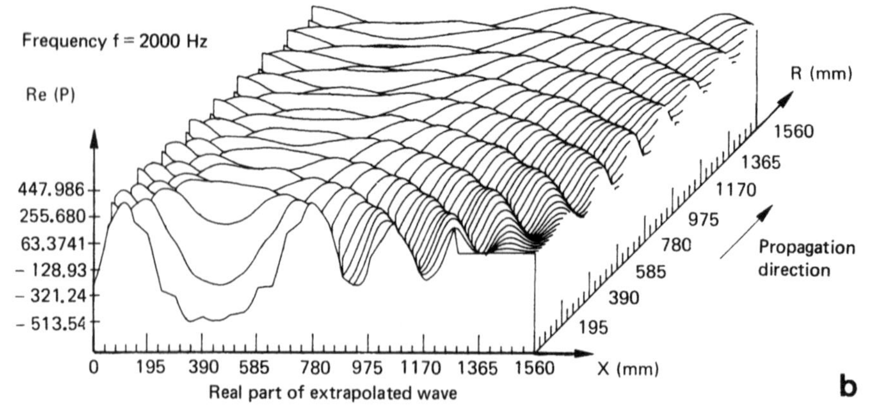

Fig. 5 - Extrapolated wave field from data recorded in the experiment of Fig. 4. The recording array comprises 27 transducers. The driver unit is at $r_o = 4.11\lambda$. The array length is $d = 7.64\lambda$. (a) Modulus of the field. (b) Real part.

3 - SPECTRAL METHODS

Spectral methods are extensively used in fluid mechanics and most notably in direct simulations of turbulence. Methods based on spectral techniques have also found many applications in acoustics and seismic data processing. Calculations based on Fourier representations are performed in seismic reflectivity migration, wave extrapolation and back projection determinations of vibrator nearfields. Many papers dealing with such problems are cited in Candel, 1983. A recent contribution belonging to this class is that of Stepanishen and Chen, 1984 concerning the nearfield and intensity of cylindrical vibrators. Another interesting study of Bojarski, 1984, deals with the scattering by a cylinder. This author shows that this classical problem may be solved with three FFT and a few additional calculations. This original method requires about 20 lines of computer code.
In circumstances where the wave field is axisymmetric the Fourier-Bessel (Hankel) transform advantageously replaces a two-dimensional Fourier transform. This fact has been recognized and exploited by many authors. Di Napoli and Deavenport, 1979,1980, use an asymptotic evaluation of the Fourier-Bessel transform in a direct numerical determination of the Green's function in a multilayered medium. These authors are at the origin of a code designated as the fast Field Programm (FFP) which allows fast calculations of the sound field radiated by an underwater source. The Fourier-Bessel transform is used by Frisk, Oppenheim and Martinez, 1980, in an attempt to determine the reflection coefficient of the ocean floor from pressure field data. Frisk and Lynch, 1984, define a related method, based on the same transform, which may be used to characterize the shallow water Green's function. Candel and Brun, 1982, deal with the point source radiation from a semi-infinite elastic solid loaded by a light fluid. The problem is solved with a direct numerical evaluation of the Fourier-Bessel transform of the spatial spectrum describing the field. Mook, Frisk and Oppenheim, 1984, give a detailed analysis of difficulties associated with the direct evaluation of fields based on numerical Hankel transforms. These authors propose a combination of numerical and analytic methods for the computation of fields in stratified media. Raspet et al., 1985, use the FFP technique to calculate the sound propagation in a layered atmosphere above a flat ground characterized by a constant impedance. In parallel with these studies a notable effort has been devoted to the development of fast algorithms for the calculation of the Fourier-Bessel transforms and related integrals. A summary of the main references concerned with this aspect is given in Table 1. Most studies pertain to the zero

order Fourier-Bessel (Hankel) transform

$$F_0(r) = \int_0^{\infty} \zeta f(\zeta) J_0(\zeta r) d\zeta \tag{23}$$

This is because many integral representations used in acoustics are based on this transform. However it is a simple matter to show that such integral representations are not the most appropriate in situations where the field radiates outwards; this aspect is easily understood if one considers the behavior at infinity of the transform kernel

$$J_0(\zeta r) \sim \left(\frac{2}{\pi \zeta r}\right)^{1/2} \cos\left(\zeta r - \frac{\pi}{4}\right) = \frac{1}{(2\pi\zeta r)^{1/2}}\left[e^{i(\zeta r - \pi/4)} + e^{-i(\zeta r - \pi/4)}\right] \tag{24}$$

The kernel contains an outgoing and an incoming cylindrical wave. Only the outgoing wave is actually useful in the synthesis of the radiating field. As a consequence a straight forward application of a numerical Fourier-Bessel (Hankel) transform may introduce non physical converging wavelets in the calculation of a radiating field. The interference between converging and diverging components produces a highly oscillatory result. This phenomenon is visible for instance in some of the calculations of Mook, Frisk and Oppenheim, 1984. Note however that these authors give a different interpretation of their results.

In any event it appeared to us that some advantage could be obtained by using H^1_0 integral representations in calculations of radiation problems. Such representations are formally equivalent to those based on the J_0 transform but they are much easier to handle in direct numerical calculations. The H^1_0 - transform is defined by

$$F_H(r) = \int_{-\infty}^{+\infty} \zeta f(\zeta) H_0^1(\zeta r) d\zeta \tag{25}$$

A dual algorithm has been developed to calculate this integral. The method parallels those used in the Fourier-Bessel transform case. It is fast and sufficiently accurate for most practical purposes. The details of this algorithm may be found in Candel et al., 1985a,b. This algorithm is here illustrated with an example typifying situations arising in certain wave propagation problems.

Consider the function

$$f(\zeta) = a/(\zeta - a) \tag{26}$$

This function has a real pole located on the integration path. This feature is commonly encountered in many guided wave problems.

Now, the H^1_0 - transform of $f(\zeta)$ is easily evaluated analytically:

$$F_H(r) = i\pi a^2 H_0^1(ar) \qquad r > 0 \tag{27}$$

Numerically however a major difficulty arises because the pole is on the integration path. It is necessary to calculate the integral as a principal value. We found that if the pole is located at the midpoint of a sampling period the numerical integration indeed provides the principal value (see

Candel, 1983b, for a numerical illustration of this aspect). Now, if the pole is not at the center of a sampling period, an accurate estimate is still obtained if the sampling points surrounding the pole are shifted around the pole and placed symmetrically around that point.

Let us assume here that $a = (M + \frac{1}{2})\Delta\zeta$ i.e. that the pole is the midpoint of the M-th sampling period. The numerical estimates of the H_o^1 transform, obtained in this case, are shown in Figure 6.

Estimates are displayed as discrete symbols while the exact values of $F_H(r)$ are represented by a solid line (actually the plot gives the sampled and scaled values of the transform $\tilde{F}_H(l) = F_H(l\,\Delta r)/(\Delta\zeta)^2$). The numerical estimates are very close to the exact values. This agreement is extremely encouraging and indicates that the H^1_o transform may be used in guided wave problems.

To prove this point we now describe some results obtained by us in collaboration with a team from SINTRA-Alcatel (Candel, Boussarie, Loesch and Lelarge, 1985c). The test problem is sketched in Figure 7. A point source radiates acoustic waves in a shallow water channel. We assume that the water surface is flat and may be treated as a pressure release boundary

$$p(r,0) = 0 \tag{28}$$

The ocean floor is plane and characterized by an infinite impedance

$$\frac{\partial p}{\partial n}(r,h) = 0 \tag{29}$$

The geometry and boundary conditions are the simplest possible. In this situation the sound field is easily expressed as a modal series:

$$p(r,z) = (i/2h)\sum_{n=0}^{\infty} \Psi_n(z_o)\Psi_n(z)\,H_o^1(k_{n\parallel} r) \tag{30}$$

where $\Psi_n(z) = \sin(k_{n\perp} z)$ is the n-th transverse eigenfunction, $k_{n\perp} = (n + \frac{1}{2})\pi/h$ designates the transverse wavenumber of the n-th mode and $k_{n\parallel}^2 = k^2 - k_{n\perp}{}^2$ is the corresponding longitudinal wavenumber.

The pressure field may also be written as a H_o^1 transform

$$p(r,z) = \int_{-\infty}^{+\infty} g(\zeta,z,z_o)\,H_o^1(\zeta r)\,\zeta\,d\zeta \tag{31}$$

where

$$\begin{aligned} g(\zeta,z,z_o) &= \frac{1}{2\pi}\,\frac{\cosh\gamma(z_o-h)}{\gamma\cosh(\gamma h)}\,\sinh\gamma z\ , \quad 0 \leq z \leq z_o \\ &= \frac{1}{2\pi}\,\frac{\sinh\gamma z_o}{\gamma\cosh(\gamma h)}\,\cosh\gamma(z-h)\ , \quad z_o \leq z \leq h \end{aligned} \tag{32}$$

and $\gamma = |\zeta^2-k^2|^{\frac{1}{2}}$ for $|\zeta|>k$, $\gamma = -i\,|k^2-\zeta^2|^{\frac{1}{2}}$ for $|\zeta| < k$

The function $g(\zeta,z,z_o)$ is the Green's function of the transverse propagation problem. According to the last expressions the pressure field may be obtained by calculating the Green's function $g(\zeta,z,z_o)$ corresponding to each horizontal wavenumber ζ and integrating these functions as a H_o^1 - transform. It is worth noting that representations of this type

are quite general and may be used in multilayered configurations.
In such cases analytic expressions for the Green's function are not available. The Green's function is then obtained by solving a transverse two-point boundary value problem or by making use of a transmission matrix formulation.
Returning to the problem at hand we first observe that the Green's function (32) has an infinite number of poles. These poles are such that

$$\cosh(\gamma h) = 0 \qquad \text{or equivalently such that}$$
$$\zeta_n^2 = k^2 - [(n+\tfrac{1}{2})\tfrac{\pi}{h}]^2 \tag{33}$$

The poles exactly correspond to the modes involved in the modal series (30). A finite number of these poles are located on the integration path. Figure 8 displays the typical structure of the transverse Green's function. In this case the function has 8 poles on the real axis and it exhibits 8 peaks. Direct numerical transformation of this function requires a proper sampling around each pole. The result plotted as a transmission loss is displayed in Figure 9. The same figure also shows values of this quantity determined from the modal series expansion (30). The two methods agree except in the vicinity of the source where the modal series is inaccurate.
The calculation demonstrates that a direct numerical H_0^1 - transformation may be used to evalutate the pressure field. However the calculation is successful only because the sampling points around each real pole have been shifted and placed to form a symmetrical pattern around each pole.

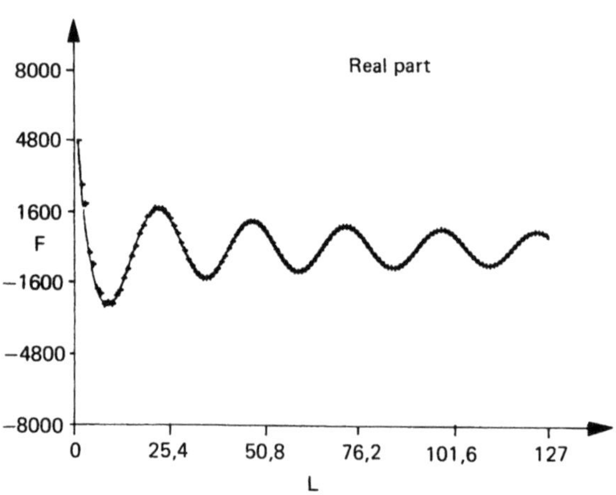

(a)

(b)

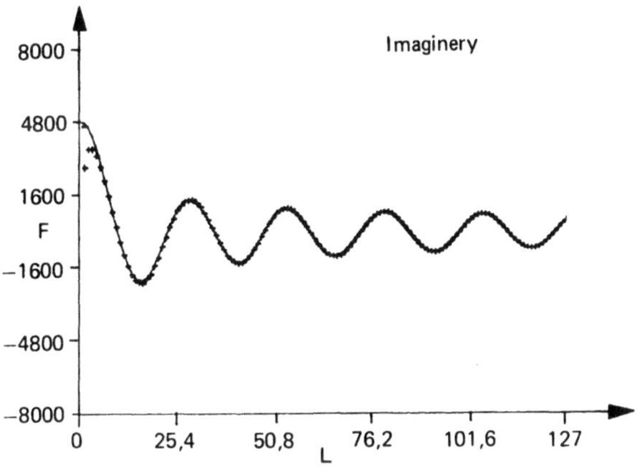

Fig. 6 - H_o^1 - transform of the test function defined by expression (26). The pole is at a = (M + ½)ΔS where M = 40. Numerical estimates are represented by discrete symbols. The exact transform values are displayed as solid lines.

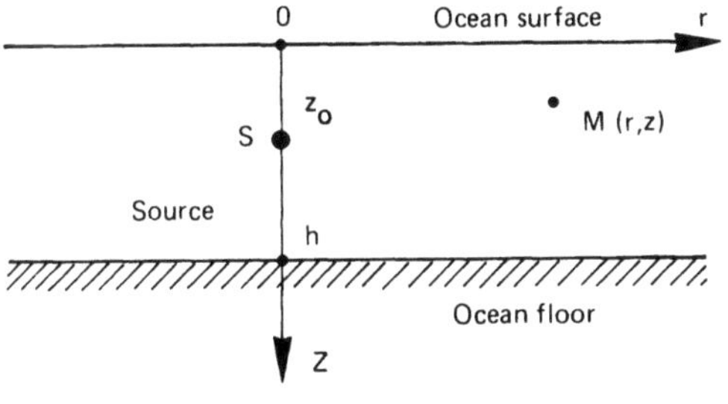

Fig. 7 - A point source radiates acoustic waves in a shallow water channel. The ocean surface is flat, the ocean bottom is flat and has an infinite impedance.

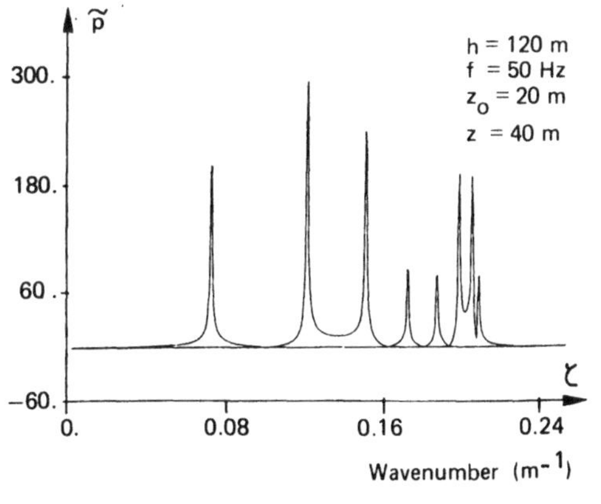

Fig. 8 - Typical structure of the transverse Green's function. The modulus of g is displayed as a function of the horizontal wavenumber ξ .

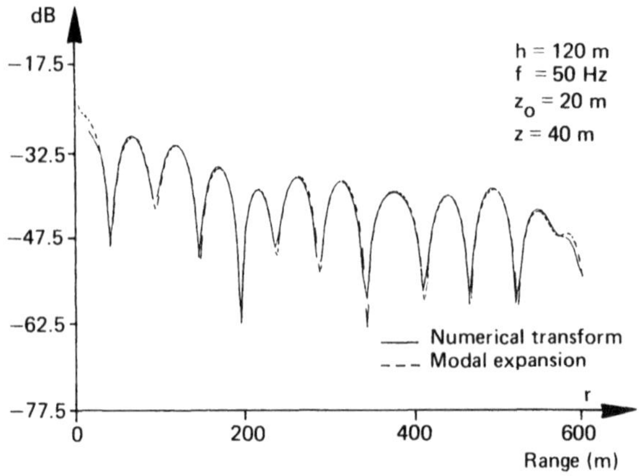

Fig. 9 - Transmission loss calculated with the H_0^1-transform method and with the modal series expression.

4 - BOUNDARY INTEGRAL METHODS

Boundary integral methods have found numerous applications in the fields of elasticity, electromagnetic wave propagation and acoustics. A brief review of the acoustics literature dealing with BIM is contained in Candel, 1983, and will not be repeated here. Recent additions to the literature are due to Tobocman 1984,1985, and concern wave scattering problems treated by means of a direct integral equation method. Seybert et al., 1984,1985, consider simple radiation problems and use a similar approach but base the calculation of the influence integrals on a boundary element formulation. An important and elegant contribution to the field is that of Hamdi, 1982. The method proposed by this author relies on an indirect formulation in terms of single and double layer potentials. The potentials are obtained as the stationary solutions of a variational equation. This formulation leads to a symmetric linear system of equations which may be efficiently solved numerically. This technique has been used by Hamdi and his co-workers to treat acoustic radiation problems and some acousto-elastic coupled situations.

The method is here illustrated with an analysis of scattering problems. We consider a diffracting object bounded by a closed surface S (Figure 10). Let p_i designate an incident acoustic field radiated by sources located outside S in the domain Ω and let p_s represent the field scattered by the surface S. The total pressure field outside S is the sum of these two fields

$$p_T = p_i + p_s \tag{34}$$

Now we consider for simplicity that S is a rigid surface. Then,

$$\frac{\partial p_T}{\partial n} = 0 \tag{35}$$

The boundary value problem for the scattered field is then described by the following set of equations

$$\nabla^2 p_s + k^2 p_s = 0 \qquad \text{in} \quad \Omega \tag{36}$$

$$\frac{\partial p_s}{\partial n} = -\frac{\partial p_i}{\partial n} \qquad \text{on} \quad S \tag{37}$$

$$\lim_{r \to \infty} r\left(\frac{\partial p_s}{\partial n} - ikp_s\right) = 0 \tag{38}$$

The field p_s is solution of the Helmholtz equation outside S, it satisfies Sommerfeld's radiation condition at infinity and its normal derivative on S is specified by (37).

Now this problem may be replaced by an integral equation formulation. There are many possible ways to obtain such a formulation and the literature on the subject is extensive.
We adopt a double layer potential representation of the pressure field. At a point M belonging to Ω the field is expressed as

$$p_s(\underset{\sim}{r}_M) = \int \mu(\underset{\sim}{r}_P) \frac{\partial G}{\partial n_P}(\underset{\sim}{r}_M, \underset{\sim}{r}_P)\, dS_P \tag{39}$$

where

$$G(\underset{\sim}{r}_M, \underset{\sim}{r}_P) = \frac{1}{4\pi} \frac{e^{ik|\underset{\sim}{r}_M - \underset{\sim}{r}_P|}}{|\underset{\sim}{r}_M - \underset{\sim}{r}_P|} \tag{40}$$

is the free space Green's function and μ (r_P) is a double layer potential. It is easy to show that expression (39) satisfies the Helmholtz equation (36) and the radiation condition (38) at infinity. Application of the boundary condition (37) then yields an integral equation for the potential :

$$-\frac{\partial p_i}{\partial n_M} = ⨍ \mu(\underset{\sim}{r}_P) \frac{\partial^2 G}{\partial n_M \partial n_P}\, dS_P \tag{41}$$

where M now belongs to the surface S and ⨍ designates the principal value of the surface integral. This principal value may be obtained as the following limit

$$⨍ \mu(\underset{\sim}{r}_P) \frac{\partial^2 G}{\partial n_M \partial n_P}\, dS_P = \lim_{\substack{M \to M_0 \\ M_0 \in S}} \frac{\partial}{\partial n_M} \int \mu(\underset{\sim}{r}_P) \frac{\partial G}{\partial n_P}\, dS_P \tag{42}$$

Equation (42) may be solved directly but it is possible to avoid the calculation of the principle value and obtain a symmetrical problem by making use of the method developed by Hamdi, 1982. Equation (41) is replaced by a equivalent variational formulation. The potential μ is obtained as the stationary solution of the variational equation

$$\delta \mathcal{L}(\mu) = 0 \qquad \text{where}$$

$$\begin{aligned} \mathcal{L}(\mu) = \frac{1}{2} \int_S \int_S \{ & k^2 (\underset{\sim}{n}_M \cdot \underset{\sim}{n}_P)\, \mu(\underset{\sim}{r}_M) \mu(\underset{\sim}{r}_P) \\ & - \underset{\sim}{n}_M \times \nabla_M \mu(\underset{\sim}{r}_M) \cdot \underset{\sim}{n}_P \times \nabla_P \mu(\underset{\sim}{r}_P) \} \, G \, dS_M \, dS_P \\ & + \int_S f(\underset{\sim}{r}_M) \mu(\underset{\sim}{r}_M) \, dS_M \qquad , \quad M \in S \end{aligned} \tag{43}$$

and

$$f(\underset{\sim}{r}_M) = - \partial p_i / \partial n_M$$

and M, P are the gradient operators at points M and P.

Once μ is determined the scattered field $p_S(\underset{\sim}{r}_M)$ in the exterior domain Ω is obtained from (39). At points belonging to S expression (39) must be modified and the scattered field is given by

$$p_S(\underset{\sim}{r}_M) = \frac{1}{2} \mu(\underset{\sim}{r}_M) + \int \mu(\underset{\sim}{r}_P) \frac{\partial G}{\partial n_P} (\underset{\sim}{r}_M, \underset{\sim}{r}_P) \, dS_P \tag{44}$$

In practice, the search of μ is conducted in discrete form. A surface mesh of N triangular pannels comprising M nodes is defined and the potential μ is evaluated at the nodes of this mesh. The functional $\mathcal{L}$ may then be written in discrete form

$$\mathcal{L}(\mu) = \frac{1}{2} \{\mu\}^T D \{\mu\} - \{\mu\}^T \{V\} \tag{45}$$

where D is an (MxM) complex symmetric influence matrix and V is an (Mx1) source vector. The potential μ which makes $\mathcal{L}$ stationary is then solution of the linear system

$$D\{\mu\} = \{V\} \tag{46}$$

Further details on this method are contained in Hamdi, 1982.

We shall now illustrate this method by considering two examples of calculations. The first case provides a verification of the method. A plane wave propagating in the positive x direction is scattered by a hard sphere (Figure 11). The modulus of the total field divided by that of the incident field $|p_T| / |p_i|$ is shown in the form of contours in Figure 12. The same quantity obtained from a series expansion solution of the problem is shown in Figure 13 (this exact calculation is due to

Waterhouse, 1984). The numerical and exact solutions nearly coincide. The main differences are observed in a very thin layer adjacent to the diffracting sphere. In this region special precautions must be taken in the evaluation of the pressure field. The difficulty is related to the estimation of integral (39) which becomes singular when the point M approaches the surface of the diffracting objects.
The calculation is otherwise satisfactory. Theoretical arguments and numerical experiments indicate that the solution remains accurate if the typical size h of the surface pannels is less or equal than a quarter of the wavelength $h \leqslant \lambda/4$. If this criterion is not satisfied large errors may occur. A second source of difficulties is the problem of "irregular frequencies" of the internal Helmholtz problem. Special techniques are required to eliminate this problem. This aspect is extensively discussed in the literature dealing with boundary integral methods (see Candel, 1983, for a list of references on this particular subject).
The second scattering problem is of direct technical interest. It concerns the structure of the acoustic field in the vicinity of the Ariane launcher. It is known that modern spacecraft launchers produce intense sound fields at lift-off. The sound field radiated by the launcher propulsion system and exhaust gases is diffracted by the outer structure of the rocket and this may result in local augmentations of the sound level. A detailed analysis of this problem is being performed at ONERA in relation with the Ariane IV program. Figure 14 shows a typical surface mesh used to perform the calculations. This mesh represents the upper portion of the launcher. Figure 15 exhibits the field structure in the vicinity of the rocket. The calculation is performed at a frequency f = 125 Hz and for an incidence angle $\theta = 29°$ characteristic of waves radiated by noise sources in the jets when the launcher is on the pad. On the incidence side the direct and diffracted fields interfere and the sound field near the structure is enhanced. On the other side diffraction produces a reduced sound field. Another view of the sound field structure is given in Figure 16. The plot now corresponds to a plane perpendicular to the rocket axis and it is located at a small distance from the base of the rocket equipment bay. The pressure field is enhanced on the incidence side and in the two quadrants adjacent to this direction. This field structure is strongly influenced by the presence of the base surface.

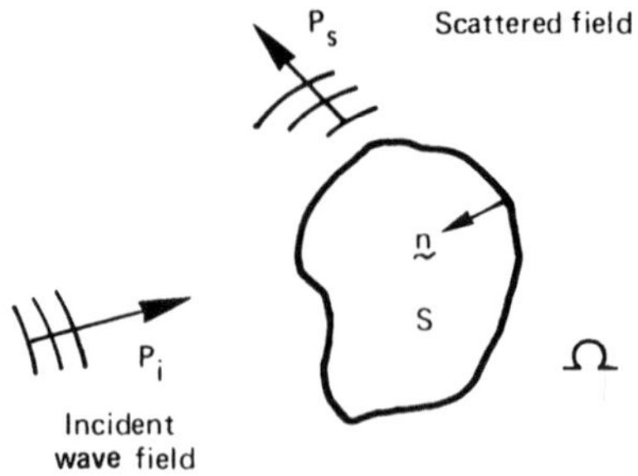

Fig. 10 - Geometry of the scattering problem.

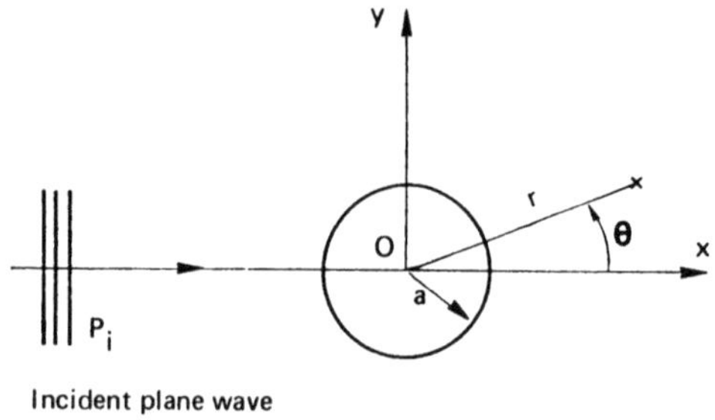

Fig. 11 - A plane wave is incident on a hard sphere of radius a.

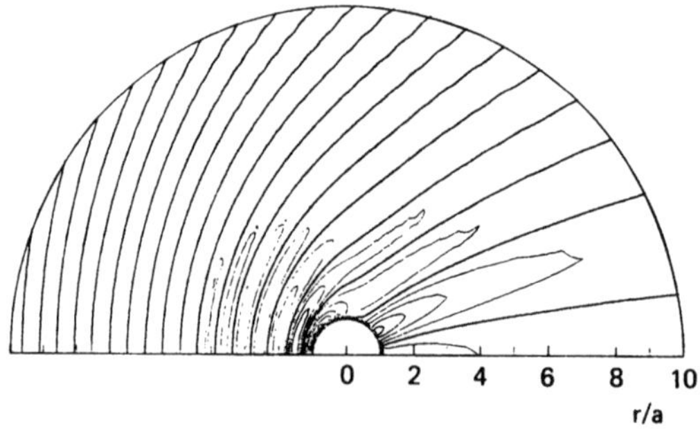

Fig. 12 - Diffraction of a plane wave by a hard sphere. The incident wave propagates from left to right and its wavevector is parallel to the x axis. The normalized pressure modulus $|p_T| / |p_i|$ is displayed in the form of constant contour lines. Thick contours correspond to a value of 1. The difference between successive contours is 0.1. ka = 3. Numerical results.

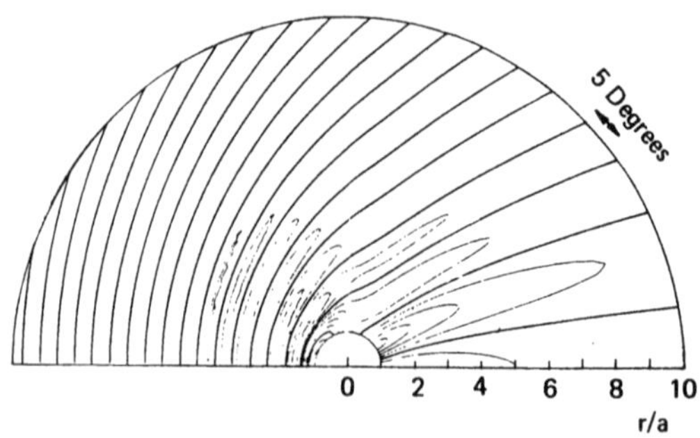

Fig. 13 - Same legend as Fig. 12. Results obtained by Waterhouse, 1984. ka = 3.

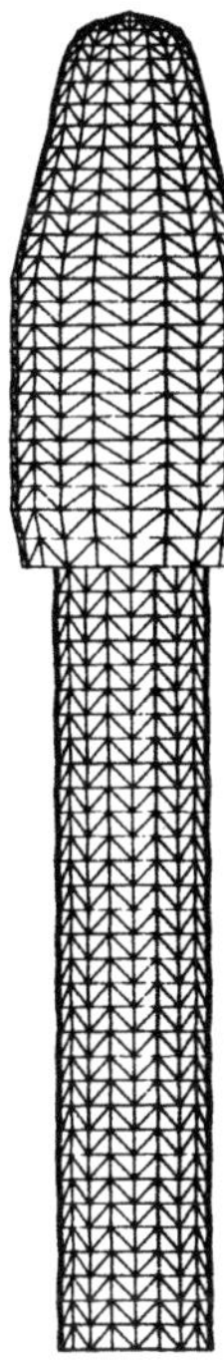

Fig. 14 - The surface mesh used to represent the upper part of the Ariane IV launcher.

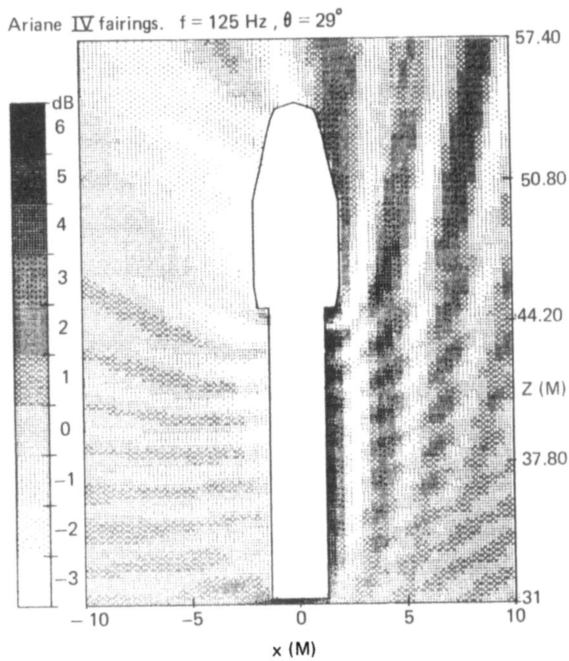

Fig. 15 - Field structure in the vicinity of the upper portion of the Ariane IV launcher. The plot corresponds to an axial plane containing the incident wavevector. The reflection diffraction index RDI = 20 $\log_{10} |p_T|/|p_i|$ is displayed on a scale of grey levels. f = 125 Hz, θ = 29°

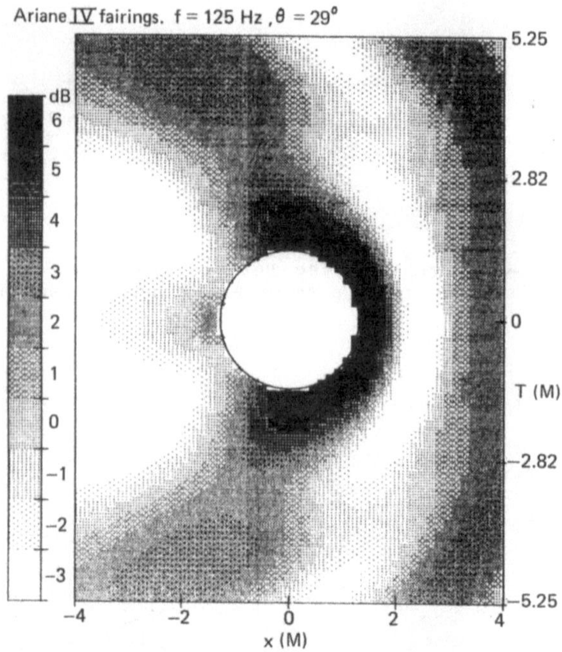

Fig. 16 - Field structure in the vicinity of the upper portion of the Ariane IV launcher. The plot corresponds to a plane perpendicular to the rocket axis. This plane is located at z = 44,519 m at a distance of from the base of the equipment bay. The reflection diffraction index RDI is displayed on a scale of grey levels. f = 125 Hz, θ = 29°.

Reference	Transform type	Method	Qualitative features
Siegman, 1977	$\int_0^\infty \rho f(\rho) J_0(\rho r)\, d\rho$	Gardner transformation and convolution evaluation with three FFT	Fast, but non uniform sampling of function and transform arguments.
Talman, 1978	"	"	"
Johansen and Sorensen, 1979	$\int_0^\infty \rho f(\rho) J_k(\rho r)\, d\rho$	Gardner transformation and linear filtering	"
Mook, 1983	$\int_0^\infty \rho f(\rho) J_0(\rho r)\, d\rho$	Abel transform evaluated as a convolution followed by a FFT	Fast, but non uniform sampling of data and result.
Anderson, 1982	$\int_0^\infty \rho f(\rho) J_k(\rho r)\, d\rho$	Linear filtering	"
Brund and Chavel, 1977	$\int_0^\infty \rho f(\rho) J_0(\rho r)\, d\rho$	Sampling theorem for rotationally invariant functions	Slow, accurate
Cavanagh and Cook, 1979	$\int_0^\infty \rho f(\rho) J_0(\rho r)\, d\rho$	Expansion in Gaussian-Laguerre polynomials	Requires a large number of terms to achieve convergence
Nachamkin and Maggiore, 1980	"	Series expansion of the Bessel function	Slow convergence
Miraglia, Piacentini and Rivarola, 1980	$\int_0^\infty \rho f(\rho) J_k(\rho r)\, d\rho$	Expansion of $f(\rho)$ in a series of Laguerre polynamials	Slow
Di Napoli and Deavenport 1979, 1980	$\int_0^\infty \rho f(\rho) J_0(\rho r)\, d\rho$	Asymptotic evaluation of Bessel function followed by an FFT	Fast provides asymptotic estimales
Oppenheim, Frisk and Martinez, 1980	$\int_0^\infty \rho f(\rho) J_0(\rho r)\, d\rho$	Projection slice theorem	Slow but accurate
Candel, 1981a	"	FFT followed by summation of selected components	Intermediate execution speed. Accurate uniform sampling
Candel, 1981b	"	Dual algorithm combining an asymptotic calculation with the previous method	Fast and accurate, uniform sampling
Candel, 1981c	$\int_0^\infty \rho f(\rho) J_k(\rho r)\, d\rho$ $k = 1, 2 \ldots, M$	FFT followed by Fourier series calculations	Fast and accurate, uniform sampling
Candel, 1981d	$\int_0^\infty \rho f(\rho) J_k(\rho r)\, d\rho$ k even k odd (simultaneous evalu ation)	"	"
Candel, Boussarie, Loesch and Lelarge, 1985a,b	$\int_0^\infty \rho f(\rho) H_0^1(\rho r)\, d\rho$	Dual algorithm combining an asymptotic calculation with an algorithm for the low order components	Fast and accurate, uniform sampling

ACKNOWLEDGEMENTS

The numerical calculations used as illustrations in Section 2 have been obtained with Daniel Blacodon. Work on the H_0^1-transform and its applications was performed with Eric Boussarie, Jean-Marc Loesch and A. Lelarge. Part of that work was carried at SINTRA-Alcatel, Arcueil, under contract from DRET. Numerical calculations of scattering were obtained with Patrice Malbêqui using the REDSTAR Code. The kernel of this code is due to M.A. Hamdi from the University of Compiègne. The calculations pertaining to the Ariane IV launcher were performed under a contract from CNES, Evry, monitored by R. Bec. Permission to use Figure 13 was granted by R.V. Waterhouse. It is a pleasure to acknowledge the contributions of my coworkers, many helpful discussions with Prof. Harold Levine during his sabbatical period at ONERA and the financial support provided by DRET and CNES for various parts of this research.

REFERENCES

ANDERSON W.L., 1982, ACM Transactions on Mathematical Softwave, 8, 344-368. Fast Hankel transforms using related and lagged convolutions.

BLACODON D. and CANDEL S.M., 1984, 9ème Colloque d'Acoustique Aéronautique Compiègne, 14-16 Nov. 1984. Extrapolation radiale de champs axisymétriques par une méthode spectrale. II.

BOJARSKI N.N., 1984, J. Acoust. Soc. An., 75, 320-323. Scattering by a cylinder: a fast exact numerical solution.

BRUNOL J. and CHAVEL P., 1977, Proceedings IEEE, 1089-1090. Fourier transformation of rotationally invariant two-variable functions: computer implementation of Hankel transform.

CANDEL S.M., 1981(a), Computer Physics Communication, 23, 343-353. An algorithm for the Fourier-Bessel transform.

CANDEL S.M., 1981(b), IEEE Transactions on Acoustics Speech and Signal Processing, ASSP-29, 963-972. Dual algorithms for fast calculation of the Fourier-Bessel transform.

CANDEL S.M., 1981(c), J. Computat. Phys., 44, 243-261. Simultaneous calculation of Fourier-Bessel transforms up to order N.

CANDEL S.M., 1981(d), La Recherche Aérospatiale, (English Edition), 1981-5, 23-24. Algorithms for the computation of even and odd Fourier-Bessel transforms.

CANDEL S.M. and BRUN Y., 1982, Revue du CETHEDEC, 19, 207-219. Rayonnement d'une source à proximité d'un interface solide-fluide. Résolution par transformation de Fourier-Bessel numérique.

CANDEL S.M., 1983, Proc. of Int. Symp. on Recent Advances in Aerodynamics and Adroacoustics, Aug. 22-26, 1983. A review of numerical methods in acoustic wave analysis. Also ONERA TP 1983-93.

CANDEL S.M. and CHASSAIGNON C., 1984, J. Acoust. Soc. Am., 76, 1823-1828. Radial extrapolation of wave fields by spectral methods.

CANDEL S.M., BOUSSARIE E., LOESCH J.M. and LELARGE A., 1985(a), submitted to J. Computat. Phys. A dual algorithm for fast calculation of the H_o^I - transform.

CANDEL S.M., BOUSSARIE E., LOESCH J.M. and LELARGE A., 1985(b), Colloque GRETSI, Nice - 20-24 mai 1985. Un algorithme dual pour le calcul rapide de la transformée H_o^1.

CANDEL S.M., BOUSSARIE E., LOESCH J.M. and LELARGE A., 1985(c). In preparation. Analysis of guided wave motion in shallow water channels by fast H_o^1 - transform methods.

CAVANAGH E. and COOK B., 1979, IEEE Transactions on Acoustics Speech and Signal Processing, ASSP-27, 361-366. Numerical evaluation of Hankel transforms via Gaussian Laguerre Polynomial expansions.

CHIN R.C.Y., HEDSTROM G. and THIGPEN L., 1984, J. Computat. Phys., 54, 18-56. Numerical methods in seismology.

DI NAPOLI F.R. and DEAVENPORT R.L., 1979, In Ocean Acoustics, J.A. de Santo, editor, Springer Verlag, Berlin, 79-157. Numerical models of underwater acoustic propagation.

DI NAPOLI F.R. and DEAVENPORT R.L., 1980, J. Acoust. Soc. Am., 67, 92-105.

FRISK G.V., OPPENHEIM A.V. and MARTINEZ D.R., 1980, J. Acoust. Soc. Am., 68, 602-612. A technique for measuring the plane-wave reflection coefficient of the ocean bottom.

FRISK G.V. and LYNCH J.F., 1984, J. Acoust. Soc. Am., 76, 205-216. Shallow water waveguide characterization using the Hankel transform.

GUYOMAR D., 1982, Thesis, University of Compiègne. Modes vibratoires d'un crystal piézoélectrique et méthodes de reconstruction de l'état de vibration.

HAMDI M.A., 1892, Doctoral thesis, Université de Compiègne. Formulation variationnelle par équations intégrales pour le calcul de champs acoustiques linéaires proches et lointains.

JOHANSEN H.K. and SORENSEN K., 1979, Geophysical Prospecting, 27, 876-901 Fast Hankel transforms.

MEDEIROS A.F. and STEPANISHEN P.R., 1984, J. Acoust. Soc. Am., 75, 1732-1740. The forward and backward projection of acoustic fields from axisymmetric ultrasonic radiators using impulse response and Hankel transform techniques.

MIRAGLIA J.E., PIACENTINI R.D. and RIVAROLA R.D., 1980, Computer Physics Communications, 19, 299-303. Computational methods for eikonal differential cross sections.

MOOK D.R., 1983, IEEE Transactions on Acoustics Speech and Signal Processing, ASSP-31, 979-985. An algorithm for the numerical evaluation of the Hankel and Abel transforms.

MOOK D.R., FRISK G.V. and OPPENHEIM A.V., 1984, J. Acoust. Soc. Am., 76, 222-243. A hybrid numerical/analytic technique for the computation of wave fields in stratified media based on the Hankel transform.

NACHAMKIN J. and MAGGIORE C.J., 1980, J. Computat. Phys., 37, 41-55. A Fourier-Bessel transform method for efficiently calculating the magnetic field of solenoids.

OPPENHEIM A.V., FRISK G.V. and MARTINEZ D.R., 1980, J. Acoust. Soc. Am., 68, 523-529. Computation of the Hankel transform using projections.

RASPET R., LEE S.W., KUESTER E., CHANG D.C., RICHARDS W.F., GILBERT R. and BONG N., 1985, J. Acoust. Soc. Am., 77, 345-352. A fast field program for sound propagation in a layered atmosphere above an impedance ground.

SEYBERT A.F., SOENARKO B., RIZZO F.J. and SHIPPY D.J., 1984, Transactions of the AS ME, 106, 414-420. Application of the BIE method to sound radiation problems using an isoparametric element.

SEYBERT A.F., SOENARKO B., RIZZO F.J. and SHIPPY D.J., 1895, J. Acoust. Soc. Am., 77, 362-368. An advanced computational method for radiation and scattering of acoustic waves in three dimensions.

SIEGMAN A.E., 1977, Opt. Lett., 1, 13-15. Quasi fast Hankel transform.

STEPANISHEN P.R. and CHEN H.W., 1984, J. Acoust. Soc. Am., 76, 942-948. Nearfield pressures and surface intensity for cylindrical vibrators.

TALMAN J.D., 1978, J. Computat. Phys., 29, 35-48. Numerical Fourier and Bessel transforms in logarithmic variables.

TOBOCMAN W., 1984, J. Acoust. Soc. Am., 76, 599-607. Calculation of acoustic wave scattering by means of the Helmholtz integral equation.I.

TOBOCMAN W., 1985, J. Acoust. Soc. Am., 77, 369-374. Comparison of the T-matrix and Helmoltz integral equation methods for wave scattering calculations.

WATERHOUSE R.V., 1984, J. Acoust. Soc. Am., 75, 695-706. Near field diffraction patterns at a sphere or cylinder.

Computational Aeroacoustics – Present Status and Future Promise

J. C. HARDIN and S. L. LAMKIN

NASA Langley Research Center
Hampton, Virginia

Kentron International
Hampton, Virginia

ABSTRACT

The field of computational aeroacoustics (CAA), in which sound generation by unsteady flows is calculated from first principles, has recently emerged as a viable discipline. This paper describes one CAA technique which has been validated for low Mach number, low Reynolds number, two-dimensional flows. Results of its application to geometries producing both dipole and quadrupole type sound sources are presented along with theoretical and experimental comparisons. In addition, the future promise of this and other CAA techniques in removing the present limitations is evaluated.

INTRODUCTION

The calculation of sound generation by an unsteady flow, much as an aerodynamicist would compute the pressure distribution of an airfoil section, has long been a goal of the aeroacoustics community. However, there are many difficulties inherent in such a computation.

The first technique one might consider would be to calculate the total compressible flow field from which that portion of the pressure field which is traveling at the local speed of sound could be identified as the acoustic radiation. However, to perform such a computation on a finite grid requires either the development of a non-reflecting boundary condition at the outer boundary, such that outgoing radiation will not be reflected and invalidate the calculation, or termination of the computation before acoustic waves can reach the outer boundary. Further, the identification problem is difficult.

Direct solutions of the governing equations also assume that the flow is essentially laminar. If the Reynolds number of the flow is high enough, the flow will be turbulent which implies that such solutions, if they exist at all, are merely single sample functions of the random turbulent field produced by critical dependence upon initial and boundary conditions. In addition, very fine grids must be employed to resolve the scales comprising the turbulent field.

Aero- and Hydro-Acoustics
IUTAM Symposium Lyon 1985
Editors: G. Comte-Bellot and J. E. Ffowcs Williams

Another approach that one might explore would be to embrace the turbulence and average the equations. However, to do so and still retain the time dependent properties of the flow necessary for CAA, higher levels of closure assumptions must be employed than have been utilized in the past. Thus, this approach must await further developments in the state-of-the-art and may, in fact, never be viable.

A third approach to CAA, which is the one adopted in this study, is based on the work of Crow[1] who showed that, for low Mach number, CAA solutions could be achieved in a two part calculation. The time-dependent incompressible flow field, which greatly reduces the problem of reflecting boundaries, is computed first. Then the sound radiation is obtained by integrating over the flow field. Since sound is a second order by-product of flows, this approach also offers the attraction of less concern with computational accuracy than would be necessary to extract sound from a totally compressible solution.

SOLUTION TECHNIQUE

In viscous, two dimensional, incompressible flow, the stream function, ψ, satisfies the Poisson equation

$$\nabla^2\psi = -\Omega \tag{1}$$

where Ω is the vorticity within the field. The vorticity further satisfies the Helmholtz equation

$$\frac{D\Omega}{Dt} = \nu\nabla^2\Omega \tag{2}$$

where ν is the kinematic viscosity of the fluid and D/Dt is the convective derivative. The use of these variables in description of the flow is attractive for CAA due to the relative simplicity of the equations and the intimate connection between acceleration of vorticity and radiation of sound.

This coupled set of partial differential equations has been solved numerically using a variety of standard computational fluid dynamic (CFD) techniques. The stream function is constant on solid boundaries and is assumed known from the analogous potential flow solution in benign regions of the outer boundary or extrapolated from the interior in outflow regions. The vorticity is similarly either taken as zero or extrapolated on the outer boundary. On solid surfaces, the no-slip condition is imposed which acts as a source for vorticity. Imposition of this condition greatly increases the resolution requirements and thus limits the Reynolds number at which computations can be achieved. For this reason, with bodies present in the field, solutions thus far have been restricted to low Reynolds number, basically laminar, flows. However, the appeal of this approach is that the computation is relatively straightforward with a minimum of assumptions employed.

After the time-dependent stream function and vorticity fields have been computed, the acoustic radiation from the flow may readily be obtained through the theory developed by Powell[2] who showed that the source of sound within such a flow may be taken as the Coriolis acceleration $\vec{L}=\vec{\Omega}\times\vec{v}$ where $\vec{\Omega}=[0,0,\Omega]$ and $\vec{v}=[\psi_y,-\psi_x,0]$ are the two dimensional vorticity and velocity fields respectively.

For a free vortical flow of finite volume V surrounded by an ambient medium, the source type is quadrupole and the acoustic pressure radiation to a farfield observer location $\vec{x}$ may be calculated from the integral

$$p(\vec{x},t) = -\frac{\rho}{4\pi a^2}\frac{x_i x_j}{x^3}\left(\frac{d^2}{dt^2}\int_V y_i L_j d\vec{y}\right)^* \tag{3}$$

where ρ and a are the density and speed of sound in the ambient medium, $x=|\vec{x}|$, and the asterisk indicates evaluation at the retarded time $t-x/a$.

When a surface is present in the flow, the sound radiation is generally more efficient and dipole in character. For this case, assuming that the characteristic wavelength of the sound radiation greatly exceeds the body length scale and that the source field is within a wavelength of the body, Howe[3] extended the theory to the case of uniform motion of the surrounding medium and showed that the sound could be computed from the elegant low-frequency Green's function approach

$$p(\vec{x},t) = -\frac{\rho}{4\pi a x(1+\vec{M}_o\cdot\frac{\vec{x}}{x})}\frac{\partial}{\partial t}\int_V \left[\vec{L}\cdot\nabla\{(\frac{\vec{x}}{x}-\vec{M}_o)\cdot\vec{Y}\}\right]^* d\vec{y} \tag{4}$$

where $\vec{M}_o$ is the Mach number vector of the uniform motion of the surrounding medium, $\vec{Y}$ is a vector of potentials such that Y_i is the potential of incompressible flow about the body such that the flow at great distances from the body is of unit speed in the i direction, and the asterisk indicates evaluation at the retarded time $t-(x-\vec{M}_o\cdot\vec{x})/a$. Note that, in this approach, no surface integrals of pressure distributions appear.

APPLICATIONS

The first application of this technique was to sound generation by a cylinder in a uniform flow[4]. This was an ideal test case since the sound source is dipole in nature and the flow is well-understood, both theoretically and experimentally. For Reynolds numbers above about 60, a von Karman vortex street forms behind the cylinder with substantial noise radiation due to the alternate vortex shedding.

Computations were carried out on a Control Data CYBER 203 computer at a Reynolds number of 200. Equation 1 was solved using the semi-implicit, successive-line-overrelaxation method, while computation of equation 2 was achieved by an explicit forward-time-center-space finite difference method. The noise radiation was then obtained via equation 4. Figure 1 is a montage

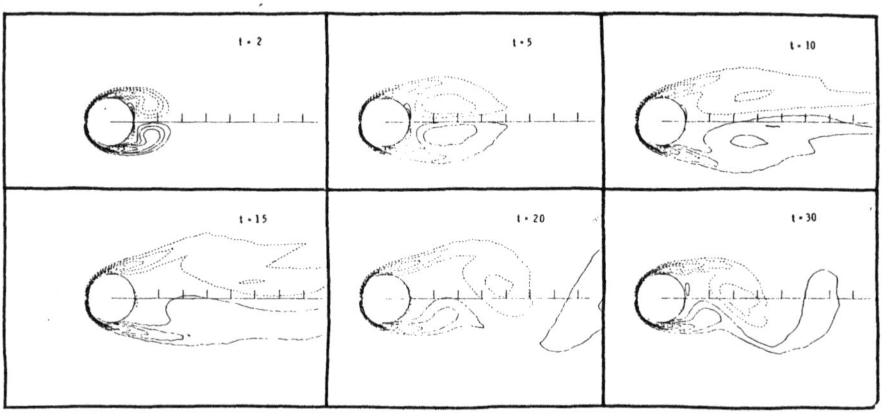

Figure 1: Contour plots of wake vorticity

of vorticity contours behind the cylinder at various nondimensional times which depicts the alternate vortex shedding characteristic of the von Karman street. A movie of this flow simulation has also been produced. Figure 2 shows the power spectral densities of the noise radiation both above and behind the cylinder. Note that, above the cylinder, the spectrum peaks at a Strouhal number of 0.2, in agreement with data, while that behind the cylinder consists of only even harmonics of St=0.2 as predicted by theory. Intensity values from this computation were also within 1.05 dB of experimental data and the directivity pattern agreed with theory.

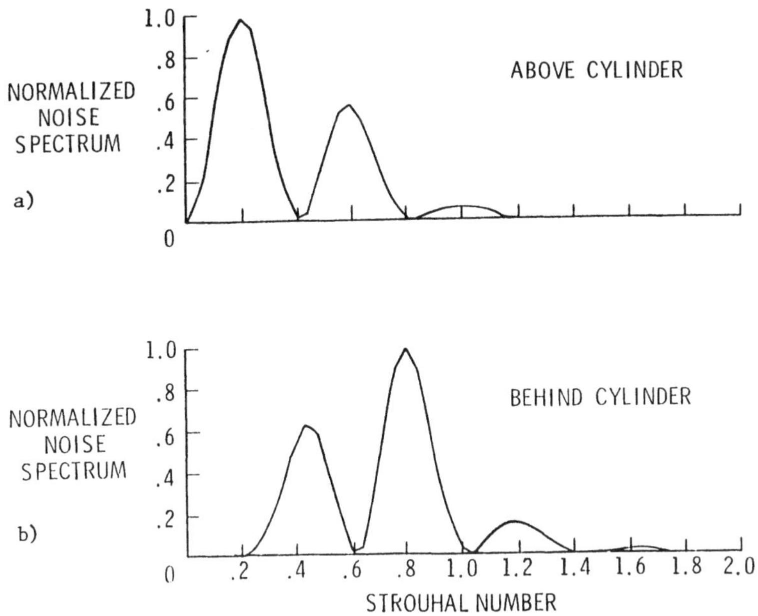

Figure 2: Noise spectrum: a) above cylinder; b) behind cylinder

A similar computational technique has also been applied to the blade/vortex interaction problem[5] via Joukowski transformation. For this problem, however, the computational Reynolds number is constrained to be much lower than the physical Reynolds number. Thus, direct comparisons could not be made with data, although qualitative features of the vortex interaction and subsequent noise radiation were reproduced.

As another test, this time of a quadrupole source, the technique has been applied to noise generation by the interaction and ultimate merging of two like-signed, distributed vortices[6] at a Reynolds number of 1000. This vortex pairing mechanism is thought by some to be a major noise source in free turbulent jet flows. The computation was initialized by two equal, Gaussian distributions of vorticity and periodic boundary conditions were imposed in both the x and y directions. Equations 1 and 2 were then transformed into the frequency domain and a spectral method employing Galerkin approximation was used to solve the transformed equations. The noise radiation was then obtained by integrating equation 3 over only one box in the periodic domain.

Figure 3 displays the normalized acoustic pressure time history for an observer location at ninety degrees to the initial separation vector of the vortices. Also superimposed on the figure are vorticity contours showing the vortices' positions at the time the sound was radiated. Note that the

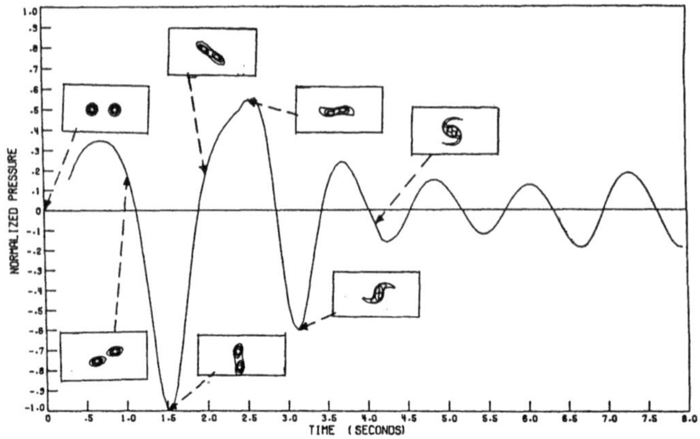

Figure 3: Time history of vortex merging noise

acoustic radiation is large during the initial phase when the vortices are rotating about one another, but dies away as the vortices begin to merge. Thus, it is not the viscous merging process, but rather the vortex acceleration due to close interaction which physically results in the noise generation. The noise radiation during the initial rotating phase has been compared with an exact analytical solution for rotating point vortices and found to agree very well as to amplitude, frequency, and phase. Thus, this CAA approach appears capable of calculating free flow quadrupole sources as well.

SUMMARY AND PROGNOSIS

This paper has presented a low Mach number CAA technique involving a two-part calculation of first the incompressible, time dependent, viscous flow and then a volume integral to calculate the resulting noise radiation. Examples of its application to both dipole and quadrupole sources in low Reynolds number, two-dimensional flows are also included. Our research presently in progress will increase the Reynolds number of this two-part CAA approach by approximating the boundary layer on bodies present in the flow as well as allow extension to simple three-dimensional flow geometries by reducing the resolution requirements. Such calculations are possible with present hardware such as the CYBER 205.

Ultimately, of course, one would hope to develop a compressible one-step CAA approach in high Reynolds number three-dimensional flows. However, it is our experience that CFD techniques cannot as yet be used as a tool by the CAA community. For example, our work has employed only analytical grids since numerical grids have not proven sufficiently accurate. Conventional aeronautics applications of CFD are to either steady-state or short time fluctuations of less critical parameters than sound. Thus, much work on highly accurate and efficient solution algorithms, grid generation schemes, non-reflecting boundary conditions, and turbulence modeling must be accomplished before such an approach is feasible. Nevertheless, given the speed at which both hardware and software are advancing, the future of computational aeroacoustics is bright.

REFERENCES

1. Crow, S. C.: Aerodynamic Sound Emission as a Singular Perturbation Problem, Stud. App. Math., Vol. XLIX, No. 1, pp. 21-44, 1970.
2. Powell, A.: Theory of Vortex Sound, Jour. Acous. Soc. Am., Vol. 36, No. 1, pp. 177-195, 1964.
3. Howe, M. S.: Contributions to the Theory of Aerodynamic Sound, with Application to Excess Jet Noise and the Theory of the Flute, JFM, Vol. 71, Pt. 4, pp. 625-673, 1975.
4. Hardin, J. C. and Lamkin, S. L.: Aeroacoustic Computation of Cylinder Wake Flow, AIAA Jour., Vol. 22, No. 1, pp. 51-57, 1984.
5. Hardin, J. C. and Lamkin, S. L.: Aeroacoustic Interaction of a Distributed Vortex with a Lifting Joukowski Airfoil, AIAA Paper No. 84-2287, 1984.
6. Lamkin, S. L. and Hardin, J. C.: Aeroacoustic Computation from Vortex-Pair Merging, SIAM 1983 Fall Meeting, Norfolk, Virginia, 1983.

The Wavenumber-Frequency Spectrum of the Wall Pressure Field Computed from a Numerical Solution of the Navier Stokes Equation

R.A. Handler and R.J. Hansen
Naval Research Laboratory, Washington, DC (USA)

R. Leighton
University of Michigan, Ann Arbor, Michigan (USA)

S.A. Orszag
Princeton University, Princeton, New Jersey (USA)

Summary

The wall pressure fluctuations in a turbulent channel flow have been calculated from a numerical solution of the full Navier Stokes equation. No closure or subgrid scale approximations have been made. The wavenumber-frequency spectra of the pressure fluctuations obtained in this work show the following characteristics of particular note: (a) a "convective ridge" associated with those components of the pressure fields propagating in the streamwise direction and at a significant fraction of the freestream velocity; (b) comparable dimensions of the wall pressure structure in the streamwise and spanwise directions; (c) a nonzero value of the pressure spectrum at zero wavenumber. This last result is shown theoretically to be a consequence of the boundary conditions associated with a channel flow.

Introduction

The wavenumber-frequency spectrum of turbulence-induced wall pressure fluctuations is of both fundamental and practical interest. This spectrum may provide insight into the fundamental structure of turbulent flow in the near-wall region. From a practical point of view, it allows prediction of the small-amplitude vibration response of a structure to turbulence-induced wall pressure fluctuations. Unfortunately, this spectrum has proven extraordinarily difficult to measure in the laboratory, because of such factors as extraneous facility noise and limitations on the number of wall pressure transducers that may be used in any one given experiment.

The present work focuses on obtaining this spectrum from a numerical solution of the full three-dimensional, time-dependent Navier Stokes equation. Such solutions have been shown to correctly predict the mean velocity profile (1) and the characteristics of the wall pressure spectrum (2) over the range of parameters where the calculations are valid. Attention in these previous studies, and the present study as well, is devoted to the fully-developed turbulent channel flow.

Numerical Method

The method of solution of the equations of fluid motion is identical to that employed in references 1 and 2. The Navier Stokes and continuity equations are written in rotation form:

$$\frac{\partial \vec{v}}{\partial t} = \vec{v} \times (\vec{\nabla} \times \vec{v}) - \vec{\nabla} (p + 1/2\, v^2) + \frac{1}{R} \nabla^2 \vec{v} + \frac{2}{R} \hat{x} ; \qquad (1)$$

$$\vec{\nabla} \cdot \vec{v} = 0 ; \qquad (2)$$

where $\vec{v}$ is the velocity vector, $\hat{x}$ is the unit vector in the flow direction, and R is the Reynolds number based on the channel half-width. The boundary condition imposed at the channel walls ($z = \pm 1$) is $\vec{v} = 0$. The pressure field at the wall is related to the velocity field through a Poisson equation:

$$\nabla^2 (p + 1/2\, v^2) = \vec{\nabla} \cdot [\vec{\nabla} \times (\vec{\nabla} \times \vec{v})] ; \qquad (3)$$

$$\frac{\partial p}{\partial z} = 0 \qquad (z = \pm 1) . \qquad (4)$$

A pseudospectral solution technique is then used to solve for the velocity and pressure fields. Velocity components are expanded in Fourier series in the streamwise and spanwise directions and Chebychev polynomials in the wall-normal direction. All spatial derivatives are evaluated in the transform domain, after which inverse fast Fourier transforms are employed to transform the solution for one time step to the physical domain. The solution is moved forward to the next time step using the Adams Bashforth time stepping algorithm. Fast Fourier transforms are then used to move this updated solution back into the transform domain, where spatial derivatives are again evaluated spectrally.

In the present work the numerical solution of the Navier Stokes equation was carried out with 32 spectral modes in each spatial direction. Hence the total number of collocation points in the flow field was $(32)^3$ and the number on the wall (i.e., the number of "pressure sensors") was $(32)^2$. In carrying out the solution the flow was assumed to be initially laminar. After time t = 0, a large unstable disturbance was added to the laminar velocity field. The establishment of fully-developed conditions required approximately 4000 time steps (computational time of $0 \leq t^+ \leq 40$), where $t^+ = \frac{tU}{h}$. Here U is the laminar centerline velocity, and h is the channel half-width. The wall pressure field at each of the next 2000 time steps ($40 \leq t^+ \leq 60$) was stored for post analysis. The total computer time for an entire $(32)^3$ run was approximately two hours on a Cray 1 computer.

Results

The large number of collocation points on the solid boundary of the channel provides a means of gathering a detailed picture of the spatial characteristics of the wall pressure field in both the physical and the transform (wavenumber-frequency) domains.

A. Temporal Evolution of Wall Pressure Contours

Figures 1 and 2 are wall pressure contours at times separated by $\Delta t^+ = 0.80$. Two characteristics of these contour plots are worthy of note. First, selected features show clear movement from left to right (in the flow direction) from the first plot to the second. In these figures x^+ and y^+ are spatial coordinates scaled on the viscous length ν/u^*, where ν is the kinematic viscosity and u^* is the friction velocity. This observation is in accordance with the experimental studies of Dinkelacker, et. al. (3), where convection of individual wall pressure patterns in the streamwise direction was also evident. Additionally, the characteristic dimensions of the pressure patterns in Figures 1 and 2 appear qualitatively consistent with those observed in reference 3.

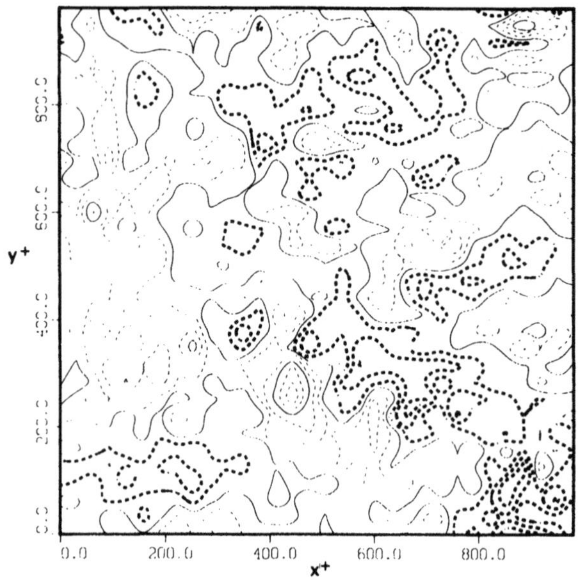

Fig. 1. Contour plot of the wall pressure at $t^+ = 40.2$. Bold dashed, solid, and light dashed lines represent positive, zero, and negative pressure contours respectively.

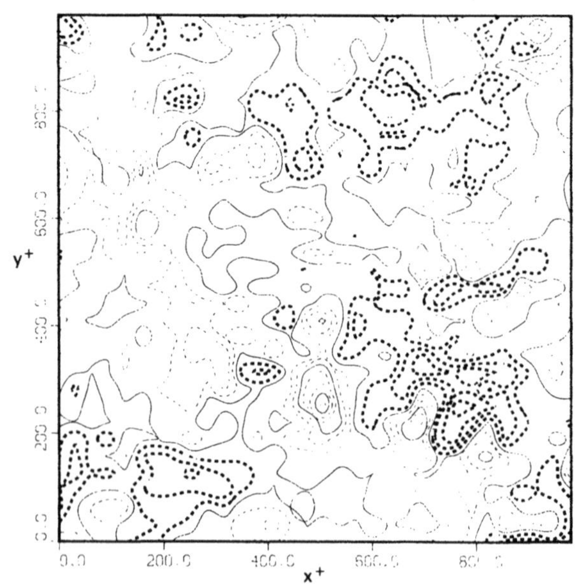

Figure 2. Contour plot of the wall pressure at $t^+ = 41.0$.

B. Wavenumber-Frequency Spectrum

The wavenumber-frequency spectrum of the wall pressure field, $P(k_1, k_3, \omega)$, is given by the expression

$$P(k_1, k_3, \omega) = \frac{2}{L_1 L_3 T} E\left\{p^*(k_1, k_3, \omega)\, p(k_1, k_3, \omega)\right\} \tag{5}$$

Here k_1 and k_3 denote the streamwise and spanwise wavenumbers, respectively, ω is the radian frequency, $p(k_1, k_3, \omega)$ is the Fourier transform of the wall pressure field, and $E\{\ \}$ denotes expected value. L_1, L_3 and T are the spatial and temporal lengths over which the Fourier transforms are computed. Two projections of $P(k_1, k_3, \omega)$ are of particular interest:

$$\hat{P}(k_1, \omega) = \int_{-\infty}^{\infty} P(k_1, k_3, \omega)\, dk_3 ; \tag{6}$$

$$\hat{\hat{P}}(k_1, k_3) = \int_{-\infty}^{\infty} P(k_1, k_3, \omega)\, d\omega . \tag{7}$$

$\hat{P}(k_1, \omega)$ is the quantity of primary interest in practical flow-induced vibration problems as well as an indicator of how the pressure field convects in the streamwise direction. $\hat{P}(k_1, k_3)$, on the other hand, is a representation of the relative length scales of the dominant structures in the wall pressure field in the streamwise and spanwise directions.

Figure 3 shows a contour plot of $\hat{P}(k_1, \omega)$. The convective ridge is the line traced out on this plot by the maximum value of $\hat{P}(k_1, \omega)$ for each $\frac{\omega h}{U}$. The convection velocity associated with the convective ridge in Figure 3 is of the order of 0.6 times the laminar centerline velocity. In addition to this dominant feature, it should be noted that $\hat{P}(k_1, \omega)$ does not approach zero as k_1 approaches zero. This result is in contrast to the well known theoretical result for the wall pressure spectrum at zero wavenumber in a turbulent boundary layer flow (4). This characteristic of $\hat{P}(k_1, \omega)$ is considered theoretically in a subsequent section of this work.

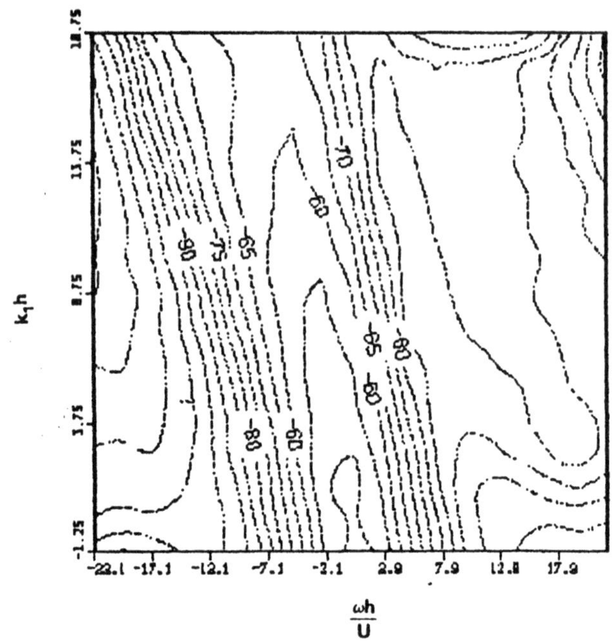

Figure 3. Contour plot of $\hat{P}$ (k_1, ω). The contour levels correspond to 10 log ϕ, where ϕ is the normalized spectrum amplitude.

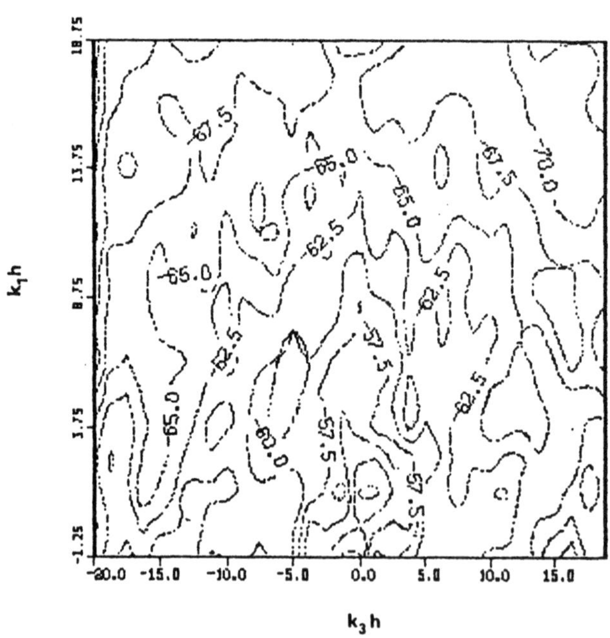

Figure 4. Contour plot of $\hat{\hat{P}}$ (k_1, k_3).

The behavior of $\hat{\hat{P}}(k_1, k_3)$ is shown in contour plot form in Figure 4. Both oridinate and abscissa are normalized by the channel half-width. This plot shows that the wall pressure field is characterized by comparable wavenumbers in the streamwise and cross flow directions. Since wavenumber is inversely proportional to wavelength, an equivalent statement is that the energy containing portion of the wall pressure spectrum has comparable characteristic wavelengths in the two orthogonal directions.

Theoretical Estimate of the Low Wavenumber Limit in Channel Flow

The low wavenumber limit of the wall pressure spectrum is obtained by taking the Fourier transform of the Navier Stokes equation (assuming homogeniety in space and time). The result for the pressure is:

$$\frac{d^2\, p(x_2, \vec{k}, \omega)}{d^2\, x_2} - k^2\, p(x_2, \vec{k}, \omega) = T\,(x_2, \vec{k}, \omega), \tag{8}$$

where $\vec{k}$ is the wavevector in the horizontal plane $\left(\vec{k} = (k_1, k_3)\right)$, $k = |\vec{k}|$, x_2 is the wall-normal coordinate, and T is the turbulence source term. The approximate boundary conditions for the channel flow are:

$$\frac{dp}{dx_2}\,(0, \vec{k}, \omega) = \frac{dp}{dx_2}\,(d, \vec{k}, \omega) = 0. \tag{9}$$

where d=2h. The solution to equation (8) with these boundary conditions is:

$$p(\vec{k}, \omega) = \frac{1}{k \sinh kd} \int_0^d T(x_2, \vec{k}, \omega) \cosh k(x_2-d)\, dx_2. \tag{10}$$

Since the source term is quardropole in character we can, for the purpose of extracting the qualitative behavior of $p(x_2, \vec{k}_1 \omega)$ on k, set

$$T\,(y, \vec{k}, \omega) = k^2\, t(y, \vec{k}, \omega)\ . \tag{11}$$

We then obtain the following limit for the pressure transform;

$$\lim_{k \to 0} p(0,\vec{k},\omega) = \frac{1}{d} \int_0^d t(x_2,\vec{0},\omega)dx_2. \tag{12}$$

Thus, the pressure spectrum near k = 0 should have some nonzero value which is in accordance with the numerical results and in contrast with the boundary layer result.

Conclusions

A numerical solution of the full Navier Stokes equation has been employed to investigate the turbulence-induced wall pressure field in a channel flow. This numerically obtained pressure field has instantaneous contours and convection characteristics reminescent of those observed in previous experimental studies. Moreover, the $k_1 - \omega$ projection of the spectrum is nonzero at zero wavenumber and the k_1-k_3 projection shows that the energy-containing patterns in the pressure field have comparable characteristic dimensions in the spanwise and streamwise directions. Previous experiments have established neither the value of the $k_1 - \omega$ spectrum near zero wavenumber nor provided a high spatial resolution $k_1 - k_3$ spectrum. In these respects the present numerical study has provided entirely new information about turbulence-induced wall pressure fluctuations in channel flows.

Acknowledgements: The work reported herein was supported by the Office of Naval Research Selected Research Opportunity in Hydroacoustics and the Naval Research Laboratory. In addition, one of us (R.L.) was on an Office of Naval Research Doctoral Fellowship during the conduct of this work.

References

1. Orszag, S.A., and Patera, A.T.: Calculation of Von Karman's constant for turbulent channel flow. Phys. Rev. Lett., 47 (1981) 832-835.

2. Handler, R.A., Hansen, R.J., Sakell, L., Orszag, S.A., and Bullister, E.: Calculation of the Wall Pressure Field in a turbulent channel flow. Phys. Fluids, 27 (1984) 579-582.

3. Dinkelacker, A., Hessel, M., Meier, G.E.A., and Schewe, G.: Investigation of pressure fluctuations beneath a turbulent boundary layer by means of an optical method. Phys. Fluids, 20 (1977) 5216-5224.

4. Kraichnan, R.H.: Pressure fluctuations in turbulent flow over a flat plate. J. Acoust. Soc. Am., 28 (1956) 378-390.

A Mixed Functional for the Numerical Resolution of Fluid-Structure Interaction Problems

M.A. HAMDI ; P. JEAN

Division Acoustique et Vibrations Industrielles
Université de Technologie de Compiègne
60200 Compiègne France

Summary

A new mixed functional is presented in terms of the displacement of the structure and the fluid loading on its boundary. It consists in coupling the classical functional of the structure with a variational formulation by integral equations of the fluid. This method has a double advantage : it avoids the difficult problem of singularities appearing when collocation technic is used and it leads, after discretisation by boundary finite elements method to an algebric symmetric system of small size which reduces considerably the calculation costs.

1 - Introduction

In many fields of applied mechanics as for example aeronautic and ships industries, structure interacts with surrounding fluid and radiates sound.Experience shows that the fluid modifies the resonance charecteristics of the structure,in fact it makes critical certain vibrational modes which leads in some unfavorable cases to the structure 's ruin.

Several analytical works have been developed to study the phenomenon of interaction between elastic and acoustic waves, but they are limited to the case of simple structures as plates and cylindrical or spherical shells(1,2,3) . Only numerical methods, as finite elements method and integral equations, are capable to predict the phenomenon in the general case of structures of arbitrary shapes(4,5,6).

However,the finite element method necessitates the discretisation of the structure and the surrounding acoustic media. This leads at high frequency range or when the fluid domain is infinite to an algebric system of large size which increases the computation costs. In addition the Sommerfeld 's radiation condition is not quite satisfied in the case of external problems.

To remove the previous disadvantages, we present a new numerical method which consists in coupling the classical functional of the structure with a variational formulation by integral equations for the fluid .

Our method has a double advantage when compared to collocation technics (7,8) : it avoids singularities and leads after discretisation by a boundary finite element method to a symetric algebric system of small size which can be solved by fast algorithms.

Aero- and Hydro-Acoustics
IUTAM Symposium Lyon 1985
Editors: G. Comte-Bellot and J. E. Ffowcs Williams

This method is suitable for solving hydro- and aeroelastic problems [1,3] . In this paper we show a comparison between numerical and analytical results in the case of a thin spherical shell immersed in a deep water. The sphere is excited by internal load or by an harmonic incident plane wave. The results are in very good agreement and shows the ability of the method for solving this kind of problems.

2 - Equations of the coupled problem

Let us consider a closed elastic stucture Ω which interacts as shown in fig. 1 with two internal and external acoustic media. The coupled system is submitted to an harmonic incident wave $p_\infty(x)e^{-i\omega t}$.

The resulting harmonic displacement $\vec{W}(x)e^{-i\omega t}$ of the structure and the complex pressure field $p(x)e^{-i\omega t}$ inside and outside the domain Ω must satisfy the following system of partial differential equations :

$$\operatorname{div}\sigma + \rho\omega^2\vec{W} = 0 \qquad \text{in } \Omega \tag{1}$$

$$(\Delta + k_\alpha^2)p = 0 \qquad \text{in } \Omega_\alpha \tag{2}$$

$$\sigma.\vec{n} + p\vec{n}\,\Big|_{S_\alpha} = 0 \tag{3}$$

$$\frac{\partial p}{\partial n} - \rho_\alpha\omega^2\, W\,\Big|_{S_\alpha} = 0 \tag{4}$$

$$\lim_{r\to\infty} r\left(\frac{\partial}{\partial n} - ik_2\right)(p - p_\infty) = 0 \qquad r = |ox| \tag{5}$$

The displacement vector $\vec{W}$ satisfies in Ω the equation of motion (1), where ρ is the mass density of the structure and σ is the stress tensor which is related to the strain tensor ε by the Hooke 's law :

$$\sigma_{ij} = C_{ijkl}\varepsilon_{kl} \qquad \varepsilon_{kl} = (W_{k,l} + W_{l,k})/2 \tag{6}$$

The pressure field in the fluid domain Ω_α satisfies the Helmholtz 's equation (2), where $k_\alpha = \omega/c_\alpha$ is the wave number and c_α is the speed of sound in Ω_α.

The pressure and the displacement vector are coupled on S_α by the momentum and kinematic boundary conditions (3) and (4), where W is the normal component of the displacement and ρ_α is the mass density of the inviscid fluid in Ω_α. Finally, far away from the structure Ω, the diffracted external pressure field must satisfy the Sommerfeld's radiation condition (5).

3 - Variational formulation of the problem

3.1 The structure

The classical variational formulation of the structure can be used to find the displacement vector $\vec{W}$ satisfying the equation of motion (1) in Ω and the boundary condition (3) for a given pressure on S_α. So we have,

$$E(\vec{W},\vec{W}') + \sum_{\alpha} \langle p,W'\rangle_{S_\alpha} = 0 \qquad \forall \vec{W}' \in (H^1(\Omega))^3 \tag{7}$$

Where,

W' is the normal component of the arbitrary vector $\vec{W}'$ on S_α

$\langle pW'\rangle_{S_\alpha} = \int_{S_\alpha} pW'dS_\alpha$ is the virtual work due to pressure loading on S_α.

$E(\vec{W},\vec{W}')$ is the virtual internal work given by

$$E(\vec{W},\vec{W}') = \langle\sigma(\vec{W}).\varepsilon(\vec{W}') - \rho\omega^2(\vec{W},\vec{W}')\rangle_\Omega \tag{8}$$

3.2 The fluid

From Helmholtz's equation (2) and taking into account the kinematic condition (4), it is easy to show by using the second Green's formula, that the pressure field in Ω_α has the integral representation :

$$p(x) = p_\infty(x) + \langle p_y \frac{\partial G_\alpha(x,y)}{\partial n_y} - \rho_\alpha\omega^2 W_y G_\alpha(x,y)\rangle_{S_\alpha} \tag{9}$$

where $G_\alpha(x,y) = e^{ik_\alpha R}/4\pi R$ is the free Green's function of the Helmholtz's equation (2),which satisfies the Sommerfeld radiation condition (5).

When x lies on S_α, we have from (9) the two integral representations :

$$p(x)/2 = \text{p.v} \langle p_y \frac{\partial G_\alpha}{\partial n_y}\rangle_{S_\alpha} - \rho_\alpha\omega^2 \langle W_y G_\alpha\rangle_{S_\alpha} + p_\infty(x) \tag{10}$$

$$W(x)/2 = \frac{\text{f.p}}{\rho_\alpha\omega^2} \langle p_y \frac{\partial^2 G_\alpha}{\partial n_x \partial n_y}\rangle_{S_\alpha} - \text{p.v} \langle W_y \frac{\partial G_\alpha}{\partial n_x}\rangle_{S_\alpha} + U_\infty(x) \tag{11}$$

where the integrals corresponding to the first and second derivative of the Green's function must be taken respectively in the sense of Cauchy principal value(v.p) and Hadamard finite part (f.p). $U_\infty = 1/\rho_\alpha\omega^2 \cdot \partial p_\infty/\partial n$ is the normal component of the acoustic displacement due to the harmonic incident wave.

For a given W,equation(11) can now be used to determine the pressure on S_α. Instead of solving (11) by collocation method,we associate to it the variational equation:

$$\omega^{-2} D_\alpha(p,p') = \langle p'_x W_y \frac{\partial G_\alpha}{\partial n_x} \rangle_{S_\alpha * S_\alpha} + \langle p'(W/2 - U_\infty) \rangle_{S_\alpha} \quad \forall\ p' \quad H^{1/2}(S_\alpha) \qquad (12)$$

where,

$$D_\alpha(p,p') = 1/\rho_\alpha \ \langle p'_x p_y \frac{\partial^2 G_\alpha}{\partial n_x \partial n_y} \rangle_{S_\alpha * S_\alpha}$$

As shown in ref(9) the symmetric bilinear form $D_\alpha(p,p')$ is also given by :

$$D_\alpha(p,p') = \langle (k_\alpha^2(p'\vec{n}_x, p\vec{n}_y) - (\vec{n}_x \wedge \nabla p', \vec{n}_y \wedge \nabla p))G_\alpha \rangle_{S_\alpha * S_\alpha} \qquad (13)$$

This transformation has the great advantage to avoid the explicit evaluation of the finite part..

Using equation (1) and adding (7) to (12) we obtain for the coupled system, the following variational formulation :

$$E(\vec{W},\vec{W}') + \sum_\alpha B_\alpha((W,p),(W',p')) = \sum_\alpha L_\alpha(W',p') \qquad \forall\ \{\vec{W}',p'\} \qquad (14)$$

Wher B_α is the bilinear symmetric form given by :

$$B_\alpha((W,p),(W',p')) = -\rho_\alpha \omega^2 \langle W_x W'_y G_\alpha \rangle_{S_\alpha * S_\alpha} - \omega^{-2} D_\alpha(p,p') +$$

$$+ 1/2 \langle W'p + Wp' \rangle_{S_\alpha} + \langle W_x p_y + W'_x p'_y) \frac{\partial G_\alpha}{\partial n_y} \rangle_{S_\alpha * S_\alpha} \qquad (15)$$

In the case of thin shell and when internal and external fluids are identical, it is more convenient to write the bilinear form B in terms of fluid load $q = p_2 - p_1$. This leads to the more simple expression :

$$B'((W,q),(W',q')) = \langle W'q + Wq' \rangle_{S_\alpha} - \omega^{-2} D(q,q') \qquad (15')$$

L_α is the linear form associate to the harmonic incident wave :

$$L_\alpha(W',p') = \langle U_\infty p' - p_\infty W' \rangle_{S_\alpha} \qquad (16)$$

Since the bilinear forms E and B_α are symmetric, the displacement vector $\vec{W}$ and the fluid loading on S_α make the functional,

$$F(\vec{W},p_1,p_2) = 1/2\{E(\vec{W},\vec{W}) + \sum B_\alpha((W,p),(W,p)) - 2L_\alpha(W,P)\} \qquad (17)$$

stationary.

4 - Discretisation by boundary finite element method

After the discretisation of the domain $\bar{\Omega}$ by finite element method, and making the discrete form of the functional (17) stationary, we obtain the two following symmetric systems. The first one corresponds to the case of thin shells.

$$\begin{bmatrix} \mathbf{K} - \omega^2\mathbf{M} & \mathbf{C} \\ \mathbf{C}^t & \mathbf{D}(k)/\omega^2 \end{bmatrix} \cdot \begin{vmatrix} \mathbf{W} \\ \mathbf{q} \end{vmatrix} = \begin{vmatrix} 0 \\ \mathbf{U}_\infty \end{vmatrix} \qquad \text{(I)}$$

$$\begin{bmatrix} \mathbf{K} - \omega^2(\mathbf{M} + \mathbf{A}1(k1) + \mathbf{A}2(k2)) & \mathbf{C}1/2 + \mathbf{B}1(k1) & \mathbf{C}2/2 + \mathbf{B}2(k2) \\ \mathbf{C}1^t/2 + \mathbf{B}1^t(k1) & \mathbf{D}1(k1)/\omega^2 & 0 \\ \mathbf{C}2^t/2 + \mathbf{B}2^t(k2) & 0 & \mathbf{D}2(k2)/\omega^2 \end{bmatrix} \cdot \begin{vmatrix} \mathbf{W} \\ \mathbf{p}1 \\ \mathbf{p}2 \end{vmatrix} = \begin{vmatrix} \mathbf{T} \\ \mathbf{U}1 \\ \mathbf{U}2 \end{vmatrix} \qquad \text{(II)}$$

Where,

$(k - \omega^2\mathbf{M})$ is the impedance of the dry structure, $\mathbf{A}_\alpha(k_\alpha)$ is the pseudo added mass on S_α, $\mathbf{C}_\alpha/2 + \mathbf{B}(k_\alpha)$ is the interaction matrix, $\mathbf{D}_\alpha(k)$ is the admittance of the acoustic fluid. Finally $\mathbf{T}$ and $\mathbf{U}_\alpha$ correspond to the acoustic load and displacement due to the harmonic wave.

When the internal and external fluids are incompressible ($k_1=k_2=0$), all the complex matrices depending on the wave numbers becames real and the condensation of fluid loading gives rise to the two added mass matrices :

$$\mathbf{M}_a^{\mathbf{I}} = \mathbf{C}^t\,\mathbf{D}^{-1}(0)\mathbf{C}$$

and

$$\mathbf{M}^{\mathbf{II}} = \sum \{\mathbf{A}_\alpha(0) + (\mathbf{C}_\alpha/2 + \mathbf{B}_\alpha(0))^t\,\mathbf{D}^{-1}(0)(\mathbf{C}_\alpha/2 + \mathbf{B}_\alpha(0))\}$$

The coupled problem consists forthis last case in solving the eigenproblem :

$$[\mathbf{K} - \omega^2(\mathbf{M} + \mathbf{Ma})] \cdot |\mathbf{W}| = |\mathbf{R}|$$

5 - Numerical results

We consider a thin spherical shell immersed in deep water as shown in figure 2. The sphere is excited by an internal load or by an harmonic incident plane wave $p_\infty = e^{-ikz}$. Numerical results are obtained by using equal axsymmetric finite elements of typical size l such that $kl \leq \pi/2$.

Figure 3 shows, in the case of internal load, the spectrum of the radial velocity of the sphere and of the fluid pressure on its external surface. Numerical and analytical solutions are in very good agreement.

Fig. 4 shows the spectrum of the retrodiffused pressure and the directivity pattern of sphere at ka = 8.377, when it is excited by the incident plane wave. The numerical and analytical results are also in very good agreement.

6 - Conclusion

A new numerical method has been developed to solving the fluid-structure interaction problems.It has the great advantage to avoid the discretisation of the fluid domain and to lead to an algebric symmetric system of small size which can be solved by fast algorithms and reduces considerably the computational costs.this method which is validated by analytical and experimental resuls,is suitable for solving aero-and hydroelastic problems where it is necessary to calculate the fluid loading and to determine the radiation efficiency of the vibrating structures.A general computer program based on this method is developed at the reserch center of the University of Technologie of Compiegne .

7 - References

1. Eatwell,G.P:Aspects of wave propagation in fluid-loaded structures.Ph D,University of Bath (1981).
2. Lauchle,G.C:Interaction of spherical acoustic wave with an elastic spherical shell. Journal of sound and vibration,44,n°1,pp.37-46 ,(1979).
3. Husson ,D:Diffraction d'une onde acoustique plane par une sphère élastique C.E.R.D.S.M , 83140 Six Foors.
4. Zienkiwiecz,O.C; Newton,R.E:Coupled vibration of structure submerged in compressible fluid. Int. Symp.of finite element tech.,Stuttgart (1969).
5. Ohayon ,R.;Morand,D:Variational formulation for the elasto-acoustics vibration problem:finite element results. Sec.Int.Symp.of F.E.M applied to flow problems.Rappalo 14-18 june 1976 Italy.
6. Mayer,W.L;Bell,W.A:Journal of Acoust.Soc.Am.,65,n°3,pp.631-638 (1979).
7. Filippi,J.F:Journal of sound and vibration,74,n°2,pp.473-500 (1977).
8. Geers,T.L;Felippa,C.A:Doubly asymptotic approximations for vibration analysis of submerged structures. J.Acoust.Soc.Am. vol 13(4) pp.1152-1159 (1983).
9. Hamdi,M.A:C.R. Acad.Sc. Paris, t.292, Série II.17 (janv.1981)
10. Hamdi,M.A; Mebarek,L: Rayonnement de structures à géometries axisymètriques 9°Colloque d'Acoust.Aéronautique, Compiegne 14-16 nov. 1984,France.

Fig 1. COUPLED SYSTEM

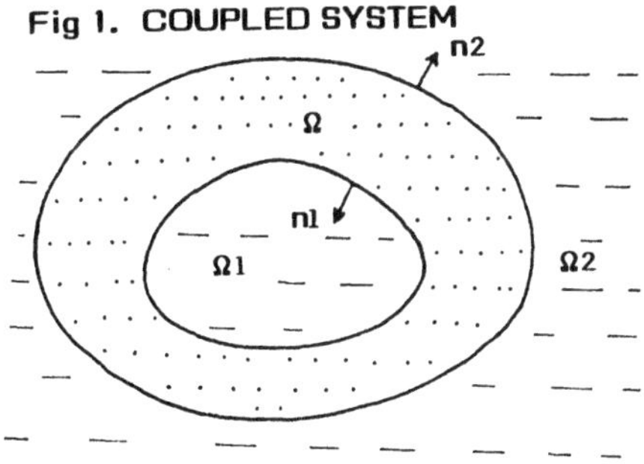

Characteristics of the shell

$c_s/c = 3.53 \qquad c_s = [E/2\rho_s(1+\nu)]^{1/2}$

$\rho_s/\rho = 7.67$

$h/a = 0.01$

$\nu = 0.3$

Fig 2. SPHERICAL THIN SHELL

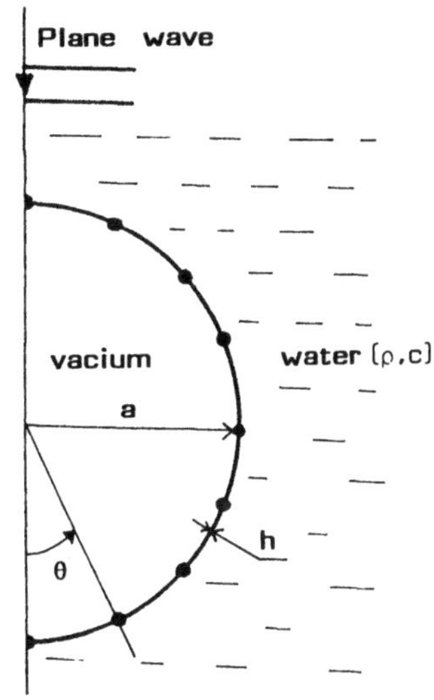

Numerical model

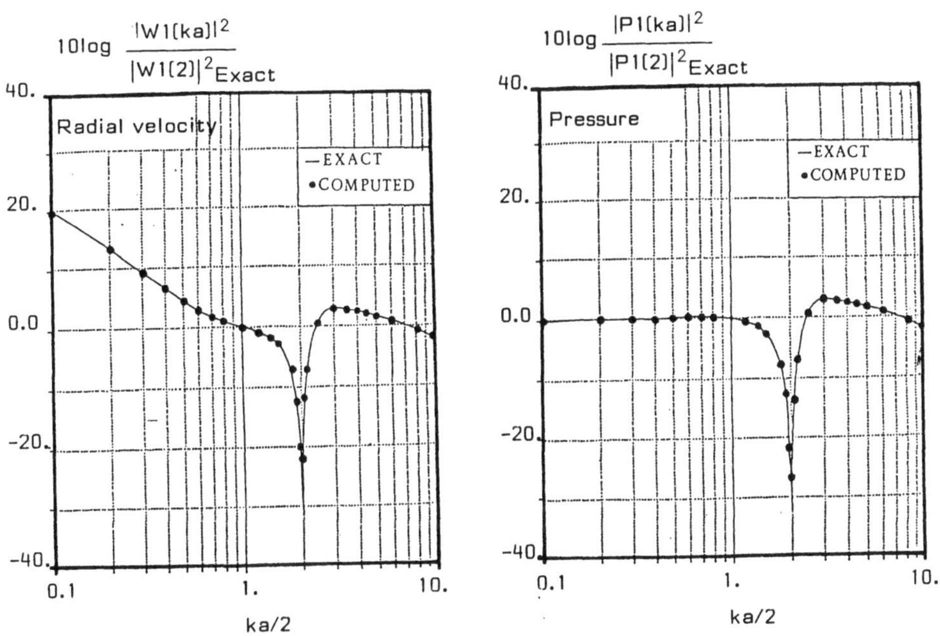

Fig 3. SPHERICAL THIN SHELL EXCITED BY INTERNAL LOADING (P=cosθ)

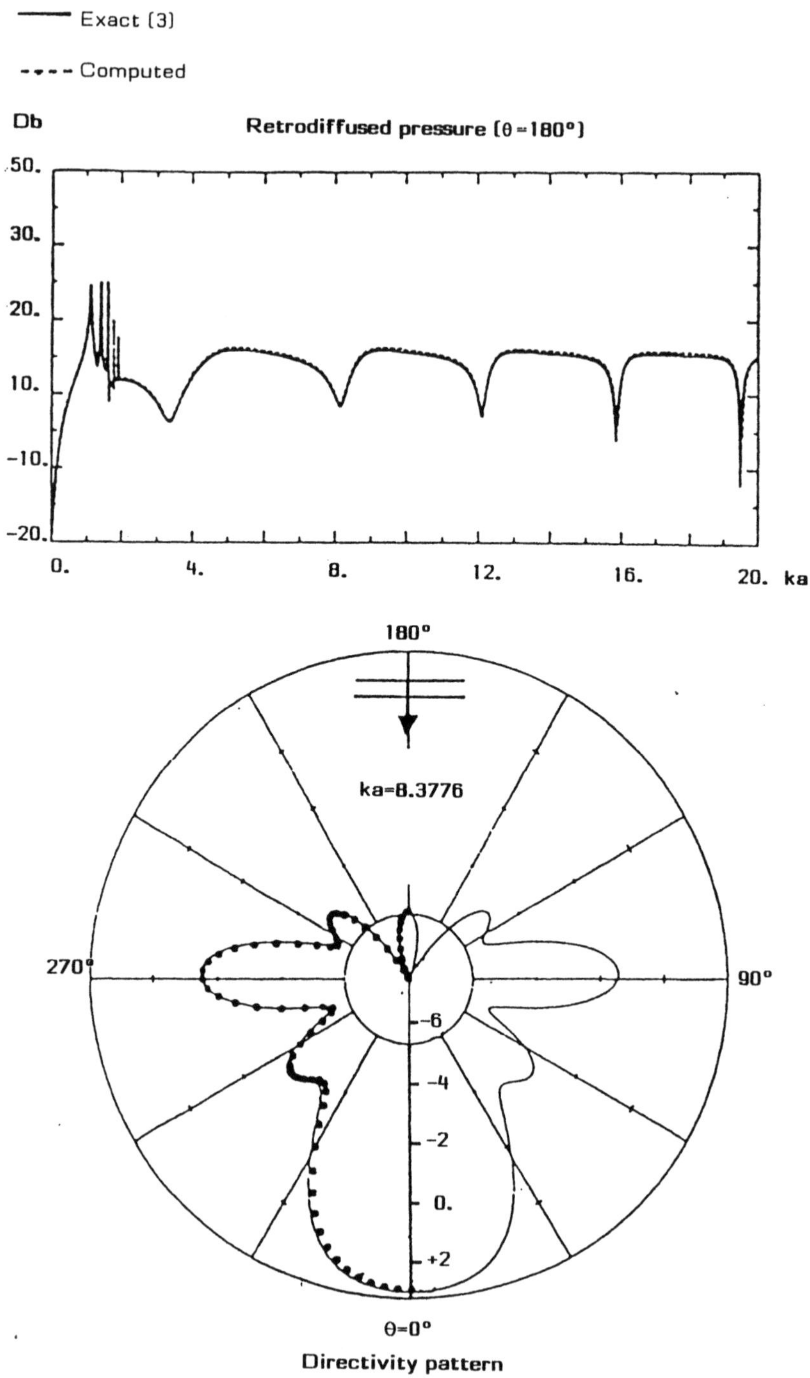

FIG. 4 SPHERICAL THIN SHELL EXCITED BY INCIDENT PLANE WAVE

Rayonnement d'une Plaque Mince, Homogène ou Raidie, Encastrée dans un Plan Parfaitement Rigide.

Paul J.T. FILIPPI

Laboratoire de Mécanique et d'Acoustique

31 chemin Joseph Aiguier, B.P. 71

13277 MARSEILLE CEDEX 9 (FRANCE)

SUMMARY : In the homogeneous case, the plate displacement is expressed by a Green formula, which involves the infinite fluid-loaded plate kernel. When stiffeners are present, they are accounted for by line forces, the densities of which are unknown functions. The boundary conditions and the continuity conditions on the stiffeners lead to a system of integral equations, the solution of which exists and is unique. This system is solved numerically by a collocation technique, in which the unknown functions are approximated by C^1-elements. Examples of directivity patterns of rectangular plates are given.

I-ENONCE DU PROBLEME

Soit Ω un domaine borné du plan $Z=0$, de frontière $\partial\Omega$ munie (presque partout) d'une normale extérieure n. Ce domaine est occupé par une plaque mince de caractéristiques suivantes : E = module d'Young ; ν = coefficient de Poisson ; h = épaisseur ; m = masse par unité de surface ; D = $Eh^3/12(1-\nu^2)$ = rigidité. Le complémentaire $\complement\overline{\Omega}$ de $\overline{\Omega}$ est une surface parfaitement rigide. La plaque est encastrée le long de $\partial\Omega$. Le demi-espace $Z>0$ est occupé par un fluide parfait homogène de caractéristiques : ρ_0 = densité ; c_0 = vitesse du son. Le demi-espace $Z<0$ est occupé par un fluide d'influence négligeable. La source d'énergie est une force harmonique $(\exp[-i\omega t])$ appliquée à la plaque, et de densité F.

Soit G la fonction de Green du problème de Neumann pour l'équation de Helmholtz, et correspondant au demi-espace $Z>0$, et Γ la fonction de Green de la plaque indéfinie couplée. Désignons par u le déplacement de la plaque. La représentation de Green de u est de la forme :

$$u = \Gamma \underset{2}{*} F + \Gamma \underset{2}{*} \{\mu_1 \boxtimes \delta_{\partial\Omega} + \mu_2 \boxtimes \delta'_{\partial\Omega}\} \tag{1}$$

où les densités de simple et double couches μ_1 et μ_2 satisfont le système d'équations intégrales de frontière suivant :

Aero- and Hydro-Acoustics
IUTAM Symposium Lyon 1985
Editors: G. Comte-Bellot and J. E. Ffowcs Williams

la poutre à une excitation f, on obtient une équation intégrale supplémentaire en exprimant que u est égal à w le long de C. On montrerait aisément que le triplet (f,μ_1,μ_2) existe, et est unique.

IV-APPROXIMATION NUMERIQUE

Soit $M_i (i=1,2,\dots,N)$ une suite de points de $\partial\Omega$, d'abscisses curvilignes s_i. On définit les fonctions $F_i(s)$ par :

$$F_i(s) = 2\left\{\left|\frac{s - s_i}{s_i - s_{i-1}}\right| - 1\right\}^2 \left\{\left|\frac{s - s_i}{s_i - s_{i-1}}\right| + \frac{1}{2}\right\} \quad \text{pour } s \in [s_{i-1}, s_i]$$

$$= 2\left\{\left|\frac{s - s_i}{s_{i+1} - s_i}\right| - 1\right\}^2 \left\{\left|\frac{s - s_i}{s_{i+\&} - s_i}\right| + \frac{1}{2}\right\} \quad \text{pour } s \in]s_i, s_{i+1}]$$

$$= 0 \quad \text{pour } s \notin [s_{i-1}, s_{i+1}]$$

On approche les fonctions $\mu_1(s)$ et $\mu_2(s)$ par des fonctions C^1 :

$$\mu_{1,2} = \sum_{i=1}^{N} \mu^i_{1,2} \, F_i(s)$$

Les $\mu^i_{1,2}$ sont déterminés en écrivant que les équations (2) sont satisfaites aux N points de collocation M_i. Lorsqu'il y a un raidisseur, on approche f d'une façon analogue, ce qui augmente d'autant le nombre d'équations et d'inconnues du système linéaire à résoudre. Enfin, il peut être utile, dans le cas de contours anguleux, d'introduire des fonctions singulières [2] : elles permettent de mieux décrire le déplacement de la plaque au voisinage des angles. Si seul le diagramme de rayonnement de la plaque à l'infini est à calculer, ce raffinement ne semble pas nécessaire.

V-EXEMPLES ET CONCLUSIONS

Nous avons calculé le rayonnement d'une plaque rectangulaire en acier, d'épaisseur 5 cm, et occupant le domaine $(-133 \text{ cm} < X < +133 \text{ cm} ; -100 \text{ cm} < Y < +100 \text{ cm})$. Elle est excitée par une force ponctuelle en X=Y=0, à 31,5 Hz. Sur la figure n°1, on présente le diagramme de directivité de la plaque homogène dans le plan X=0. La figure n°2 correspond au cas où la plaque est munie d'un raidisseur situé sur la droite X=0, de 12 cm d'épaisseur et de largeur 3 cm. La comparaison de ces deux digrammes met en valeur l'effet du raidisseur.

La méthode que nous avons développée a un champ d'application restreint à l'étude des plaques bafflées. Toutefois, elle a l'avantage de

$$(2)\quad \begin{cases} \mathrm{Tr}\ \Gamma_2 * \{\mu_1 \otimes \delta_{\partial\Omega} + \mu_2 \otimes \delta_{\partial\Omega}\} = \mathrm{Tr}(-\Gamma_2 * F) \\ \mathrm{Tr}\ \partial_n [\ \Gamma_2 * \{\mu_1 \otimes \delta_{\partial\Omega} + \mu_2 \otimes \delta_{\partial\Omega}\}\ ] = \mathrm{Tr}(-\partial_n [\Gamma_2 * F]) \end{cases}$$

La pression acoustique rayonnée par la plaque a pour expression :

$$(3)\qquad p = \omega^2\ \rho_0\ G_3 * (\ u \otimes \delta_Z\ )$$

En désignant par (R,θ,ϕ) les coordonnées polaires d'un point, le comportement asymptotique de p est donné par :

$$(4)\qquad p \simeq \omega^2\ \rho_0\ \frac{e^{ikR}}{2\pi R}\ \hat{u}(k\ \sin\theta\ \sin\phi\ ,\ k\ \sin\theta\ \cos\phi\ ) + O(R^{-2}),\ k = \omega/c_0$$

où $\hat{u}(\xi_1, \xi_2)$ est la transformée de Fourier de u.

II-EXISTENCE ET UNICITE DE LA SOLUTION

Soit $H(\Delta, R^3_+)$ l'espace des fonctions f de carré sommable dans R^3_+, et telles que Δf soit également de carré intégrable. L'espace $H^{loc}(\Delta, \overline{R}^3_+)$ est l'espace des fonctions f telles que :

$$f \in H^{loc}(\Delta, R^3_+) \text{ si } \varphi f \in H(\Delta, R^3_+) \text{ pour toute } \varphi \in \mathcal{D}(R^3)$$

Soit $H(\Delta^2, \Omega)$ l'espace des fonctions $v \in L^2(\Omega)$ telles que $\Delta v \in L^2(\Omega)$. En faisant le bilan d'énergie du système physique plaque-fluide, on montre [1] qu'il existe une et une seule solution (p,u) dans $H^{loc}(\Delta, \overline{R}^3_+) X H(\Delta^2, \Omega)$. On en déduit que le système (2) possède une et une seule solution, généralement C^∞ par morceaux.

III-INFLUENCE D'UN RAIDISSEUR

Une modélisation simple du raidisseur consiste à assimiler celui-ci à une poutre. La plaque exerce sur la poutre une force f, et, réciproquement, la poutre exerce sur la plaque une force linéïque $-f \otimes \delta_C$, où C est la ligne le long de laquelle est fixé le raisisseur. L'expression (1) du déplacement de la plaque devient :

$$(5)\qquad u = \Gamma_2 * F + \Gamma_2 * \{-f \otimes \delta_C + \mu_1 \otimes \delta_{\partial\Omega} + \mu_2 \otimes \delta'_{\partial\Omega}\}$$

La fonction f est une nouvelle inconnue. En désignant par w la réponse de ne demander qu'un volume de calculs assez faible, comparativement aux autres méthodes. Elle permet, de ce fait, de faire l'étude numérique détaillée de cas types, sans nécessiter des moyens lourds. Elle pourrait être pratique pour développer des méthodes d'optimisation visant à déterminer le traitement acoustique optimal à faire subir à un élément de structure plan, en vue d'obtenir une réduction, fixée a priori, de l'énergie acoustique rayonnée. Notons enfin que ce type de technique est applicable à d'autres problèmes : par exemple, le calcul du rayonnement d'une coque cylindrique de longueur finie, prolongée par deux cylindres semi-infinis et parfaitement rigides.

BIBLIOGRAPHIE

1. - Paul J.T. FILIPPI, 1985, "Boundary integral equations for sound radiation by a harmonically vibrating baffled plate", in "Singularities and Constructive Methods for their Treatment", Lecture Notes in Mathematics n°1121, Springer Verlag.

2. - Pierre GRISVARD, 1984, "Boundary value Problems in non smooth domains", Pitman.

3. - Paul J.T. FILIPPI, 1984, "Rayonnement acoustique des structures", Rapport D.R.E.T. n°82/302.

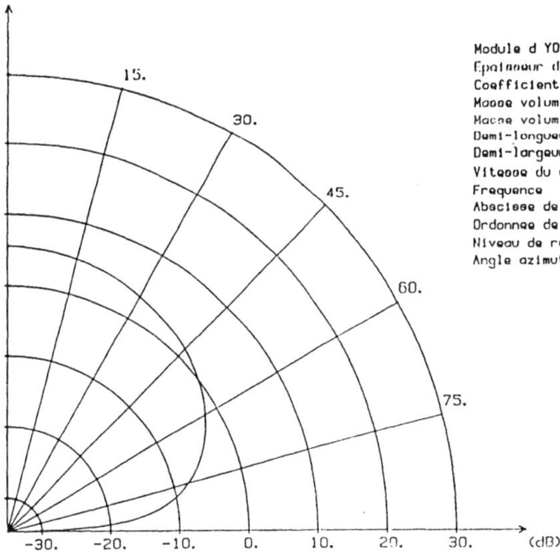

Module d YOUNG	(cgs)	E	=	0.216E+13
Epaisseur de la plaque	(cm)	H	=	5.000
Coefficient de POISSON		PNU	=	0.276
Masse volumique de la plaque	(g/cm3)	ROP	=	7.840
Masse volumique du fluide	(g/cm3)	ROF	=	1.000
Demi-longueur de la plaque	(cm)	FLONG	=	133.000
Demi-largeur de la plaque	(cm)	FLARG	=	100.000
Vitesse du son	(cm/s)	VC	=	0.149E+06
Frequence	(Hz)	FREQ	=	31.500
Abscisse de la source	(cm)	XSO	=	0.000
Ordonnee de la source	(cm)	YSO	=	0.000
Niveau de reference	(dB)	REF	=	-118.327
Angle azimutal	(rd)	PHI	=	1.571

PLAQUE RECTANGULAIRE ENCASTREE figure n°1

CNRS equipe P.A.P.E.

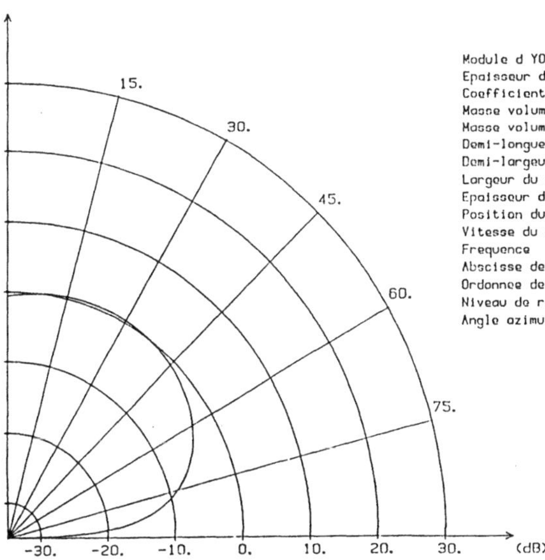

Module d YOUNG	(cgs)	E	=	0.216E+13
Epaisseur de la plaque	(cm)	H	=	5.000
Coefficient de POISSON		PNU	=	0.276
Masse volumique de la plaque	(g/cm3)	ROP	=	7.840
Masse volumique du fluide	(g/cm3)	ROF	=	1.000
Demi-longueur de la plaque	(cm)	FLONG	=	133.000
Demi-largeur de la plaque	(cm)	FLARG	=	100.000
Largeur du raidisseur	(cm)	FETIL	=	3.000
Epaisseur du raidisseur	(cm)	PETIH	=	12.000
Position du raidisseur	(cm)	POSIT	=	0.000
Vitesse du son	(cm/s)	VC	=	0.149E+06
Frequence	(Hz)	FREQ	=	31.500
Abscisse de la source	(cm)	XSO	=	0.000
Ordonnee de la source	(cm)	YSO	=	0.000
Niveau de reference	(dB)	REF	=	-118.327
Angle azimutal	(rd)	PHI	=	1.571

PLAQUE RECTANGULAIRE ENCASTREE RAIDIE figure n°2

CNRS equipe P.A.P.E.

Hydroacoustics

Topics in Hydro-acoustics

MAURICE M. SEVIK

David Taylor Naval Ship R&D Center
Bethesda, Maryland 20084 USA

ABSTRACT

The characteristics of three sources of flow-induced vibrations and noise which commonly occur in hydroacoustic applications are discussed. The first, a monopole-like source, concerns cavitation noise. Recent data obtained on weakly cavitating jets and vortices which develop at the tip of lifting surfaces suggests the existence of a universal spectrum function. An expression for this function is provided and its domain of validity is defined. The second source is dipole-like and concerns the sound field from a propeller rotating in a spatially non-uniform flow field. If the wavelength of the sound is much larger than a characteristic length of the source, the sound field reduces to that of a dipole. The source strength, however, involves some complexities because of hydroelastic interactions between the structure and the fluid. Theoretical predictions based on unsteady lifting surface theory are compared with experimental data and indicate that the dynamic response of the blades significantly affects the sound pressure levels. A third source of noise of importance in hydroacoustics arises from pressure fluctuations within turbulent boundary layers. Recent measurements of wave number spectra on smooth and rough surfaces are reported.

1. UNIVERSAL SPECTRUM FORMS OF CAVITATION NOISE

Cavitation is the most intense source of noise in fluid machinery and underwater acoustics. It occurs in a flow field whenever the local pressure falls below a critical value. The conditions under which cavitation incepts, the subsequent development of a cavitation zone and the sound emitted by the intense, interactive processes that take place within that zone are extremely complex. Theoretical analyses are almost exclusively restricted to isolated spherical bubbles containing varying amounts of non-condensible gas. Although these analyses have made valuable contributions to the fundamental understanding of cavitation noise, the problem of predicting sound pressure levels in practical applications still remains largely unsolved. In marine engineering, for example, a common approach is to

Aero- and Hydro-Acoustics
IUTAM Symposium Lyon 1985
Editors: G. Comte-Bellot and J. E. Ffowcs Williams

measure the cavitation noise of a model propeller in a water tunnel or towing tank and then attempt to deduce the fullscale levels that would occur at sea. One of the difficulties with this procedure is that the environmental conditions that are found at sea differ significantly from those that exist in laboratories, in addition to the fact that cavitation, Froude and Reynolds scaling criteria cannot be simultaneously achieved. Here, we shall review the factors which influence cavitation noise and describe results of some large scale measurements which may be useful in "calibrating" test facilities and scaling techniques.

As is well known, cavitation voids develop around nucleation sites such as suspended microbubbles or small, gas-filled crevices in hydrophobic particles. The critical pressure depends upon the number and size of nuclei per unit volume of liquid. Measurements made in the ocean reveal that the density distribution of nuclei is a function of location, depth, and time of year. Data cited by Shen, Growing, and Pierce(1), shown in Figs. 1 and 2 illustrate the variation of the density distribution function of nuclei at sea and in a water tunnel.

From a static equilibrium theory, the critical pressure P_c for cavitation is

$$P_c(r) = P_v - \left(\frac{3\gamma-1}{2\gamma}\right)\frac{4S}{3R_c} \qquad \text{for } r > R$$

where P_v is the vapor pressure of the liquid and R_c is the critical bubble radius. The theory assumes the nuclei are spherical bubbles of radius R immersed in a liquid with surface tension S. The bubbles contain a gas whose ratio of specific heats is γ. A bubble may become unstable at a critical pressure which is less than the vapor pressure and this pressure difference is greater the smaller the bubble radius. In practice, microbubbles whose radius are of the order of 10^{-3} cm or larger are sufficient to trigger cavitation at vapor pressure.

When cavitation occurs, the gas contained in the original nucleus expands into a much larger volume, and its partial pressure is greatly exceeded by the vapor pressure of the liquid. This is called "vaporous cavitation" in contrast to "gaseous cavitation" which occurs when growth is due to the diffusion of dissolved gas into the bubble from the surrounding liquid. The bubble growth rates associated with vaporous cavitation are many orders of magnitudes larger than those of gaseous cavitation. When the rarefaction is replaced by a compression, calculations by Hickling and Plesset(2) show that the termination phase of the bubble collapse is controlled by the gas content and by the compressibility of the liquid.

Since this phase contributes most of the sound energy, the possibility of reducing cavitation noise by the addition of gas to the cavitation zone appears to be a promising idea. The acoustic effects of gas entrained in a bubble are illustrated in Fig. 3 which is due to Baiter[3]. Here, Baiter defines an acoustic efficiency η_{ac} as the ratio of acoustic energy radiated E_{ac} to the potential energy $p_o V_M$ initially available for the collapse to be set in motion

$$\eta_{ac} = \frac{E_{ac}}{p_o V_M}$$

where V_M is the maximum bubble volume and p_o is a constant value assumed by the ambient pressure p_a after a step rise initiating bubble collapse. This acoustic efficiency is normalized by a compressibility factor $m = [p_o/(\rho c^2]^{1/2}$ which has a value of 7×10^{-3} in water at $p_o = 1$ bar. The gas content parameter $P(R_M)/p_a$ is defined as the ratio of the partial pressure of the non-condensible gas contained in the bubble when its radius is maximum to the ambient pressure p_a. Baiter's calculations clearly show the noise-reducing effects of noncondensible gas.

In hydrodynamic cavitation, the pressures surrounding the cavitation nuclei are affected by the details of the flow which, in turn, depend upon the Reynolds number. Tip and hub vortex cavitation inception on marine propellers, for example, are strongly influenced by Reynolds number since the radius of the vortex core is governed by the boundary layer in the neighborhood of the tips and roots of the blades. A similar statement applies to cavitation inception in turbulent flows such as regions of flow separation or submerged jets. On the other hand, cavitation inception at the boundary of two- or three-dimensional bodies can be predicted with good accuracy from scale-model tests or calculations, providing the surfaces are smooth and free of protrusions.

Recent studies of bubble dynamics were performed by Hentschel and Lauterborn[4]. They used a holographic technique in order to observe the dynamics of a swarm of micro-bubbles generated in the rarefaction region of an acoustic sound field. An examination of the reconstructed holograms obtained with an ultra-high-speed camera shows the complexity of the interactions which occur within the cavitation zone, as a function of time. It is interesting to note, for example, the formation of small jets which penetrate coalescing bubbles. This phenomenon appears to be more generally applicable and is illustrated in Fig. 4, from Sevik and Park[5]. Here, bubbles of air were emitted from a small pipe placed at the center of a jet of water which was discharged into a large reservoir. In spite of the differences in length and time scales involved in the experiments of references 4 and 5, the hydrodynamics of coalescing bubbles appears to be quite similar. Remarkable photographs published by Brinkmeyer(6) and later also by Hentschel(7) clearly show the shock waves generated by a collapsing cavity and their effects on adjacent bubbles.

The prediction of cavitation noise in practical applications is evidently a problem of some complexity. For example, if it is based on measurements of the noise radiated by scale models in a water tunnel a large number of hydrodynamic and hydroacoustic conditions must be met if an accurate extrapolation to full-scale is to be achieved. The facility requirements, measurement techniques, and scaling laws which are needed for this purpose were outlined by Blake and Sevik(8). In order to "calibrate" such an approach it is desirable to possess full-scale data of the sound radiated by various forms of cavitation by performing a set of "canonical" experiments in the environment in which the machine operates. For example, Cumming and Morgan(9) have observed various types of cavitation on marine propellers such as tip and hub vortex cavitation, leading edge attached sheet cavitation and mid-chord bubble and cloud cavitation. Frequently, these various forms of cavitation occur simultaneously and the contribution of each form to the total sound energy cannot be distinguished.

For this reason, measurements were made at sea on a set of hydrofoils on which tip vortex cavitation could be induced as the only form of cavitation. The Reynolds number, based on chord length, was on the order of 10^7. Ambient and background noise levels were controlled in such a manner that incipient stages of cavitation could be detected by acoustic means. Fig. 5 shows examples of the noise measured in the far field with an array of hydrophones, as a function of time. Here, the sound intensity in a 1/3 octave band centered at several kHz is displayed. Fig. 5a shows intermittent spikes of noise which corresponds to a condition close to incipient cavitation, whereas in Fig. 5b the cavitation zone is more extensive. The cavitation inception condition is determined from data exemplified in Fig. 6. If the background noise at the hydrophones is sufficiently low, a sudden rise in level is observed, as cavitation incepts. The corresponding ratio of static to dynamic head defines the incipient cavitation index, σ_i. As the velocity of the flow is increased, the cavitation noise becomes more intense and continuous, although, as shown in Fig. 5b, high level, intermittent bursts are still present. Typical spectra of the radiated noise are shown in Fig. 7, for two values of σ/σ_i. These spectra have similar charactristic features. They contain a peak level $\overline{p_m^2}\ (f_m, \Delta f)$ at a frequency f_m and exhibit similar decay rates with frequency.

There are cases in which it is desirable to detect the onset of cavitation in its early, incipient stages. For this purpose, it is useful to know the frequency f_m and the maximum sound pressure level to be expected at that frequency. A simple dimensional analysis due to Blake[8] allows an empirical prediction of these quantities. Since cavitation is a monopole-like source, the sound pressure level $p(\underline{r},t)$ at a point $\underline{r}$ distant from the cavitation zone is proportional to the net volumetric acceleration of this zone at any instant of time

$$p(\underline{r},t) = \rho \int_{vol} \frac{\ddot{q}(\underline{y},\, t-r/c)}{4\pi r}\, dV(\underline{y})$$

where q is the time-dependent cavity volume per unit volume of fluid.

Characteristic length and time scales $\mathcal{L}$ and τ describe the growth of the cavitation zone after inception:

$$\mathcal{L} = L\,(\sigma_1 - \sigma)^{1/2}$$

and

$$\tau = L\,(\sigma_1 - \sigma)^{1/2}\,\rho^{1/2}\,p_\infty^{-1/2}$$

where L is a length scale of the body generating the cavities, as for example, the chord of a hydrofoil or the diameter of a jet.

Based on the choice of these scales, the mean-square pressure p^2 in a frequency band Δf will vary as

$$\frac{\overline{p^2}\,(f,\Delta f)}{p_\infty^2\,(L/r)^2} = \left[\frac{\Delta f\, L\,(\sigma_1-\sigma)^{1/2}}{p_\infty^{1/2}\rho^{-1/2}}\right](\sigma_1-\sigma)\;\phi\left(\frac{f\, L\,(\sigma_1-\sigma)^{1/2}}{p_\infty{}^{1/2}\rho^{-1/2}},\ \sigma_1, R_L\right)$$

where R_L is the Reynolds number. At the frequency f_m, the theory of isolated bubble cavitation noise suggests that the function ϕ would have a maximum value when

$$\frac{f_m\, L\,(\sigma_1-\sigma)^{1/2}}{p_\infty{}^{1/2}\rho^{-1/2}} = \alpha$$

For that

$$\frac{p_\infty{}^{1/2}\rho^{-1/2}}{f_m L} = \frac{(\sigma_1-\sigma)^{1/2}}{\alpha}$$

where α is a constant that depends on the type of flow. In proportional frequency bands Δf = const. x f, the maximum spectrum level varies as

$$\frac{\overline{p_m^2}}{p_\infty^2\,(L/r)^2} \sim (\sigma_1 - \sigma)\ \phi\ (\alpha, \sigma_1, R_L)$$

For a given form of cavitation in a given type of flow, the function ϕ should only vary with f/f_m. This function can be conveniently found from the ratio

$$\left(\frac{f}{f_m}\right)\frac{\phi\left(\frac{f}{f_m}\ \alpha, \sigma_1, R_L\right)}{\phi(\alpha, \sigma_1, R_L)} = \frac{\overline{p^2(f,\Delta f)}}{\overline{p_m^2}}$$

Figs. 8 and 9 show the dependence of $\overline{p_m^2}$ and f_m on $(\sigma_1 - \sigma)^{1/2}$ for tip vortex cavitation. The choice of characteristic length and time scales for the cavitation zone appears to be valid up to a critical value of $(\sigma_1 - \sigma)^{1/2}$. It is probable that above this value more advanced cavitation significantly alters the structure of the flow. The spectrum functions which are plotted in Fig. 10 indicate the existence of a universal spectrum form.

To provide further support for this hypothesis, the same approach was applied to the noise measurements made by Jorgensen(10) on cavitating jets. The frequency and pressure factors differ from those obtained previously because shear layer instead of tip vortex cavitation is involved. Fig. 11 shows the spectrum function which has a remarkable resemblance to Fig. 10.

It appears that an empirical prediction of cavitation noise produced by a given type of cavitation is possible and requires the knowledge of the spectrum function $\phi\left(\frac{f}{f_m}\ \alpha, \sigma_1, R_L\right)$ as well as $\overline{p_m^2}$ and f_m. A collection of data from carefully conducted large scale experiments under realistic operating conditions would provide a useful data base for "calibrating" model test facilities and model-scale test results.

2. DIPOLE-LIKE SOURCES IN HYDROACOUSTICS

In the absence of monopole-like sources such as cavitation or pulsating gas bubbles, dipole-like sources are next in importance. In most hydroacoustic applications, these sources are compact and the Mach numbers of the flow are close to zero. In this important class of problems, the radiated sound power is related to the statistics of the alternating forces acting on the bodies by a simple relationship. For an acoustically compact, rigid body, the sound intensity is directly proportional to the time derivative of the mean square fluctuating force F(t)

$$I(\underline{x}) = \frac{1}{16\pi^2 \rho c^3} \frac{\cos^2\theta}{r^2} \left[\frac{\partial \overline{F(t)}}{\partial t} \right]^2$$

where θ is measured from the direction of the fluctuating force vector.

A marine propeller is an example of such a radiator. A modern design, manufactured by the Bird-Johnson Co. Marine Division, is illustrated in Fig. 12. It is clear that it operates in an incident velocity field which contains spatial and temporal non-uniformities. The response of each blade is proportional to the product of the intensity of the flow disturbances and of the blade's hydroelastic admittance function. If structural resonances are encountered, a possible enhancement of the response must be considered as this will influence the level of the radiated noise as well as the fatigue life of the blades.

The hydrodynamic acceptance function of a propeller has maxima at wavelengths which are integral numbers of the blade spacing. The resultant sound is proportional to the integrated value of the fluctuating force components which act on the propeller and energy is concentrated at nearly discrete values of the circumferential wave number $k_\theta = \pm m \frac{B}{R_o}$ where m is an integer B the number of blades, and R_o the propeller radius.

Although the radiation problem of an acoustically compact propeller is trivial, the prediction of the source strength is a hydroelastic problem of some complexity. The approach is to decompose the incident flow field into a steady and a randomly varying pattern and calculate the response to each separately.

The prediction of forces acting on a propeller which rotates with angular velocity Ω in a steady, but circumferentially non-uniform incident flow requires the application of a three-dimensional unsteady lifting surface theory because of the low aspect ratio of the blades. The unknown pressure difference Δp at a point x^1 on a blade is related to the self-induced velocity $w(\underline{x},\overline{t})$ by an integral equation

$$w(\underline{x},t) = -\int_S \Delta p(\underline{x}^1)\, K(\underline{x}, \underline{x}^1, t)\, ds$$

where s represents the area of the blades. The kernel function K represents the velocity induced at $\underline{x}$ by an oscillatory pressure of unit amplitude located at $\underline{x}^1$.

The boundary conditions require that $w(\underline{x},t)$ exactly cancel velocity components normal to the blade surfaces. If these do not deform under load, then

$$w(\underline{x},t) = v(R) \exp iq(\Omega t - \gamma_q)$$

where $\frac{2\pi R}{q}$ is the wavelength of the disturbance and v(R) is the velocity component of the blade at a radial position R. The phase angle γ_q is illustrated in Fig. 15. Modern designs, however, have complex geometries and several resonances may occur within the operating range. In this case $w(\underline{x},t)$ must also cancel a velocity component due to blade vibrations. The linearized boundary condition in this case is given by

$$w(\underline{x},t) = \left\{ \frac{\partial}{\partial t} + U \cot\beta(R) \frac{\partial}{R\partial\phi_0} \right\} \delta(\underline{x},t)$$

where $\beta(R)$ is the pitch angle of the blade at the radial position R and $\delta(\underline{x},t)$ is the normal deflection at (R,ϕ_0)

and U is the volume mean velocity over the propeller disk. In addition, the Kutta condition at the trailing edge applies at all instants of time.

Tsushima and Sevik(11) solved the integral equation, subject to these boundary conditions. To verify the predictions, experiments were performed on a "flexible" and on a "rigid" propeller of identical geometries, shown in Fig. 13. The blades had a skew angle of 120 degrees. The "flexible" propeller was made of an epoxy resin and the "rigid" propeller of aluminum alloy. The resonances of the epoxy propeller were within the operating rpm; those of the aluminum propeller were outside that range. The lowest resonance frequency of the epoxy propeller when submerged in water was 46 Hz, whereas that of the aluminum one was higher by a factor of 2.34. During the experiments the exciting frequency was limited to 55 Hz. The epoxy propeller had a loss factor of 2.5×10^{-2}.

The propellers were mounted in the water tunnel of the Applied Research Laboratory at the Pennsylvania State University, as shown in Fig. 14. An electric motor varied the rev/min so as to maintain a constant advance ratio over a range of tunnel flow velocities. A screen upstream of the propeller generated a strong three cycle wake, as shown in Fig. 15. A simple dynamometer, consisting of a piezoelectric crystal embedded in the propeller shaft, was capable of measuring only a single force component, namely the unsteady thrust.

The theoretical calculations of the unsteady forces included the first five modes of vibration of the blades. The dynamic response of the blades gave rise to significant differences in the magnitude and phase of the forces, as illustrated in Fig. 16. Since the simple dynamometer used in the experiments only measured the fluctuating thrust, theoretical as well as experimental results are plotted in Fig. 17 as the ratio of the root-mean-square value of the force acting on the flexible propeller $|T_F|$ to that acting on the rigid propeller $|T_R|$. Although the agreement between theory and experiment is not precise, the general features of the response are evident in both. Additional experiments and analyses performed by Brooks[12] confirmed earlier results.

These results can be scaled linearly, providing the geometry and the incident flow field are similar and the loss factor is the same. For example, a propeller made of bronze, ten times larger than the model, would have a fundamental frequency of approximately 11 Hz and the peak amplitude of the unsteady thrust force would occur at about 180 rev/min. Near resonance, the intensity of the radiated sound would be 15 dB higher than that of an ideally "rigid" propeller.

3. SOME RECENT MEASUREMENTS OF BOUNDARY LAYER NOISE

The noise radiated by turbulent boundary layers, the wall motions induced by fluctuating pressures and the resulting acoustic field have received much attention in hydroacoustics. The early, classical papers of Powell[13] and Ffowcs Williams[14] treated turbulent flows contiguous to rigid or flexible boundaries which were infinite in extent and devoid of inhomogeneities. They showed that the radiation is quadrupole in character. However, at the Mach numbers of interest in underwater acoustics, this source of radiation has an extremely low acoustic efficiency. It was evident that quadrupole theory could not explain the much higher levels observed in practice.

Two sources contribute to the sound field in practical applications, namely direct radiation from the flow and radiation from structures that are excited by the fluctuating wall pressures. Real structures, of course, contain numerous impedance discontinuities, and actual boundary layers are seldom spatially homogeneous. Accordingly, there exist many acoustic scattering mechanisms which convert the intense reactive energy of turbulent eddies into more efficient acoustic sources. Powell, for example, showed that although pressure dipoles vanish over a surface, they need not vanish near its edges. More recently, Howe[15] examined the problem of sound generation by scattering of the near field of low Mach number boundary layer turbulence by a rough, rigid wall. He concluded that the sound intensity varies as the sixth power of the main stream velocity, in accordance with experimental data due to Hersh[16]. This scattering mechanism affects the direct radiation from the flow.

A wave number conversion process which influences the radiation from structures was analyzed by Crighton[17]. The problem involved the long range coupling between a quadrupole eddy which injects energy into an elastic structure and surface inhomogeneities, such as ribs or frames. This energy remains trapped in the surface and an adjacent fluid layer and propagates in subsonic surface wave modes until a discontinuity is encountered. Crighton showed that at low frequencies and under conditions of heavy fluid loading, this process efficiently converts some of the near field reactive energy of the quadrupole into radiative energy. In practical applications, estimates of the structural response and of the resulting sound radiation require that the wave number spectrum of turbulent boundary layer pressures be known. The acceptance of structures of interest has maxima in a range of wave numbers

$$\frac{\omega}{c} < k << \frac{\omega}{U_c}$$

where ω/c and ω/U_c are the sonic and convective wave numbers, respectively. Accordingly, in hydroacoustic applications, the wave numbers of greatest interest are those in the sonic and supersonic range for direct radiation from the flow, and those in the low, subsonic wave number range for radiation from flow induced vibrations.

Existing models of the wave number-frequency spectrum of the wall pressure field beneath a turbulent boundary layer assume that it is spatially homogeneous and temporally stationary. The cross-correlation function for the pressure at two arbitrary points on the wall is therefore given by the ensemble average with respect to time

$$R(\underline{\xi}, t) = \langle p(\underline{x}, t)\, p(\underline{x} + \underline{\xi}, t + \tau)\rangle$$

where $\underline{\xi} = (\xi_1, \xi_3)$ is the separation distance between two points and τ is a time delay. The distances in the streamwise and crossflow directions are ξ_1 and ξ_3, respectively.

The wave vector-frequency spectrum is the Fourier trans form of $R(\underline{\xi}, \tau)$

$$\Phi_p(\underline{k},\omega) = \left(\frac{1}{2\pi}\right)^3 \int_{-\infty}^{\infty}\int_{-\infty}^{\infty}\int_{-\infty}^{\infty} R(\underline{\xi}, \tau) \exp\left[i(\underline{k}\,\underline{\xi} - \omega\tau)\right] d^2\underline{\xi}\, d\tau$$

where ω is the frequency and $\underline{k} = (k_1, k_3)$ the wave vector.

The stream wise and crossflow components are denoted by k_1 and k_3, respectively. The cross-spectral density is the partial Fourier transform over time

$$\Psi(\underline{\xi}, \omega) = \frac{1}{2\pi} \int_{-\infty}^{\infty} R(\underline{\xi}, \tau) \exp(-i\omega\tau)\, d\tau$$

The autospectral density function is given by

$$\Phi(\omega) = \Psi(0,\omega) = \frac{1}{2\pi} \int_{-\infty}^{\infty} R(0,\tau) \exp(-i\omega\tau) d\tau$$

and

$$\Phi(\omega) = \int_{-\infty}^{\infty} \int_{-\infty}^{\infty} \Phi_p(\underline{k},\omega)\, d^2\underline{k}$$

The first model of the wave number frequency spectrum was based on similarity principles proposed by Corcos[18]. Wall pressure measurements by Willmarth and Wooldridge[19], and Bull[20] suggested a separable representation for the cross-spectral density function in the form

$$\Psi(\xi_1, \xi_3, \omega) = \Phi(\omega)\, A\left(\frac{\omega\xi_1}{U_c}\right) B\left(\frac{\omega\xi_3}{U_c}\right) \exp\left(-i\, \frac{\omega\xi_1}{U_c}\right)$$

The "Corcos model", which is the Fourier transform of this function, has several limitations. For example, it does not account for the effects of compressibility which control the sonic and supersonic phase velocity range of the spectrum. It also violates the $|k|^2$ dependence of the spectrum at low wave numbers, for perturbations which propagate at low subsonic phase velocities. Note that the convection velocity U_c is a function of the frequency: larger values of U_c are associated with smaller values of $\frac{\omega\delta^*}{U_\infty}$ where δ^* is the displacement thickness of the boundary layer and U_∞ is the free stream velocity.

Ffowcs Williams[21] extended the Corcos model to make it applicable to the low wave number elements of the spectrum. Starting with Lighthill's acoustic analogy and his own earlier paper[22], he proposed the following representation for the wall pressure spectrum

$$\Phi_p(\underline{k},\omega) = \rho^2 U_\infty^3 \delta^{*3}\, \Phi_0\left(\frac{\omega\delta^*}{U_\infty}\right) A_0\left(1 - \frac{k_1 U_c}{\omega}\right) B_0\left(\frac{k_3 U_c}{\omega}\right) \times$$

$$\left\{a_0\left(\frac{U_c|\underline{k}|}{\omega}\right)^2 + a_1 M^2 + a_2 M^4 \ln(R/\delta^*)\,\delta\left[\left(\frac{U_\infty|\underline{k}|}{\omega}\right)^2 - M^2\right]\right\}$$

where ρ is the density of the fluid, δ the Dirac delta function, and $M = \frac{U_\infty}{c}$. The functions Φ_0, A_0, B_0 and the constants a_0, a_1 and a_2 must be determined by experiments. An objective of the present paper is to use experimental data from which function $\Phi_0\left(\frac{\omega\delta^*}{U_\infty}\right)$ and the product $\left[A_0(1)B_0(0)\; a_1\right]$ can be found.

The last term in Ffowcs Williams' expression represents the acoustically co-incident elements of the spectrum. Here, R denotes the effective extent of the turbulence zone which contributes to the energy at $k^2 = \frac{\omega^2}{c^2}$.

A third, recent model by Chase[23] for the hydrodynamic range of wave numbers accounts for both the linear mean shear and the nonlinear turbulence-turbulence interaction terms. In other words, the model includes the sum of interactions of each scale of motion with the mean flow as well as the self-interactions of eddies and the coupled interactions between different scales of motion. The spectrum is of the form

$$\frac{\Phi_p(\underline{k},\omega)\, U_\infty}{\tau_w^2} = \frac{U_\infty}{U_\tau}\left(C_M k_1^2 K_M^{-5} + C_T\left(k_1^2 + k_3^2\right) K_T^{-5}\right)$$

where

$$K_i^2 = \frac{\left(\omega - U_c k_1\right)^2}{h_i^2 U_\tau^2} + (k_1^2 + k_3^2) + (b_i\delta)^{-2};\; i = M, T$$

and the subscripts M and T refer to either the mean shear or the turbulence interaction. The shear stress at the wall τ_w, the friction velocity U_τ , and the dimensionless coefficients C_M, C_T, h_i and b_i depend upon the surface finish of the wall. For smooth walls, Chase recommends the following values:

$$h_M \simeq h_T \simeq 3 \ ; \ C_T = 0.0474 \ ; \ C_M = 0.0745$$

$$b_T = 0.378 \ ; \ b_M = 0.756$$

With this choice of coefficients, the convective pressure levels agree well with experimental values and the wave number spectrum displays the proper $|k|^2$ dependence in the low wave number domain. Coefficients for rough walls are given by Blake[24] in his recent book.

With the advent of large computers, the possibility of calculating wall pressure fluctuations numerically from the full Navier-Stokes equations is being explored. A recent paper by Handler, Hansen, Orszag et al[25] provides a comparison of numerical and experimental results for a fully developed channel flow. Although numerical techniques show promise, further developments are needed if information of practical value is to be provided.

In order to measure the weak signals in the low wave number and acoustic wave number domains, two essential requirements must be met. First, a very quiet test facility must be used such as an anechoic wind tunnel or a buoyant body in a region of water with very low ambient noise levels, as in the experiments of Haddle and Skudrzyk[26]. Secondly, an instrument must be available which selectively responds to the spatial scales of interest and rejects others - such as convective disturbances - which may be more intense by a factor as great as 10^5. Jameson[27] and Martin[28] used vibrating plates and membranes as spatial filters. By selecting specific modes of vibrations, the force on the panel applied by the matching scales of the turbulence in the boundary layer is calculated. This technique applies to the hydrodynamic range of wave numbers. Maidanik and Jorgensen[29] devised an array of pressure transducers - a wave vector filter - which measures the spectral density of the wall pressures in terms of k and ω. The array is designed to respond to a bandwidth of specific wave numbers only. When coupled to a frequency filter, the resultant output is a measure of the spectral density. Formally, the array measures a mean square pressure $\overline{p^2}$ (ω, Δω) at a frequency ω and bandwidth Δω in terms of the following filter functions

$$\overline{p^2}\ (\omega,\ \Delta\omega) = \int_{-\infty}^{\infty}\int_{-\infty}^{\infty}\int_{-\infty}^{\infty} |H(\omega)|^2\ |S(\underline{k})|^2\ |A(\underline{k})|^2\ \Phi_p(\underline{k},\omega)\ d^2\underline{k}d\omega$$

where $|H(\omega)|^2$ is the frequency filter function providing a narrowband output of the signal which is further subjected to the spatial response kernel of each transducer forming the array $|S(k)|^2$ and the array function $|A(k)|^2$. Wave vector filter measurements have been performed by Geib and Farabee[30] in an anechoic wind tunnel. An array of six microphones mounted flush in the wall of the tunnel was used, as shown in Figure 18. Measurements of the spectral density of wall pressure fluctuations were made on smooth and rough walls, for low wave number and supersonic elements of the spectrum. In addition, the experiments performed by Haddle and Skudrzyk were repeated with improved methods and equipment.

Data from various sources in the hydrodynamic range of wave numbers are compared by Blake[24] with the Corcos and Chase models in Fig. 19. The Corcos model applies in the neighborhood of the convective region, whereas the Chase model agrees well with the data over a broader range of wave numbers. This conclusion was also reached by Hwang and Geib[31] who used a linear regression model for estimating the normalized wave vector-frequency spectrum with a set of measured levels. The convenient form of the Corcos model, however, motivates attempts to extend its range of validity to lower wave numbers as more experimental data are obtained.

The supersonic elements of the spectrum are shown in Figs. 20 and 21. These figures contain data from experiments performed on a buoyant body in water as well as those performed in an anechoic wind tunnel. The Mach numbers during the water experiments were less than 0.01 and the momentum thickness Reynolds numbers exceeded 10^5. The wall surface was smooth. In the anechoic wind tunnel experiments, Mach numbers ranged from 0.03 to 0.15 and the momentum thickness Reynolds numbers were on the order of 3×10^4. Smooth, as well as rough walls were investigated. In terms of the equivalent roughness height of the surface irregularities k_s, values of $\frac{k_s U_\tau}{\nu}$ ranged from less than 10 to 1000. The wave vector filter used in the anechoic wind tunnel had a bandwidth marked by the -3dB sensitivity of $0<k\,\delta^* < 0.3$, whereas in water this bandwidth was smaller by a factor of approximately 10. The collapse of the data taken over such a wide range of parameters suggests the possibility that direct radiation from the boundary layer may have been measured. However, the cautionary words that usually accompany such statements must not be forgotten.

The data can be approximated by the relation

$$10 \log \frac{\Phi_p(0,\omega)\, U_\infty}{M^2 \tau_w^2 (\delta^*)^3} = 7.5 - 45 \log \left(\frac{\omega \delta^*}{U_\infty} \right) \quad \text{in the range:}$$

$$3 < \frac{\omega \delta^*}{U_\infty} < 30$$

or alternatively

$$\frac{\Phi_p(0,\omega)\, U_\infty}{M^2 \tau_W^2 (\delta^*)^3} = 5.62 \left(\frac{\omega\delta^*}{U_\infty}\right)^{-4.5}$$

As $\underline{k} \to 0$, Ffowcs Williams' spectrum function reduces to

$$\Phi_p(0,\omega) = \rho^2 U_\infty^3 (\delta^*)^3 \;\Phi_0\left(\frac{\omega\delta^*}{U_\infty}\right) A_0(1) B_0(0)\, a_1\, M^2$$

In terms of the dynamic head $q = \frac{1}{2}\rho\, U_\infty^2$

$$\frac{\Phi_p(0,\omega)\, U_\infty}{M^2 q^2 (\delta^*)^3} = 4\, A_0(1) B_0(0)\, a_1\, \Phi_0\left(\frac{\omega\delta^*}{U_\infty}\right)$$

The data in Figure 21 are plotted in accordance with these parameters. The collapse of the data does not appear to be as good as in Figure 20; points from different experiments have different slopes, for example. If q^2 were replaced by the shear stress τ_W^2 as in Figure 20, then

$$\frac{\Phi_p(0,\omega)\, U_\infty}{M^2 \tau_W^2 (\delta^*)^3} = \frac{4\, A\,(1) B\,(0) a}{c_f^2}\; \Phi_0\left(\frac{\omega\delta^*}{U_\infty}\right)$$

where c_f denotes the skin friction coefficient

$$c_f = 2\left(\frac{U_\tau}{U_\infty}\right)^2$$

In this case, over the range of frequencies shown in Fig. 20,

$$A_0(1) B_0(0) a_1 = 1.41\, c_f^2$$

and

$$\Phi_0\left(\frac{\omega\delta^*}{U_\infty}\right) = \left(\frac{\omega\delta^*}{U_\infty}\right)^{-4.5}$$

Levels at $k^2 = \left(\frac{\omega}{c}\right)^2$ were observed to be higher than at supersonic wave numbers, with most of the energy arriving at grazing incidence from an up stream direction. Although there are several possible explanations for this observation the source has not been positively identified at this time.

References

1. Shen, Y.T., S. Growing, R. Pierce, "Cavitation Susceptibility Measurements by a Venturi," International Symposium on Cavitation Inception, A.S.M.E., 1984

2. Hickling, R. and M.A. Plesset, "Collapse and Rebound of a Spherical Bubble in Water," Phys. Fluids, Vol. 7, pp. 7-14, 1964

3. Baiter, H.J., "Estimates of the Acoustic Efficiency of Collapsing Bubbles," International Symposium on Cavitation Noise, A.S.M.E., 1982

4. Hentschel, W., W. Lauterborn, "Jet Formation and Bubble Interaction Studied with Laser-Produced Cavitation Bubbles", Mecanique Materiaux Electricite, pp. 365-366, 1980

5. Sevik, M. and S.H. Park, "The Splitting of Drops and Bubbles by Turbulent Fluid Flow," Trans. A.S.M.E. Journal of Fluids Engineering, Vol. 95, pp. 53-60, 1973

6. Brinkmeyer, E., "Koharent-Optische Verfahren zu Untersuchung kurzer Stosswellen bei Akustisch Erzeugter Kavitation," Dissertation, Gottingen, 1976

7. Hentschel, W., "Hochfrequenzholografische Untersuchungen zum Verhalten von Kavitationsblasen in Druckfeldern," Dissertation, Gottingen, 1983

8. Blake, W.K. and M. Sevik, "Recent Developments in Cavitation Noise Research", Internationl Symposium on Cavitation Noise, A.S.M.E., 1982

9. Cumming, W.E. and W.B. Morgan, "Propeller Design Aspects of Large High-Speed Ships", Proceedings Symposium on High Powered Propulsion of Large Ships, NSMB Publication No. 490, 1974

10. Jorgensen, D.W., "Noise from Cavitating Submerged Jets", Journal of Acoustical Society of America, Vol. 33, pp. 1334-1338, 1961

11. Tsushima, H. and M. Sevik, "Dynamic Response of Marine Propellers to Nonuniform Flowfields", Journal of Hydronautics, Vol. 7, No. 2, pp. 71-77, 1973

12. Brooks, J.E., "Vibrations of a Marine Propeller Operating in a Nonuniform Inflow," Dissertation, Catholic University, Washington, D.C., 1980

13. Powell, A., "Aerodynamic Noise and the Plane Boundary," Journal of Acoustical Society of America, Vol. 32, pp. 982-990, 1960

14. Ffowcs Williams, J.E., "Sound Radiation from Turbulent Boundary Layers Formed on Compliant Surfaces", Journal of Fluid Mechanics, Vol. 22, pp. 347-358, 1965

15. Howe, M.S., "On the Generation of Sound by Turbulent Boundary Layer Flow Over a Rough Wall", Proceedings of Royal Society, London, A 395, pp. 247-263, 1984

16. Hersh, A.S., AIAA paper No. 83-0786, 1983

17. Crighton, D.G., "Long Range Acoustic Scattering by Surface Inhomogeneities Beneath a Turbulent Boundary Layer", International Symposium on Turbulence-Induced Vibrations and Noise of Structures, pp. 107-121, 1983

18. Corcos, G.M., "The Resolution of Pressures in Turbulence", Journal of Acoustical Society of America, Vol. 35, pp. 192-199, 1963

19. Willmarth, W.W. and C.E. Wooldridge, " Measurements of the Fluctuating Pressure at the Wall Beneath a Thick Turbulent Boundary Layer", Journal of Fluid Mechanics, Vol. 14, pp. 187-210, 1962

20. Bull, M.K., "Wall-Pressure Fluctuations Associated with Subsonic Turbulent Boundary Layer Flow", Journal of Fluid Mechanics, Vol. 28, pp. 719-754, 1967

21. Ffowcs Williams, J.E., "Boundary Layer Pressures and the Corcos Model: a Development to Incorporate Low Wave Number Constants", Journal of Fluid Mechanics, Vol. 125, pp. 9-25, 1982

22. Ffowcs Williams, J.E., "Surface Pressure Fluctuations Induced by Boundary-Layer Flow at Finite Mach Number", Journal of Fluid Mechanics, Vol. 22, pp. 507-519, 1965

23. Chase, D.M., "Modeling the Wave-Vector Frequency Spectrum of Turbulent Boundary Layer Wall Pressure", Journal of Sound Vibration, Vol. 70, pp. 29-67, 1980

24. Blake, W.K., "Aero-Hydroacoustics for Ships", DTNSRDC Report 84/010, 1984

25. Handler, R.A., R.J. Hansen, S.A. Orszag et al., "Calculation of the Wall Pressure Field in a Turbulent Channel Flow", Physics of Fluids, Vol. 27, No. 3, pp. 579-582, 1984

26. Haddle, G.D. and E.J. Skudrzyk, "The Physics of Flow Noise", Journal of Acoustical Society of America, Vol. 56, pp. 130-157, 1969

27. Jameson, P.W., "Measurement of the Low-Wavenumber Component of Turbulent Boundary Layer Pressure Spectral Density", Proceedings of 4th Symposium on Turbulence in Liquids, 1975

28. Martin, N.C., "Wavenumber Filtering by Mechanical Structures", M.I.T. Ph.D. Thesis, 1976

29. Maidanik, G. and D.W. Jorgensen, "Boundary Wave Vector Filters for the Study of the Pressure Field in Turbulent Boundary Layer", Journal of Acoustical Society of America, Vol. 42, pp. 494-501, 1967

30. Geib, F.E. and T.M. Farabee, "Measurement of Boundary Layer Pressure Fluctuations at Low Wavenumber on a Smooth Wall," Paper presented at 91st Meeting, Acoustical Society of America, 1976

31. Hwang, Y.F. and F.E. Geib, "Estimation of the Wavevector-Frequency Spectrum of Turbulent Boundary Layer Wall Pressure by Multiple Linear Regression", A.S.M.E. Symposium on TurbulenceInduced Vibration and Noise of Structures, 1983

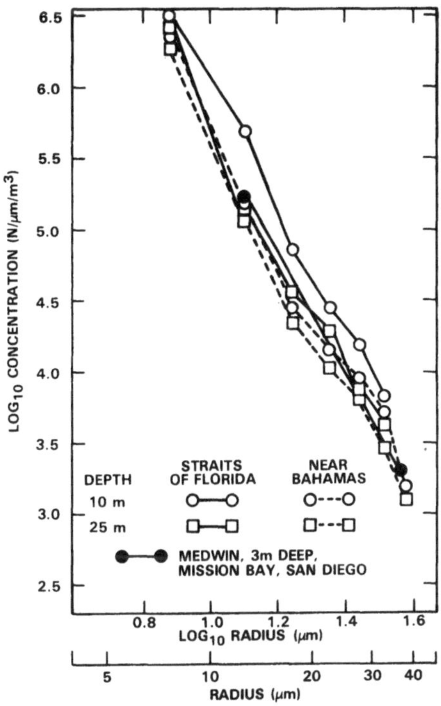

Fig. 1 Measured Bubble Spectra At Sea (Ref. 1)

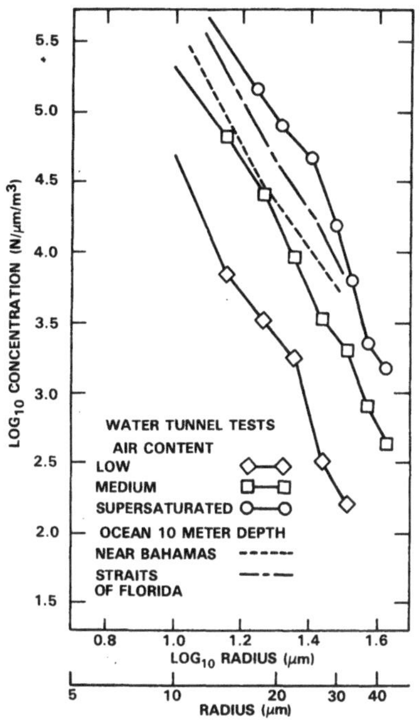

Fig. 2 Bubble Concentration in 12-inch Water Tunnel at DTNSRDC (Ref. 1)

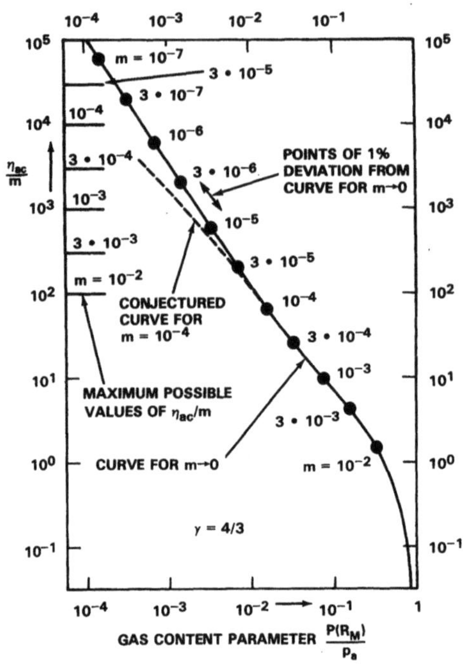

Fig. 3 Acoustic Efficiency for Weak Collapse With Indications of Deviations From This Relation in Case of Moderately Strong Collapse and Maximum Possible Values of η_{ac}/m in Case of Strong Collapse Depending on the Compressibility Parameter $m = \sqrt{p_0/\varrho c^2}$ (Ref. 3)

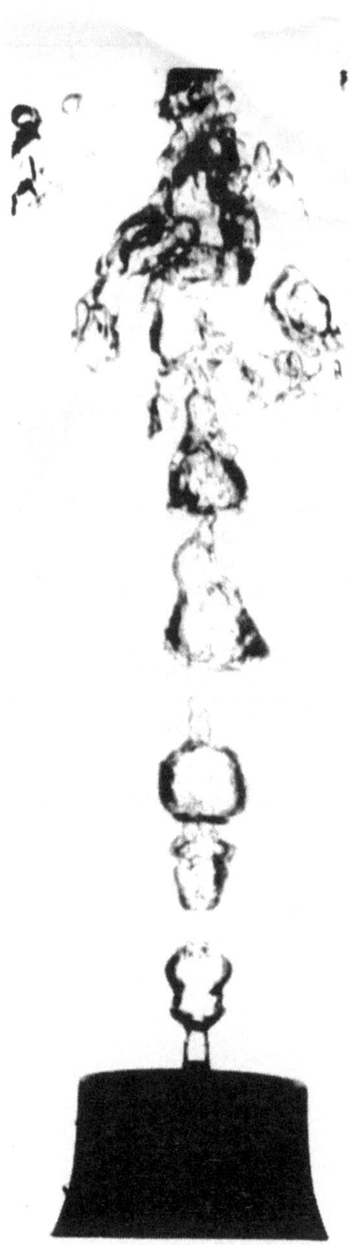

Fig. 4 Two Phase Flow Pattern (Ref. 5)

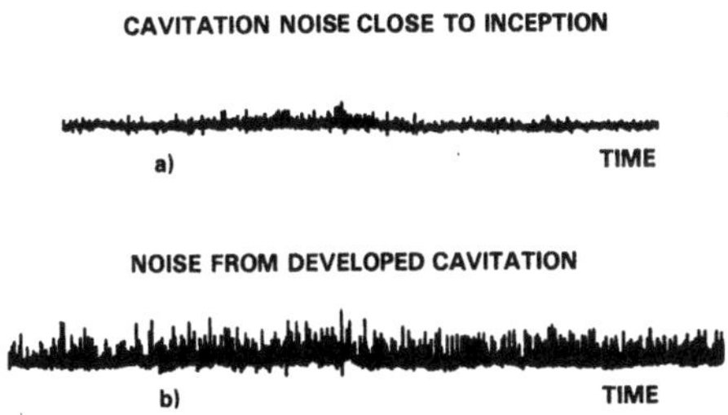

Fig. 5 Temporal Variations of Tip Vortex Cavitation Noise

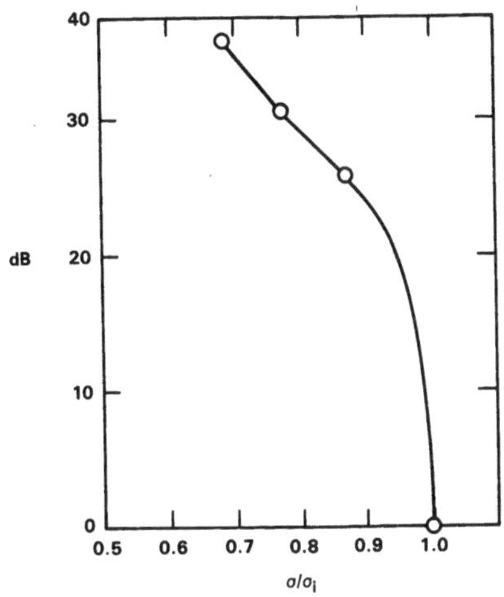

Fig. 6 Noise Level Differences Above Background, at High Frequencies

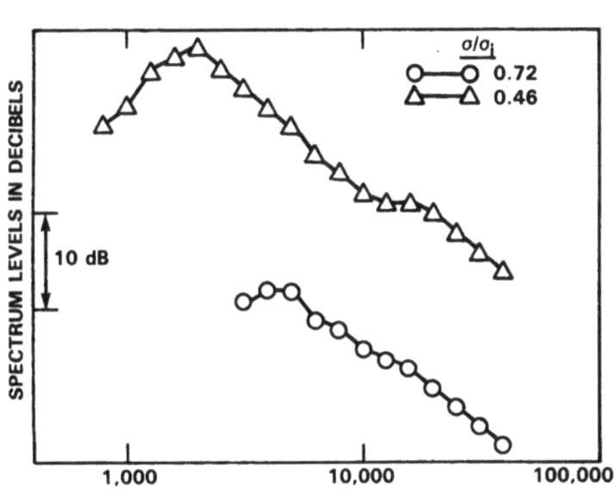

Fig. 7 Typical Spectra of Tip Vortex Cavitation Noise

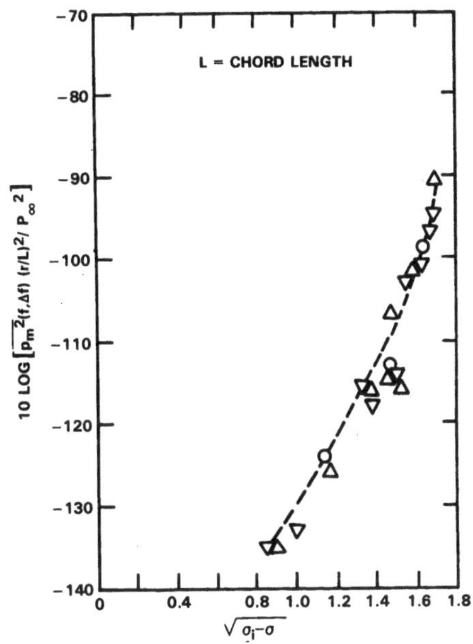

Fig. 8 Pressure Factor for Sound Radiated From a Tip Vortex on Large Hydrofoils. Various Points Refer to Different Test Conditions and Static Pressures ($R_L \sim 10^7$)

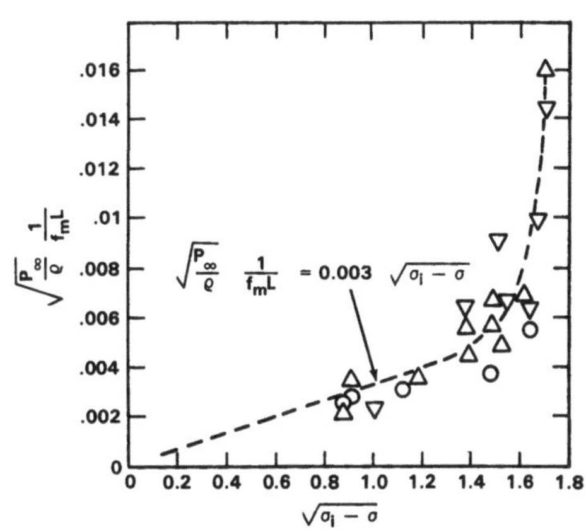

Fig. 9 Frequency Factor for Sound Radiated From a Tip Vortex on Large Hydrofoils. Various Points Refer to Different Test Conditions and Static Pressures ($R_L \sim 10^7$)

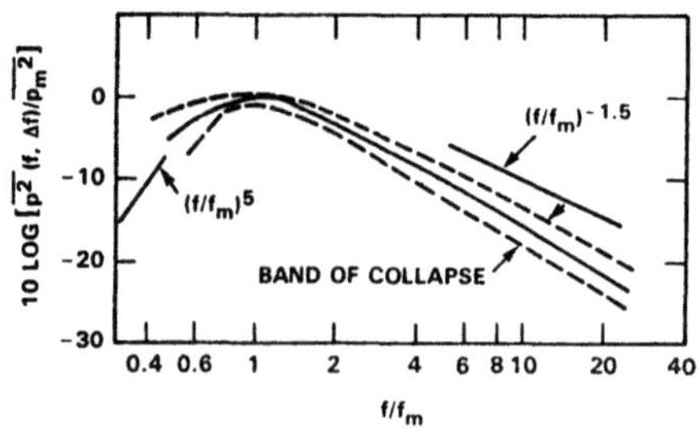

Fig. 10 Universal Spectrum Function For Sound From Tip Vortex Cavitation on Hydrofoils

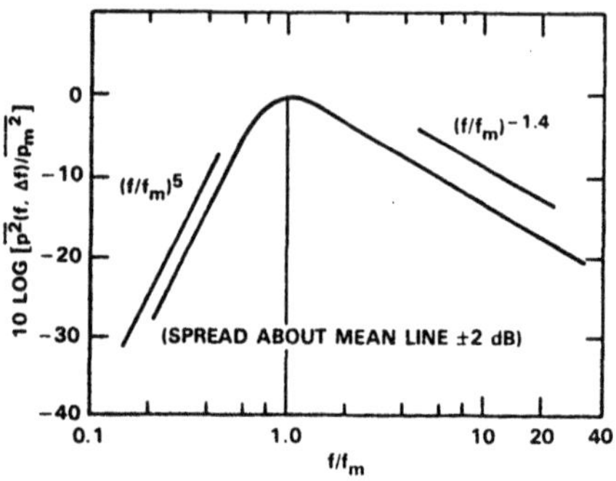

Fig. 11 Universal Spectrum Function for Sound From Cavitating Jets (Jorgensen Data)

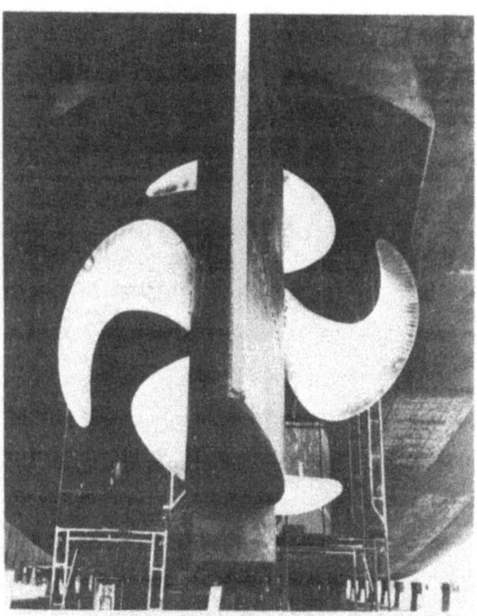

Fig. 12 Propeller for Great Lakes Ship "Charles E. Wilson" Courtesy Bird-Johnson Co. Marine Division, Walpole, MA., USA

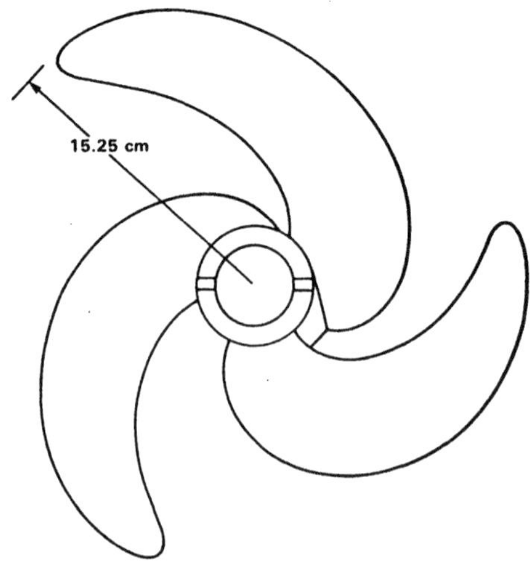

Fig. 13 Model Propeller Geometry (Ref. 11)

Fig. 14 Installation of Screen and Propeller in the Water Tunnel (Ref. 11)

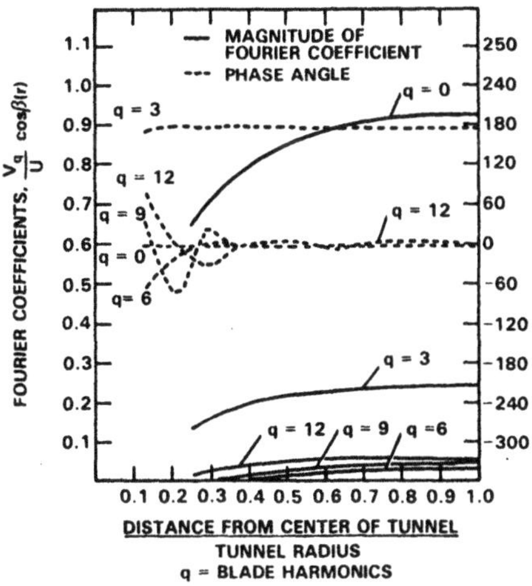

Fig. 15 Wake Harmonics Due to Screen (Ref. 11)

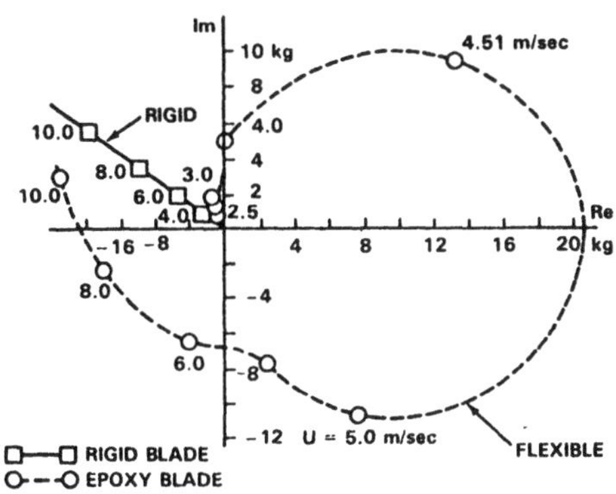

Fig. 16 Predicted Magnitudes and Phases of Unsteady Thrust Force for Rigid and Flexible Propellers (Ref. 11)

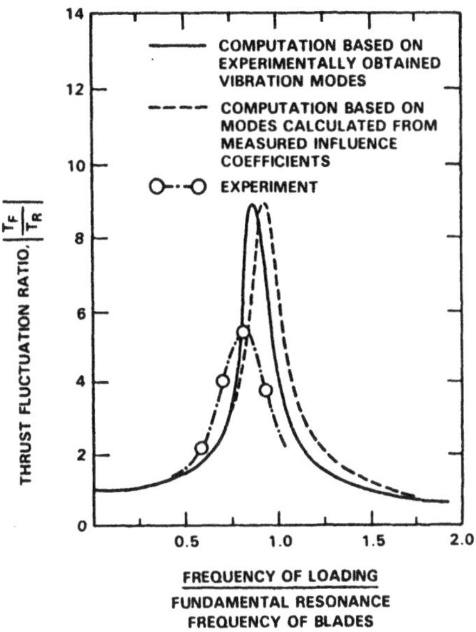

Fig. 17 Effect of Blade Resonances on Fluctuating Force

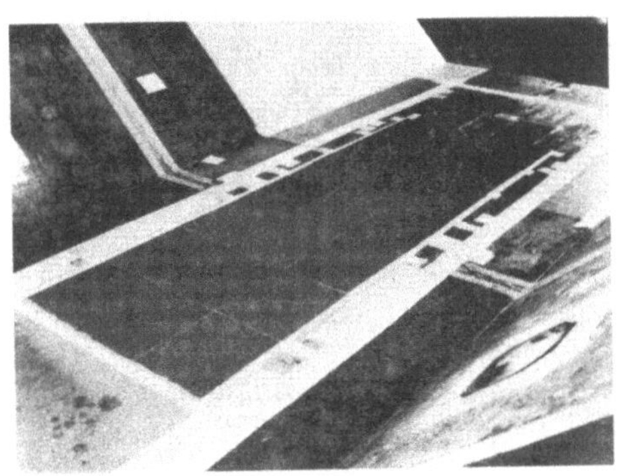

Fig. 18 Experimental Setup in Anechoic Wind Tunnel

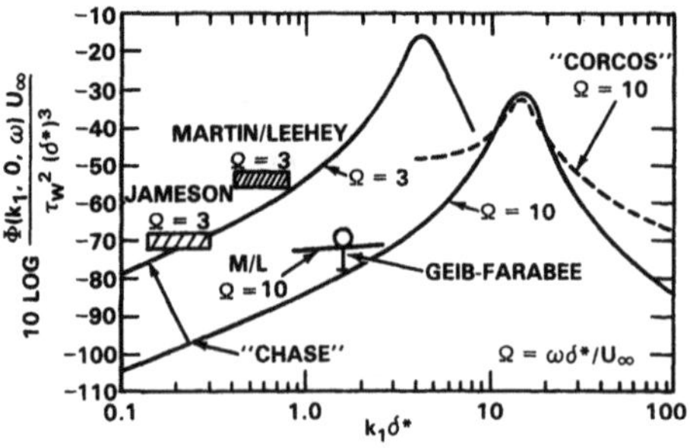

Fig. 19 Wall Pressure Spectra in Purely Hydrodynamic Range of Wave Number, M <0.01 (Ref. 24)

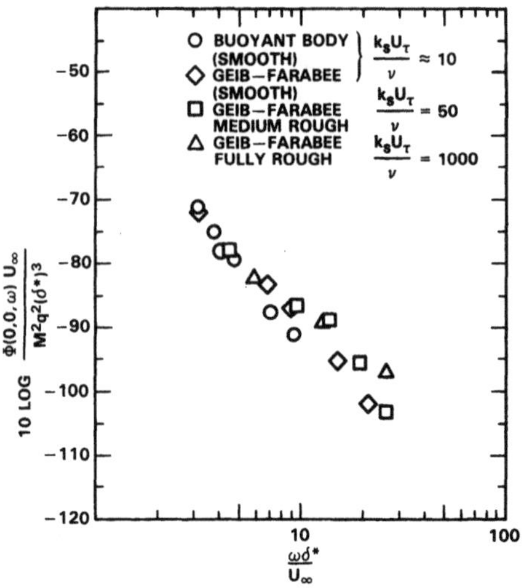

Fig. 21 Data Normalized to Dynamic Head.

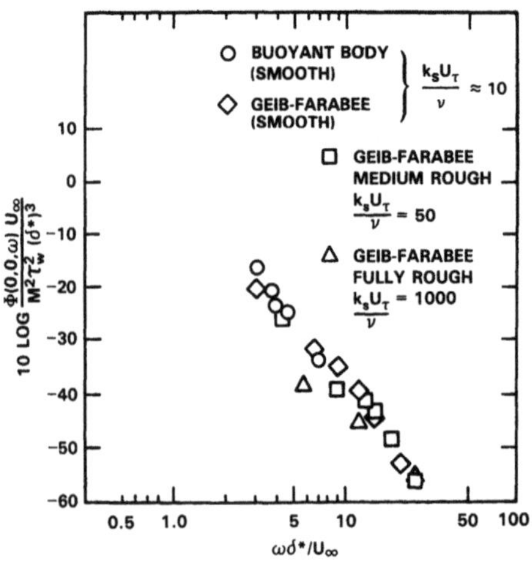

Fig. 20 Variation of Boundary Layer Noise With Dimensionless Frequency

Interaction Hydroelastique entre une Plaque et un Ecoulement

F. LEMONNIER et R. PERRET

Section Hydroacoustique

NEYRTEC - Etablissement de Grenoble d'ALSTHOM
B.P. 61 X
38041 X GRENOBLE

RESUME

Le chant d'hélice est la conséquence d'une intéraction hydroélastique entre un écoulement et une pale d'hélice. Ce phénomène a été étudié dans une configuration simplifiée, plaque plane sans incidence. La forme du bord de fuite est le paramètre le plus important; il influe à la fois sur la fréquence de lâcher de tourbillons à l'aval de la plaque et sur le niveau de vibrations induites. Le coefficient d'amortissement est la seule caractéristique modale des plaques à évoluer avec la vitesse de l'écoulement.

1. INTRODUCTION

L'étude a eu pour cadre le phénomène de chant d'hélice. Ce phénomène qui se traduit pour une émission de bruit intense à une fréquence précise est dû aux intéractions hydroélastiques entre une pale de l'hélice et l'écoulement autour de celle-ci, et se produit pour des conditions de fonctionnement très particulières.

La seule technique totalement empirique que l'on connaisse pour réduire les risques d'apparition du chant d'hélice consiste à pratiquer un chanfrein sur le bord de fuite des pales. Mais elle n'est pas sytématiquement efficace. L'élimination totale de tout risque passe nécessairement par une meilleure connaissance du phénomène.

On a pu mettre en évidence que le chant d'hélice est dû à une coïncidence entre une fréquence de résonance de la pale et la fréquence du lâcher de tourbillons qui se produit au bord de fuite (Réf. 1).

La fréquence de ce lâcher de tourbillons permet de construire un nombre de Strouhal basé sur l'épaisseur de la plaque.

Aero- and Hydro-Acoustics
IUTAM Symposium Lyon 1985
Editors: G. Comte-Bellot and J. E. Ffowcs Williams

$$S = f \frac{D}{U}$$

S = nombre de Strouhal

D = épaisseur du profil

U = vitesse de l'écoulement

f = fréquence du lâcher de tourbillons

L'influence de la forme du bord de fuite sur ce nombre adimensionnel et sur l'amplitude des vibrations a déjà été remarqué par divers auteurs (réf. 3, 4, 5).

L'étude qui a été menée a eu pour objet d'analyser à l'aide d'essais en tunnel hydrodynamique sur des configurations simplifiées les intéractions hydroélastiques entre un écoulement et une plaque plane, pour différents types de fixation de la plaque et diverses formes de bord de fuite.

2. DESCRIPTION DU MOYEN D'ESSAIS

Le moyen d'essais utilisé est le tunnel hydrodynamique THV de 600 l/s qui équipe le Laboratoire de Mécanique des Fluides de NEYRTEC. Une veine transparente de 200 x 200 mm. a été réalisée spécialement afin qu'elle puisse être équipée des plaques à analyser. Le montage prévu permet de monter des plaques encastrées sur un bord ou fixées par des barres de torsion. Une zone d'eau morte est réservée sur le côté de la veine; elle a été prévue pour permettre l'équipement des plaques (accéléromètres, excitation par pot vibrant) sans gêner l'écoulement (voir figure n° 1).

Les dimensions des plaques et des barres de torsion ont été calculées afin que leurs fréquences propres puissent être excitées par les lâchers de tourbillons dans la gamme de vitesse 1 - 15 m/s, en se basant, a priori, sur un nombre de Strouhal moyen de 0,25.

Quatre angles de bord de fuite ont été réalisés : 30, 45, 60 et 90°.

3. ESSAIS REALISES

3.1. Essais en air et en eau hors écoulement

Les réponses propres des différentes plaques définies ci-dessus ont été mesurées dans l'air et dans l'eau. Pour chacun des six premiers modes, ont été déterminés : la fréquence, l'amortissement et la déformée.

3.2. Essais en tunnel hydrodynamique

Les essais des plaques excitées par un écoulement se sont déroulés sur le tunnel hydrodynamique THV de 600 l/s. Ils ont consisté :

- à mesurer les vibrations des plaques en fonction de la vitesse d'écoulement, autour des différentes fréquences propres
- à mesurer les variations du nombre de Strouhal dans la gamme de vitesse 1 à 15 m/s;
- à mesurer, sur les plaques encastrées, les caractéristiques modales (fréquence propre, amortissement) pour différentes vitesses d'écoulement (de 2 à 12,5 m/s) par superposition d'une excitation par pot vibrant.

Quatre formes de bord de fuite ont été étudiées dans le cas des plaques en torsion. Seules les deux angles extrèmes (30° et 90°) ont été retenus pour les plaques encastrées.
Une visualisation du lâcher de tourbillons derrière les plaques a pu être réalisée en faisant caviter l'écoulement.

4. PRINCIPAUX RESULTATS

Les principaux résultats obtenus lors de l'étude peuvent être analysés de la façon suivante :

a/ La valeur du nombre de Strouhal dépend de la forme du bord de fuite.Elle est d'autant plus grande que le profil est plus effilé. La coïncidence avec une fréquence de résonance donnée se produira donc à plus basse vitesse avec un bord de fuite plus fin. Au passage des fréquences propres, le phénomène d'accrochage entraîne une modification du nombre de Strouhal (voir figure n° 2).

b/ La forme du bord de fuite a un effet sur le niveau de vibration à la fréquence de Strouhal (-10 dB d'écart entre un bord de fuite à 90° et un bord à 30°) (voir figure n° 3).

c/ Le phénomène d'accrochage entre le lâcher de tourbillons et les premières fréquences propres a été constaté pour toutes les formes de bord de fuite étudiées et quel que soit le montage de la plaque.

d/ Afin de tenir compte le mieux possible de l'obstruction apportée par la plaque à l'écoulement, il est nécessaire de corriger l'épaisseur de la plaque par l'épaisseur de déplacement de la couche limite. Pour une plaque donnée, cette correction permet d'obtenir un nombre de Strouhal pratiquement indépendant du nombre de Reynolds. Elle est, par contre, insuffisante pour bien rendre compte d'une variation d'épaisseur. La notion de nombre de Strouhal universel nécessiterait la connaissance de l'écoulement à l'aval du bord de fuite et, en particulier, de la configuration de l'allée de tourbillons.

e/ Nous n'avons pas constaté de zone de Reynolds critique comme dans le cas des cylindres, dans toute la gamme explorée : $1,9 \times 10^5$ à $2,9 \times 10^6$ (nombre de Reynolds calculé sur la corde du profil).

f/ Il n'y a pas de modification des fréquences propres des plaques en fonction de la vitesse d'écoulement; par contre, on constate une augmentation de l'amortissement des modes (quand on a pu le mesurer) avec la vitesse et l'amortissement est d'autant plus grand que la fréquence propre est plus basse.

g/ Il n'y a émission acoustique, à l'accrochage, que pour des modes élevés.

5. CONCLUSION

La connaissance précise des caractéristiques modales des plaques mesurées hors écoulement dans l'air, puis dans l'eau, a permis d'effectuer une étude approfondie de l'influence de l'angle de bord de fuite sur les vibrations de ces plaques excitées.

L'adoption de bords de fuite effilés est favorable à la réduction des vibrations, d'une part, parce que l'énergie des lâchers de tourbillons est plus faible et, d'autre part, parce que le phénomène d'accrochage entre la fréquence de Strouhal et une fréquence propre se produit à plus faible vitesse.

L'amélioration de la prédiction des fréquences des lâchers de tourbillons doit passer par une meilleure connaissance de l'écoulement à l'aval du bord de fuite.

- mesures de vitesse dans le sillage
- visualisation

La connaissance des conditions d'émission acoustique d'une plaque ou d'une pale excitée sur un mode propre devrait permettre de mieux préciser les risques d'apparition du phénomène du chant d'hélice, dans la mesure où l'on a constaté que tout mode propre est excité par les lâchers de tourbillons et que l'émission acoustique est, par contre, exceptionnelle. Afin de se rapprocher du cas plus complexe des hélices on étudiera ultérieurement sur des profils toujours bidimensionnels mais non forcément plans :

- l'influence de l'incidence,
- l'influence de la non synchronisation des lâchers de tourbillons due, soit au fait que le profil n'est pas perpendiculaire à l'écoulement, soit à une géométrie du profil non constante sur l'envergure,
- l'influence du tourbillon marginal.

Cette étude a été financée par le Ministère de la Défense, Direction des Recherches et Etudes Techniques (DRET)

Références

(1) G.H. TOEBES; P.S. EAGLESON (1961)
"Hydroelastic Vibration of Flat Plates, related to Trailing Edge Geometry"
ASME Paper n° 61 HYD 16

(2) G. HESKESTAD, D.R. OLBERTS (1952)
"Influence of Trailing Edge Geometry on Hydraulic - Turbine-Blade Vibration Resulting from Vortex Excitation
ASME Paper n° 59 Hyd 7

(3) W.K. BLAKE, L.J. MAGA (1975)
"On the Flow Excited Vibrations of Cantilever Struts in Water.I : Flow Induced Damping and Vibration"
J.A.S.A. 57 n° 3

(4) W.K. BLAKE, L.J. MAGA, G. FINKELSTEIN (1977)
"Hydroelastic variable, Influencing Propeller and Hydrofoil Singing"
ASME Proceding of the Symposium on Noise and Fluid Angineering Winter Neeting.

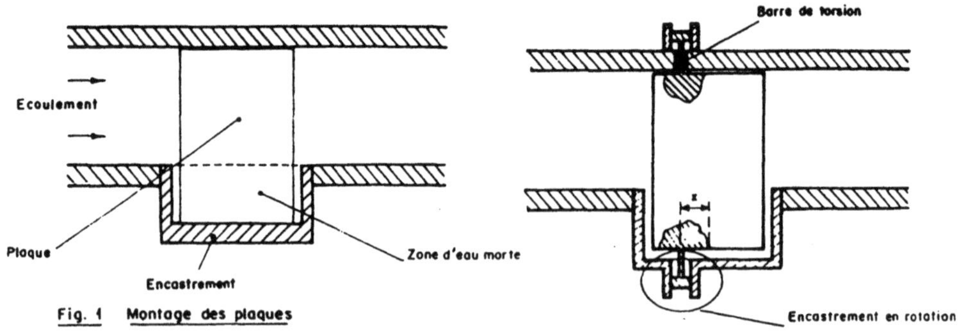

Fig. 1 Montage des plaques

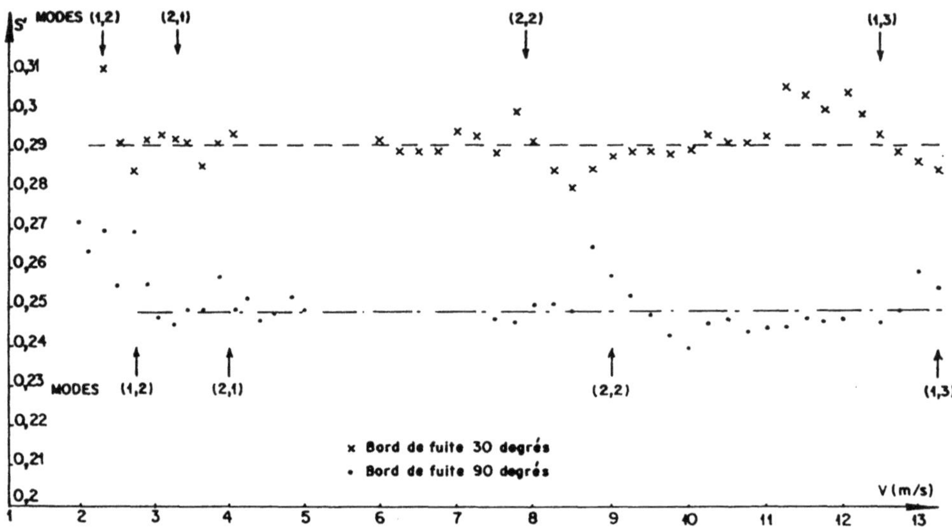

Fig. 2 Nombre de Strouhal corrigé par l'épaisseur de déplacement en fonction de la vitesse d'écoulement

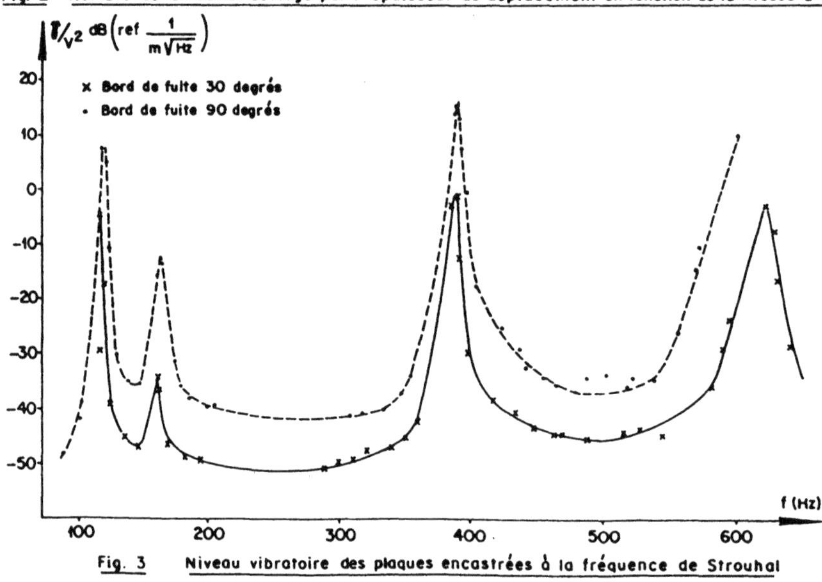

Fig. 3 Niveau vibratoire des plaques encastrées à la fréquence de Strouhal

The Acoustics of Disturbed Steady Motions

D. C. HILL

Cambridge University Engineering Department,
Trumpington Street,
Cambridge,
England.

Summary

This paper bears on the fundamental scattering processes taking place on a wavebearing surface which has an edge, by describing a simple model. The role of the form drag in balancing the momentum flow in the near and far fields is shown. The distinct difference in nature between a surface supported wave travelling towards an edge, as against away from it, is apparent from both energetic and drag considerations.

Introduction

The mode of transfer of energy from bounding surfaces and their nearfield to the farfield, in the presence of an abrupt edge is studied. We look at two particular situations in which an infinite plane surface, bounded on one side by an inviscid fluid, supports harmonic waves of fixed amplitude. A fixed region of the boundary plane is in motion; elsewhere the surface is undisturbed. At the edge of the region there is a discontinuity which is important to the production of sound. The form drag is a particularly important feature caused by the scattering of momentum into the sound field.

The first problem concerns a semi-infinite plane wave travelling towards an edge beyond which the surface is flat, while in the second example the wave is driven away from the edge.

Case 1.

Case 2.

The Pressure Field

The outgoing pressure field produced by motion of an infinite plane surface is given by

$$p(\underline{x},t) = \frac{\rho_0}{2\pi}\int \left[\frac{\partial v_n}{\partial t}\right] \frac{dS}{r} \qquad (2.1)$$

where v_n is the normal velocity and $[\;]$ denotes retarded time $t-r/c$, r being the distance between the source and field points.

The prescribed normal velocity in the first problem is

$$v_n = vH(-x')e^{i(\omega t-kx')}, \omega/k<c, \qquad (2.2)$$

the y=0 plane being the boundary surface, with the fluid occupying the region y>0.

Then

$$p(\underline{x},t) = \tfrac{1}{2}\rho_0 v\omega\int_{-\infty}^{0} e^{i(\omega t-kx')} H_0^{(2)}\left(\tfrac{\omega}{c}\sqrt{(x-x')^2+y^2}\right)dx'. \qquad (2.3)$$

If there were no factor $H(-x')$ the pressure would be the evanescent wave

$$E(\underline{x},t) = \rho_0 vcm(1-m^2)^{-\frac{1}{2}}ie^{i(\omega t-kx)}e^{-ky(1-m^2)^{\frac{1}{2}}}, m=\omega/kc. \qquad (2.4)$$

Then (2.3) can be rearranged as

$$p(\underline{x},t) = H(-x)\{\, E(\underline{x},t) + \tfrac{1}{2}\rho_0 v\omega C_-(\underline{x})e^{i\omega t}\}$$

$$+ H(x)\tfrac{1}{2}\rho_0 v\omega C_+(\underline{x})e^{i\omega t}, \qquad (2.5)$$

where

$$C_-(\underline{x}) = -\int_0^{\infty} e^{-ikx'} H_0^{(2)}\left(\tfrac{\omega}{c}\sqrt{(x-x')^2+y^2}\right)dx',$$

$$C_+(\underline{x}) = \int_{-\infty}^{0} e^{-ikx'} H_0^{(2)}\left(\tfrac{\omega}{c}\sqrt{(x-x')^2+y^2}\right)dx'.$$

In order to find the structure of the correction terms C_- and C_+ which supplement the evanescent field the Hankel functions are expanded using the "Summation Theorem" as stated in Gradshteyn and Ryzhik [2], p.979, §8.531.

$$H_0^{(2)}\left(\tfrac{\omega}{c}\sqrt{(x-x')^2+y^2}\right) = J_0(\tfrac{\omega}{c}x')H_0^{(2)}(\tfrac{\omega}{c}r) + 2\sum_{n=1}^{\infty} J_n(\tfrac{\omega}{c}x')H_n^{(2)}(\tfrac{\omega}{c}r)\cos n\theta, \qquad (2.8)$$

where $r=\sqrt{x^2+y^2}$, and θ lies in the interval $\pi/2$ to π, with $x'>0$.

Noting that

$$\int_0^\infty J_n(\frac{\omega}{c}x')e^{-ikx'}dx' = -\frac{1}{k}\frac{1}{\sqrt{1-m^2}}\frac{m^n}{(1+\sqrt{1-m^2})^n}ie^{-in\pi/2}, \quad (2.9)$$

and that

$$x=r\cos\theta$$
$$y=r\sin\theta \quad (2.10)$$

defines a set of polar coordinates, θ being measured from the positive x-axis, then for θ lying in the interval π/2 to π

$$C_-(r,\theta)=\frac{i}{k}\frac{1}{\sqrt{1-m^2}}H_0^{(2)}(\frac{\omega}{c}r)+\frac{2i}{k}\frac{1}{\sqrt{1-m^2}}\sum_{n=1}^{\infty}\frac{m^n e^{-in\pi/2}}{(1+\sqrt{1-m^2})^n}H_n^{(2)}(\frac{\omega}{c}r)\cos n\theta. \quad (2.11)$$

In a similar fashion it can be found that C_+ is identical in form to C_-.

To find the farfield approximation for large r the asymptotic form for $H_n^{(2)}$ can be substituted in (2.11)

$$H_n^{(2)}(z) \sim \sqrt{2/\pi}\frac{1}{z^{\frac{1}{2}}}e^{-i(z-\pi/4)}e^{in\pi/2}. \quad (2.12)$$

Then

$$C_-(r,\theta)\underset{r\to\infty}{\sim}\frac{i}{k}\sqrt{2c/\pi\omega}\frac{1}{r^{\frac{1}{2}}}\;e^{i(\pi/4-\omega r/c)}\times$$

$$\times\frac{1}{\sqrt{1-m^2}}\{1+2\sum_{n=1}^{\infty}\frac{m^n\cos n\theta}{(1+\sqrt{1-m^2})^n}\}. \quad (2.13)$$

The infinite series is convergent, and using the result of §1.447(2) on p.40 of Gradshteyn and Ryzhik [2]

$$\frac{1}{\sqrt{1-m^2}}\{1+2\sum_{n=1}^{\infty}\frac{m^n\cos n\theta}{(1+\sqrt{1-m^2})^n}\}=\frac{1}{1-m\cos\theta}. \quad (2.14)$$

This gives the farfield pressure as being

$$p(\underline{x},t)\underset{r\to\infty}{\sim}H(-x)E(\underline{x},t)+\frac{A}{r^{\frac{1}{2}}}\frac{1}{1-m\cos\theta}ie^{i\omega(t-r/c)}e^{i\pi/4}, \quad (2.15)$$

where $A=\rho_0 v(c/2\pi\omega)^{\frac{1}{2}}\omega/k$.

The evanescent surface wave travels in towards the edge where it is scattered into a radially outgoing sound wave. There is also a directional dependance in the form of a "Doppler" factor $1/(1-m\cos\theta)$. The waves are scattered most strongly in the forward direction as seen in figure 1.

As the phase velocity of the surface wave becomes near sonic there is a strong forward beam of sound formed from the scattered evanescent wave.

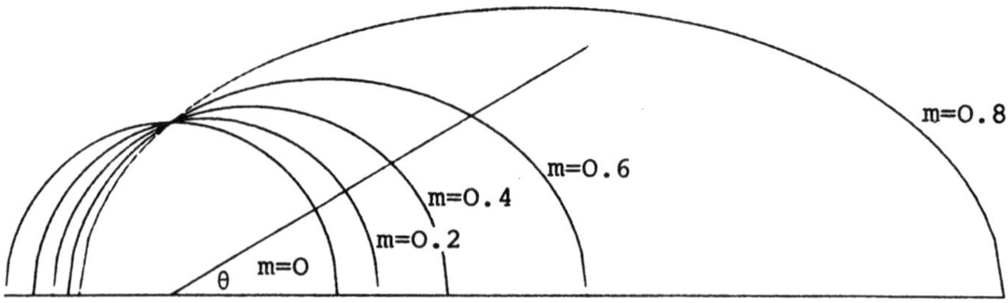

Fig.1. Amplitude of the scattered wave as a function of direction in the form of a polar diagram.

To study a wave moving away from the edge we consider

$$v_n = vH(x')e^{i(\omega t-kx')}, \omega/k<c, \tag{2.16}$$

The pressure field follows almost immediately by noting that the sum of (2.16) and (2.2) produces the field $E(\underline{x},t)$. This gives the field due to (2.16) as

$$p(\underline{x},t) = \frac{1}{2}\rho_0 v\omega\int_0^\infty e^{i(\omega t-kx')}H_0^{(2)}\left(\frac{\omega}{c}\sqrt{(x-x')^2+y^2}\right)dx', \tag{2.17}$$

with asymptotic form

$$p(\underline{x},t) \underset{r\to\infty}{\sim} H(x)E(\underline{x},t) - \frac{A}{r^{\frac{1}{2}}}\,\frac{1}{1-m\cos\theta}\,ie^{i\omega(t-r/c)}e^{i\pi/4}. \tag{2.18}$$

The Energetics Of The Motion

The rate of energy radiation from the surface can be calculated by integrating the time averaged intensity $\overline{Re(p)Re(v_n)}$ on the surface over the activated part of the plane.

When the wave is incident on the edge, from (2.5), the pressure on the surface is

$$Re(p(x,0,t)) = Re(E(x,0,t))$$

$$-\frac{1}{2}\rho_0 v\omega\int_0^\infty\{\cos(\omega t-kx')J_0(\tfrac{\omega}{c}(x'-x))+\sin(\omega t-kx')N_0(\tfrac{\omega}{c}(x'-x))\}dx', \tag{3.1}$$

where $x<0$.

Using v_n from (2.2)

$$\overline{Re(p)Re(v_n)} = -\frac{1}{4}\rho_0 v^2 c\int_{-\frac{\omega}{c}x}^\infty(\cos(\eta/m)J_0(\eta)-\sin(\eta/m)N_0(\eta))d\eta. \tag{3.2}$$

Writing $\overline{m}=1/m$ then

$$\overline{\mathrm{Re}(p)\,\mathrm{Re}(v_n)} = -\frac{1}{4}\rho_0 v^2 c \frac{\partial}{\partial \bar{m}} \int_{-\frac{\omega}{c}x}^{\infty} (\sin \bar{m}\eta J_0(\eta) + \cos \bar{m}\eta N_0(\eta)) \frac{d\eta}{\eta}. \tag{3.3}$$

Then the total rate of energy radiation for a wave incident on the edge is

$$\phi_i = \int_{-\infty}^{0} \overline{\mathrm{Re}(p)\,\mathrm{Re}(v_n)}\, dx$$

$$= -\frac{1}{4}\rho_0 v^2 c \frac{\partial}{\partial \bar{m}} \int_0^{\infty} (\sin \bar{m}\eta J_0(\eta) + \cos \bar{m}\eta N_0(\eta))\, d\eta$$

$$= 0. \tag{3.4}$$

Using a similar method with expression (2.17) for p, in order to describe the case of a wave moving away from the edge, the energy radiation rate can be calculated to be

$$\phi_a = \frac{1}{2} \frac{\rho_0 v^2}{k} \frac{1}{(1-m^2)^{3/2}} \frac{\omega}{k}. \tag{3.5}$$

This is the sum of the energy which travels outwards in the cylindrical wave, calculable from (2.15), and the energy travelling away from the edge, parallel to the surface in the evanescent field, calculable from (2.4).

When the wave is incident on the edge no work is done by the surface, which plays a passive role. All the energy in the incoming evanescent wave is scattered into the cylindrical wave.

When the wave travels away from the edge the surface does work on the fluid at a rate ϕ_a to produce the energy for both the evanescent and cylindrical waves.

The Effect Of Form Drag

Using the momentum conservation law in integral form

$$\int_{\sigma(t)} (p\delta_{ij} + \rho v_i v_j)\, dS_j = \int_{V(t)} \frac{\partial}{\partial t}(\rho v_i)\, dV \;, \tag{4.1}$$

allows the time averaged form drag on the surface $\overline{\mathrm{Re}(p)\,\mathrm{Re}(\frac{\partial \xi}{\partial x})}$ to be related to the mean forces on the fluid in the near and far fields.

Since the drag is of second order in v, it is necessary to know p to second order. L.V. King [1] gives the second order correction as being

$$\frac{p^2}{2\rho_0 c^2} - \frac{1}{2}\rho_0 v_i^2 \tag{4.2}$$

where p, and v_i, are the first order pressure and velocity respectively, p being defined by (2.1). For the cylindrical sound waves in the farfield, since they are almost plane, (4.2) is zero, but for the evanescent wave, (4.2) is non-zero.

The total mean form drag on the surface is

$$D_f = \int_{-\infty}^{\infty} \overline{\mathrm{Re}(p)\,\mathrm{Re}\left(\frac{\partial\xi}{\partial x}\right)}\,dx \tag{4.3}$$

with p evaluated at y=0.

For the case of a wave incident on the edge

$$\mathrm{Re}\left(\frac{\partial\xi}{\partial x}\right) = -\frac{k}{\omega}\mathrm{Re}(v_n) - \frac{v}{\omega}\sin(\omega t - kx)\,\delta(x). \tag{4.4}$$

Then

$$D_f = -\frac{k}{\omega}\phi_i - \frac{v}{\omega}\,\overline{\mathrm{Re}(p)\sin(\omega t-kx)}\,\Big|_{x=0}$$

$$= -\frac{v}{\omega}\,\overline{\mathrm{Re}(E(x,0,t))\sin(\omega t-kx)}\,\Big|_{x=0}$$

$$+ \frac{1}{2}\rho_0 v^2 \int_0^{\infty} \{\sin kx' J_0(\tfrac{\omega}{c}x') + \cos kx' N_0(\tfrac{\omega}{c}x')\}\,dx'$$

$$= \frac{1}{2}\,\frac{\rho_0 v^2}{k}\,\frac{1}{(1-m^2)^{\frac{1}{2}}}. \tag{4.5}$$

This constitutes a force on the fluid, parallel to the surface, acting against the incoming wave.

A similar calculation can be made for the case of the wave being driven away from the edge for which

$$D_f = -\frac{1}{2}\,\frac{\rho_0 v^2}{k}\,\frac{1}{(1-m^2)^{3/2}}\,. \tag{4.6}$$

The departing surface wave experiences a force accelerating it into its steady evanescent state.

In both cases these surface forces are balanced by forces associated with the evanescent and scattered wavefields arising due to

Reynold's stress, and steady forcing due to the second order pressure terms.

Conclusion

An evanescent wave which is driven towards an edge is scattered into a cylindrical outgoing wave, whose amplitude drops off like $r^{-\frac{1}{2}}$. There is also a directional dependance in the amplitude of $1/(1-m\cos\theta)$ (see fig.1), m being the ratio of the surface phase speed to the sound speed. No work is done by the surface on the fluid, so that all the energy incident in the nearfield is transferred to the sound field. The surface exerts a force on the fluid parallel to the surface which acts in the opposite direction to the approaching evanescent wave.

When an attempt is made to drive an evanescent wave away from an edge, a cylindrical wave identical to the previous case, except for a phase difference π, is set up. The surface now does work on the fluid at a rate sufficient to supply energy for both the evanescent and cylindrical waves, while exerting a force on the fluid acting in the direction of the departing evanescent wave.

References

1. L.V. King. On the acoustic radiation pressure on spheres. Proc. Roy. Soc. London. Ser. A,147 (1934) p.212.

2. Gradshteyn,I.S.; Ryzhik,I.M.; Table of integrals, series, and products. Corrected and enlarged edition. Academic Press (1980).

Temporal Variability in Shallow Water Acoustic Transmission and its Correlation with the Environment

H.B. ALI, T. AKAL, O.F. HASTRUP

SACLANT ASW Research Centre Viale San Bartolomeo 400, I-19026 La Spezia, Italy

ABSTRACT

The ocean is a complex, highly variable medium of acoustic propagation. Medium-induced variability in the acoustic index of refraction (sound speed) can have significant effects on acoustic propagation. Using the results of SACLANTCEN measurements in two diverse geographical areas, the present paper discusses the correlation between the temporal variability of ocean parameters and the fluctuations in acoustic propagation. Following a discussion of the relevant spatial and temporal scales, brief comments are made on the theoretical approaches to propagation in a random medium. This is followed by a discussion of selected results of measurements in the Mediterranean and in the North Atlantic. It is shown that fluctuations in acoustic transmission loss closely correlate with ocean variability resulting from internal waves, tides, tidally-advected changes in water masses, and ocean fine-structure.

1. INTRODUCTION

The propagation of sound in the ocean is inevitably accompanied by fluctuations in the amplitude and phase of an acoustic signal received at large distances from the source. The fluctuations manifest not only changing patterns of interaction with the bottom and surface, particularly important in shallow water propagation, but also passages of the wave through time-varying inhomogeneities in the ocean medium. The resulting fluctuations, or scintillations, in acoustic intensity are analogous to the twinkling of stars arising from multiple scattering of lightwaves in the irregular layers of the upper atmosphere. The variability in acoustic propagation can be considered to arise from variations in the index of refraction, or sound velocity, of the medium, which, in turn, are induced by a variety of ocean processes covering a wide range of temporal and spatial scales.

Aero- and Hydro-Acoustics
IUTAM Symposium Lyon 1985
Editors: G. Comte-Bellot and J. E. Ffowcs Williams

2. SPATIAL AND TEMPORAL SCALES

Depending on the temporal and spatial scales involved, the mechanisms can be considered either deterministic or random <1>. The general circulation of the ocean ("ocean climate") and its associated current systems (Gulf Stream, Kuroshio, etc.) are characterized by horizontal scales of variability limited only by the size of the basin, vertical scales of a few hundred metres, and temporal scales from a few days to seasonal. These are deterministic structures. The intermediate scales of variability, including ocean motions such fronts and eddies, can also be considered to be deterministic perturbations from the mean structure. The associated scales of variabilities are of the order of 100 to 1000 km in the horizontal, to ocean depth in the vertical, and days to months in time.

Smaller scale phenomena comprise internal waves, fine-structure, and microstructure. These phenomena are considered random. Internal waves are characterized by scales from 100 m to 10 km or more in the horizontal, 1 to 100 m in the vertical, and from about 10 min to 1 day in time. Because they owe their existence to the restoring forces due to the density gradient and the Coriolis force, the frequency spectra of internal waves are bounded by the inertial frequency at the low end and by the buoyancy frequency (Brünt-Väisälä) at the high end. Internal wave-induced variability has been found to be a very significant source of sound scattering and has received considerable attention in recent years <2,3,4>. In essence, this increased activity arises from two complementary facts. On the one hand the spatial scales of internal waves match the acoustic wavelength over a broad frequency range, and thereby affect the acoustic field; on the other hand the internal wave is one of the few phenomena for which a reasonably effective statistical model (Garrett-Munk) is available. Fine- and microstructure variability involve scales from several metres to hundreds of metres in the horizontal and 1 cm to about 10 m in the vertical. It is of the order of milliseconds in time. Such variability would be expected to affect sound propagation in the frequency range from approximately 1 to tens of kHz. Table 1 summarizes the temporal scales of environmentally-induced variations encountered in acoustic data.

3. THEORETICAL APPROACHES

Attempts to understand the non-deterministic component of intensity fluctuations have been centered around the theory of propagation through a

PERIOD	PRINCIPAL ENVIRONMENTAL PARAMETER	PRINCIPAL ACOUSTIC EFFECTS
ms	FINE AND MICROSTRUCTURE	AMPLITUDE, PHASE AND PROPAGATION PATH FLUCTUATIONS; HIGH FREQUENCY SCATTERING
5 s	SURFACE WAVE SIGNATURES	AMPLITUDE MODULATION; SCATTERING
10 min – 24 h	INTERNAL WAVE SIGNATURE	AMPLITUDE AND PHASE FLUCTUATION; SIGNIFICANT, REFRACTION OF NEARLY HORIZONTAL PATHS
4 h	SHALLOW WATER TIDES	NOISE-LIKE AMPLITUDE FLUCTUATIONS; PHASE FLUCTUATIONS SIMPLY CORRELATED WITH TIDAL VARIATIONS. GREATEST EFFECTS IN SHALLOW WATER
12 h	SEMIDIURNAL TIDES	
24 h	DIURNAL TIDES	
4 day	MESOSCALE PHENOMENA (EDDIES, FRONTS)	FLUCTUATIONS IN PROPAGATION PATH, ACOUSTIC INTENSITY; DISTORTION OF SIGNAL SHAPE.
15 day		
1 month		
	OCEAN CLIMATE	FLUCTUATION IN AMPLITUDE, PHASE AND PROPAGATION PATH
1 year		

Table 1

Temporal variations observed in acoustic data

random medium. The only medium descriptor needed in such a theory is the correlation function of the fluctuations (index of refraction) or its equivalent (structure function, spectrum). Because the pertinent stochastic wave equation cannot be solved deterministically, solutions are obtained in terms of statistical moments of the acoustic pressure, such as the expected value (ensemble average), various correlation functions, and higher order moments. The fourth moment is of particular interest since this gives the intensity fluctuations.

Earlier theories of intensity fluctuations were based on either the Born approximation or the Rytov Method of Small Perturbations. These approaches are valid for the weak or single scattering cases, which are adequate for high frequencies and/or short ranges. In particular, the Born approximation is valid only for small fluctuations in both the intensity and the phases; the Rytov method is valid for small intensity fluctuations, but has less stringent requirements in phase.

Much of the new work in extending the range of validity to large intensity fluctuations rising from multiple scattering is based on the parabolic equation (PE) method. The PE method is a simplification of the wave equation for the case of scattering that is mostly in the forward direction. The two main methods of solving the PE are the Feynman path integral (used by Flatte, et al <3>) and the moment-equation (used by Uscinski, et al <4>). Both methods have had some recent success in attributing measured intensity fluctuations to the action of internal waves <3,4>. As alluded to earlier, this has been possible because of the availability of the required statistical description of the stochastic medium, viz, the Garrett-Munk empirical model.

4. EXPERIMENTAL RESULTS

From the gamut of environmentally-induced fluctuations, a few representative samples relevant to low frequency acoustic propagation will be discussed. In particular, using the results of SACLANTCEN measurements in the Mediterranean and in the North Atlantic, the effect on transmission loss of long-period internal waves (inertial oscillations), diurnal and semi-diurnal tides, and tidally-advected changes in water masses will be demonstrated.

Generally, both time-dependent and range-dependent propagation were of interest in these measurements but the discussions will focus on time-dependency. For such measurements, the experimental configuration consisted of a source and a receiver (generally deployed from two ships) separated by a fixed distance which was different for each area. The sources (explosives) were broad-band and deployed at hourly intervals. The receiver was usually a vertical array of hydrophones. Simultaneous samplings of the pertinent oceanographic parameters (sound speed, temperature, salinity and density) were taken at both the source and receiver locations.

4.1 Mediterranean Measurements

For the Mediterranean measurements, the fixed distance between source and receiver was 35 km; the water depth ranged from about 40 to 85 m along the propagation path. A typical example of temporal variability of the sound speed profile is shown in Fig. 1a, which shows the variation of the sound speed with depth over a period of 25 hours. From the surface down to about 20 m the region is essentially isovelocity with sound speed of approximately 1538 m/s. The depth from approximately 25 to 35 m comprises the steepest portion of the thermocline. Fluctuations in sound speed are quite evident, particularly at a depth of about 25 m within the thermocline. The plot clearly indicates an oscillation in the thickness of the mixed layer (surface duct). Further, "kinks", typical features in the curve, appear to migrate vertically with time — a motion often suggesting the presence of internal waves <Gregg, 5>. Of particular note is the fine-structure evident in the expanded portion of the profile (Fig. 1b). Generally, the fine-structure was characterized by vertical dimensions of the order of a few centimetres to one or two metres. This fine structure tends to scatter high frequency sound.

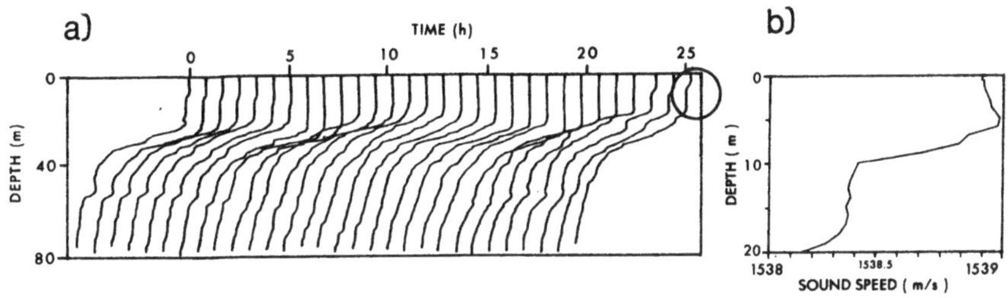

Fig. 1 Sound speed profiles
a) Temporal variation over approx. 25 hours
b) Enlargement of circled profile in a)

The frequencies of these oscillations are of particular interest because they provide clues to the responsible mechanisms. Figure 2 compares the spectra of the transmission losses at 200 Hz and 1600 Hz, at a depth of 40 m, with those of the sound speed at a depth of 25 m. Several features should be noted. The dominant fluctuations occur in the frequency range from 0.06 to 0.08 cycle/h, which coincide with a range of periods of 17 to 12 hours, respectively. Within this broad response are included both the semi-diurnal and inertial periods (approximately 17 hours for this area). A longer time series than was used here would probably have enabled the spectral separation of the two responses. From the results obtained at other depths it does appear, however, that the relative significance of the semi-diurnal tidal and inertial effects depends on depth. Apparently there are fluctuations of inertial period in the surface layers, and, as the depth increases, the dominant fluctuations are semidiurnal. Inertial oscillations may occur in connection with a sudden change of the wind speed (i.e., a wind gust of short duration) and rapid changes in barometric pressure <6>. The whole Mediterranean is dominated by meteorological forcing; therefore it is likely that inertial oscillations are present in the study area. A comparison of the spectra reveals a good correlation

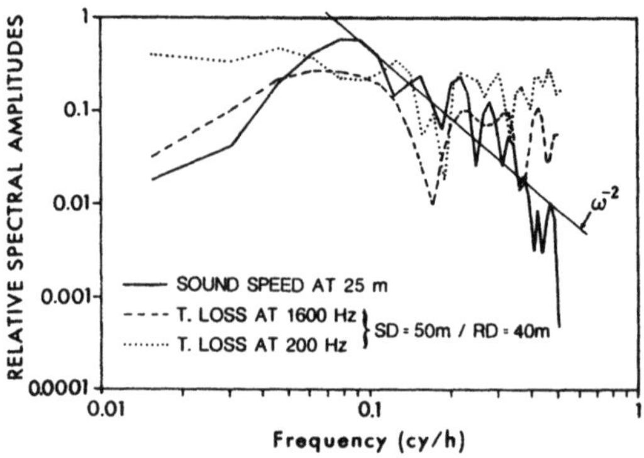

Fig. 2
Comparison of transmission loss and sound speed spectra for Mediterranean measurements.

between the spectrum of environmental variability and the higher frequency transmission loss spectrum.

Note that the drop-off rate (sloping line) of the sound speed spectrum is approximately ω^{-2}, a slope consistent with that predicted by the Garrett-Munk model for internal waves.

4.2 North Atlantic Measurements

The experimental configuration for the North Atlantic measurements is shown in Fig. 3. The fixed distance between source (S) and receiver (R) was 60 km in water changing in depth from 300 to 75 m. A large part of the track was over a sand, silt and clay bottom. The receiver was located in a hard-bottom (sand) region. From the point of view of temporal variability, a significant aspect of the propagation path, SR, is its proximity to the Polar Front. In particular, SR is directly affected by movements of the Front, the water above 20 m changing from Polar to Atlantic with an approximately 12 h periodicity, while waters below 20 m remain of the Polar type.

In effect, therefore, the Polar Front serves as a demarcation line between two different propagation areas. To the north of the Front, in isothermal water, an important part of the propagation will be in shallow water under upward refracting conditions, resulting in small total transmission losses as a result of less (acoustic) energy-bottom interaction. On the other hand, in waters affected by the Front, propagation will be under strong downward refracting conditions, as a result of the steep gradients, and thereby subject to higher losses, particularly at low frequencies.

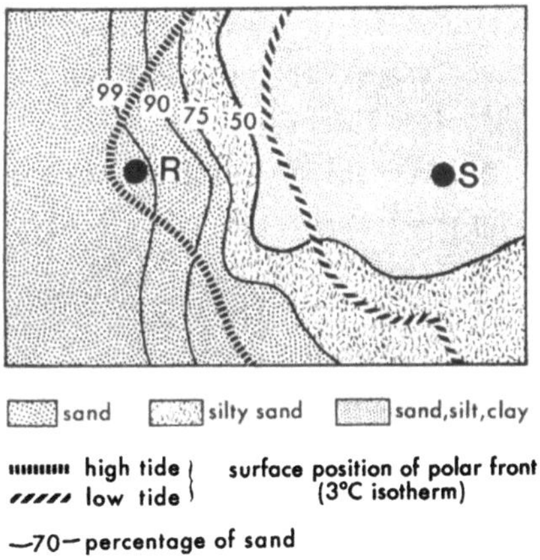

Fig. 3
Environment at the North Atlantic measurements site.

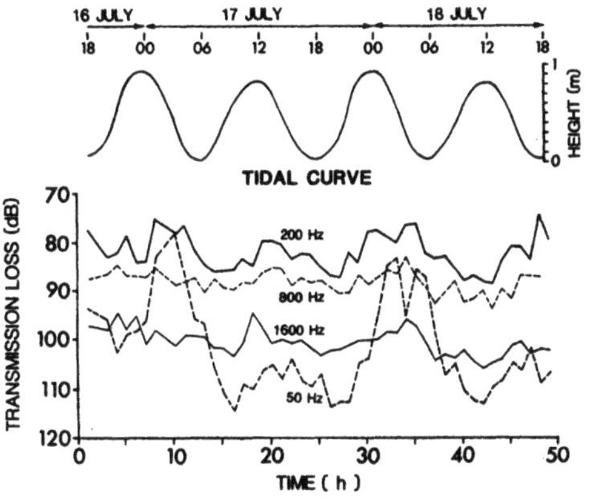

Fig. 4
Temporal variation of transmission losses and predicted tidal curve.

Figure 4 provides a typical example of the fluctuations in measured transmission loss at site R for several frequencies, along with the predicted tidal curve for the area. Obviously the fluctuations in transmission loss correlate quite well with tidal periodicity, particularly for the lower frequencies. This is consistent with the assumption that the lower frequencies are more affected by the tidally-induced change in downward-refracting conditions resulting from advected changes in water masses. Both diurnal and semi-diurnal periodicities are evident, the diurnal periodicity seemingly the more significant. Furthermore there is a correspondence, albeit inexact, between high tide and high transmission loss, consistent with the fact that the front reaches site R during periods of high tide. Although the curves pertain to a particular source and receiver depth combination (240 and 50 m respectively), the observed trends represent fairly all the data obtained at this site.

Figure 5 compares the transmission loss spectra at 25 and 1600 Hz (at a depth of 20 m) with the sound speed spectrum (at a depth of 18 m). As seen, the correlation between the sound speed spectrum and that of the low-frequency transmission loss is good. Note again that the fluctuations in low frequency transmission loss are much more marked than those at the higher frequency. Moreover, the fluctuations reveal more clearly the presence of diurnal and semi-diurnal effects. This selective frequency effect induced by the medium contrasts with that seen in the Mediterranean results. This is apparently due to the fact that (acoustic) energy-bottom interaction is the primary cause of the transmission losses in the North Atlantic measurements. However, for the Mediterranean measurements apparently natural periodic mechanisms such as internal waves and tides are at work.

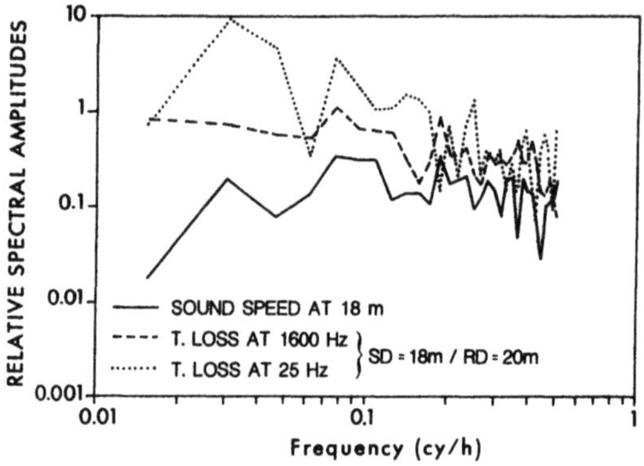

Fig. 5
Comparison of transmission loss and sound speed spectra for North Atlantic measurements.

CONCLUSIONS

The results discussed in this paper lead to the general conclusion that medium-induced variability in the acoustic index of refraction can significantly affect the acoustic propagation. In particular, for the two areas (Mediterranean and N. Atlantic) investigated, fluctuations in acoustic transmission losses closely correlate with ocean variability at diurnal, semi-diurnal and inertial periods, as well as with the variation of ocean fine-structure. The principal mechanisms involved appear to be internal waves, tides, tidally advected changes in water masses, meteorological forcing, and ocean fine-structure.

References

1. ALI, H.B. Spatial and temporal variabilities in underwater acoustic transmission: an analytical review, SACLANTCEN SM-166. La Spezia, Italy, SACLANT ASW Research Centre, 1983. [AD 129 959].

2. DESAUBIES, Y. Statistical aspects of sound propagation in the ocean. In: Urban, H. ed. Adaptive Methods in Underwater Acoustics, Dordrecht, Reidel, The Netherlands, 1985.

3. FLATTE, S.M., DASHEN, R., MUNK, W.H., WATSON, K.M. and ZACHARIASEN, F. eds. Sound Transmission through a Fluctuating Ocean. Cambridge, Cambridge University Press, 1979.

4. USCINSKI, B.J., MACASKILL, C. and EWART, T.E. Intensity fluctuations, Part I: theory, Part II: comparison with the Cobb experiment. Journal of the Acoustical Society of America 74, 1983: 1474-1499.

5. GREGG, M.C. Finestructure: oceanic "kinks and wiggles." Naval Research Reviews 34, 1982: 52-72.

6. POLLARD, R.T. On the generation of inertial waves in the ocean. Deep-Sea Research 17, 1970: 795-812.

Underwater Sound Generation by Waterhammer

A. LAAKE , G.E.A. MEIER , A.P. SZUMOWSKI

Max-Planck-Institut für Strömungsforschung
Böttingerstraße 4 - 8
D - 3400 Göttingen
Fed. Rep. of Germany

Summary

If an accelerated water column is abruptly stopped, waterhammer occurs. This generates a pressure wave, which is radiated into the surrounding liquid. A delta pulse-like pressure signal with large bandwidth and amplitude is achieved.

Application of Sound Waves to Geophysical Prospecting

The past years have shown the decline of natural resources of oil, gas and minerals on all continents. As a result recent geophysical prospecting mainly using the seismic reflection method concentrates on the shelf areas. Acoustic waves, generated in water, propagate through the water into the sea bottom. They are reflected and refracted back to the sea surface where they are detected by an array of hydrophones. One characteristic seismic parameter to be determined is the damping factor Q^{-1}. During the last years Q^{-1} was found to be strongly dependent on frequency [1]. Therefore further investigations require sound generators for signals with large bandwidth. This paper is concerned with the generation of delta shaped pulses by mechanical means (Fig. 1).

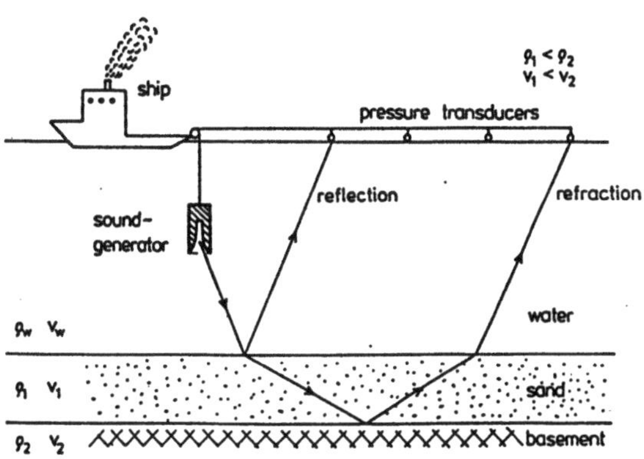

Fig. 1. Application of sound waves to geophysical prospecting

Aero- and Hydro-Acoustics
IUTAM Symposium Lyon 1985
Editors: G. Comte-Bellot and J. E. Ffowcs Williams

Theory of Sound Generation by Waterhammer

In a pipe waterhammer occurs, if an accelerated liquid column of the typical length L is stopped abruptly within the time Δt [2] .

Joukowski Condition $$\Delta t < \frac{2L}{a} \tag{1}$$

where a is the velocity of sound in the liquid. The momentum of the liquid flow, which moves with the velocity v, is transferred to a pressure wave. This transformation can be treated by a one-dimensional approach [3] , [4] and has to fulfill the

Euler equation $$\frac{\partial v}{\partial t} + v\frac{\partial v}{\partial x} + \frac{1}{\rho}\frac{\partial p}{\partial x} = 0 \tag{2}$$

and the

Continuity equation $$\frac{\partial \rho}{\partial t} + \frac{\partial (\rho v)}{\partial x} = 0 \tag{3}$$

where p is the pressure in the liquid and ρ is the density of the liquid. These two equations form a system, which has the following characteristics

Characteristics $$\frac{dx}{dt} = v \pm a . \tag{4}$$

Additionally we have the compatibility equations for the moving liquid column and the generated pressure wave.

Compatibility equations $$dp \pm \rho\, a\, dv = 0 \tag{5}$$

Experimental Arrangement

In a heavy steel cylinder a water column covered by a piston is accelerated by the pressure difference between pressure p_n at the horn and lower pressure p_v above the piston. After the piston has been stopped at the stop face, it is driven back by supply pressure from above. The pressure signals at the mouth of the horn are measured by a pressure transducer (Fig. 2).

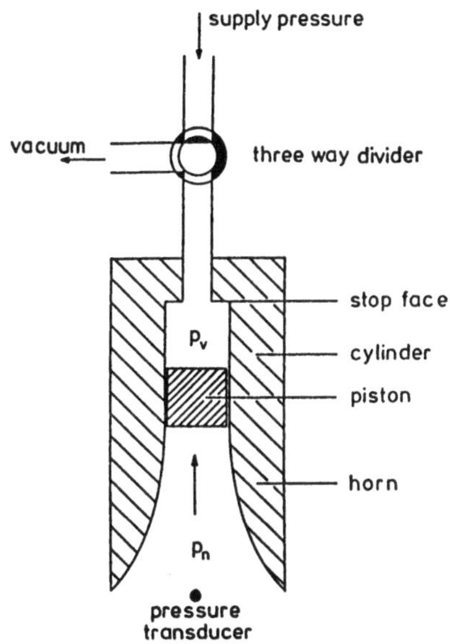

Fig. 2. Experimental arrangement

Wave Pattern

When the piston is suddenly stopped, the inertia of the water column compresses the water at the piston. The generated pressure wave follows the characteristics (4) out of the horn. At the open end of the horn one part of the wave is reflected as a rarefaction wave propagating into the cylinder. Thereby the pressure in the cylinder is reduced to the pressure p_n. At the hard wall of the piston the reflection of the rarefaction wave causes a reduction of the pressure to a value below vapour pressure. The result is an area of cavitation with strongly reduced sound velocity. The acoustically generated cavitation as well as the following annihilation of the cavitation results in a considerable consumption of acoustic energy and consequently reduced amplitudes. The resulting damped reflections now propagate again with the normal sound velocity (Fig. 3).

Signature of pressure waves:

———— compressional wave

- - - - - - rarefaction wave

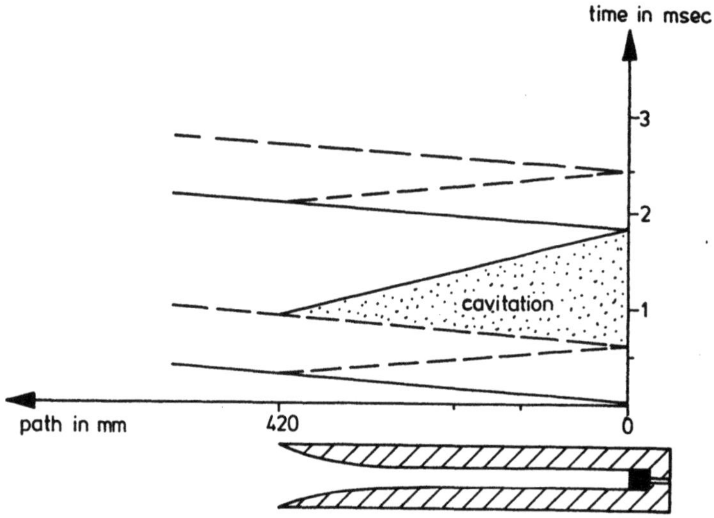

Fig. 3. Wave pattern of the sound generator

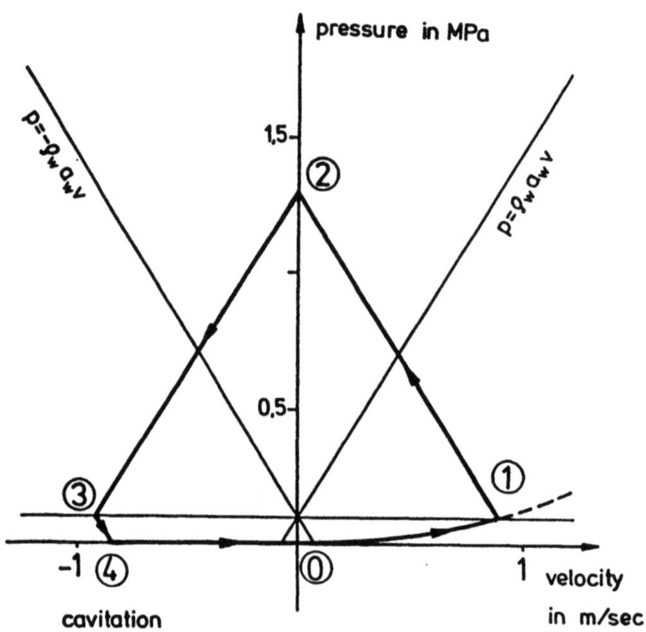

Fig. 4. Phase diagramm of the sound generator

Phase Diagram

The cycles of the reflected waves can be shown in a phase diagram of sound pressure p and the velocity of the water column v. The motion starts at the origin of the coordinate system (Fig. 4 point 0). The acceleration of the water column leads to point 1. The generated pressure wave follows the compatibility equation (5) until it is reflected at point 2. The rarefaction wave reduces the pressure and velocity to point 3. When the reflected rarefaction wave reduces the pressure to a value below vapour pressure (point 4) the compatibility equations are not fulfilled any longer, and cavitation occurs. The pressure remains nearly zero until the reflected pressure wave cancels cavitation (Fig. 4).

Pressure Signal

A typical signal measured at the mouth of the horn shows a peak shaped like a delta pulse with an amplitude of 1.3 MPa and a half width of 0.3 msec. Cavitation causes the successive under pressure for a time span of 1.2 msec. Subsequently damped reflections (1-3) of pressure waves and rarefaction waves are observed (Fig. 5).

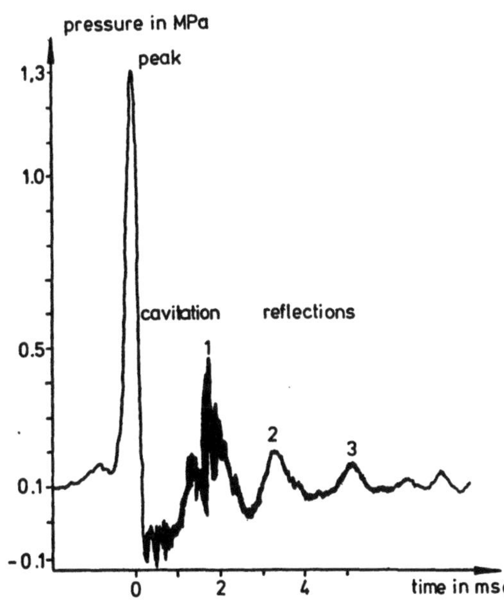

Fig. 5. Pressure signal at the mouth of the horn

References

1. Meißner, R., Theilen, F., Proc. of the 11th World Petroleum Congress, London 1983

2. Joukowski, N.E., Mem. Acc. Sc., St. Petersbourg 1889

3. Meier, G.E.A., Szumowski, A.P., Ber. d. MPI f. Strömungsforschung 15/1980, Göttingen 1980

4. Szumowski, A.P., Ber. d. MPI f. Strömungsforschung 10/1980, Göttingen 1980

5. Laake, A.: Diploma Thesis. Ber. d. MPI f. Strömungsforschung 107/1984, Göttingen 1984

Hydroacoustics (C'nd)

Finite Amplitude Sound Beams Resulting from Nonlinear Vibration of a Circular Membrane Undergoing Axisymmetric Resonant Excitation

J. H. GINSBERG

School of Mechanical Engineering
Georgia Institute of Technology
Atlanta, Georgia

Summary

This paper analyses the interaction between a vibrating circular membrane contained in an infinite baffle and the resulting sound field radiated into a fluid medium contained in the half-space above the membrane. The case of resonant excitation of the membrane leads to nonlinear coupling between transverse and in-plane displacement. Nonlinearity within the fluid medium is described by a recent general treatment of finite amplitude sound beams resulting from boundary motion. The vibratory response of the membrane is evaluated in a perturbation technique based on the modes of free vibration. The results give amplitude-frequency relations for the plate that account for the inertial and damping impedances of the fluid, as well as expressions that may be solved for the pressure signal in the fluid.

Finite Amplitude Effects in Acoustic Radiation

Consider a circular membrane of radius a that is fixed at its edges to an infinite baffle. An excitation at frequency ω close to a natural frequency is applied on one side, and a fluid medium occupies the half-space on the other side. The speed of sound and density of the fluid at ambient conditions are c_0 and ρ_0, respectively. Dimensional cylindrical coordinates are R/k and z/k within the fluid which occupies $z > 0$, and dimensional time is t/ω, where $k = \omega/c_0$. A recent study described the effects of material and convective nonlinearity on the acoustic radiation resulting from an arbitrary motion on the boundary [1]. Suppose that the normal velocity on the boundary is

$$v_z\Big|_{z=0} = \frac{1}{2i}\,\varepsilon\, c_0\, f(R)\, \exp(it) + \text{c.c.} \tag{1}$$

where $\varepsilon \ll 1$ is the acoustic Mach number and $f(R)/R^{1/2}$ is bounded for large R. Then the pressure at any (R, z), omitting a mean value correction term, was found to be described by

Aero- and Hydro-Acoustics
IUTAM Symposium Lyon 1985
Editors: G. Comte-Bellot and J. E. Ffowcs Williams

$$p = \frac{1}{2}\,\varepsilon\rho_0 c_0{}^2 \int_0^\infty \frac{mV}{\mu}\,\exp(it - \mu\xi)\,J_0(m\alpha)\,dm + c.c. \tag{2}$$

In general, c.c. shall denote the complex conjugate of all preceding terms. The parameters m and μ are transverse and axial wavenumbers, respectively, $J_0(\ )$ denotes the Bessel function of zero order, and V is the Hankel transform of the spatial pattern f(R).

$$\mu = i(1 - m^2)^{1/2} \text{ if } m < 1 \quad \& \quad \mu = (m^2 - 1)^{1/2} \text{ if } m > 1$$

$$V = \int_0^\infty R\,f(R)\,J_0(mR)\,dR \tag{3}$$

The parameters (α, ξ) are strained coordinates defined in implicit form by

$$z = \xi - \pi\varepsilon\beta_0\xi\,\left\{(mV/\mu)\,\exp(it)\,\mathrm{erfc}\left[(\mu\xi)^{1/2}\right] + c.c\right\}\,J_0(n\alpha)$$

$$R = \alpha + \pi\varepsilon\beta_0\xi\,\left\{(mV/\mu)\,\exp(it)\,\mathrm{erfc}\left[(\mu\xi)^{1/2}\right] + c.c\right\}\,J_1(n\alpha) \tag{4}$$

The evaluation of the response of the fluid-membrane system requires that eqs. (3) and (4) be interfaced with the equations governing the membrane.

Equations of Motion for the Membrane

An elastic membrane undergoing finite deformation due to a resonant excitation was studied by Chobotov and Binder [2]. Several shortcomings of the earlier work shall be corrected here. Small errors associated with using an assumed mode function shall be addressed by employing the exact Bessel function mode. Furthermore, the study here shall describe the situation for resonance of any mode, rather than only the fundamental. The last matter is that the resistive and reactive portions of the fluid impedance will be derived analytically. In Ref. [2] the acoustic impedance was based on a low frequency approximation that did not account for diffraction.

Chobotov and Binder began with a derivation of displacment equations of motion that accounted for in-plane deformation and geometrical nonlinearity. Examination of these equations reveals that the membrane displacements are scaled such that

$$\text{transverse displacement} = \varepsilon\,\frac{w}{k}\ ;\ \text{in-plane displacement} = \varepsilon^2\,\frac{u}{k} \tag{5}$$

Based on these definitions and the fact that the acoustic Mach number ε in eq. (1) is very small, the equation for in-plane motion may be rewritten from its original form in [2] as

$$\frac{\partial}{\partial R}\left[\frac{1}{R}\frac{\partial}{\partial R}(Ru)\right] + \frac{1}{2}\frac{\partial}{\partial R}\left(\frac{\partial w}{\partial R}\right)^2 + \frac{(1-\nu)}{2R}\left(\frac{\partial w}{\partial R}\right)^2 = \frac{c_0^{\,2}}{c_u^{\,2}}\frac{\partial^2 u}{\partial t^2} \tag{6}$$

The corresponding equation for transverse motion is

$$\frac{1}{R}\frac{\partial}{\partial R}\left\{R\frac{\partial w}{\partial R} + \frac{\varepsilon^2}{e_0}\left[R\frac{\partial u}{\partial R}\frac{\partial w}{\partial R} + \nu\, u\frac{\partial w}{\partial R} + \frac{1}{2}R\left(\frac{\partial w}{\partial R}\right)^3\right]\right\}$$
$$+ \frac{\Gamma}{\varepsilon ka}\left[Q(R)\cos(t) - \frac{p|_{z=0}}{\rho_0 c_0^{\,2}}\right] = \frac{c_0^{\,2}}{c_w^{\,2}}\frac{\partial^2 w}{\partial t^2} \tag{7}$$

In the foregoing $\Gamma = \rho_0 c_0^{\,2} a/\sigma_0 h$ (h is the thickness of the membrane), Q(R) is the radial profile of the excitation applied to the membrane, and

$$c_u = [E/\rho(1-\nu^2)]^{1/2}\ ;\ c_w = (\sigma_0/\rho)^{1/2}\ ;\ e_0 = (c_w/c_u)^2 \tag{8}$$

Note for later use that $c_w \ll c_u$. In addition, the magnitude of Q is required to be sufficiently small to induce a transverse displacement whose peak velocity actually is a small fraction of c_0.

<u>Vibratory Response</u>

The eigenfunctions for a membrane whose edges are fixed are

$$\Phi_j = J_0(\eta) \quad \text{where} \quad \eta = \lambda_j R/ka \quad \text{and} \quad J_0(\lambda_j) = 0 \tag{9}$$

Some approximate values are $\lambda_1 = 2.405$, $\lambda_2 = 5.520$, $\lambda_3 = 8.654$. Proximity to resonance is specified by $\omega \approx \lambda_j c_w/a$.

In general, the transverse displacement may be expanded in a series of modes. When the excitation is close to one of the natural frequencies, the corresponding mode may be expected to dominate the response. Since the nonlinear terms in eq. (7) are $O(\varepsilon^2)$, the transverse motion in this case should satisfy

$$w = \Phi_j\, w_j(t) + \varepsilon^2 \hat{w}(R,t) \tag{10}$$

where $\hat{w}(R,t)$ is orthogonal to Φ_j over the interval $0 \le R \le ka$.

Substitution of eq. (10) into eqs. (6) and (7) shows that all derivatives have comparable magnitudes. The largeness of c_u relative to c_w, in combination with the resonance of w, makes it permissible to neglect the inertia of in-plane motion. An expression for u may then be found analytically. Toward this end the O(1) term for w in eq. (10) is substituted into eq. (6). Using the chain rule to replace R by η results in an inhomogeneous ordinary differential equation for u.

$$\frac{\partial}{\partial\eta}\left[\frac{1}{\eta}\frac{\partial}{\partial\eta}(\eta u)\right] = -\frac{\lambda_j}{2a} w_j^2 \left\{\frac{\partial}{\partial\eta}\left[J_1(\eta)^2\right] + \frac{(1-\nu)}{\eta}\left[J_1(\eta)^2\right]\right\} \tag{11}$$

This equation may be integrated twice, after which the constants of integration are selected so as to satisfy the condition that u = 0 at R = 0 and R = ka. Repeated application of the recursion relations for Bessel functions yields the following expression.

$$u = \frac{\lambda_j}{ka} w_j^2 U$$

$$U = \frac{1}{4}(1+\nu)\, J_0(\eta)\, J_1(\eta) + \frac{1}{4}\nu\, \eta \left[J_1(\lambda_j)^2 - J_0(\eta)^2 - J_1(\eta)^2\right] \tag{12}$$

Amplitude-Frequency Relations

The equation for $w_j(t)$ is obtained by substituting eqs. (10) and (12) into eq. (7), then applying orthogonality of $\hat{w}$ with respect to Φ_j. The result is

$$\frac{1}{2}\left(\frac{\lambda_j}{ka}\right)^2 \left[\lambda_j\, J_1(\alpha_j)\right]^2 \left[\left(\frac{\omega a}{\lambda_j c_w}\right)^2 \frac{\partial^2 w_j}{\partial t^2} + w_j\right] + \frac{\varepsilon^2}{\varepsilon_0}\left(\frac{\lambda_j}{ka}\right)^4 \delta_j\, w_j^3$$

$$+ \frac{\Gamma}{\varepsilon ka}\int_0^{\lambda_j} \eta\, \frac{p|_{z=0}}{\rho_0 c_0^2}\, J_0(\eta)\, d\eta = \frac{\Gamma}{\varepsilon ka}\cos(t)\int_0^{\lambda_j} \eta\, Q\, J_0(\eta)\, d\eta \tag{13}$$

where δ_j is a coefficient of nonlinear elasticity that is found to be

$$\delta_j = \int_0^{\lambda_j}\left\{U\left[(1+\nu)\, J_1(\eta)^2 - 2\nu\, J_0(\eta)\, J_1(\eta)\right] + \frac{1}{2}\eta\, J_1(\eta)^4\right\} d\eta \tag{14}$$

values of this parameter for the first three resonances when $\nu = 0.29$ are $\delta_1 = 0.08206$, $\delta_2 = 0.11996$, $\delta_3 = 0.14069$. Chobotov and Binder used the approximate fundamental mode $\Phi_1 = 1 - (R/ka)^2$ to obtain $\delta_1 = 0.08424$.

Because of the smallness of the nonlinear term in eq. (13), w_j will be harmonic in the first order approximation. Let A be the complex amplitude, and let $\bar{w}$ denote the dimensional transverse displacement found from eq. (5).

$$w_j = \frac{A}{2} \exp(it) + c.c. \; ; \; \bar{w} = \frac{\varepsilon A}{2k} J_0(\eta) \exp(it) + c.c. + O(\varepsilon^2) \qquad (15)$$

The Hankel transform V of the normal velocity at the membrane may be evaluated by comparing the (dimensional) time derivative of $\bar{w}$ to eq. (1). This defines the shape factor f(R) for R < ka. For R > ka, f(R) = 0 because the baffle is stationary. The transform is found in this manner to be

$$V = - A\,G \; ; \quad G = \int_0^{ka} R\, J_0(\lambda_j R/ka)\, J_0(mR)\, dm \qquad (16)$$

The coefficient G, which may be evaluated in closed form, leads to an expression for p according to eq. (3). The acoustic loading applied to the membrane is readily found from that expression, because $\xi = 0$ and $\alpha = R$ at z = 0, see eqs. (4).

$$p|_{z=0} = -\frac{1}{2} \rho_0 c_0^2 \,\varepsilon A \int_0^{\infty} \frac{nG}{\mu} \exp(it - \mu z)\, J_0(mR)\, dm + c.c. \qquad (17)$$

The final step leading to the amplitude-frequency relations is to use the method of harmonic balance to equate all terms in the equation of motion that are proportional to exp(it). (This procedure is equivalent to removing secular terms in singular perturbation schemes.) The equation obtained in this manner relates the amplitude εA to the frequency ω and the generalized modal amplitude Q_j for the excitation.

$$\left[1 + \Gamma\,(\gamma_d i - \gamma_m) - \left(\frac{\omega a}{\lambda_j c_w}\right)^2\right] \varepsilon A + \frac{3\delta_j \varepsilon^3 A^2 A^*}{2e_0[kaJ_1(\lambda_j)]^2} = \Gamma \frac{ka}{\lambda_j^2} Q_j \qquad (18)$$

where εA^* denotes the complex conjugate. The coefficients γ_m and γ_d are reactance (added mass) and resistance (damping) impedance coefficients, respectively, resulting from the acoustic radiation.

$$\gamma_d = \frac{2}{ka\, \lambda_j^2 J_1(\lambda_j)^2} \int_0^1 \frac{m\, G^2}{(1 - m^2)^{1/2}}\, dm$$

$$\gamma_m = \frac{2}{ka\ \lambda_j^{\,2}\ J_1(\lambda_j)^2} \int_1^\infty \frac{m\ G^2}{(m^2 - 1)^{1/2}}\, dm \tag{19}$$

Equivalent expressions were derived by Bouwkamp [3] by using a scattering integral.

The real and imaginary parts of eq. (18) may be solved for the magnitude and phase of εA; the latter represents the phase lag of the response relative to the excitation. Then the acoustic signal may be evaluated by substituting the value of εA into eqs. (16) to form the Hankel transform V.

Results

Figures 1 and 2 depict the resistance and reactance coefficients as a function of ka. The peaks in γ_m are centered around $ka = \lambda_j$ for the respective resonances, whereas γ_d rises almost linearly beyond those locations. Suppose that the fluid is water, in which case $c_w/c_0 \ll 1$ ($\sigma_0 \ll \rho_0 c_0^{\,2}$ because σ_0 cannot exceed the yield stress). Furthermore, note that $ka = \lambda_j\ c_w/c_0$ when $\omega = \omega_j$, and that $ka \gg 2\pi$ for closely confined sound beams. It follows that resonant excitations at high ka values for beam forming only arise for very high order modes.

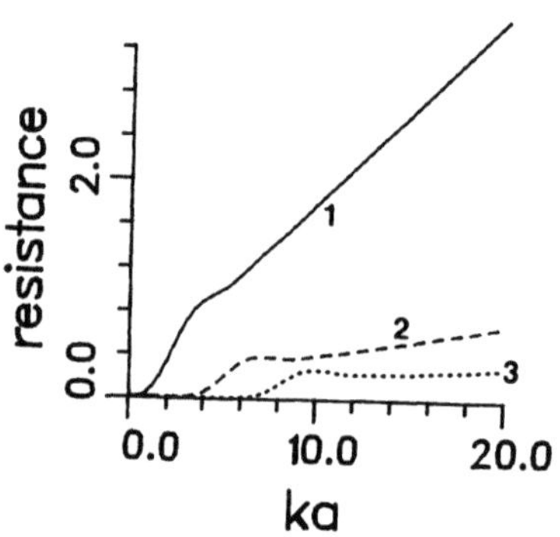

Fig. 1. Resistance γ_d for the three lowest modes.

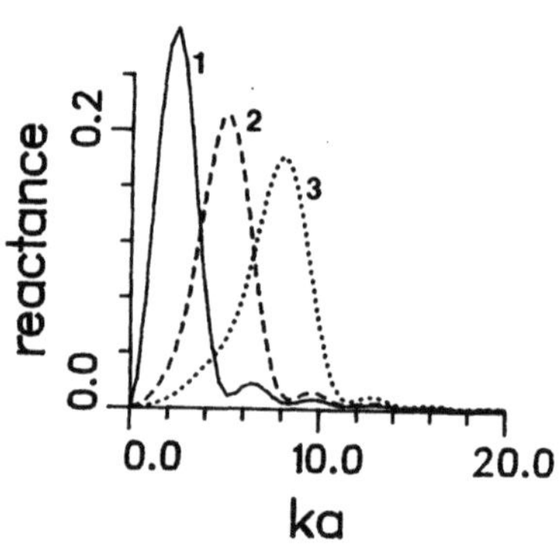

Fig. 2. Reactance γ_m for the three lowest modes.

The dependence of the harmonic amplitudes forming the pressure waveform along the axis of a sound beam is depicted in Figure 3. Only the fundamental frequency, marked #1, is treated by a linearized analysis.

The corresponding result for a piston (uniform velocity) is shown in Figure 4. In both cases ka = 40 and ε = 0.00102, which produces a maximum sound pressure level in the piston case of 250 dB re 1 µPa. The fluctuations in the piston case are due to diffraction effects that alternately reinforce and cancel the fundamental frequency in the near field, $z < (ka)^2/2\pi$. Diffraction effects are much less significant for the membrane because the particle velocity is continuous across the edge, so the propagation curves are smoother, and the peak values are lessened.

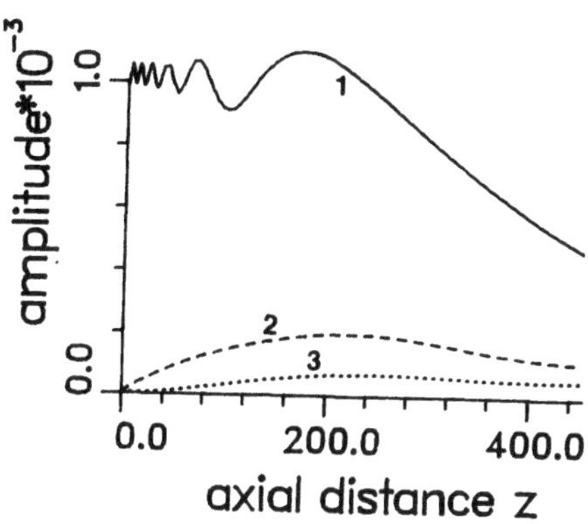

Fig. 3. Range dependence of the amplitude of the three lowest harmonics in a sound beam generated by a membrane, as a fraction of $\rho_0 c_0^2$.

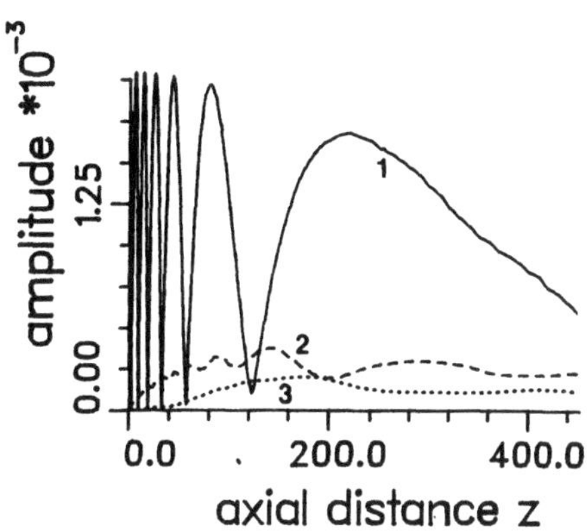

Fig. 4. Range dependence of the amplitude of the three lowest harmonics in a sound beam generated by a piston, as a fraction of $\rho_0 c_0^2$.

References

1. Ginsberg, J. H.: Nonlinear King integral for arbitrary axisymmetric sound beams at finite amplitudes. II. Derivation of uniformly accurate results. J. Acoust. Soc. Am. 76 (1984) 1208 - 1214.

2. Chobotov, V. A. and Binder, R. C.: Nonlinear response of a circular membrane to sinusoidal excitation. J. Acoust. Soc. Am. 36 (1964) 59 - 73.

3. Bouwkamp, C. J.: A contribution to the theory of acoustic radiation. Phillips Research Reports 1 (1946) 251-277.

Acknowledgement

This work was supported by the Office of Naval Research, code 425-UA.

Theoretical Prediction for the Reduction of Flow-Induced Noise on a Plane Surface

SUNG H. KO

Department of the Navy
Naval Underwater Systems Center
New London, Connecticut 06320 USA

Summary

A study was made of a theoretical model for evaluating the reduction of the noise induced by turbulent flow over a plane surface. The flow-induced noise received by a rectangular hydrophone embedded in a layer of elastomer was calculated to evaluate the noise reduction relative to the turbulent boundary layer noise on a hard surface. The formulation of the problem is based on the Corcos model representing a turbulent wall pressure spectrum, a transfer function representing a wavevector filter, and the rectangular hydrophone function. The major results presented in this paper are noise reduction contour plots for 200 Hz and 1,000 Hz.

Introduction

The analytical model is a plane elastomer layer backed by a plane surface; the other side of the elastomer layer is exposed to turbulent flow. In order to evaluate the noise reduction relative to the turbulent boundary layer noise on a hard surface, the flow-induced noise received by a rectangular hydrophone embedded in the elastomer was calculated. The flow-induced noise level is characterized by the frequency spectral density expressed in dB//$\mu Pa^2 Hz^{-1}$. The turbulent boundary layer forcing function used in this study is a turbulent wall pressure spectrum given by Corcos [1].

The calculation of the flow-induced noise received by a circular hydrophone beneath turbulent boundary layers has been made by Chase and Stern [2], and by Ko [3]. In their calculations, both Chase and Stern [2] and Ko [3] used the angular averaged wall pressure spectrum for the convective domain as a turbulent boundary layer forcing function. The turbulent wall pressure

Aero- and Hydro-Acoustics
IUTAM Symposium Lyon 1985
Editors: G. Comte-Bellot and J. E. Ffowcs Williams

spectrum used for the subconvective domain is based on the convenient assumption of isotropy in the plane of the wall [2].

Theoretical Analysis

The physical model (Fig. 1) considered in the present analysis is a plane elastomer layer of an infinite extent backed by a plane surface; the other side of the elastomer is exposed to turbulent flow. The desired frequency spectral density sensed by a rectangular hydrophone embedded in the elastomer layer is written as

$$Q(\omega) = \int_{-\infty}^{\infty}\int_{-\infty}^{\infty} P(k_x,k_y;\omega)\ S(k_x,k_y)\ T(k_x,k_y;\omega)\ dk_x dk_y \quad , \tag{1}$$

where $Q(\omega)$ is the frequency spectral density, $P(k_x,k_y;\omega)$ is the turbulent wall pressure spectrum, $S(k_x,k_y)$ is the rectangular hydrophone function, $T(k_x,k_y;\omega)$ is the transfer function, ω is the angular frequency, and k_x and k_y are the wavenumbers in the flow (x) and cross-flow (y) directions, respectively.

Modeling of a turbulent wall pressure spectrum [1], [4], [5], has been a subject for many years, but at the present time no explicit model is available for the full wavevector-frequency spectrum of wall pressure fluctuations beneath turbulent boundary layers. A model developed by Corcos [1] gives a relatively high level in the low wavenumber domain, but it is a convenient expression for describing pressure fluctuations induced on a flat surface by boundary-layer turbulence. In his study, Chase [4] has shown a model describing the wavevector-frequency spectrum of wall pressure for the convective domain, and he has suggested a convective white model for the subconvective domain. Ffowcs Williams [5] has recently shown a Corcos-like description of the wall pressure spectrum with yet undetermined constants, which must be obtained from free field experiments. Among these representative models, the following Corcos model has been selected for this study.

$$P(k_x,k_y;\omega) = \frac{a_o\rho_o^2 v_*^4}{\pi^2\omega}\left\{\frac{\alpha_1 k_c}{(k_x - k_c)^2 + (\alpha_1 k_c)^2}\right\}\left\{\frac{\alpha_3 k_c}{k_y^2 + (\alpha_3 k_c)^2}\right\} , \tag{2}$$

where ρ_o is the fluid density, $k_c = \omega/u_c$ is the convective wavenumber, u_c is the convective flow velocity, v_* is the friction velocity, $a_o = a_+ (1+\gamma)$, $a_+ = 0.766$, $\gamma = 0.389$, $\alpha_1 = 0.09$ and $\alpha_3 = 7\alpha_1$.

A hydrophone function is a spectral filtering function acting over the hydrophone area which depends on the wavenumber. The hydrophone function used for a rectangular hydrophone is written as

$$S(k_x,k_y) = \left\{\frac{\sin(k_xL_x/2)}{(k_xL_x/2)} \cdot \frac{\sin(k_yL_y/2)}{(k_yL_y/2)}\right\}^2 , \tag{3}$$

where L_x and L_y are the dimensions of the rectangular hydrophone in the x- and y-directions, respectively.

The transfer function $T(k_x,k_y;\omega)$ is a measure of the wavevector filtering. The elastomer layer is assumed to be perfectly bonded to the rigid backing surface. The propagation of elastic waves in the elastomer layer is described by the vector differential equation, the solution of which is the vector displacement:

$$\mu\nabla^2\vec{u} + (\lambda + \mu)\,\nabla(\nabla \cdot \vec{u}) = \rho_c \frac{\partial^2\vec{u}}{\partial t^2} , \tag{4}$$

where $\vec{u}$ is the vector displacement, λ and μ are the Lamé constants, and ρ_c is the material density. The vector displacement $\vec{u}$ can be written in terms of a scalar function ϕ and a vector potential function $\vec{\psi}$ as follows:

$$\vec{u} = \nabla\phi + \nabla x\vec{\psi} , \tag{5}$$

if ϕ and $\vec{\psi}$ are solutions to the equations

$$\nabla^2\phi = \frac{1}{c_c^2}\frac{\partial^2\phi}{\partial t^2} , \tag{6}$$

$$\nabla^2\vec{\psi} = \frac{1}{c_s^2}\frac{\partial^2\vec{\psi}}{\partial t^2} , \tag{7}$$

$$\cdot \;\nabla \cdot \vec{\psi} = 0 , \tag{8}$$

where $c_c = [(\lambda+2\mu)/\rho_c]^{1/2}$ is the compressional wave speed and $c_s = (\mu/\rho_c)^{1/2}$ is the shear wave speed. Using the functions ϕ

and $\vec{\psi}$, stress tensors and displacement components were obtained and were used for matching pertinent boundary conditions. The transfer function obtained for this study is

$$T(k_x,k_y;\omega) = \left| T_{zz}(k_x,k_y;\omega)\right|^2 \quad , \qquad (9)$$

where T_{zz} is the Fourier transform of the normal stress $\tau_{zz}(x,t)$ evaluated at the stand-off distance d.

Numerical Results and Discussion

The calculation of the turbulent boundary layer noise reduction requires the evaluation of the integral shown in Eq. 1. The integrand of Eq. 1 is the product of the turbulent wall pressure spectrum, the hydrophone function, and the transfer function. Fig. 2 shows the integrand of Eq. 1 plotted as a function of k_x in the positive direction with $k_y = 0$ for 500 Hz. The baseline data used in the calculation are as follows: standoff distance d = 0, elastomer layer thickness h = 2.54 cm, hydrophone dimension $L_x = L_y = 5.08$ cm, loss factor of elastomer $\zeta_s = 0.3$, shear speed $c_t = 16.66$ m/sec, compressional wave speed $c_c = 2000$ m/sec, elastomer density $\rho_c = 1200$ kg/m^3, sound speed in water $c_o = 1500$ m/sec, water density $\rho_o = 1000$ kg/m^3, free stream velocity U = 20 knots, convective flow velocity $u_c = 0.6U$ and friction velocity $v_* = 0.035$ U. The ordinate in Fig. 2 is expressed in dB//(dynes/cm^2)^2·cm^2·sec. The frequency spectral density is shown in Fig. 3. The dashed line is the calculated result for the flush mounted point hydrophone (d = 0, h = 0, and $L_x = L_y = 0$). The solid line is the calculated result for the 5.08 cm square hydrophone embedded in a 2.54 cm thick layer. The difference between these two curves is the turbulent boundary layer noise reduction.

Fig. 4 shows the effect of the square hydrophone dimension for various layer thicknesses on the noise reduction for 1,000 Hz. Fig. 5 shows the effect of the layer thickness for various hydrophone dimensions on the noise reduction for 1,000 Hz. By cross-plotting the calculated results shown in Figs. 4 and 5, the contour levels of the noise reduction as functions of the layer thickness and the square hydrophone dimension for 1,000 Hz were obtained (Fig. 6). As can be seen in the figure, more noise reduction is achieved as the square hydrophone dimension increases for a given layer thickness, and as the layer thickness increases for a given square hydrophone dimension. Note that substantial noise reductions are observed at zero layer thickness as the square hydrophone dimension increases. These reductions are mainly attributed to the spatial averaging of the hydrophone. However, as noted in the figure, the amount of the noise reduction is less noticeable for a large hydrophone than for a small hydrophone as the layer thickness increases. Fig. 7 shows a similar contour plot for 200 Hz.

Concluding Remarks

Turbulent boundary layer noise reduction was investigated using a simple model consisting of a layer of plane elastomer backed by a rigid surface; the other side of the elastomer layer was exposed to turbulent flow.

Contour plots of the noise reduction as functions of the layer thickness and the square hydrophone dimension were presented for 200 Hz and 1,000 Hz. The study shows that the effect of the square hydrophone spatial averaging diminishes as the value of the layer thickness becomes large.

References

1. Corcos, G.M.; The structure of the turbulent pressure field in boundary-layer flows. J. Fluid Mech., 18 (1964).

2. Chase, D.M., and Stern, R; Turbulent-boundary-layer pressure transmitted into an elastomer layer. BBN Technical Memorandum No. 382 (1977).

3. Ko, S.; Turbulent boundary layer noise reduction by means of an elastomer layer. NUSC Technical Memorandum No. 821094 (1982).

4. Chase, D.M.; Modeling the wavevector-frequency spectrum of turbulent boundary layer wall pressure. J. Sound Vib., 70(1) (1980).

5. Ffowcs Williams, J.E.; Boundary-layer pressures and the Corcos model: a development to incorporate low-wavenumber constraints. J. Fluid Mech., 125 (1982).

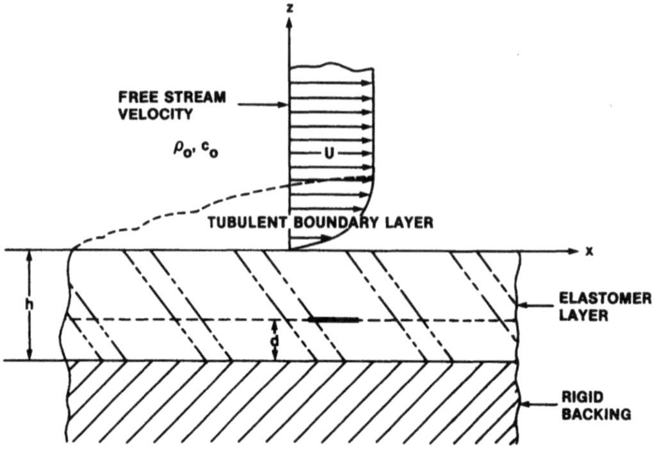

Fig. 1. Theoretical model

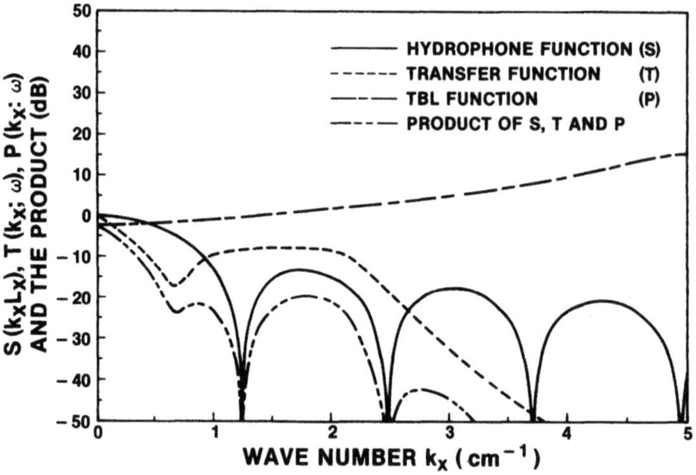

Fig. 2. The integrand of equation (1) as a function of the wavenumber k_x

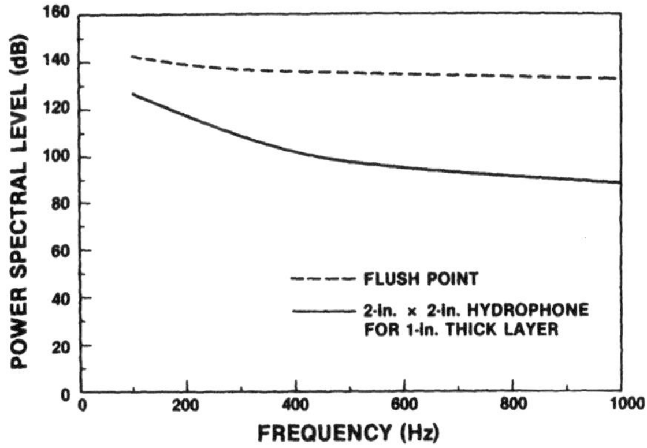

Fig. 3. Frequency spectral density

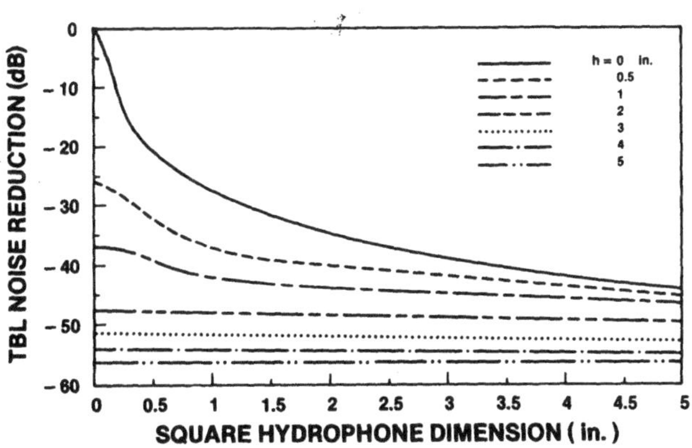

Fig. 4. Effect of the hydrophone dimension (1,000 Hz)

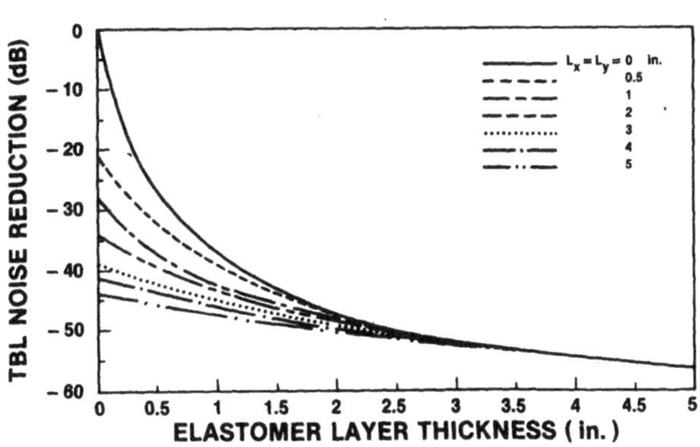

Fig. 5. Effect of the layer thickness (1,000 Hz)

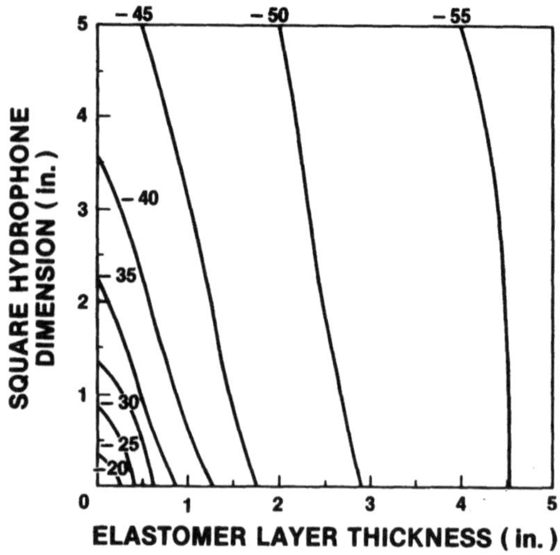

Fig. 6. Contour plot for 1,000 Hz

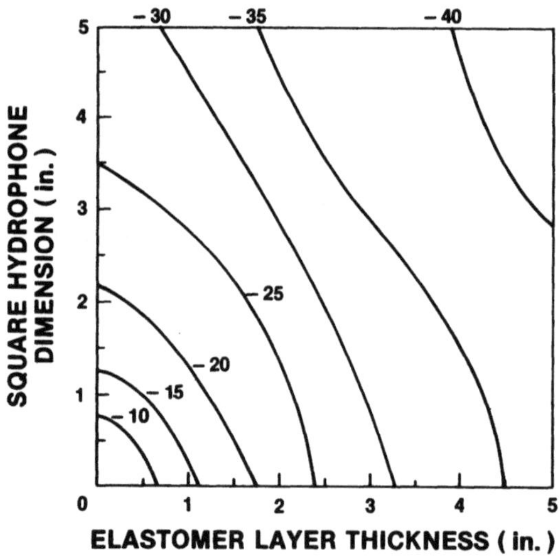

Fig. 7. Contour plot for 200 Hz

The Vibration of a Periodically Ring-Stiffened, Fluid-Loaded Cylinder

C. H. HODGES, J. POWER AND J. WOODHOUSE

Topexpress Ltd.,
13-14 Round Church St.,
Cambridge CB5 8AD, U.K.

Summary
Fluid loading is incorporated into previously-reported detailed modelling of ribbed cylinder vibrational behaviour at low frequencies. Preliminary experiments are under way, and are showing promising agreement with the theory described here.

Basic modelling

Theoretical modelling of the modal structure and transmission properties of a fluid-loaded cylindrical shell, braced periodically with circular, T-section ribs, will be described. The work is the natural extension to incorporate fluid loading (without mean flow) of modelling which has already proved very successful in describing the modal structure of such a cylinder in vacuo [1,2,3]. The theory aims at predicting the travelling Floquet waves on an infinite periodic cylinder using an approach which makes the incorporation of fluid loading easy. The treatment also emphasises the importance of realistic modelling of the ribs, and includes a description of internal degrees of freedom of the ribs.

Each of the displacement components (axial, radial and circumferential) of a travelling wave in a particular waveguide mode on our periodic cylinder satisfies a relation of the form

$$\xi(z,t) = \exp(iqz - i\omega t)\ \exp(in\theta)\ f(z) \qquad (1)$$

where q is the Bloch or Floquet wavenumber and f(z) is a periodic function of the axial coordinate with the periodicity of the structure. The separate waveguide modes are identified by values n=0,1,2,.. Our previous modelling of a ribbed cylinder [2] used this equation, with the function f(z) expressed in terms of a suitable set of quantities to be

Aero- and Hydro-Acoustics
IUTAM Symposium Lyon 1985
Editors: G. Comte-Bellot and J. E. Ffowcs Williams

described shortly. To calculate the travelling waves in vacuo, an extension of the variational principle to complex displacements was then used with these quantities as variational parameters. This leads to an eigenvalue problem of the usual form $(S - \omega^2 T)\,\underline{c} = 0$ where S and T are matrices representing the potential and kinetic energies, and $\underline{c}$ is the vector of variational parameters.

The contributions of the shell (assumed thin) to S and T have been calculated by Arnold and Warburton [4], once the cylinder displacement components are described by truncated Fourier series. The Fourier coefficients form the bulk of our variational parameters. However, it is also necessary to go beyond the usual "passive" modelling of ribs [5], and to include the effect of internal degrees of freedom, i.e. rib motions which can occur when the shell is held fixed. At low frequencies the most important of these is a cantilever-like resonance involving axial motion of the ribs. (This is true at least for the geometry we have studied, with ribs having a thin web and relatively massive flange as is usual in such structures.)

To incorporate such deformations in our model the radial variation of the rib web axial displacement is represented as a polynomial $\sum b_\alpha x^\alpha$, where x is the radial distance from the centre of the shell. Two of these coefficients b_0 and b_1 are determined by the shell displacement via compatibility conditions at the root of the web. The remaining ones b_2, b_3, etc. are treated as free variational parameters. For our particular geometry and the frequency range considered here (up to three times the ring frequency), convergence tests have shown that five Fourier coefficients in each shell displacement component give sufficient accuracy, and that the web axial displacement may be truncated at the cubic term (as suggested by static elasticity theory). As a result we need seventeen variational parameters, 3×5 Fourier coefficients plus b_2 and b_3, which gives a 17×17 matrix eigenvalue problem to solve. At higher frequencies, of course, one needs larger matrices in order not to lose accuracy. Not only must the truncation limits be extended, but further rib internal degrees of freedom enter the calculation - shear resonances, compressional resonances and flange bending resonances for example.

Incorporation of fluid loading

The Fourier representation of the shell displacement used in the present approach limits its application to periodic systems and introduces small truncation errors (depending on the frequencies of interest). However, it has the compensating advantage that the modifications required to incorporate fluid loading are very simple. All that is needed is a correction to the diagonal elements of the kinetic energy matrix T which correspond to radial motion of the shell. For the mth Fourier component, this correction has the form (per unit length of cylinder)

$$2\pi\rho a^2 \left| \frac{H_n(z)}{zH_n'(z)} \right|_{z=z_m} \tag{2}$$

where ρ is the fluid density, a the cylinder radius, prime denotes differentiation, and the argument of the Hankel function H_n is given by

$$z_m = a\left[\frac{\omega^2}{c^2} - \left(q + \frac{2\pi m}{d}\right)^2\right]^{1/2} . \tag{3}$$

The presence of fluid loading necessitates two changes to the in vacuo calculations. Firstly, if any of the Fourier components is acoustically fast there is a real z_m and a complex diagonal matrix element in T, which is therefore non-Hermitian. To deal with this case it is therefore necessary to use an eigenvalue routine which copes with non-Hermitian matrices. Secondly, the correction to the matrix T is frequency-dependent, and to obtain a self-consistent solution of the eigenvalue problem we must use some kind of iterative procedure. This slows down the numerical computation, but has not caused any great difficulties. The only place where problems might be anticipated is where any branch of the dispersion curves would have crossed the fluid sound-speed line, since a singularity then appears in the fluid-loading correction. In practice, it has not proved hard to cope with this adequately in the computations.

One further general point should be noted. In our approach it is convenient to deduce complex frequencies for real wavenumbers q. For spatial attenuation along the cylinder for realistic driving, one needs to know the complex wavenumber for real frequency. It is not difficult to deduce spatial attenuation rates from our results provided $\mathrm{Im}(\omega)$ is not too large. To lowest order one finds that $\mathrm{Im}(q) = \mathrm{Im}(\omega)/c_g$, which just expresses the fact that the spatial attenuation is the temporal decay rate divided by the group velocity $c_g = \mathrm{Re}(\partial\omega/\partial q)$.

We have already reported extensive experimental results on a ribbed cylinder in air [3], and we are now undertaking a programme of measurements on the same ribbed cylinder with fluid loading, to validate the modelling described here. Preliminary results from these measurements show very encouraging agreement with the theoretical predictions of pass-band positions as a function of frequency and waveguide angular order n. These measurements were made by driving the cylinder impulsively and observing the response at some distance along it with an array of accelerometers distributed uniformly around the circumference. The measured sets of time series responses were processed to separate the different cylinder waveguide modes of propagation, as has been described previously for in-air measurements [1-3].

Once the modelling with fluid loading has been validated, there is scope for generalising the treatment in many ways. For example, the model could be extended to higher frequencies, or the effect of structural complications such as damping treatment could be incorporated.

Acknowledgement

This work has been carried out with the support of the Procurement Executive, Ministry of Defence, U.K.

<u>References</u>

1. Hodges, C.H.; Power, J.; Woodhouse, J. The use of the sonogram in structural acoustics, and an application to the vibration of cylindrical shells. J. Sound Vib. in press (1985)
2. Hodges, C.H.; Power, J.; Woodhouse, J. The low-frequency vibration of a ribbed cylinder Part 1: theory. J. Sound Vib. in press (1985)
3. Hodges, C.H.; Power, J.; Woodhouse, J. The low-frequency vibration of a ribbed cylinder Part 2: observations and interpretation. J. Sound Vib. in press (1985)
4. Arnold, R. N.; Warburton, G. B. Flexural vibration of the walls of thin cylindrical shells having freely supported ends. Proc. Roy. Soc A197 (1949) 238-256. Reprinted in Kalnins, A.; Dym, C. L. (eds.) Benchmark Papers in Acoustics vol. 8. Dowden, Hutchinson and Ross, Stroudsberg, Pennsylvania 1976.
5. Cremer, L.; Heckl, M.; Ungar, E. E. Structure-borne Sound. Springer-Verlag, Berlin 1973. See chapter V.6.

Surface Impedance Loading of Panels

G. MAIDANIK and L.J. MAGA

David Taylor Naval Ship Research and Development Center
Washington, D.C. 20084, USA

ABSTRACT

In the endeavor to control the aero- and hydro-acoustical behavior of a complex dynamic system, the analysis of its response is of essential interest. Often practical complexes can be modelled in terms of a multitude of interwoven plane structural and fluid elements. Some of the structural elements in these models are panels. The analysis of the response of a panel in a complex is not only of intrinsic significance but is also central to the analysis of the response of the complex as a whole. (This is particularly true if the panel happened to be ribbed.) To ascertain the response of a panel one may need to ascertain the surface impedance loadings perceived by the surface of the panel in situ. The manner of ascertaining the surface impedance loadings on a panel which constitutes a structural element in a complex is the subject of this paper.

TEXT

The formalism of the response of a dynamic system interwoven of structural and fluid components is central to problems in aero- and hydro-acoustics. Often in such problems the structures are modelled as plane parallel boundaries and the fluids are considered to occupy the spaces between these boundaries as typified in Figure 1. Further, it is often idealized in this modelling that the dynamic system is stationary in time and possesses no surface impedance nonuniformities in planes parallel to the boundaries. This idealization yields considerable simplification. Since the impulse response is the analytical tool of this paper this simplification is demonstrated for the impulse response formalism as outlined in Figure 1. Indeed, the application of Fourier transformation renders the equation essentially one-dimensional. However, even under this idealization, the impulse response formalism is straightforward only if the external drives lie within the confines of the fluids and the response in the fluids only is of interest. If the external drives lie within a boundary and/or there is interest in the response of a structural member of a boundary, special provisions need to be made in the formalism to accommodate these external drives and/or to yield response estimates of a structural member in situ. With these provisions instituted one may also estimate, as a by-product, the blocked drives and the surface impedance loadings that a structural member in situ perceives on its surfaces. Such estimates may be used to extend the impulse response formalism to accommodate situations in which a structural member in a boundary may not obey the idealization imposed on it with respect to the temporal and spatial uniformity of its surface impedance, e.g., a panel in a boundary may be ribbed.

Aero- and Hydro-Acoustics
IUTAM Symposium Lyon 1985
Editors: G. Comte-Bellot and J. E. Ffowcs Williams

The significance of this accommodation is obvious to those who have investigated acoustical and hydrodynamical phenomena employing idealized uniform (unribbed) and nonuniform (ribbed) panels.

The paper outlines a technique for estimating the surface impedance loadings perceived by a panel in situ in a boundary of a dynamic system typified in Figure 1. The development of this specific technique, however, relies on the impulse response matrix formalism previously developed by the authors and on recent extensions made in this formalism [1-3]. In that sense the paper is of a broader content; the estimation of the surface impedance loadings is presented merely as an example.

A brief recounting of the impulse response matrix formalism is first presented, Figures 1 through 6. The extension to this formalism is presented in Figures 7 and 8. The estimates of the surface impedance loadings on a chosen panel is then presented in Figures 8 and 9.

It is convenient to divide the boundaries alternately into two groups, (r) and (q), as illustrated in Figure 1. Using this division one may cast the impulse response formalism in terms of two components, one in reference to those external drives that are directed toward the boundaries in the one group and those toward the boundaries in the other group, as indicated in Figure 1. In this figure $\hat{\alpha}$ is a unit vector directed toward the boundaries in group α; α = r or q.

Each form of propagation in a fluid is modelled by a basic one-dimensional dynamic system (BODDS). A typical BODDS is defined in Figure 2. The equations that define the wavenumber k_h and the propagator function $t_h(x_h|x_h')$ embody the wave equation and the radiation condition at infinity for the (h)th form of propagation.

The dynamic system exemplified in Figure 1 may then be modelled by an ensemble of BODDSs, as sketched in Figure 3. The impulse response formalism presented in Figure 1 can then be cast in terms of the impulse response matrix formalism; the casting procedure is stated in Figure 3.

The composition of the impulse response matrix is sketched in Figure 4. With the assistance of this sketch one may readily derive the form of the impulse response matrix $\underset{\approx}{g}(\underset{\sim}{x}|\underset{\sim}{x}')$ and its component $\underset{\approx}{g}_\alpha(\underset{\sim}{x}|\underset{\sim}{x}')$; α = r and q.

Assuming that each liquid admits to a single propagation form, the dynamic system in Figure 1 may be modelled as indicated in Figure 5. A composition of a typical boundary is also indicated in Figure 5. The constituent structural members are considered to be alternately panels and compliant layers and are defined in terms of the surface impedances Z_p; $2 \leq p \leq N - 1$, and Z_c^p; $1 \leq p \leq N - 1$, respectively. The surface impedances of the fluids are defined by $Z_1 = Z^j = (\rho_j\omega/k_j)$, and $Z_N = Z^i = (\rho_i\omega/k_i)$. The partial junction (boundary) matrix $\underset{\approx}{\Lambda}_{\alpha hk}$ of the typical (hk)th boundary is of rank two; the diagonal elements $\Lambda_{\alpha hh}$ and $\Lambda_{\alpha kk}$ are the reflection coefficients at the interfaces at $x_{\alpha h}$ and $x_{\alpha k}$, respectively. The off-diagonal elements $\Lambda_{\alpha hk}$ and $\Lambda_{\alpha kh}$ are the transmission coefficients from the (k)th to the (h)th and from the (h)th to the (k)th fluids, respectively. The junction matrix $\underset{\approx}{\Lambda}_\alpha$ is composed of the proper placements, in a matrix, of the partial junction matrices. To estimate the partial junction matrices, e.g., $\underset{\approx}{\Lambda}_{rji}$, and whence $\underset{\approx}{\Lambda}_r$, one may employ the equivalent circuit diagram of the chosen boundary as shown in Figure 6.

It is observed that the surface impedances Z_p are placed in series and Z_c^p in parallel. These placements define the physical properties of the individual structural member in the boundary. In particular, a compliant layer by definition allows for different velocities to exit on its two surfaces. The employment of the equivalent circuit diagram to ascertain the elements of a partial junction matrix, e.g., $\underset{\approx}{\Lambda}_{rji}$, is prescribed in Figure 6.

Inserting the junction matrices so ascertained in the matrix formalism outlined in Figures 3 and 4 would enable one to estimate the responses in the fluids induced by external drives that are applied within the fluids [1,2]. Provisions are now outlined that are made to accommodate situations in which one may seek to ascertain the responses of structural members within a chosen boundary and/or introduce external drives that may be applied within a chosen boundary.

A chosen boundary, e.g., the (ji)th, divides the dynamic system into two stacks (parts), stack (ν) and stack (ξ), as indicated in Figure 7. The interfaces at x_{rj} and x_{ri} of that boundary are rendered blocked and the stack impulse response matrix $\underset{\approx}{g}_{(\zeta)}$, and/or its component $\underset{\approx}{g}_{(\zeta)\alpha}$, for each stack is estimated by the procedure for ascertaining the impulse response matrix $\underset{\approx}{g}$, and/or its component $\underset{\approx}{g}_{\alpha}$, previously outlined in Figures 1 thorugh 6. The blocked conditions are prescribed in Figure 7. Also prescribed in this figure is the manner of determining the stack surface impedance $Z^h_{(\zeta)}$; $\zeta = (\nu)$ with $h = j$, and $\zeta = (\xi)$ with $h = i$, perceived by the interface at x_{rj} facing stack (ν) and the interface at x_{ri} facing stack (ξ), respectively. Similarly, using the stack impulse response matrices one may readily prescribe procedures for ascertaining the blocked drive $P_{b(\zeta)}$; $(\zeta) = (\nu)$ and (ξ), perceived by the interface at x_{rj} facing stack (ν) and the interface at x_{ri} facing stack (ξ), respective [3]. The structural members in the chosen boundary and their definitions in terms of surface impedances are indicated in Figure 7.

The dynamics of the chosen boundary may be estimated employing the equivalent circuit diagram shown in Figure 8. (It is noted that in Figure 6, Z_1 and Z_N are the fluid surface impedances $Z_1 = Z^j$ and $Z_N = Z^i$, whereas in Figure 8 they are the stack surface impedances $Z_1 = Z^j_{(\nu)}$ and $Z_N = Z^i_{(\xi)}$, respectively.) The dynamics exhibited in Figure 8 may be used to extend the impulse response matrix formalism to accommodate situations in which one may desire to ascertain the response of a structural member within a boundary and/or to ascertain the response of the dynamic system to an external drive that is applied within a boundary.

The dynamics of the equivalent circuit diagram exhibited in Figure 8 may be employed, in addition, to estimate the surface impedance loadings on a panel in situ in a boundary. The prescription for such estimates is given in this figure.

A specific simple example of a surface impedance loading on a panel is cited in Figure 9. Some asymptotic forms and a generalization of the specific example are also indicated in this figure.

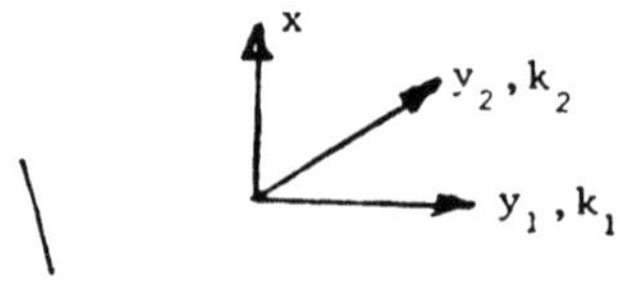

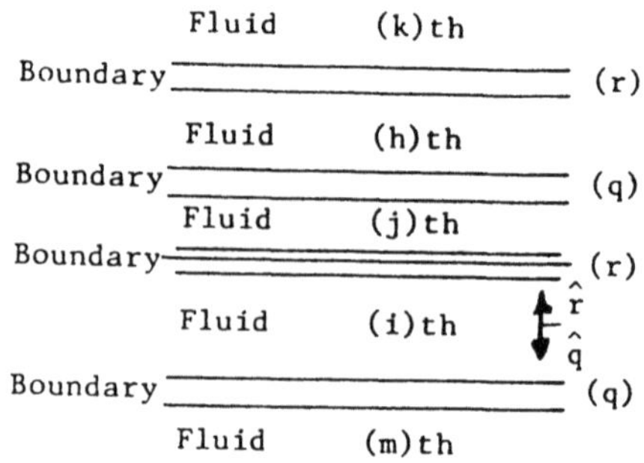

$\{\underset{\sim}{k},\omega\}$ Fourier conjugate of $\{\underset{\sim}{y},t\}$

$\underset{\sim}{\omega} = \{\underset{\sim}{k},\omega\}$

$d\underset{\sim}{y} = dy_1 \, dy_2$

$$p'(\underset{\sim}{y},x,t) = \int g'(\underset{\sim}{y}|\underset{\sim}{y}',x|x',t|t') \, d\underset{\sim}{y}' \, dx' \, dt' \, p_e'(\underset{\sim}{y}',x',t')$$

p' response, g' impulse response function, p_e' external drive

If dynamic system uniform in $\{\underset{\sim}{y},t\}$ then

$$p(x,\underset{\sim}{\omega}) = \int g(x|x',\underset{\sim}{\omega}) \, dx' \, p_e(x',\underset{\sim}{\omega}) \; ; \; p(x) = \int g(x|x') \, dx' \, p_e(x').$$

$$\underset{\sim}{g}(x|x') = \sum_{r,q} g_\alpha(x|x') \, \hat{\alpha} \; ; \; \underset{\sim}{p}_e(x) = \sum_{r,q} \hat{\alpha} \, p_{e\alpha}(x) \; ; \; [\hat{\alpha}\hat{\beta}] = \delta_{\alpha\beta}$$

$$p(x) = \sum_{r,q} p_\alpha(x) \; ; \; p_\alpha(x) = \int [g_\alpha(x|x') \, dx' \, p_{e\alpha}(x')] \; ; \; \alpha = r \text{ or } q.$$

Figure 1

Basic One-Dimensional Dynamic System (BODDS) Modelling

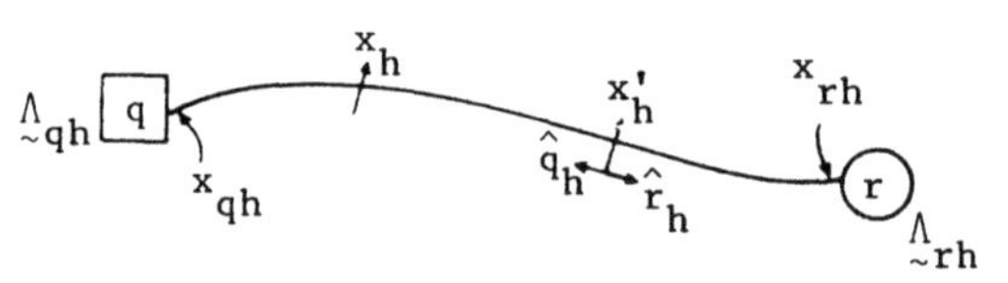

Two terminal positions $x_{\alpha h}$ at two boundaries $\alpha = r$ and q.

Two junction vectors $\underset{\sim}{\Lambda}_{\alpha h}$ to describe reflection and interactions with other BODDSs at interfaces $x_{\alpha h}$.

Two unit vectors $\hat{\alpha}_h$ directed toward interfaces at $x_{\alpha h}$.

Single wavenumber:

$$k_h = (\chi_h)^{1/2} \, U(\chi_h) - i(-\chi_h)^{1/2} \, U(-\chi_h) \; ; \; \chi_h = (\omega/c_h)^2 - k^2$$

Propagator function $t_h(x_h|x_h') = \exp[-ik_h|x_h - x_h'|]$.

Span function $S_{\alpha h}(x_h|x_h') = |U(x_h - x_{\alpha h}) - U(x_h - x_h')|$.

Drive applied at position x_h', observation at position x_h

Density ρ_h and speed of propagation (sound) c_h.

Figure 2

Ensemble of Basic One-Dimensional Dynamic Systems

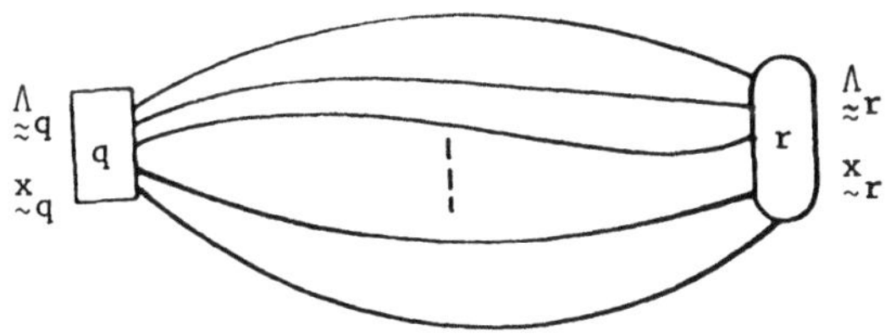

$$\Lambda_\alpha = \{\Lambda_{\alpha h}\} = (\Lambda_{\alpha hk}) \; ; \; x_\alpha = \{x_{\alpha h}\} \; ; \; x = \{x_h\}$$

$$p(x) = \sum S_{\alpha h}(x_h | x_{\beta h})\, p_h(x_h) \; ; \; p_e(x) = \sum S_{\alpha k}(x_k | x_{\beta k})\, P_{ek}(x_k)$$

$$g(x|x') = \sum \sum S_{\alpha h}(x_h | x_{\beta h})\, g_{hk}(x_h | x'_k)\, S_{ak}(x'_k | x_{\beta k})$$

$$p(x) \int g(x|x')\, dx'\, p_e(x') \; ; \; dx = (dx_h\, \delta_{hk})$$

$$p(x) = \{ph(x_h)\} \; ; \; p_e(x) = \{p_{ek}(x_k)\} \; ; \; g(x|x') = (g_{hk}(x_h | x'_k))$$

$$S_\alpha(x|x) = (S_{\alpha h}(x_h | x'_h)\, \delta_{hk}) \; ; \; S_{\alpha h}(x_h | x_{\beta h}) = S_{\beta h}(x_h | x_{\alpha h}$$

Figure 3

Derivation of Impulse Response Matrix

$$p_e = \sum_{r,q} \alpha\, p_{e\alpha}(x) \; ; \; g(x|x') = \sum_{r,q} g_\alpha(x|x')\, \alpha$$

$$p(x) = \sum_{r,q} p_\alpha(x)$$

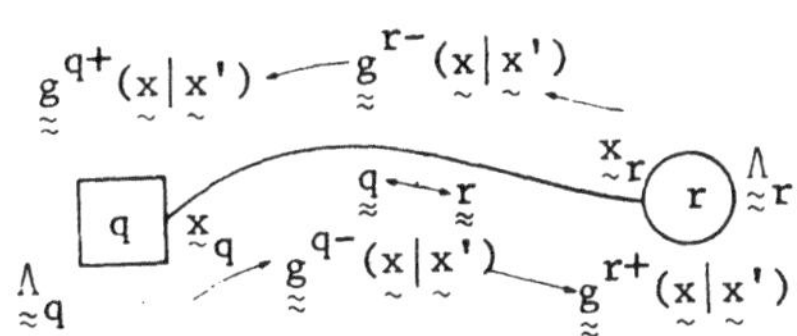

$$[\alpha\ \beta] = (\delta_{\alpha\beta}\, \delta_{hk})$$

$$g^{\alpha+}(x|x') = g_{d\alpha}(x|x') + g^{\beta-}(x|x')$$

$$g(x|x') = \sum_{r,q} g^{\alpha+}(x|x')$$

$$g^{\alpha-}(x|x') = t(x|x_\alpha)\, \Lambda_\alpha\, g^{\alpha+}(x_a|x')$$

$$g_{d\alpha}(x|x') = S_\alpha(x|x')\, t(x|x')\, \alpha$$

$$g^{\alpha+}(x_\alpha|x') = (I - \overline{\Lambda}_\beta\, \overline{\Lambda}_\alpha)^{-1}\, \{g_{d\alpha}(x_\alpha|x') + \overline{\Lambda}_\beta\, g_{d\beta}(x_\beta|x')\}$$

$$t(x|x') = (t_h(x_h|x'_h)\, \delta_{hk}) \; ; \; \overline{\Lambda}_\alpha = t(x_\alpha|x_\beta)\, \Lambda_\alpha$$

$$t_h(x_{\alpha h}|x_{\beta h}) = \exp[-ik_h | x_{\alpha h} - x_{\beta h})] = t_h(x_{\beta h}|x_{\alpha h})$$

Figure 4

Uncompounded Boundaries and the Composition of a Typical Boundary

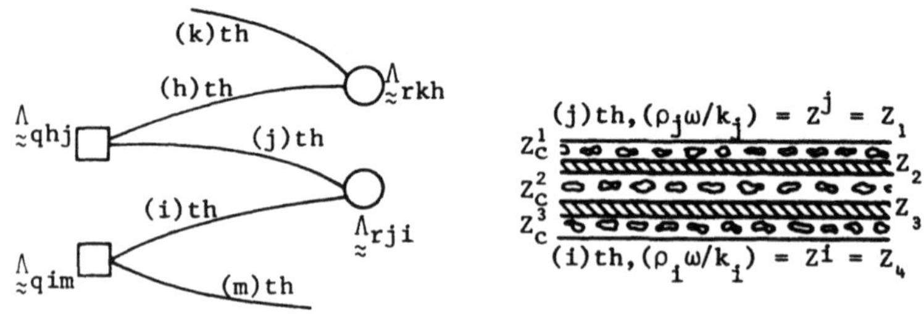

$$\Lambda_{\alpha hk} = \begin{pmatrix} \Lambda_{\alpha hh} & \Lambda_{\alpha hk} \\ \Lambda_{\alpha kh} & \Lambda_{\alpha kk} \end{pmatrix} ; \quad \Lambda_{\alpha} = (\Lambda_{\alpha hk}) \; ; \; h = \begin{cases} k \\ k-1 \end{cases}$$

Absence of a compliant layer $Z_c^p \to \infty$

Absence of a panel (or a fluid) $Z_p \to 0$

$\Lambda_{\alpha hh}$ and $\Lambda_{\alpha kk}$ reflection coefficients at the boundary

$\Lambda_{\alpha hk}$ and $\Lambda_{\alpha kh}$ transmission coefficients through the boundary

Figure 5

Equivalent Circuit Diagram to Estimate $\Lambda_{\alpha ji}$

$V = Y P_e \; ; \; Y = (Z)^{-1}$

$Z_c^o = Z_c^N = 0$

$Z_1 = Z^j \; ; \; Z_N = Z^i$

Z^h fluid surface impedance

$Z^h = (\rho_h \omega / k_h)$

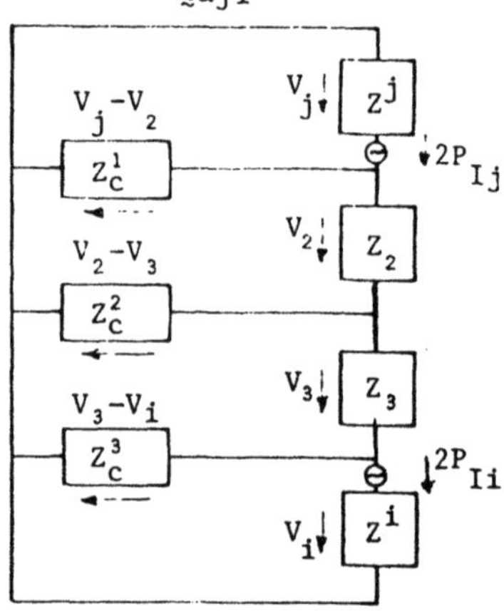

$$Z = \left((Z_c^{p-1} + Z_p + Z_c^p)\, \delta_{p,p} - Z_c^p(\delta_{p+1,s} + \delta_{p,s-1})\right) \; ; \; 1 \le p \le N$$

$$\Lambda_{\alpha jj} = 1 - 2Z^j Y_{ji}$$

$$\Lambda_{\alpha ji} = 2Z^j Y_{ji}$$

Figure 6

Dynamic system divided at chosen (ji)th boundary into two parts: stack (ν) and stack (ξ). Interfaces at x_{rj} and x_{ri} are rendered blocked in estimating the impulse response matrix: $\underset{\approx}{g}_{(\zeta)}(\underset{\sim}{x}|\underset{\sim}{x}') = \sum_{r,q} \underset{\approx}{g}_{(\zeta)\alpha}(\underset{\sim}{x}|\underset{\sim}{x}')\, \underset{\approx}{\alpha}$

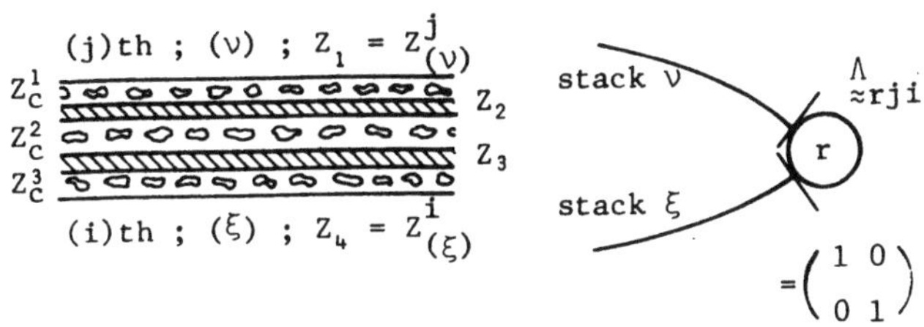

Surface impedance $Z^h_{(\zeta)}$ of stack (ζ) perceived by the interface at x_{rh} that faces that stack

$$Z^h_{(\zeta)} = Z^h\, g_{(\zeta)qhh}(x_{rh}|x_{rh}).$$

Figure 7

Equivalent Circuit Diagram for chosen (ji)th boundary

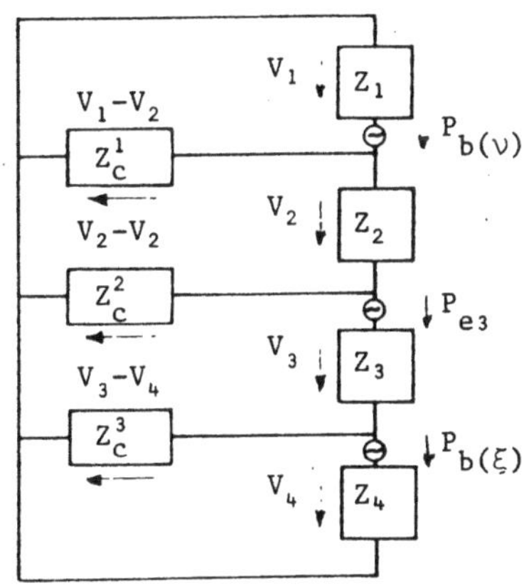

$$\underset{\sim}{V} = \underset{\approx}{Y}\,\underset{\sim}{P}_e \ ; \ \underset{\approx}{Y} = (\underset{\approx}{Z})^{-1}$$

$$Z^o_c = Z^N_c = 0$$

$$Z_1 = Z^j_{(\nu)} \ ; \ Z_N = Z^i_{(\xi)}$$

$$\underset{\approx}{Z} = \begin{pmatrix} \underset{\approx}{Z}_{(\nu)} & -Z^{n-1}_c & \\ -Z^{n-1}_c & (Z^{n-1}+Z_n+Z^n_c) & -Z^n_c \\ & -Z^n_c & \underset{\approx}{Z}_{(\xi)} \end{pmatrix}$$

$$\left.\begin{matrix}\underset{\approx}{Z}_{(\nu)} \\ \underset{\approx}{Z}_{(\xi)}\end{matrix}\right\} = \left((Z^{p-1}_c+Z_p+Z^p_c)\,\delta_{p,p} - Z^p_c(\delta_{p+1,s}+\delta_{p,s-1})\right) \begin{cases} 1 \le p \le n-1 \\ n+1 \le p \le N \end{cases}$$

In particular, defining $\underset{\approx}{Y}_{(\zeta)} = \underset{\approx}{Z}_{(\zeta)}^{-1}$ one obtains

$$\left.\begin{matrix} Z_{(\nu)n} = Z^{n-1}_c\left(1-Z^{n-1}_c\, Y_{(\nu)n-1,n-1}\right) \\ Z_{(\xi)n} = Z^n_c\left(1-Z^n_c\, Y_{(\xi)n+1,n+1}\right) \end{matrix}\right\}$$ Surface impedance loadings on the (n)th panel in the (ji)th boundary

Figure 8

$$\underset{\approx}{Z}(\nu) = \begin{pmatrix} z^{j}_{(\nu)}+Z^{1}_{c} & -Z^{1}_{c} \\ -Z^{1}_{c} & Z^{1}_{c}+Z_{2}+Z^{2}_{c} \end{pmatrix}$$

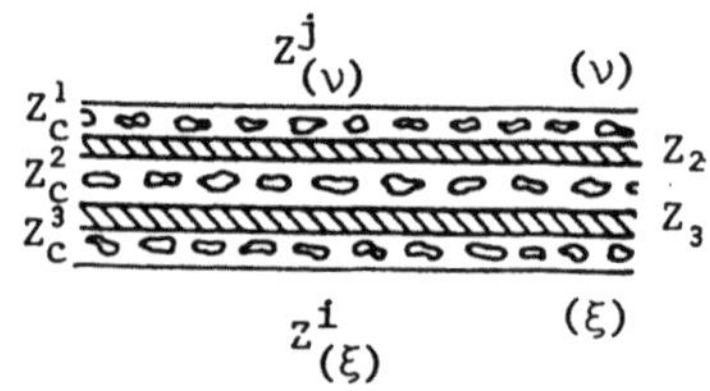

$$\underset{\approx}{Y}(\nu) = |\underset{\approx}{Z}(\nu)|^{-1} \begin{pmatrix} Z^{1}_{c}+Z_{2}+Z^{2}_{c} & Z^{1}_{c} \\ Z^{1}_{c} & z^{j}_{(\nu)}+Z^{1}_{c} \end{pmatrix}$$

$$Y_{(\nu)2,2} = \left(Z_{2}+Z^{2}_{c}+z^{1c}_{(\nu)}\right)^{-1} \quad ; \quad Z_{(\nu)3} = Z^{2}_{c}\left(Z_{2}+z^{1c}_{(\nu)}\right)\left(Z^{2}_{c}+Z_{2}+z^{1c}_{(\nu)}\right)^{-1}$$

$$z^{1c}_{(\nu)} = Z^{1}_{c}\, z^{j}_{(\nu)}\left(Z^{1}_{c}+z^{j}_{(n)}\right)^{-1} \quad ; \quad Z^{o}_{c} = 0$$

Asymptotics

$Z^{1}_{c} \to \infty \qquad Z_{(\nu)3} = Z^{2}_{c}(Z_{2}+z^{j}_{(\nu)})(Z^{2}_{c}+Z_{2}+z^{j}_{(\nu)})^{-1}$

$Z_{2} \to 0 \qquad Z_{(\nu)3} = Z^{2}_{c}\, z^{j}_{(\nu)}(Z^{2}_{c}+z^{j}_{(\nu)})^{-1}$

$Z^{2}_{c} \to \infty \qquad Z_{(\nu)3} = z^{j}_{(\nu)}$; Stackless $z^{j}_{(\nu)} \to z^{j}$

Generalization

$$Z_{(\nu)n} = z^{(n-1)c} = Z^{n-1}_{c}\left(Z_{n-1}+z^{(n-2)c}\right)\left(Z^{n-1}_{c}+Z_{n-1}+z^{(n-2)c}\right)^{-1}$$

Figure 9

REFERENCES

1. L.J. Maga and G. Maidanik 1983 Journal of Sound and Vibration 88, 473-488. Response of Multiple coupled dynamic systems.
2. G. Maidanik, L.J. Maga, and T.J. Eisler 1983 Journal of Sound and Vibration 89, 31-37. Response of Multiple layered media.
3. G. Maidanik and L.J. Maga 1985 (to be published). Response of coupled basic one-dimensional dynamic systems.

Flow Noise

Excitation of Instability Waves

D. W. BECHERT

Deutsche Forschungs- und Versuchsanstalt für Luft- und Raumfahrt e.V.
Abteilung Turbulenzforschung
Müller-Breslau-Str. 8
1000 Berlin 12, FRG

Summary

A brief survey is given on artificial excitation of instability waves in parallel flows. There are basically two categories of unstable flows:
(i) convectively unstable and
(ii) absolutely unstable flows.
The perturbation level in convectively unstable flows, e.g., boundary layers and most free shear flows, depends on the excitation only. On the other hand, absolutely unstable flows, e.g., wake flows and hot jets, do maintain fluctuations independent of any exterior excitation. Finally, effects of non-parallel mean flow and of varying boundary conditions on instability wave excitation are discussed.

Introduction

Shear flows such as boundary layers, jets and wakes exhibit periodic or quasi-periodic instability waves in the transition regime from laminar to turbulent flow conditions. In some cases we see also large scale periodic fluctuations in fully established turbulent flows, e.g., in the wake of cylinders. The question as to how these instability waves are excited is addressed here. Are there only forced instability waves, like in a controlled jet or boundary layer experiment? Or, are there also situations where the existence of instability waves is independent of exterior perturbations? Why do some instabilities like wake flows exhibit fluctuations at one frequency, whereas other flows, like jets, produce a broad band spectrum of the fluctuations? As it will turn out, the answer to these questions is closely connected to the issue of spatial versus temporal stability theory. But when do we have to apply which? These questions are not only of academic interest, as will be shown in the following example: To keep a flow laminar on an aircraft reduces drag. However, if even apparently spurious interactions between sound and little inhomogeneities of the surface do control the magnitude of the instability waves (i.e., the Aizin-Polyakov-Mechanism [1,2]), or if little protrusions, e.g., caused by dirt, produce self-sustained local oscillations, the practicality of some suggested designs be-

Aero- and Hydro-Acoustics
IUTAM Symposium Lyon 1985
Editors: G. Comte-Bellot and J. E. Ffowcs Williams

comes indeed questionable. Ignoring these mechanisms could cause wide discrepancies between model test data and full scale results on aircraft under operational conditions.

There are also other aspects of instability, like, e.g., the peculiar periodicity of a wake flow (as opposed to the broad band fluctuations in a jet), that have become clear due to very recent progress in stability theory. The present short survey is ment to highlight some basic ideas of recent papers on the excitation of instability waves.A significant aspect in that context is how receptive flows are to sound, i.e., we will deal with the 'receptivity' problem, a term suggested by Morkovin [3]. In the reference list of this paper the numerous previous investigations are not listed where artificial excitation has been merely observed and used, but we focus on more recent papers which are relevant to the understanding of this interaction.

A Model Case: Parallel Flow

Some essential aspects of instability waves as well as the effects of artificial excitation can be shown with a parallel mean flow. In addition, we restrict our analysis to subsonic flows at low Mach number*). At low Mach number, the wavelength of the instability waves is very small compared with the wavelength of a sound wave at the same frequency. Consequently, the interaction usually takes place within a small fraction of an acoustical wavelength. Already Helmholtz [4] had found, that a fluctuating flow field within a region small compared with the sound wavelength can be analyzed by incompressible flow theory. Thus, we take also advantage of this latter simplification. In most known cases, acoustical excitation is a linear process. Therefore, we use the linearized equations of an unsteady incompressible flow with a parallel mean flow U in the x-direction:

$$\frac{\partial u}{\partial t} + U\frac{\partial u}{\partial x} + v\frac{\partial U}{\partial y} + \frac{1}{\rho}\frac{\partial p}{\partial x} - \nu\Delta u = 0 \qquad (1)$$

$$\frac{\partial v}{\partial t} + U\frac{\partial v}{\partial x} + \frac{1}{\rho}\frac{\partial p}{\partial y} - \nu\Delta v = 0 \qquad (2)$$

$$\frac{\partial w}{\partial t} + U\frac{\partial w}{\partial x} + \frac{1}{\rho}\frac{\partial p}{\partial z} - \nu\Delta w = 0 \qquad (3)$$

*) It is possible to include compressibility, but the essential interaction mechanism can be shown already with an incompressible flow.

$$\frac{\partial u}{\partial x} + \frac{\partial v}{\partial y} + \frac{\partial w}{\partial z} = 0 \tag{4}$$

By taking the derivatives with respect to x, y and z of equations (1) ÷ (3) and by using the continuity equation (4) we find

$$\frac{\partial v}{\partial x} = - \frac{\Delta p}{2\rho U'} \tag{5}$$

with $U'=\partial U/\partial y$ and $\Delta = [\frac{\partial^2}{\partial x^2} + \frac{\partial^2}{\partial y^2} + \frac{\partial^2}{\partial z^2}]$. Equation (5) can be inserted into the x-derivative of eq. (2) and we obtain an equation which contains only the fluctuating pressure p as a variable:

$$[\frac{\partial}{\partial t} + U\frac{\partial}{\partial x}]\ \Delta p - 2U'\ \frac{\partial^2 p}{\partial x \partial y} - \nu U' \cdot \Delta [\frac{\Delta p}{U'}] = 0 \tag{6}$$

A Fourier transform in x, z and t would be the Orr-Sommerfeld-equation for instability waves, written with p as a variable.

We will now deal with the acoustical excitation situation. For the propagation of sound waves in a fluid at rest we have the acoustic wave equation. In an incompressible fluid, which we consider here, this equation degenerates into the Laplacean equation $\Delta p = o$. We introduce an impulse perturbation $\delta(t)$ at the origin $x=y=z=o$ and we obtain

$$\Delta p = \delta(t) \cdot \delta(x) \cdot \delta(y) \cdot \delta(z) \tag{7}$$

with the solution

$$p = - \frac{\delta(t)}{4\pi r} , \tag{8}$$

where r is the radial distance from the origin. The temporal pulse pattern $\delta(t)$ is propagated immediately to the environment, as it should in an incompressible fluid.

Also equation (6) of our instability problem can be used in a fluid at rest. There, we are left with $\partial \Delta p/\partial t = o$, the time derivative of the Laplace equation. If we want to keep eq. (6) as it is and want to introduce a pressure pulse, we have to perturb this equation on the right hand side by the derivative $\delta'(t)$ of a delta function in time. Thus, for a pressure pulse perturbation at an arbitrary location $x=x_o$; $y=y_o$; $z=z_o$ we have to write:

$$[\frac{\partial}{\partial t} + U\frac{\partial}{\partial x}]\ \Delta p - 2U'\frac{\partial^2 p}{\partial x \partial y} - \nu U' \cdot \Delta[\frac{\Delta p}{U'}] = \delta'(t)\cdot\delta(x-x_0)\cdot\delta(y-y_0)\cdot\delta(z-z_0) \qquad (9)$$

The solution of this equation is the Green's function of the instability problem. One method to determine the solution of eq. (9) is to take Fourier transforms in t, x and z to obtain an equation for the transformed quantity P (ω,α,y,β). The equation for P is no longer a partial differential equation, but a normal fourth order differential equation in y. This equation for P can be solved numerically using the methods developed previously for the Orr-Sommerfeld-equation [5]. In order to obtain the Green's function, i.e., the pressure p (t,x,y,z) produced by a pressure pulse, we have to subsequently perform the inverse Fourier transform [6,7]. We find wave packets in the shear layer triggered by the pressure pulse (see Fig. 1).

The interesting thing is now this:

(i) There are parallel flow configurations where the wave packet is swept away with the flow. There is no perturbation left at the location where the excitation took place. These flows are convectively unstable (see Fig. 1).

(ii) Another class of flows shows an amplification of fluctuations at the location of the excitation. These flows are absolutely unstable. (see Fig. 1).

There is no way to guess in advance what flow configuration belongs to which group. In order to identify the flow behaviour, however, it is not necessary to perform the whole inverse Fourier transform operation. It is already possible from the numerical solution of the transformed equation for P to conclude whether a flow is convectively or absolutely unstable. One criterion of absolute instability is the coexistence of amplified waves propagating both in the downstream and in the upstream direction. For more details see the work of Huerre & Monkewitz [7], Koch [8] and Gaster [9]*).

*) In these latter investigations, sometimes equations with different flow variables have been used. Our preference of the pressure p as a variable is caused by the advantage of really being able to introduce an acoustic perturbation and, in addition, maintaining the true 3-dimensional problem.

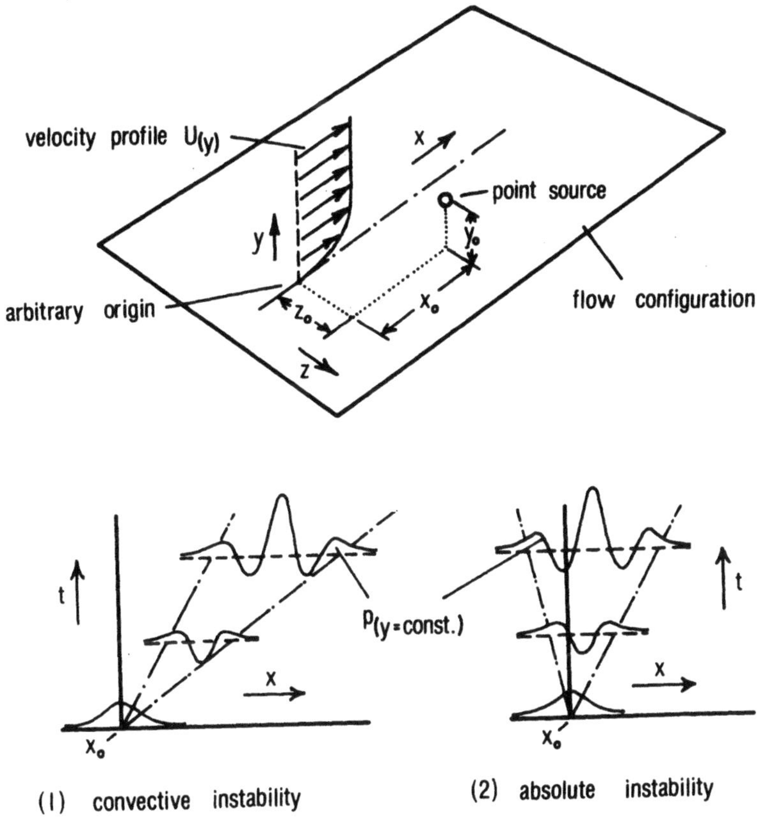

Fig. 1. Flow configuration and wave packet development

Considering the answer of the flow to an impulse excitation (i.e., the Green's function) is not as restrictive as it looks at first glance. Knowing the Green's function allows us to construct solutions for more complex excitation conditions. This can be done by arranging distributions of Green's functions so that the particular initial and boundary conditions are fulfilled. In addition, if we keep only the Fourier transform in t or, in other words, choose a δ function in space but periodic in time, we can also model such experimental situations where a periodical excitation exists.

In Fig. 2 we give a survey on calculated cases from the very recent literature. Flows can be either convectively or absolutely unstable. In Fig. 2 the conditions are indicated under which the flow becomes absolutely unstable. In addition to the examples in Fig. 2, it has been found by Gaster [6,9,10] that a Blasius boundary layer is also convectively unstable. We call a resonance a situation where fluctuations of one frequency dominate the spectrum. Such a case occurs in a wake of a cylinder or a flat plate with a blunt edge. It has been already suggested by Mattingly and Criminale [11] that the coexistence of an upstream propagating and a downstream propagating wave are necessary for the maintenance of this resonance. Koch [8] considers in particular that downstream location in the mean velocity profile where the absolute instability of the near wake changes into the convective instability of the wake farther downstream. He suggests that this location where the group velocity is zero, does control the frequency of the whole wake instability. Indeed, the agreement between this prediction and the experimental data is impressive [8].
The term "resonance" is used in stability theory in a related but slightly different sense, i.e., in a situation where two different instability waves of the same frequency and the same phase velocity (in the above situation, the phase velocity of the two waves is opposite) interact with each other. In such a case the fluctuations can be dominated by just one type of motion. The "low speed streak" motion in the viscous sublayer of a turbulent boundary layer [13] has been recognized recently as such a phenomenon. Both, the observed lateral wave length and the observed longitudinal vortex structure could be predicted by the nonlinear resonance theory by Jang, Benney and Chen [14].
Since resonances do dominate also other flow properties like, e.g., drag and sound radiation, it is often desirable to suppress flow resonances. For the reduction of the longitudinal streaks in a viscous sublayer of a turbulent boundary layer, various nonplanar surfaces have been suggested [15,16,17],

Configuration	Stability
(1) Free shear layer [7] 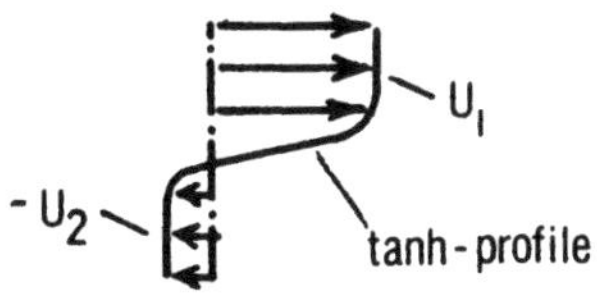	absolutely unstable for: $U_2 < -0.136\ U_1$
(2) Smooth two-dimensional wake [7,11] 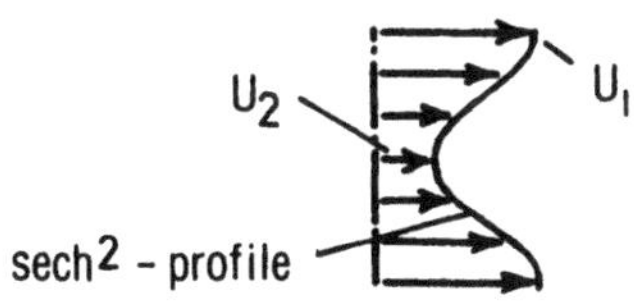	absolutely unstable for: $U_2 < 0.06\ U_1$ (asymmetrical mode)
(3) Two-dimensional wake with steep shear layers [8] 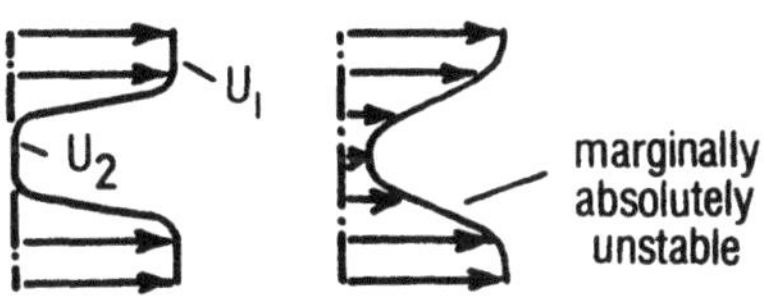	absolutely unstable for: deep wake with $U_2 = 0$; point with marginal absolute instability at downstream location with shallower wake determines frequency of wake oscillations (asymmetrical mode)
(4) Asymmetrical wake [8] 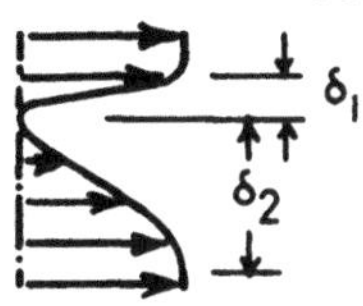	absolutely unstable for: $\delta_2 < 8\,\delta_1$ (asymmetrical mode)
(5) Lip wake [8] 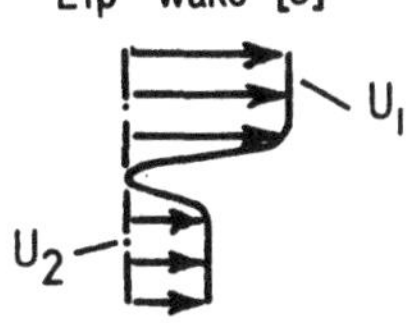	absolutely unstable for: $U_2 > 0.5\ U_1$ (asymmetrical mode)
(6) Hot circular jet with thin shear layer [7,12] 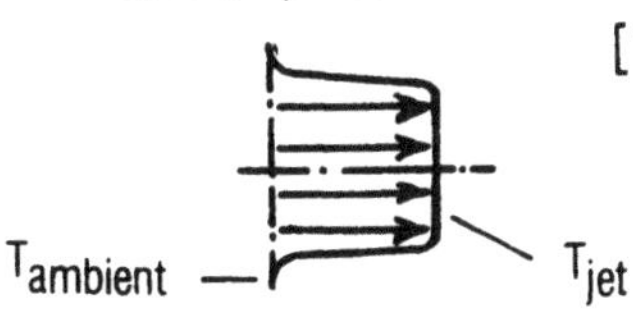	absolutely unstable for: $T_{jet} > 1.52\ T_{ambient}$ (axisymmetrical mode)

Fig. 2. Conditions for absolute instability of some flow configurations

and some of them have been tested thoroughly enough to be sure that there is real drag reduction.

For the resonance of a wake of a cylinder, the mean velocity profile and the boundary conditions are most essential for the instability. Strykowski and Srinivasan [18] have shown recently, that a little additional wire at one side of the cylinder wake causes an inflow into the near wake which changes the mean flow profile. With this additional wire, therefore, the Kármán vortex street (or wake resonance) could be suppressed completely for a certain Reynolds number range. To suppress the cylinder wake resonance, also other devices like, e.g., a wake splitter plate or a wire spiral on the circumference of the cylinder (Screwton spiral) have been proved to work. The various jet noise suppressors developed thus far are likely to suppress also the hot jet resonance, as suggested by Huerre & Monkewitz [7].

Effects of Nonparallel Mean Flow, Effects of Varying Boundary Conditions in the Streamwise Direction

Convectively unstable flows, such as boundary layers and free shear layers, depend on exterior excitation to maintain the fluctuations. An excitation by a localized source like the impulsive pressure source on the right hand side of eq. (9) contains all frequencies and wave numbers. Real excitations, however, are often not localized as a point source is. Obviously, the excitation is most efficient if the Fourier transform of the excitation field, say, in the variables ω (frequency) and α (the streamwise wave number) coincides with the range of unstable waves in ω and α of the flow. Therefore, e.g., a plane sound wave with its long wavelength has almost zero interaction with the comperatively short wave length of boundary layer instability waves. This situation would only change at high Mach numbers where sound waves and instability waves have comparable wavelength. A small interaction, however, occurs also at low Mach number. This interaction is due to the fact that the boundary layer flow is slightly nonparallel, caused by the streamwise growth of the boundary layer thickness. Then, the slowly varying instability wave field has Fourier constituents in α which match the long sound waves, as has been shown by Tam [19]. Another weak interaction mechanism for the excitation of boundary layer instability waves has been suggested by Goldstein [20]: The plane sound wave interacts with the non-parallel leading edge flow in a region where the flow is still stable. There, an attenuated vorticity is produced and is convected downstream into a region where the Tollmien-Schlichting waves can grow, and in this way, the instability waves are excited. In contrast to the above mentioned analytical theories, a purely numerical approach was chosen by Murdock [21]. This describes the interaction between a uniformly pulsating streamand the laminar boundary layer flow over an infinitely thin plate.

The above mentioned theories consider certain conceivable excitation situations and were compared with the very preliminary experimental data of Shapiro [22]. A wide discrepancy between experiment and theory was found [23]. Gedney [24] subsequently extended Shapiro's experiments on the same facility. He identified a strong surging (diffraction) sound field around the leading edge of the plate caused by asymmetries in the plate environment. Since this surging excitation field is highly nonhomogeneous in space, it can easily couple directly to the Tollmien-Schlichting waves. However, computed data on this latter interaction are not yet available.*)

Since nonhomogeneities play such an important role, it is worthwhile mentioning another excitation mechanism of major practical importance. If a laminar flat plate boundary layer is exposed to plane sound waves, the Tollmien-Schlichting wave excitation is dramatically increased if a tiny change in surface geometry is applied locally. For instance, a strip of Scotch tape (having a thickness of only a very small fraction of the boundary layer thickness) causes a small distortion of the mean velocity profile. The small profile distortion provides a strong coupling between sound waves and Tollmien-Schlichting waves. This powerful mechanism has been found experimentally by Aizin & Polyakov [1,2]. A first analysis on the effects of changes of mean flow properties on receptivity has been devised recently by Goldstein [24]. This latter theoretical investigation confirms the comparatively high receptivity due to small mean velocity changes. Thus, the problem of keeping a laminar boundary layer "clean" does not only imply eliminating surface roughness to avoid local (absolute) instabilities which act as turbulators, but very small discontinuities in the surface would also enhance and control receptivity to sound, which is ubiquitous in particular on jet aircraft.

The receptivity of a s e p a r a t e d f l o w on an airfoil has been investigated recently by Ahuja [25] and Goldstein [26]. Here, the flow is by no means parallel anymore, but the nonhomogeneity of the flow at the separa-

*) This example highlights the poor state of knowledge on the issue. Maybe, this situation could be overcome by a close and long-term cooperation between experimentalists and theoreticians.

tion point is again crucial for the excitation mechanism of the instability waves in the separated shear layer. Because the mixing in the shear layer is enhanced by the excited shear layer waves, excitation of these waves can lead to reattachment of the separated flow on the airfoil [25].

Nonhomogeneities of the boundary conditions produce nonhomogeneous excitation fields which, due to their broad wave number α content, do excite instability waves very efficiently. One solved example of such an interaction is the excitation of a thin free shear layer leaving a trailing edge. For the "thin" shear layer case, where the instability wavelength is large compared with the shear layer thickness, a simple theory has been developed [27,28,29,30]. For this case, it has been proved theoretically [28] and confirmed experimentally [31] that a Kutta condition defines the flow at the trailing edge. However, this simplification of having a Kutta condition is no more valid if the stability waves become comparable in their dimensions with the shear layer thickness [32]. A survey on the issue of the unsteady Kutta condition has been given recently by Crighton [33].

Acknowledgment

I appreciate the careful review of this paper by Dr. W. F. King, DFVLR, Berlin.

References

1. Nishioka, M.; Morkovin, M.V.: Boundary-layer receptivity to unsteady pressure gradients: experiments and overview. To be published in the J. Fluid Mech.

2. Aizin, L.B.; Polyakov, N.F.: Acoustic generation of Tollmien-Schlichting waves over local unevenness of surfaces immersed in streams. In Russian. Reprint 17 (1979) Akad. Nauk, USSR, Siberian Div., Inst. Theor. & Applied Mech., Novosibirsk.

3. Obremski, H.; Morkovin, M.V.; Landahl, M.: A portfolio of stability characteristics of incompressible boundary layers. AGARDograph 134 (1969). §2.2. Free stream disturbances and boundary-layer receptivity.

4. Helmholtz, H.v.: Theorie der Luftschwingungen in Röhren mit offenen Enden. Crelle-Borchardt, Journal für die reine und angewandte Mathematik, Bd. LVII (1860) 1-72.

5. Betchov, R.; Criminale, W.O.: Stability of parallel flows. New York: Academic Press 1967.

6. Gaster, M.: A theoretical model of a wave packet in the boundary layer on a flat plate. Proc. Roy. Soc. A 347 (1975) 271-289.

7. Huerre, P.;Monkewitz , P.A.: Absolute and convective instabilities in free shear layers. To be published in the J. Fluid Mech.

8. Koch, W.: Local instability characteristics and frequency determination of self-excited wake flows. J. Sound & Vibr. 99 (1985) 53-83.

9. Gaster, M.: On transition to turbulence in boundary layers. In: Transition and turbulence. Meyer, R.E. (ed.). Proc. Univ. of Wisconsin Meeting. New York: Academic Press 1981.

10. Gaster, M.; Grant, I.: An experimental investigation of the formation and development of a wave packet in a laminar boundary layer. Proc. Roy. Soc. A 347 (1975) 253.

11. Mattingly, G.E.; and Criminale, W.O.: The stability of an incompressible two-dimensional wake. J. Fluid Mech. 51 (1972) 233-272.

12. Michalke, A.: Instabilität eines kompressiblen runden Freistrahls unter Berücksichtigung des Einflusses der Stahlgrenzschichtdicke. Z. Flugwiss. 19 (1971) 319-328.

13. Kline, S.J.; Reynolds; W.C., Schraub, F.A.; Runstadler, P.W.: The structure of turbulent boundary layers. J. Fluid Mech. 30 (1967) 741-773.

14. Jang, P.S.; Benney, D.J.; Chen, Y.M.: Origin of streamwise vortices in turbulent boundary layers. Bull. Am. Phys. Soc., Ser. II Vol. 29 No. 9 Paper BB8 (1984) 1528.
Also: Jang, P.S.; Benney, D.J.; Gran, R.L.: On the origin of streamwise vortices in a turbulent boundary layer. Report March 1985, Dynamics Technology, Inc., 21311 Hawthorne Blvd., Suite 300, Torrance, Cal.90503, USA.

15. Bushnell, D.M.: Turbulent drag reduction for external flows. AIAA 21st Aerospace Sciences Meeting, Reno, Nevada, January 10-13, 1983, AIAA-Paper, No. 83-0227.

16. Walsh, M.J.: Turbulent boundary layer drag reduction using riblets. AIAA 20th Aerospace Sciences Meeting, Orlando, Florida, January 11-14, 1982, AIAA-Paper, No. 82-0169.

17. Bechert, D.W.; Hoppe, G.; Reif, W.-E.: On the drag reduction of the shark skin. AIAA Shear Flow Control Conference, Boulder, Col., March 12-14, 1985.

18. Strykowski, P.J.; Srinivasan, K.R.: Control of vortex shedding behind bluff bodies. 5th Symposium on turbulent shear flows, Ithaca, N.Y., Cornell University, August 7-9, 1985.

19. Tam, C.K.W.: The excitation of Tollmien-Schlichting waves in low subsonic boundary layers by free-stream sound waves. J. Fluid Mech. 109 (1981) 483-501.

20. Goldstein, M.E.: The evolution of Tollmien-Schlichting waves near a leading edge. J. Fluid Mech. 127 (1983) 59-81.

21. Murdock, J.W.: The generation of a Tollmien-Schlichting wave by a sound wave. Proc. Roy. Soc. A 372 (1980) 517-534.

22. Shapiro, P.: The influence of sound upon laminar boundary layer instability. Massachusetts Institute of Technology: Acoustics and Vibration Lab. (1977) Rep. No. 83458-83560-1.

23. Goldstein, M.E.: Scattering of acoustic waves into Tollmien-Schlichting waves by small streamwise variations in surface geometry. To be published in the J. Fluid Mech.

24. Gedney, C.J.: The response of a laminar boundary layer to sound and wall vibration. Massachusetts Institute of Technology: Acoustics and Vibration Lab. (1983) Rep. No. 83560-3.

25. Ahuja, K.K.; Whipkey, R.R.; Jones, G.S.: Control of turbulent boundary layer flows by sound. AIAA 8th Aeroacoustics Conference, Atlanta, Georgia, April 11-13,1983 AIAA-Paper 83-0726.

26. Goldstein, M.E.: Generation of instability waves in flows separating from smooth surfaces. J. Fluid Mech. 145 (1984) 71-94.

27. Bechert, D.W.; Michel, U.: The control of a thin free shear layer with and without a semi-infinite plate by a pulsating flow field. ACUSTICA 33 (1975) 287-307.

28. Bechert, D.W.: Excited waves in shear layers. DFVLR-FB 82-23 (1982). See also: A model of the excitation of large scale fluctuations in a shear layer. AIAA-Paper 83-0724 (1983).

29. Michalke, A.: On the motion of weakly disturbed plane vortex sheets. Submitted to the J. Fluid Mech.

30. Michalke, A.: Survey on jet instability theory. Prog. Aerospace Sci. 21 (1984) 159-199.

31. Bechert, D.W.; Stahl, B.: Shear layer excitation, experiment versus theory. DFVLR-FB 84-26 (1984).

32. Bechert, D.W.; Pfizenmaier, E.: Optical compensation measurements on the unsteady exit condition at a nozzle discharge edge. J. Fluid Mech. 71 (1975) 123-144.

33. Crighton, D.G.: The Kutta condition in unsteady flow. Ann. Rev. Fluid Mech. 17 (1985) 411-445.

On Noise Generated by Turbulence in a Pipe Downstream of a Discontinuous Area Enlargement

B. STAHL and A. MICHALKE*)

Deutsche Forschungs- und Versuchsanstalt für Luft- und Raumfahrt
Abteilung Turbulenzforschung, 1000 Berlin 12, FRG

*) Hermann-Föttingen-Institut für Thermo- und Fluiddynamik
Technische Universität Berlin, 1000 Berlin 12, FRG

Summary

Experimental results of pressure fluctuations measured at the wall of a circular pipe downstream of a discontinuous area enlargement are presented. Close to the nozzle where the fluid enters the pipe like a jet, the wall pressure is generally dominated by turbulence, but farther downstream, under certain conditions, by sound which is sensitive to the acoustic pipe end conditions. By comparison with a pipe inlet flow it is found that the noise is mainly generated by the jet flow region. The agreement with results of acoustic duct theory based on uniform mean flow is good.

Introduction

The noise generated by turbulence in circular pipes is the subject of the present experimental investigation. For a pipe inlet flow, turbulence develops in the wall boundary layer which eventually leads to the fully developped turbulent pipe flow far downstream. Even stronger turbulence can exist in the pipe past valves or downstream of discontinuous area enlargements. Hence these pipe elements are strong noise sources. Once the noise is generated, the propagation of sound is influenced by the acoustic properties of the pipe as described by duct acoustics. In order to investigate the turbulence and noise in the pipe, measurements of the wall-pressure fluctuations have been made in a similar way as Kerschen & Johnston [1] did.

Experimental setup

The pipe with a diameter D = 10 cm (Fig. 1) was fitted on the plenum chamber of a jet facility with the air injected into the pipe by a nozzle of a diameter d = 3 cm (discontinuous area enlargement) or d = D (pipe inlet flow). The pipe length was L = 26 D. Additionally, an anechoic termination could be fitted at the pipe end, in order to change the acoustic properties of the pipe. The nozzle Mach number $M_o = U_o/a_o$ was in general equal to 0.3. For the

Aero- and Hydro-Acoustics
IUTAM Symposium Lyon 1985
Editors: G. Comte-Bellot and J. E. Ffowcs Williams

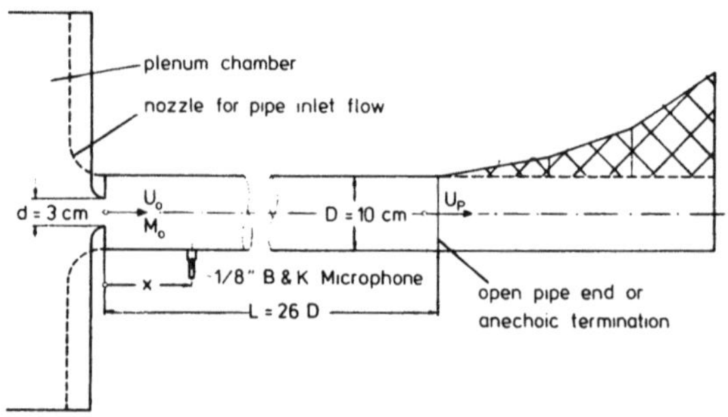

Fig. 1. Experimental setup

discontinuous area enlargement (d/D = 0.3) we have a pipe-jet flow region close to the nozzle with an increase of the static pressure in x-direction which is followed farther downstream by a pipe flow region with reduced mean velocity $U_p=U_o(d/D)^2$, approximately. For the pipe inlet flow (d/D=1) we have $U_p=U_o$, and the static pressure decays slightly along the pipe due to wall friction.

The wall-pressure fluctuations have been measured by 1/8" B&K-microphones mounted flushly into the pipe wall. The microphone position could be varied in downstream direction x (measured from the nozzle exit plane) and along the periphery ϕ. By means of a Digital-Signal-Analyzer HP 5420A the power spectral density $\tilde{p}^2_\omega$ of the microphone signal or the cross-spectral density $C_\omega + iQ_\omega$ of two signals has been derived with a bandwidth of 25 or 12.5 Hz.

Experimental results

The rms-wall-pressure $\tilde{p}$ as function of the distance x downstream of the nozzle is shown in Fig. 2 for various nozzle-to-pipe diameter ratios d/D and a nozzle Mach number M_o=0.3. We see that the results depend strongly on the diameter ratio d/D. In the following only the cases with d/D=0.3 and 1 are considered. For d/D=0.3 a peak of the rms-value occurs at x/D=2.4 which is about 6 times the value for the pipe inlet flow. Farther downstream the rms-value decays and is at x/D=10 below the value of the pipe inlet flow because of the lower pipe mean velocity U_p.

Unfortunately, microphones pick up pressure fluctuations independently of

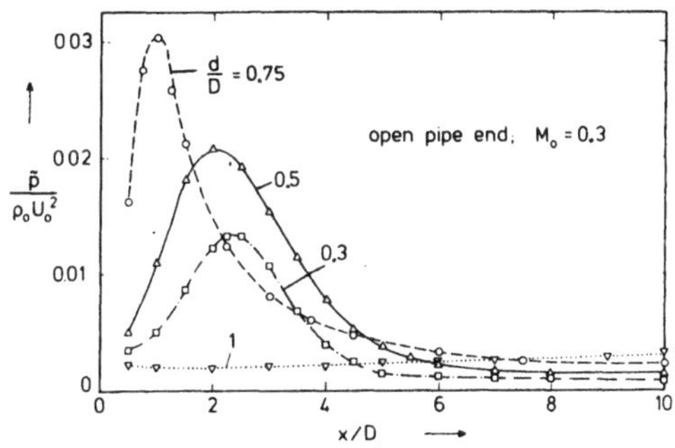

Fig. 2. Overall wall-pressure $\tilde{p}$ vs. axial distance x for various nozzle-to-pipe diameter ratios d/D. Nozzle Mach number M_0=0.3, open pipe end.

whether they correspond to turbulence or to noise. In order to get more information, let us look at the spectra. For d/D = 0.3 and open pipe end, the spectra $L_{p\omega} = 20\log(\tilde{p}_\omega/p_{ref})$ are shown in Fig. 3 at 3 different x-positions. In the pipe flow region at x/D=12 strong peaks occur at certain frequencies. These agree surprisingly well with the cut-on frequencies of higher acoustic modes due to duct theory, although in the theory a uniform pipe mean velocity is assumed. The peaks below the first cut-on frequency indicate acoustic plane wave resonances due to the open pipe end condition. At x/D = 2.4(maximum of the rms-value) in the pipe-jet flow region the spectrum is much higher and decays monotonically except for small peaks at the cut-on frequencies. At first sight, the results suggest that in the pipe-jet flow region the spectrum is dominated by turbulence, but in the pipe flow region by sound waves, although turbulence should there be present, too, which possibly could be triggered by acoustic resonances. Hence more information are needed.

Since acoustic waves depend on the pipe end conditions, additional measurements have been made with an anechoic termination, in order to avoid wave reflections. For x/D=11 the comparison of the spectra is shown in Fig. 4 for d/D=0.3 and M_0=0.3. It is obvious that the plane wave resonances are suppressed by the anechoic termination, while the resonance at the first cut-on frequency is not affected, as duct theory predicts. In order to check whether the spectrum is determined by the local turbulence or by sound generated by the jet turbulence close to the nozzle, a comparison with the spectrum in the equivalent pipe inlet flow has been made for the same pipe mean velocity and hence for the same wall turbulence, approximately. It is obvious that for the diameter ratio d/D=0.3 the level is about 20 dB higher

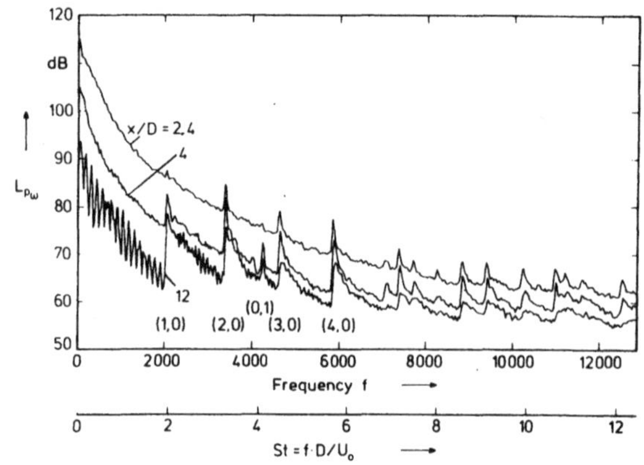

Fig. 3. Wall-pressure spectra for various axial positions x downstream of a discontinuous area enlargement d/D=0.3, M_0=0.3, open pipe end

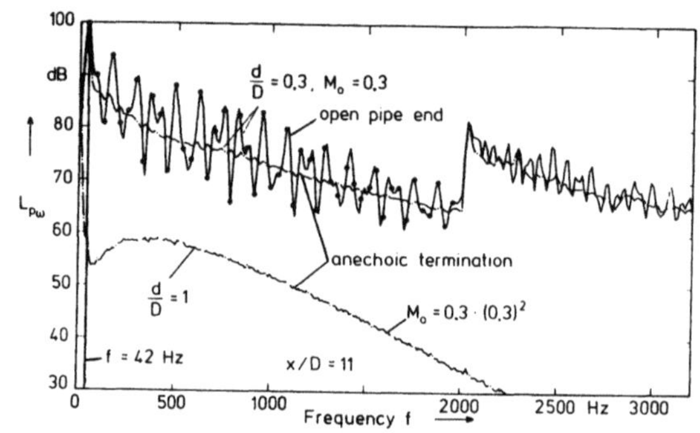

Fig. 4. Wall-pressure spectra at x/D=11 for d/D=0.3 (discontinuous area enlargement) and d/D=1 (pipe inlet flow) at the same pipe mean velocity U_p and different pipe end conditions

than for the pipe inlet flow and thus cannot be determined by the local turbulence, but by sound waves generated by the strong turbulence in the jet flow region close to the nozzle.

A clear distinction between turbulence and sound waves can, however, only be found by phase measurements. Therefore, the cross-spectral density with two axially displaced microphones has been measured with the fixed microphone placed at the rms-maximum (x/D=2.4). A plane wave resonace below the first cut-on frequency was selected with a Strouhal number St=fD/U_0=0.49 for M_0=0.3. The pressure level and the phase of the complex cross-spectral density are shown in Fig. 5 as function of the position x of the moving probe

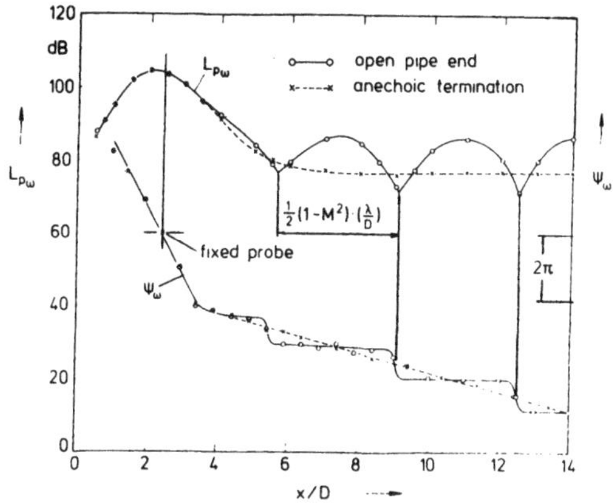

Fig. 5. Wall-pressure level $L_{p\omega}$ and phase ψ_ω of the cross-spectral density vs. downstream distance x of moving probe for different pipe end conditons. Fixed probe at x/D=2.4. Strouhal number St=0.49 (plane wave resonance), d/D=0.3, M_0=0.3.

for d/D=0.3 and both pipe end conditions. For x/D<4 the results are independent of the pipe end conditions, and the linear phase variation yields a phase velocity c_{ph}=0.5 U_0, as one would expect for jet turbulence. For x/D > 6 and anechoic termination, the pressure level becomes constant, as one would expect for plane sound waves. The phase variation here corresponds to a phase velocity c_{ph}=a+U_p (a=sound speed), as duct acoustics predicts for downstream propagating plane sound waves. Moreover, for an open pipe end, the typical standing wave pattern of acoustic plane wave resonances can be observed in this flow region. Similar results have been obtained also for other frequencies. Hence we can conclude that for the diameter ratio d/D=0.3 the wall-pressure fluctuations are dominated by turbulence up to x/D=4, but farther downstream by sound waves.

Additional information can be obtained by two microphones displaced azimuthally by $\Delta\phi$, both being at the same x-position. The quad-spectrum Q_ω is then found to be zero indicating that there is no preferred peripheral propagation. From the real part C_ω of the cross-spectral density the azimuthal pressure components can be derived by Fourier expansion with respect to $\Delta\phi$. Fig. 6 shows the azimuthal components $C_{\omega m}$ for St=0.49 as function of the azimuthal order m at x/D=2.4 and 11. We see that in the pipe-jet flow region there are various azimuthal components present, as typical for jet turbulence. In the pipe flow region, however, the energy is almost complete-

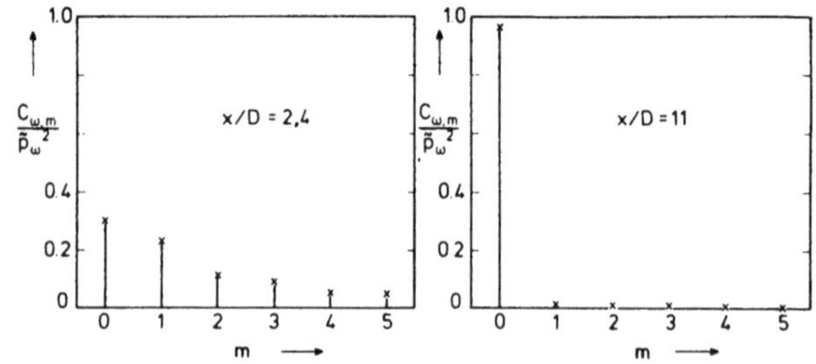

Fig. 6. Azimuthal components $C_{\omega m}$ of the co-spectrum vs. azimuthal order m. St=0.49, d/D=0.3, M_o=0.3.

ly concentrated in the axisymmetric component m=0, as one would expect for a dominating plane sound wave. It can be shown theoretically by means of the Lighthill equation that the plane waves below the first cut-on frequency are generated only by a very small fraction of the turbulence, namely by that part of the axisymmetric component of the axial velocity component which is uniform across the pipe cross-section in the source region. The result is different for the first cut-on frequency (St=1.97), as shown in Fig. 7. Here we see that even in the jet flow region the first azimuthal component contains more than 50 % energy. Hence in the source region already a strong acoustic wave is superimposed on the turbulence because of the resonance. For the first cut-on frequency at small pipe Mach numbers, it can again be shown by theory that the dominating noise contribution is generated by a fraction of the first azimuthal turbulence component of the radial and circumferential velocity components. For the pipe inlet flow the situation is quite different. In Fig. 8 the wall-pressure level and the phase are shown for a

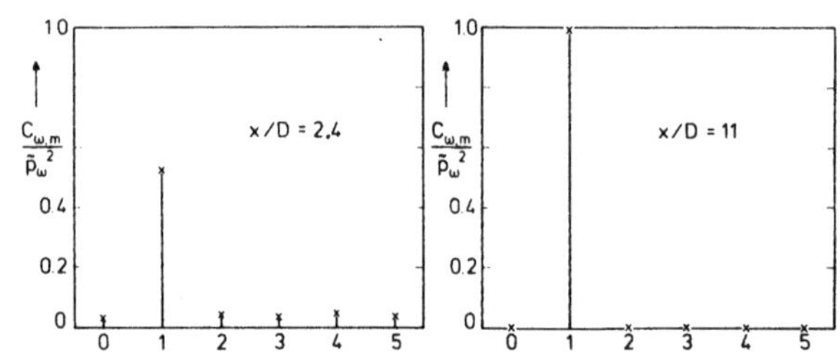

Fig. 7. Azimuthal components $C_{\omega m}$ of the co-spectrum vs. azimuthal order m. St.=1.97 (first cut-on frequency), d/D=0.3, M_o=0.3

plane wave resonance (St=0.45) and M_o=0.3. For the open pipe end condition, the pressure level indicates the standing wave pattern of a plane acoustic wave resonance. The phase variation corresponds to an upstream travelling sound wave. This is reasonable, since in this case the reflexion coefficient of the open pipe end is greater than unity. With anechoic termination the acoustic resonance is suppressed. The pressure level increases slightly with x, and the phase variation indicates a downstream travelling wave with the phase velocity c_{ph}=0.8 U_p, which is typical for wall turbulence. As a consequence, the wall pressure is here determined by the local wall turbulence unless the sound waves are strongly amplified by resonances. This is also confirmed by the composition of the azimuthal components at this position

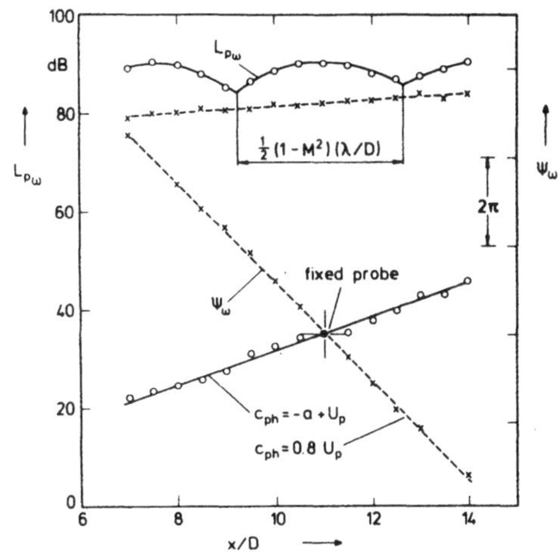

Fig. 8. Wall-pressure level $L_{p\omega}$ and phase ψ_ω of cross-spectral density vs. downstream distance x of moving probe for the pipe inlet flow (d/D=1). -o- open pipe end, -x- anechoic termination. Fixed probe at x/D=11, St=0.45 (plane wave resonance), M_o=0.3

and shown in Fig. 9. For the open pipe end the axisymmetric component clearly dominates due to the plane sound wave at resonance, but with the anechoic termination we have a composition of azimuthal components typical for turbulence. In comparison with the results for d/D=0.3, these results show clearly that the turbulent wall boundary of the inlet flow is much less effective in generating noise than the jet flow region downstream of a discontinuous area enlargement.

Finally, the Mach number dependences of the overall wall pressure level is shown in Fig. 10 for the diameter ratio d/D=0.3 and anechoic termination. In the pipe jet flow region (x/D=2.4) a U_o^4-law exists which corresponds to turbulence. In the pipe flow region (x/D=20), however, the law changes from

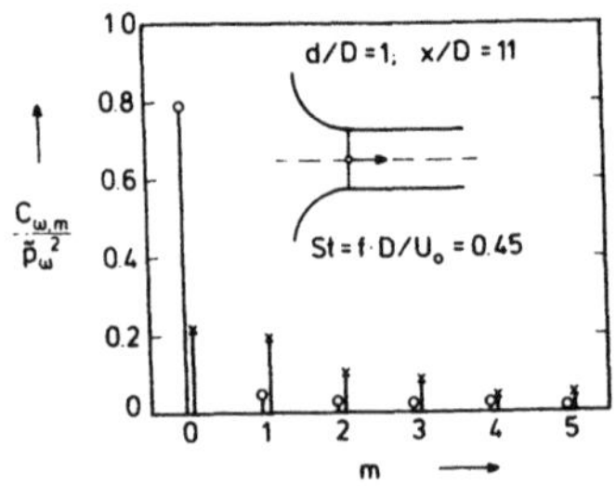

Fig. 9. Azimuthal components $C_{\omega m}$ of the co-spectrum vs. azimuthal order. Pipe inlet flow. -o- open pipe end. -x- anechoic termination

U_o^4 to U_o^6. The latter is, due to the theory of Davies & Ffowcs Williams [2], typical for the noise generated by turbulence in a duct. But even at the very low Mach numbers, the pressure fluctuations have been identified as being sound. This indicates a dipole-type contribution to the sound by the fluctuating efflux from the nozzle.

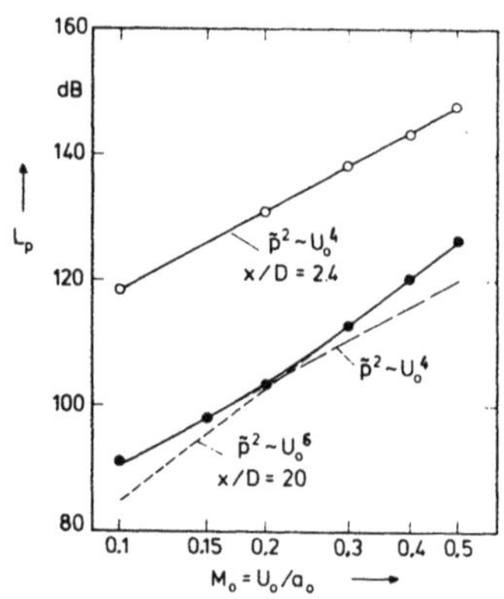

Fig. 10. Wall-pressure level L_p vs. nozzle Mach number M_o at x/D=2.4 and 20 downstream of a discontinuous area enlargement d/D=0.3, anechoic termination

This investigation has been supported by the Deutsche Forschungsgemeinschaft, Bad Godesberg.

References

1. Kerschen, E.J.; Johnston, J.P.: Modal content of noise generated by a coaxial jet in a pipe. J. Sound Vibr. 76 (1981) 95-115.

2. Davies, H.G.; Ffowcs Williams, J.E.: Aerodynamic sound generation in a pipe. J. Fluid Mech. 32 (1968) 765-778.

Acousto-elasticity of a Panel Backed by a Rectangular Cavity

S. NARAYANAN*

Institut für Technische Akustik

Technische Universität Berlin

D. PREMACHANDRAN PILLAI
Regional Engineering College
Calicut, India

Summary

The problem of low frequency sound transmission and the structural response of a panel backed by a rectangular cavity is solved by a modal analysis. The matrix equations possessing gyroscopic coupling is reduced to a standard symmetric eigenvalue problem, the solution of which yields the natural frequencies and modes of the coupled acousto-elastic system. Special orthogonal relations of the eigenvectors of the gyroscopic system facilitate the response analysis.

Introduction

Several methods like modal analysis, finite elements, statistical energy analysis and integral methods are available in the literature for the analysis of the interaction problem of internal sound fields and the flexible walls of an enclosure surrounding it. Such problems arise in the context of interior cabin noise in aircrafts and other transportation vehicles like automobiles and ships. The walls may be excited by sources like boundary layer, engine or propeller noise. For low frequency applications the modal analysis where the acoustic pressure inside the cavity and the structural motion are expressed respectively in terms of rigid wall cavity and invacuuo structural natural modes is an effective way of obtaining the coupled natural frequencies and modes which can then be used for response and pressure calculations. This paper demonstrates the efficacy of such an approach where the eigenvalue problem corresponding to a skew-symmetric matrix is reduced to a standard eigenvalue problem of a symmetric matrix by a method due to Meirovich [1].

*) AvH fellow on leave from Department of Applied Mechanics
Indian Institute of Technology. Madras 600036

Aero- and Hydro-Acoustics
IUTAM Symposium Lyon 1985
Editors: G. Comte-Bellot and J. E. Ffowcs Williams

Special orthogonal relations of the eigenvectors with respect to the skew-symmetric matrix enable response analysis and pressure evaluations. The method is applied to the example of a simple geometry of a rectangular cavity.

Coupled Acoustic Structural Equations

Consider a rectangular cavity of volume V, surrounded by rigid surfaces on all sides except one where it is closed by a flexible rectangular panel of surface area A. The acoustic pressure inside the cavity is governed by the wave equation

$$\nabla^2 p_c - (1/c^2)\ \partial^2 p_c/\partial t^2 = 0 \tag{1}$$

with

$$\partial p_c/\partial n = \begin{cases} -\rho\ \partial^2 w/\partial t^2 & \text{on } A \\ 0 & \text{otherwise} \end{cases} \tag{2}$$

In (1) and (2) w is the flexible wall deflection along the normal n, ρ and c are respectively the equilibrium fluid density and acoustic velocity in the cavity.

$\nabla^2 = (\partial^2/\partial x^2 + \partial^2/\partial y^2 + \partial z^2)$ is the Laplatian operator in Cartesian coordinates and t denotes time.
The motion of the panel is governed by the equation

$$D\nabla^4 w + \mu\ \partial^2 w/\partial t^2 = p_c - p_e \tag{3}$$

where $D = Eh^3/12\ (1 - \nu^2)$ is the flexural rigidity of the panel, $\nabla^4 = (\partial^2/\partial x^2 + \partial^2/\partial y^2)^2$, $\mu = \rho_s h$ the mass per unit area of the panel, p_e is the external pressure excitation on the panel. E is young's modulus ν the Poisson ratio, ρ_s the density and h the thickness of panel.
Let p_c be expressed in terms of the acoustic potential ϕ as

$$p_c = -\rho\ \partial\phi/\partial t \tag{4}$$

Expanding ϕ as a series of rigid wall cavity modes ξ_j

$$\phi = \sum_j a_j\ \xi_j \tag{5}$$

and expanding w as a series of invacuuo structural natural modes ψ_r

$$w = \sum_r q_r\ \psi_r \tag{6}$$

in the absence of damping or absorbent walls (1),(2) and (3) can be reduced to a set of ordinary differential equations of the form [2]

$$Vm_{jA}\,(\ddot{a}_j + \omega^2{}_{jA}\,a_j) = c^2 A \sum_r L_{jr}\,q_r \quad j = 0,1,2.... \tag{7}$$

$$m_{rS}\,(\ddot{q}_r + \omega^2_{rS}\,q_r) = -\rho A \sum_j L_{jr}\,a_j + Q^e{}_{rs} \quad r = 1,2... \tag{8}$$

where a_j and q_r respectively are j^{th} acoustic and r^{th} structural generalized coordinates, ω_{jA} and ω_{rs} are corresponding cavity and structural natural frequencies the overdots representing the order of differentiation with respect to t and the subscripts A and S referring to acoustic and structural systems. The modal coefficients occurring in (7) and (8) are given by the following integrals.

$$m_{jA} = (1/V) \int_V \xi_j^2\, dV;\; m_{rS} = \int_A \mu\, \psi_r^2\, dA;\; L_{jr} = (1/A) \int_A \xi_j \psi_r dA$$

$$\text{and} \quad Q^e_{rs} = \int_A p_e \psi_r dA \tag{9}$$

Limiting the number of cavity modes to n and structural modes to m in the analysis a finite set of equations (7) and (8) can be expressed in matrix form as

$$\begin{bmatrix} M_A & 0 \\ 0 & M_S \end{bmatrix} \begin{Bmatrix} \ddot{a} \\ \ddot{q} \end{Bmatrix} + \begin{bmatrix} 0 & -L_{AS} \\ L^T_{AS} & 0 \end{bmatrix} \begin{Bmatrix} \dot{a} \\ \dot{q} \end{Bmatrix} + \begin{bmatrix} K_A & 0 \\ 0 & K_S \end{bmatrix} \begin{Bmatrix} a \\ q \end{Bmatrix} = \begin{Bmatrix} 0 \\ Q_S \end{Bmatrix} \tag{10}$$

where (M_A, K_A) and M_S, K_S are diagonal matrices of orders n and m respectively with diagonal elements

$$M_{Aj} = Vm_{jA}/(c^2 A);\; K_{Aj} = M_{Aj}\,\omega^2_{jA};\; M_{Sr} = m_{rS}/(\rho A) \text{ and } K_{Sr} = M_{Sr}\omega^2_{rS}$$

L_{AS} is the acoustic structural coupling matrix of order nxm with elements L_{jr} and $\{Q_S\}$ is the generalized force vector of order m with elements $-Q^e_{rs}/(\rho A)$.

Equations (10) passess gyroscopic coupling in view of the skew symmetry of the coupling matrix. On inspection of the generalized acoustic stiffness matrix K_A, we find that its first element is zero as $\omega_{oA} = 0$ making it singular. This presents a problem in further analysis which can be solved in the following way.

To this end, consider the free vibration problem for $p_e = 0$. The first equation of the matrix set (10) is

$$\{Vm_{oA}/(c^2A)\}\ \ddot{a}_o = \sum_{r=1}^{m} L_{or}\ \dot{q}_r \tag{11}$$

Assuming harmonic dependence of the generalized coordinates

$a_j = \tilde{a}_j\ e^{i\omega t}$ and $q_r = \tilde{q}_r\ e^{i\omega t}$ we obtain

$$a_o = -\ ic^2A/(\omega Vm_{oA}) \sum_{r=1}^{m} L_{or}\ q_r \tag{12}$$

Substituting (12) in the rest of the equations (10), it can be shown that the resulting matrix equation is reduced by one order to (n-1)+m while retaining the original gyroscopic nature of the coupling. Denoting the reduced matrices by an asterisk, M_A^*, L_{AS}^* and K_A^* retain their original form and content except that their first row and column corresponding to the generalized coordinate a_o have been elimated. M_S^* is identically equal to M_S, while the diagonal matrix K_S is modified to a nondiagonal symmetric matrix K_S^* of the same order. The elements of K_S^* are given by

$$(K_S^*)_{rs} = K_{Sr}\ \delta_{rs} + L_{or}\ L_{os} \tag{13}$$

where δ_{rs} is the Kronecker delta.

Eigenvalue Analysis of the Acoustoelastic System

The reduced form of equations (10) after elimination of a_o can be expressed as

$$M\ \{\ddot{e}\} + L\ \{\dot{e}\} + K\{e\} = \{f\} \tag{14}$$

where $\{e\} = \begin{Bmatrix}\{a\}\\ \{q\}\end{Bmatrix}$ is a (n-1) + m dimensional vector, $\{f\} = \begin{Bmatrix}\{0\}\\ \{Q_S\}\end{Bmatrix}$

$$M = \begin{bmatrix} M_A^* & 0 \\ 0 & M_S^* \end{bmatrix} \ ;\ L = \begin{bmatrix} 0 & -L_{AS}^* \\ L_{AS}^{*T} & 0 \end{bmatrix} \ ;\ K = \begin{bmatrix} K_A^* & 0 \\ 0 & K_S^* \end{bmatrix} \tag{15}$$

Introduce a 2 {(n-1) + m} dimensional state vector {x} defined by

$$\{x\} = \begin{Bmatrix}\{\dot{e}\}\\ \{e\}\end{Bmatrix} \tag{16}$$

Equations (14) can then be expressed in state space form as

$$I\ \{\dot{x}\} + G\ \{x\} = \{F\} \tag{17}$$

$$\text{where,} \quad I = \begin{bmatrix} M & 0 \\ 0 & K \end{bmatrix}, \quad G = \begin{bmatrix} L & K \\ -K & 0 \end{bmatrix} \quad \text{and} \quad \{F\} = \begin{matrix} \{f\} \\ \{0\} \end{matrix} \tag{18}$$

with I symmetric and positive definite and G skew-symmetric. The eigenvalue problem associated with (17) for $\{F\} = \{0\}$ is

$$\lambda I\{x\} + G\,\{x\} = \{0\} \tag{19}$$

Meirovich [1] in the context of spinning structural systems has shown that the solution of (19) consists of (n-1)+m pairs of pure imaginary complex conjugates $\lambda_r = \pm\, i\omega_r$ and corresponding (n-1)+m pairs of complex conjugate eigenvectors $\{x_r\} = \{y_r\} \pm i\,\{z_r\}$. It has been further shown that the real part $\{y_r\}$ and the imaginary part $\{z_r\}$ of $\{x_r\}$ themselves satisfy equations of the form

$$\omega_r^2\, I\,\{y_r\}\; K'\,\{y_r\}, \quad \omega_r^2\, I\,\{z_r\} = K'\{z_r\} \tag{20}$$

where $K' = G^T\, I^{-I} G$ is a symmetric matrix. Thus the original eigenvalue problem of a skew-symmetric matrix has been transformed into one of a symmetric matrix thereby facilitating the use of many algorithms available for its solution and working with real quantities instead of complex quantities in the computation. The solution of (20) consists of (n-1)+m pairs of repeated eigenvalues ω_r^2 and associated pairs of eigenvectors $\{y_r\}$ and $\{z_r\}$.

The 2 {(n-1)+m} eigenvectors $\{y_r\}$ and $\{z_r\}$ are orthogonal with respect to I. If they are normalized such that

$$\{y_r\}^T\, I\,\{y_r\} = \{z_r\}^T\, I\,\{z_r\} = 1 \quad r = 1,2....(n-1)+m \tag{21}$$

the orthogonal relations can be expressed as

$$\begin{aligned} &\{y_r\}^T\, I\,\{y_s\} = \delta_{rs}; \quad \{z_r\}^T\, I\,\{z_s\} = \delta_{rs} \\ &\{y_r\}^T\, I\,\{z_s\} = 0; \quad \{z_r\}^T\, I\,\{y_s\} = 0 \end{aligned} \qquad r = 1,2,...(n-1)+m \tag{22}$$

Likewise, the eigenvectors satisfy the following orthogonal relations with respect to the G matrix

$$\begin{aligned} &\{z_s\}^T\, G\,\{y_s\} = -\,\{y_s\}^T\, G\,\{z_r\} = \omega_r\delta_{rs} \\ &\{y_s\}^T\, G\,\{y_r\} = \{z_s\}^T\, G\,\{z_r\} = 0 \end{aligned} \qquad r = 1,2,..(n-1)+m) \tag{23}$$

Response Analysis of Acoustoelastic System

Consider the linear transformation

$$\{x\} = U\,\{\eta\} \tag{24}$$

where U is the normalized modal matrix

$$U = [\{y_1\} + i\,\{z_1\},\dots\dots,\ \{y_1\} - i\,\{z_1\},\dots\dots] \tag{25}$$

Substituting (24) into (17) and premultiplyng by U^T we get

$$U^T\,I\,U\,\{\dot{\eta}\} + U^T\,G\,U\,\{\eta\} = U^T\{F\} \tag{26}$$

Using the orthogonal relations (22) and (23) it can be seen that (26) reduces to a set of decoupled equations

$$\begin{aligned}&\dot{\eta}_r - i\omega_r\,\eta_r = N_{(n-1)+m+r} \qquad\qquad r = 1,..(n-1)+m\\ &\dot{\eta}_{(n-1)+m+r} + i\omega_r\,\eta_{(n-1)+m+r} = N_r\end{aligned} \tag{27}$$

where $$\{N\} = \frac{1}{2}\,U^T\,\{F\} \tag{28}$$

is a 2 {(n-1)+m} dimensional vector. Equation (27) can be expressed in the general matrix form

$$\{\dot{\eta}\} + \Lambda\,\{\eta\} = \{N\} \tag{29}$$

where Λ is a diagonal matrix with elements $\Lambda_r = -i\omega_r$ for $r = 1,\dots(n-1)+m$ and $\Lambda_r = i\omega_r$ for $r = (n-1)+m+1,\dots,$ $2\,\{(n-1)+m\}$. The general solution of (29) is given by the convolution integral

$$\{\eta(t)\} = e^{-\Lambda t}\,\{\eta(o)\} + \int_o^t e^{-\Lambda(t-\tau)}N(\tau)\,d\tau \tag{30}$$

where $e^{-\Lambda t}$ is a diagonal matrix with elements $e^{-\Lambda_r t}$.
Substituting for $\{\eta\}$ and $\{N\}$ from (30) and (28) in (24) we obtain

$$\{x(t)\} = U\,e^{-\Lambda t}\,U^{-1}\,\{x(o)\} + \frac{1}{2}\int_o^t U\,e^{-\Lambda(t-\tau)}U^T\,\{F(\tau)\}d\tau \tag{31}$$

Equation (31) together with (16) and (4) to (6) give the structural response and the cavity pressure.

Harmonic Excitation

For the special case of harmonic excitation with

$$p_e = p_o\,(x,y)\,e^{i\omega t} \tag{32}$$

the generalized force vector $\{Q_S\}$ is given by

$$\{Q_S\} = 1/(\rho A)\, e^{i\omega t} \int_A p_o\,(x,y)\,\{\psi\}\,dA \qquad (33)$$

where $\{\psi\}$ is a vector containing the invacuuo structural natural modes. Suppressing $e^{i\omega t}$ throughout and denoting the harmonic amplitudes by a top tilda the response vector can be obtained in closed form for zero initial conditions[3]

$$\{\tilde{x}\} = \sum_{r=1}^{n-1+m} 1/(\omega^2-\omega_r^2)\,\{\omega(\{y_r\}\{y_r\}^T + \{z_r\}\{z_r\}^T)(\{\tilde{F}_I\}-i\{\tilde{F}_R\})$$

$$+ \omega_r\,(\{z_r\}\{y_r\}^T - \{y_r\}\{z_r\}^T)(\{\tilde{F}_R\} + i\,\{\tilde{F}_I\})\} \qquad (34)$$

where $\{\tilde{F}_R\}$ and $\{\tilde{F}_I\}$ respectively are the real and imaginary parts of $\{\tilde{F}\}$. It is evident from (34) that resonance condition exists when the excitation frequency ω is equal to one of the natural frequencies of the coupled acoustoelastic system.

Stationary Random Excitation

Consider the external pressure louding to be a weakly stationary random process with power spectral density function

$$\Phi_{p_e}\,(x_1,y_1;x_2,y_2;\,\omega) = \Phi(\omega)r(x_1,y_1;x_2,y_2;\omega) \qquad (35)$$

where $r(x_1,y_1;x_2,y_2;\omega)$ can be considered as the spectrum of the spatial correlation between pressures at locations (x_1,y_1) and (x_2,y_2) on the surface of the panel. The cross power spectral density function between the generalized forces Q_r and Q_s can be obtained as

$$\Phi_{Q_rQ_s}\,(\omega) = \Phi_p(\omega)\; J_{rs}\,(\omega) \qquad (36)$$

where $J_{rs}(\omega)$ is the cross acceptance function given by

$$J_{rs}\,(\omega) = 1/(\rho A)^2 \iint_{AA} r(x_1,y_1;x_2,y_2;\omega)\;\psi_r(x_1,y_1)\;\psi_s(x_2,y_2)\;dA_1\,dA_2 \qquad (37)$$

It can be shown that the cross spectral density matrix of the stationary response x(t) is given by

$$\Phi_{xx}(\omega) = U\,H(\omega)U^T\;\Phi_{FF}(\omega)\;U\,H^*(\omega)U^T \qquad (38)$$

where Φ_{FF} (ω) is the cross spectral density matrix of the excitation {F} which is filled with terms of (36) corresponding to {Q}. H(ω) is a diagonal matrix having the elements

$$H_r(\omega) = 1/i(\omega-\omega_r) \qquad r = 1,\ldots(n-1)+m$$

$$= 1/i(\omega+\omega_r) \qquad r = (n-1)+m,\ldots 2\{(n-1)+m\} \qquad (39)$$

Illustrative Example

An aluminium panel backed by an air filled cavity is considered as example. All the edges of the panel are assumed to be simply supported. The rigid wall cavity natural frequencies and modes are,

$$\omega_{jA}^2 = \omega_{ikl}^2 = \pi^2 c^2 (i^2/a^2 + k^2/b^2 + l^2/d^2) \qquad i = 0,1,2$$

$$k = 0,1,2 \qquad (40)$$

$$\phi_{jA} = \phi_{ikl} = \cos(i\pi x/a)\cos(k\pi y/b)\cos(l\pi z/d) \qquad l = o,1,2$$

The invacuuo structural natural frequencies and modes are,

$$\omega_{rs}^2 = \omega_{pq}^2 = D/\mu\ \{(p\pi/a)^2 + (q\pi/b)^2\}^2 \qquad p = 1,2$$

$$(41)$$

$$\psi_r = \psi_{pq} = \sin(p\pi x/a)\sin(q\pi y/b) \qquad q = 1,2$$

The modal coefficient of equation (9) are m_{jA} = 1, 1/2, 1/4 or 1/8 depending on whether all, two, one or none of the three subscripts corresponding to the acoustic mode are zero.
$m_{rs} = \mu ab/4$.

$$L_{jr} = L_{ik,\,pq} = \quad 4\,pq/\{\pi^2(i^2-p^2)(k^2-q^2)\} \text{ for } (i+p) \text{ and } (k+q) \text{ odd}$$

$$0 \text{ for } (i+p) \text{ or } (k+q) \text{ even} \qquad (42)$$

$$Q_{rs}^e = Q_{pq}^e = \quad 4ab/(\pi^2 pq) \text{ for } p \text{ odd and } q \text{ odd}$$

$$0 \text{ for } p \text{ or } q \text{ even} \qquad (43)$$

Material and geometric properties:

Panel-Aluminium: E = 7.24×10^{10} N/m², ν = 0.34, ρ_S = 2770 kg/m³,
Dimensions: a = 0.25 m, b = 0.5 m, h = 0.5 mm
Cavity-Air: ρ = 1.225 kg/m³, c = 330 m/s, d = 0.762 m

In the computation all structural modes and cavity modes upto 1500 rad/s have been included which corresponds to n = 2,

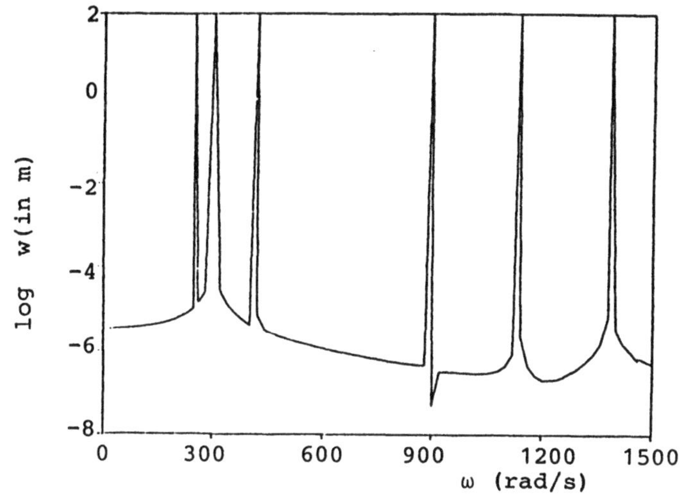

Fig. 1. Structural response (x = 0.125 m, y = 0.25 m)

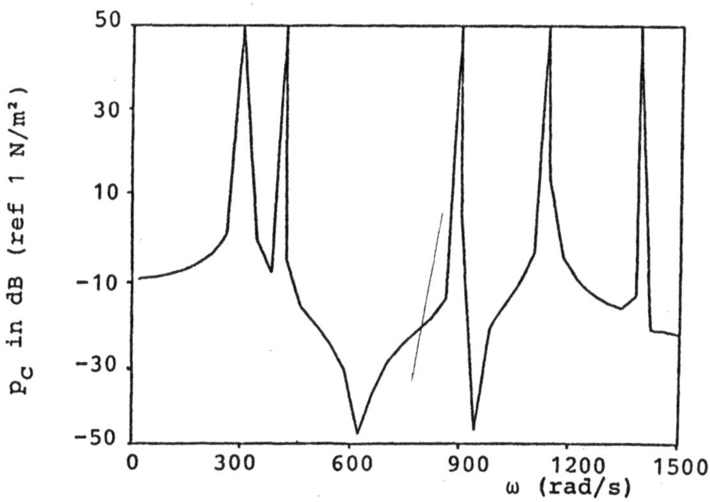

Fig. 2. Cavity pressure (x = 0.125 m, y = 0.25 m, z = 0.01 m)

and m = 14. A suitable addressing procedure in the computer program relates the single subscript used to index the cavity and structural modes to the corresponding multiple indices used in modal quantity calculations. Efficient storage schemes like the skyline storage have been adopted with significant saving in computer memory in view of the diagonal nature of many of the matrices and the large number of zeroes occurring in the coupling skew-symmetric matrix.

Figures (1) and (2) show typical plots of deflection and cavity pressure at the locations indicated for uniform unit harmonic external pressure loading ($p_e = e^{i\omega t}$). As had been observed earlier, in the absence of damping and absorbent walls the coupled acoustoelastic system behaves as an undamped oscillator with indefinitely increasing response corresponding to the coupled natural frequencies. Corresponding to structural modes for which $L_{jr} = 0$ the coupled natural frequencies do not vary from the invacuuo natural frequencies. For $L_{jr} \neq 0$ significant changes from the invacuuo values occur. For example the structural natural frequencies of 154.88 rad/s and 402.7 rad/s have been shifted to 300.28 rad/s and 418.17 rad/s in the coupled system, the increase in frequencies caused by an additional stiffness provided by the cavity to the structure mainly from the j = 0 (i = o, k = o, l = o) cavity mode. Changes in the higher frequency ranges are marginal. The method could also be used for cases with slight damping by a perturbation analysis [4].

Acknowledgement

The senior author wishes to acknowledge the Alexander von Humboldt Foundation under whose fellowship part of the computations was carried out at the Technical University, Berlin and Prof. M. Heckl and G. Ramesh with whom he had some useful discussions.

References

1. Meirovich, L.: A new method of solution of the eigenvalue problem for gyroscopic systems. AIAAJ. 12 (1974) 1337-1342.

2. Dowell, E.H.: Acoustoelasticity. Princeton University AMS Report 1280 (1976).

3. Premachandran Pillai, D.: Acoustoelasticity of a flexible panel and a rectangular cavity. M. Tech thesis. Dept. Mech. Eng. I.I.T. Madras (1984).

4. Meirovich, L.; Ryland II, G.: Response of slightly damped gyroscopic systems. J. Sound and Vib. 67 (1979) 1 - 19.

Near Field Pressure Fluctuations of a Circular Jet

Benoit Lafouasse, Anthony Chan & Chih-Ming Ho

Department of Aerospace Engineering
University of Southern California
Los Angeles, California 90089-0192

Introduction

Coherent structures in shear layers govern many dynamic aspects of flow, entrainment of mass, transfer of momentum and noise generation [1,2,3]. These structures develop from instability waves and undergo a merging process. Vortex merging reduces the time scale and increases the length scale of the shear layer. At the same time, pressure fluctuations are induced by the unsteady motion in the flow. These pressure perturbations can exert high level dynamic loading on nearby structures, [4,5,6] and radiate to far field as nose. Adjacent to the boundary of the jet, the pressure levels drop first exponentially and then algebraically. The extent of the exponential decay region was found to be scaled with the local instability wavelength [7,8].

In the present study, the relation between the generation of fluctuating pressure and the evolution of vortices is clarified. Furthermore, the transition from the near field to the far field and their length scales have been identified.

Apparatus

The jet nozzle is 5.08 cm in diameter and is connnected to the stagnation chamber with a fifth order polynomial contraction. Screens and honeycomb are used to reduce the background turbulence level. The jet is driven by compressed air which is regulated by a low noise valve. Extreme care has been applied to remove possible upstream disturbances. The origin is defined at the point on the jet axis and five diameter downstream from the nozzle exit. $r = |\vec{r}|$ is the distance away from the origin and β is the angle between $\vec{r}$ and the jet axis.

The jet is located in a anechoic chamber with 150 Hz low frequency cut-off. Multi-channel hot-wires, microphones and pressure transducers are used to record the instantaneous

Aero- and Hydro-Acoustics
IUTAM Symposium Lyon 1985
Editors: G. Comte-Bellot and J. E. Ffowcs Williams

signals. The data are digitized and processed by a PDP 11-23 microcomputer.

Results

The Hydrodynamic Pressure Contours:

For a jet at M = 0.1 and forced at its initial most amplified frequency, f_o, about one thousand measuring stations were used to survey the near field pressure in the vicinity of the jet edge. The spectrum at each station was measured. Near the nozzle exit, the spectra have a sharp peak at the most amplified frequency. At downstream locations, the peak at f_o decays as the first subharmonic, f_1, emerges. The grow-decay cycles of f_1, and higher harmonics, f_2, f_3 etc., takes place with increasing streamwise distance. The constant spectral density contours can be constructed from the measured spectra and are shown in Fig. 1. The pressure contour map clearly indicates that the pressure perturbations initiate from the corresponding vortex merging position. This result is in line with Laufer and Yen's [9] conclusion.

The Acoustic Near Field:

In previous papers, [7,8] we found that the distance from the jet edge to the point where the pressure decay curve deviates from the exponential decay was scaled with the instability wavelengths. At the far field, the length scale has to be the acoustic wavelength [10]. In the low Mach number range, the two length scales are one order of magnitude apart. In other words, there must be another transitional layer between the hydrodynamic exponential decay region and the far field. In order to examine this region, the pressure fluctuations were measured around a chosen location, r_o = 50". This location is far away from the hydrodynamic field but only a few acoustic wavelengths away from the source. The acoustic wavelength, $\lambda_a(M)$, used here is the wavelength at the preferred mode of a jet at Mach number M. The pressure intensity decreases algebraically with distance in this region as

$$p^2 \sim r^k \qquad (1)$$

The values of k are shown in Fig. 2. At β = 50° and 30°, the values of k vary with Mach number and reach the asymptotic value of -2 at M = 0.74. However, this is not a Mach number effect. It actually implies that an acoustic near field exists outside the hydrodynamic field, because r_o = 50" is about ten acoustic wavelengths away from the jet at high Mach number ($r_o/\lambda_a(M)$ = 7.48 at M = 0.75), and is only about one wavelength from a low Mach number jet ($r_o/\lambda_a(M)$ = 1.51 at M = 0.15). The angular dependence can be explained as follows: at a larger angle, e.g. β = 75°, the spectrum peak has a higher frequency than that at a lower angle. Therefore, the acoustic waves at a larger angle are few more acoustic wavelengths away than those at smaller angles.

The existence of an acoustic near field can be further substantiated by the measurements of acoustic intensity dependence on the Mach number for 0° $\lesssim$ r $\lesssim$ 50"

$$p^2 = M^N \qquad (2)$$

The values of N are plotted against $r/\lambda_a(M)$ in Fig. 3. It takes about ten acoustic wavelengths to reach the eighth power dependence on Mach number. The low Mach number measurements have low power dependence, $4 \lesssim N \lesssim 6$, because of their long acoustic wavelengths. If the set of data is plotted as the parameter of the angle with respect to the jet axis (Fig. 4), a slight angular dependence can be observed especially between β = 30° and 75°.

Directional patterns measured within the acoustic near field are shown in Fig. 5. About one or two acoustic wavelengths away from the jet, the sound waves radiate strongly in the forward direction. The directional pattern approaches an asymptotic form more than three acoustic wavelengths from the jet edge.

Conclusions

The constant pressure contours near the jet show that the evolution of coherent structures are the source of pressure perturbations. The data obtained in the vicinity of the jet indicates that two length scales govern the decay of the fluctuating pressure. A thin hydrodynamic near field extends about one instability wavelength from the jet edge and pressure decays exponentially. A thick acoustic near field takes about ten acoustic wavelengths to reach the far field. The pressure disturbances decay algebraically in this region.

Acknowledgement

This work is supported by NASA Langley Research under Contract Number NAG-1-420.

References

1. Brown, G.L. and Roshko, A.: On density effects and large structure in turbulent mixing layers. J. Fluid Mech. 64 (1974) 775-816.

2. Winant, C.D. and Browand, F.K.: Vortex pairing, the mechanism of turbulent mixing layer growth at moderate Reynolds numbers. J. Fluid Mech. 63 (1974) 237-255.

3. Ho, C.M. and Huerre, P.: Perturbed free shear layers. Ann. Rev. Fluid Mech. 16 (1984) 365-424.

4. Yu, J.C., Reddy, N.N. and Whitesides, J.L., Jr.: Noise and flow characteristics of an externally blown flap. Proc. of Second Interagency Symp. on Univ. Res. in Transportation Noise 1 (1974) 219-237.

5. Ho, C.M., Leve, H.L. and Plocher, D.A.: Surface pressure fluctuations generated by a jet impinging on a curved plate. AIAA Journal 15 (1977) 1348.

6. Ho, C.M. and Nosseir, N.S.: Dynamics of an impinging jet, Part 1, The feedback phenomenon. J. Fluid Mech. 105 (1981) 119-142.

7. Ho, C.M. and Lafouasse, B.: Hydrodynamic pressure field of an axisymmetric jet. AIAA Paper No. 84-2316.

8. Gutmark, E. and Ho, C.M.: Near field pressure fluctuations of an elliptical jet. AIAA J. 23 (1985) 354-358.

9. Laufer, J. and Yen, T.C.: Noise generation by a low Mach number jet. J. Fluid Mech. 134 (1983) 1-31.

10. Crighton, D.G. and Huerre, P.: Model problems for the generation of superdirective acoustic field by wave packets in shear layers. AIAA Paper No. 84-2295, 1984.

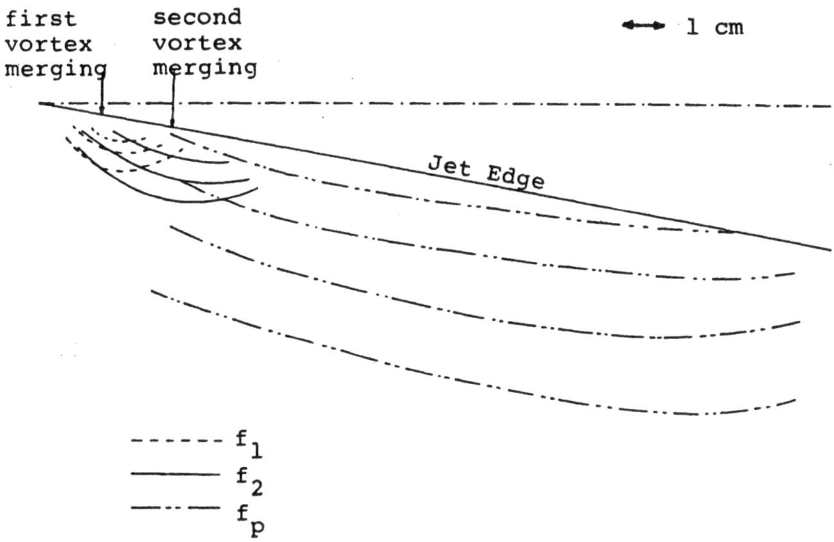

Figure 1. Constant spectral intensity contours M = 0.088
f_1 = first subharmonic f_2 = second subharmonic
f_p = preferred mode

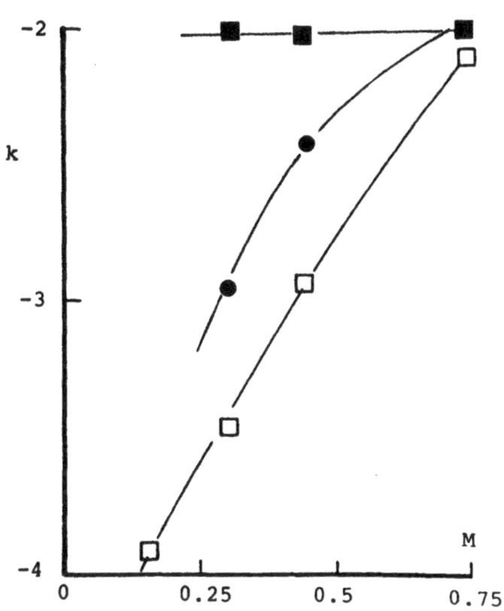

Figure 2. Variation of k with respect to Mach number
■ β = 75° ● β = 50° □ β = 30°

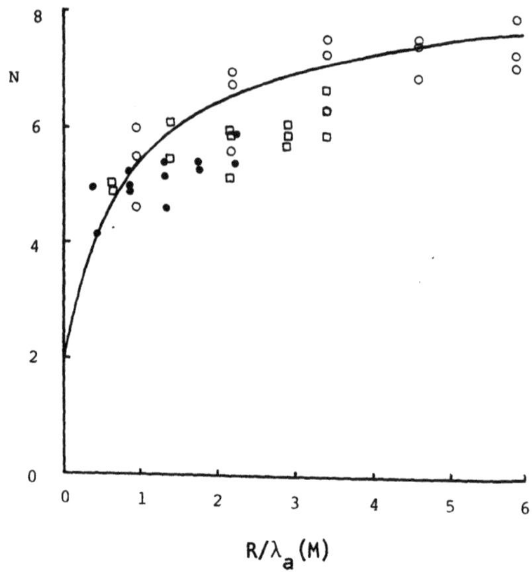

Figure 3. Variation of N with respect to normalized distance
● M = 0.15 □ M = 0.45 0 M = 0.74

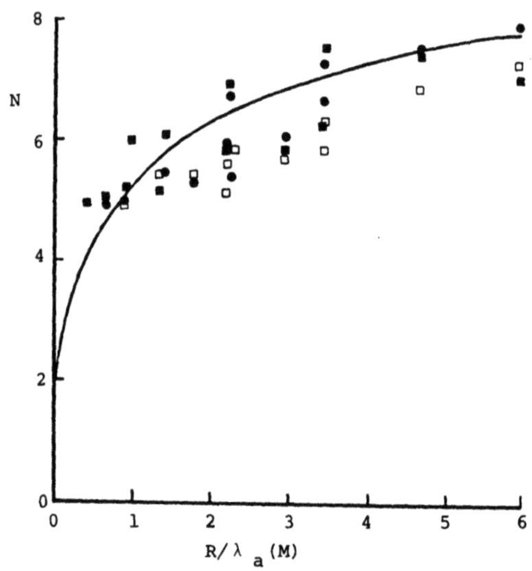

Figure 4. Variation of N with respect to normalized distance
■ β = 75° ● β = 50° □ β = 30°

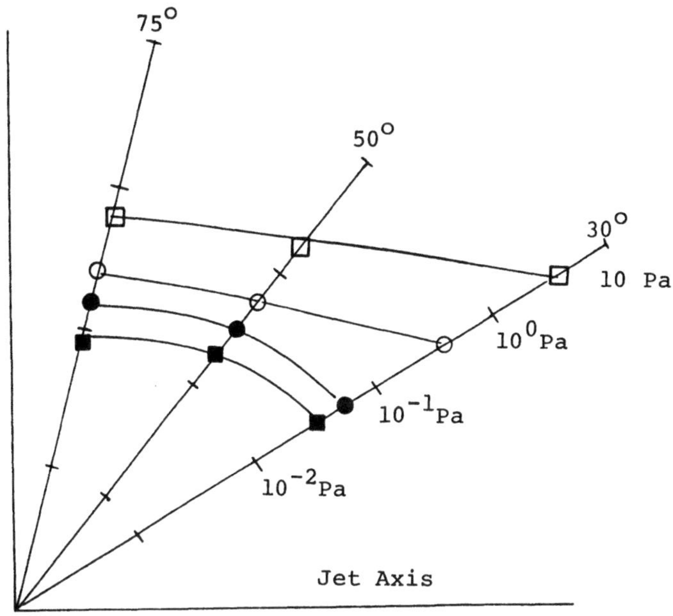

Figure 5. Directional pattern M = 0.3
□ $R/\lambda_a(M) = 1.1$ 0 $R/\lambda_a(M) = 1.8$
● $R/\lambda_a(M) = 2.4$ ■ $R/\lambda_a(M) = 3.1$

Experimental Investigation of Near and Far Acoustic Field of a Small Turbojet

J. HAERTIG, C. JOHÉ

Institut Franco-Allemand de Recherches de Saint-Louis (ISL)
68301 Saint-Louis (France)

Summary

Experimental results concerning the near and far acoustic field of a small turbojet are presented. At idle speed (subsonic jet) the acoustic field of the engine is similar to the one of a strongly excited free jet model (tone noise with high azimutal coherence). At maximum speed (supersonic subcritical jet) tone noise is more or less covered by jet noise and the far field is similar to the one of quiet jet model.

Introduction

Jet models obtained through expansion of compressed air are commonly used to investigate jet noise. It is well known that the aerodynamic and acoustic properties of such jets are modified when artificially excited by the use of loudspeaker placed upstream in the settling chamber for example [1]. This experiment was conducted to compare the noise radiated by such jets with the one of a true turboengine [2], [3].

Experimental set-up

The engine is a MICROTURBO TRS 18 provided with a single stage compressor (43 blades), a turbine (16 blades) and a nozzle with diameter D = 105 mm. At idle speed the engine is rotating at 27 000 RPM, ejection speed U and ejection gas temperature T_i are respectively about 260 m/s and 780 K. At maximum speed U = 510 m/s, T_i = 1020 K and the rotating speed is 42 000 RPM.

In order to investigate the near field, microphones (B.& K. 1/4") were placed on a ring support (radius R = 3D) surrounding the jet at different stations X downstream of the exit

Aero- and Hydro-Acoustics
IUTAM Symposium Lyon 1985
Editors: G. Comte-Bellot and J. E. Ffowcs Williams

plane of the nozzle. During far field investigations the microphones were placed along a circular arc (radius 48D) in a meridian plane.

Near field

At idle speed, noise spectra (bandwidth 36 Hz, arbitrary unit for amplitude) show high peaks at frequencies which correspond to rotating engine parts (Figure 1). Internal noise, mainly due to the turbine, gives pressure fluctuations with very high azimutal coherence (Figure 2). Moreover it has been found that the coherent fluctuations due to the turbine (7.2 kHz) mainly consist of modes $n = 0$ and $n = 1$ (amplitude ratio mode 1/mode 0 $\approx$ 1/3).

At maximum speed the jet noise increases (Figure 3) but we still can note a large azimutal coherence of the internal tone noise at $X = 1D$ (Figure 4).

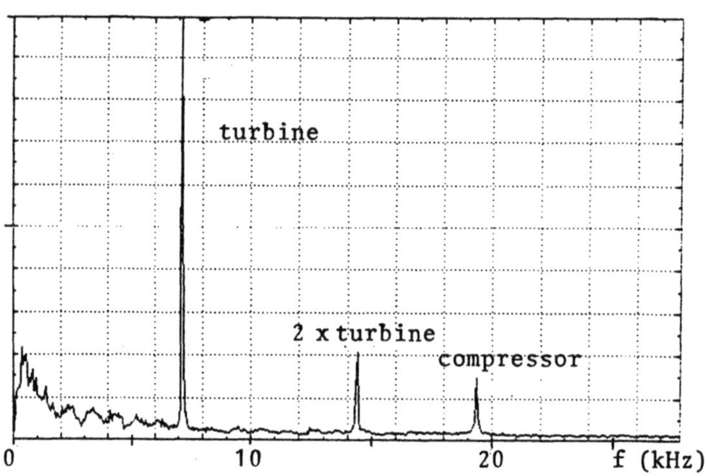

Figure 1. Near field narrow band (36 Hz) spectrum MICROTURBO TRS 18 "idle" $X/D = 1$, $r = 3D$

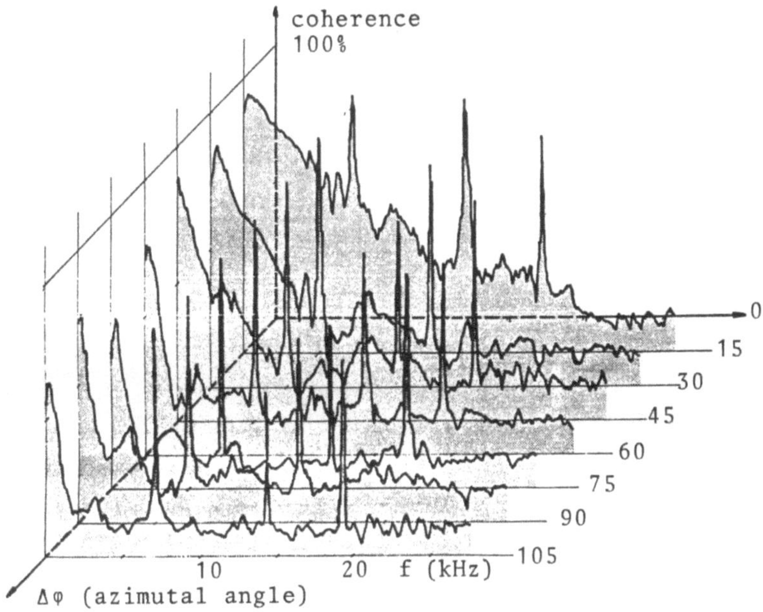

Figure 2. Azimutal coherence function (same conditions as Figure 1)

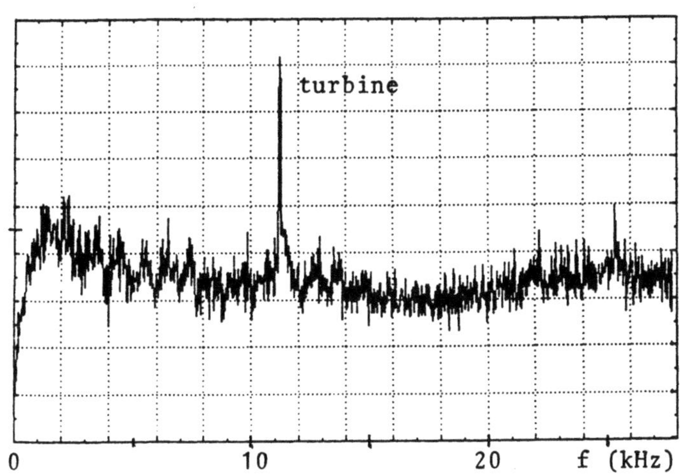

Figure 3. Near field narrow band spectrum (same scale as Figure 1)
MICROTURBO TRS 18 "maximum", X/D = 1, r = 3D

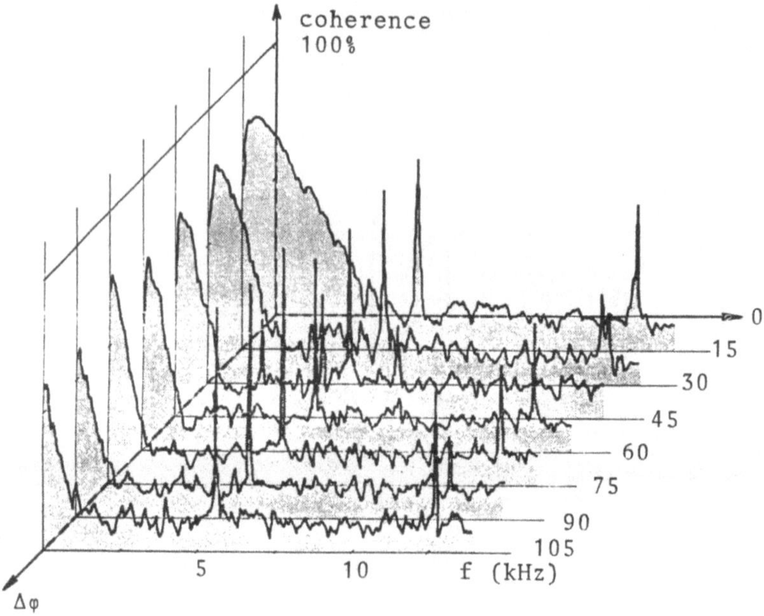

Figure 4. Azimutal coherence function (same conditions as Figure 3)

At last, the table representing the main azimutal mode components in the near field of a quiet jet model and engine jet allows a comparison of the mean azimutal structure. For jet model and engine at maximum speed, the variation of amplitude vs mode number is similar but at idle speed, the plane wave mode $n = 0$ is prevailing.

Far field

At idle speed, frequency spectra show that wide band noise is still dominated by internal tone noise (Figure 5). The mean total noise is about 35 μbar, the tone level due to the turbine ($\Delta f = 36$ Hz, $f = 7.2$ kHz) is roughly 12 μbar. At maximum speed, tone noise is dominated by wide band jet noise (Figure 6), the directivity is sharper (Figure 7) and the noise level is much larger ($\theta = 30°$; 550 μbar).

mode	free jet		MICROTURBO "idle"			MICROTURBO "maximum"		
	1D	3D	1D	3D	6D	1D	3D	6D
0	0.19	0.28	0.219	0.345	0.636	0.120	0.194	0.535
1	0.24	0.23	0.167	0.187	0.223	0.14	0.144	0.226
2	0.14	0.12	0.116	0.151	0.099	0.105	0.127	0.096
3	0.10	0.11	0.087	0.105	0.029	0.085	0.103	0.053
4	0.07	0.07	0.069	0.056	0.010	0.070	0.079	0.027
5	0.05	0.04	0.057	0.038		0.058	0.068	0.014

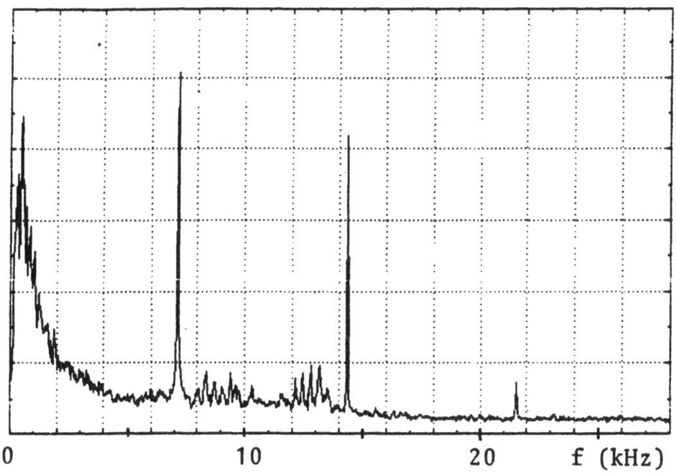

Figure 5. Far field narrow band spectrum MICROTURBO TRS 18 "idle" r/D = 48, θ = 40°

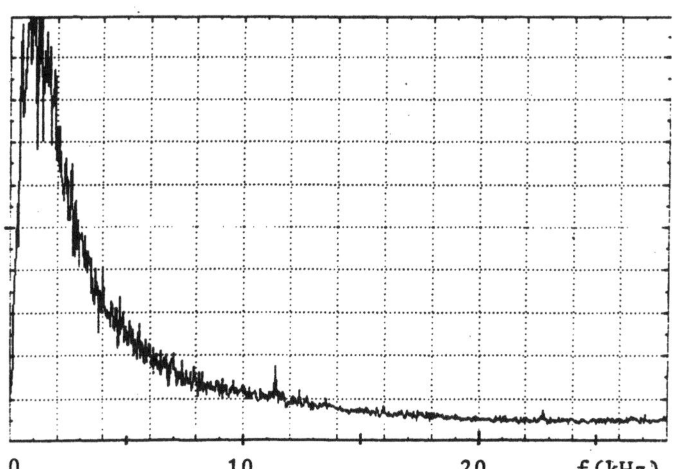

Figure 6. Far field narrow band spectrum (same scale as Fig.5) MICROTURBO TRS 18 "maximum", r/D = 48, θ = 40°

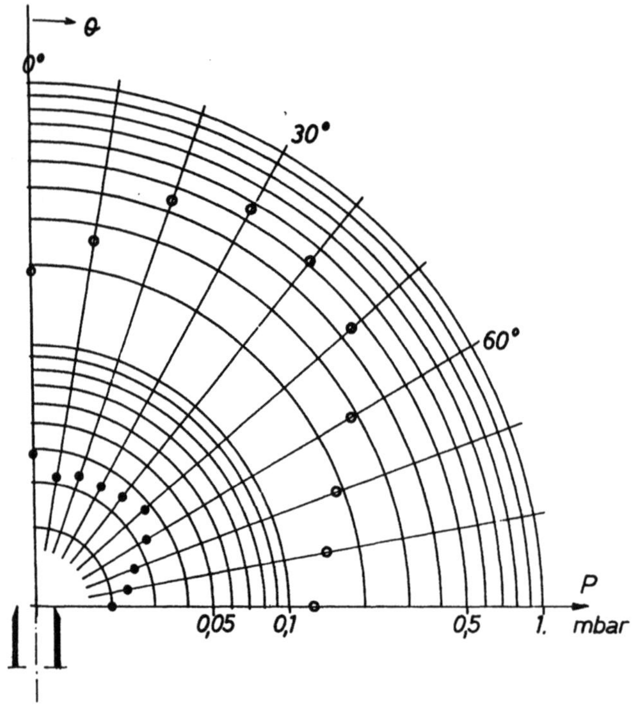

Figure 7. Far field overall noise level r/D = 48
MICROTURBO ● idle
○ maximum

Conclusions

We have found out that at idle speed the acoustic field of an engine is similar to the one of a very strongly excited jet model (Figure 8), whereas at maximum speed the internal high frequency noise is covered by the wide band jet noise especially in the far field.

The experimental investigation of both quiet and artificially excited jet models are justified because the noise radiated by engine and models is similar and, above all, acoustic annoyance of emerging tones in wide band noise is very heavy.

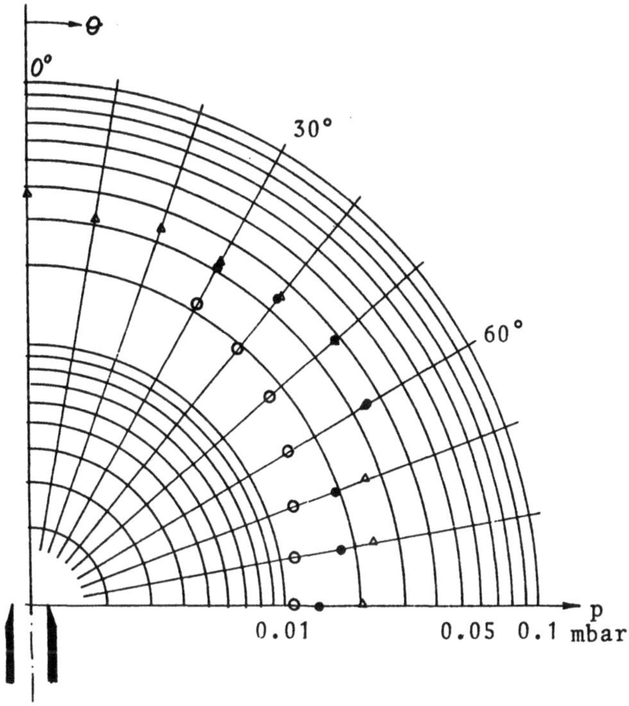

Figure 8. Far field noise level r/D = 48

o quiet jet model

● strongly excited jet model

△ MICROTURBO TRS 18 "idle"

References

[1] J. HAERTIG. Etude du couplage entre champ moyen cohérent et turbulent dans un jet libre rond excité. ISL - R 115/83

[2] L.BOBIN, G. KRONENBERGER, A. le BAS. Etude expérimentale des corrélations azimutales du champ acoustique d'un jet libre froid. ISL - CR 79/4

[3] C. JOHÉ, J. HAERTIG. Etude expérimentale des champs acoustiques proche et lointain d'un turboréacteur MICROTURBO TRS 18. ISL - R 113/84

Aeroacoustics

Wind Tunnel Investigation of Aeolian Tones from Rough Surface Cylinders

F.-R. GROSCHE

DFVLR Göttingen, Germany, and
von Karman Institute for Fluid Dynamics
Chaussée de Waterloo, 72
B - 1640 Rhode Saint Genèse - Belgium

SUMMARY

The generation of aeolian tones by a rough cylinder was investigated in a large wind tunnel at Reynolds numbers between 10^4 and 4×10^4. At low Reynolds numbers, the measured sound pressure level agrees well with predictions for smooth cylinders, but it increases less rapidly with flow velocity at higher Reynolds numbers. This can be explained by the effects of the surface roughness of the cylinder. The sound radiation could be substantially reduced by means of helical trip wires.

1. INTRODUCTION

Periodic flow separation from circular cylinders generates the well known "aeolian tones" [1,2]. The wind noise of electrical overhead lines, e.g. high tension cables, is also due to this effect. If the lines are situated in the vicinity of populated areas, the sound may be annoying and thus needs to be reduced. Wind tunnel experiments are a convenient technique to study the problem under controlled flow conditions.

This paper presents results obtained with a section of a full scale high tension cable mounted in the open test section of a large wind tunnel. In contrast to the conditions of earlier investigations reviewed in [2], the surface of the cable cannot be considered as smooth. Therefore results of a study of periodic vortex separation from rough surface cylinders [3] are compared with our test results.

2. EXPERIMENTS

The measurements were conducted in the low speed wind tunnel of DFVLR Göttingen, which has an open test section with a 3 m × 3 m square nozzle. The cable generating the aeolian tones was mounted vertically as indicated in Fig. 1. The aspect ratio of the cable was very large, $\ell/D\approx130$. Edge effects could therefore be neglected. Sound measurements were made at y=1 m lateral distance from the cable at the position x=0 shown in Fig. 1 and at x=0.5 m and x=1.0 m. The flow velocity was varied between 7.5 m/s and 25 m/s. The cable consisted of a core of steel wires surrounded by two layers of aluminium outer wires of 3.14 mm diameter; see Fig. 2. The overall diameter was D=22.4 m. After measuring the sound radiation of the "clean" cable shown in Fig. 2.,

Aero- and Hydro-Acoustics
IUTAM Symposium Lyon 1985
Editors: G. Comte-Bellot and J. E. Ffowcs Williams

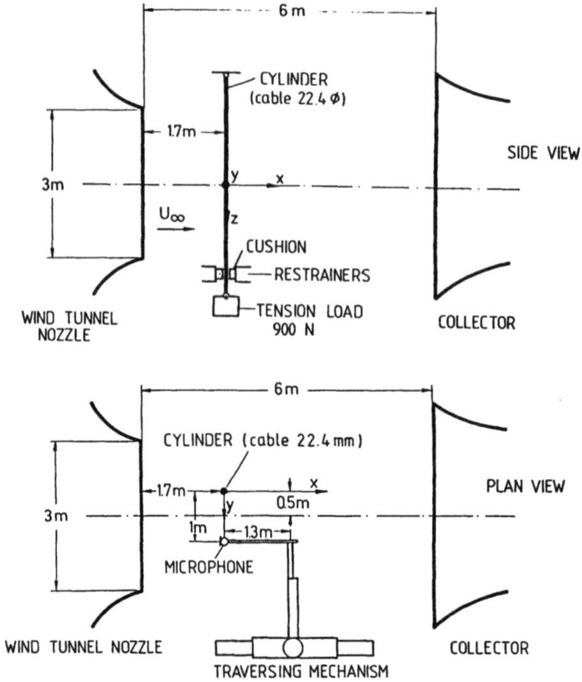

Fig. 1 - Experimental setup

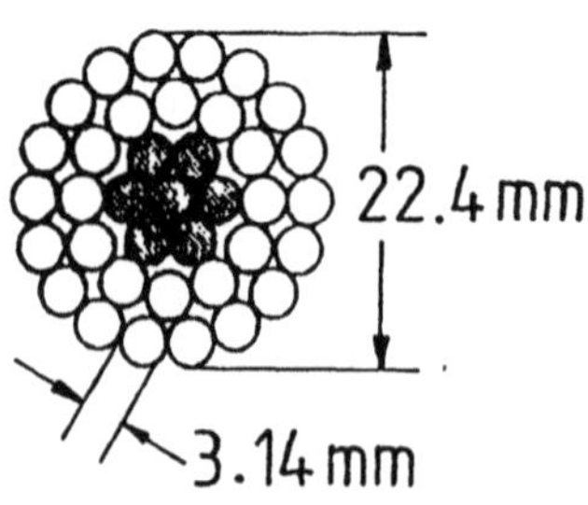

Fig. 2 - Cross section of the cable (clean config.)

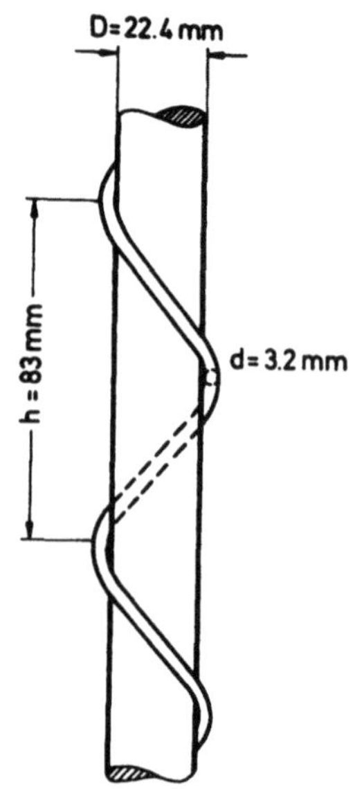

Fig. 3 - Cable with trip wire helix

configurations according to Fig. 3 were tested. A trip wire was wrapped helically around the cable to reduce the periodic vortex separation. Several configurations with different wire diameters d and pitch ratios h/D were tested in addition to the configuration of Fig. 3. Also, the effect of a double helix was investigated.

After the tests with the model installed, the background noise of the tunnel without model was measured at all wind velocities $7.5\ m/s < U_\infty < 25\ m/s$ with the microphone in the position shown in Fig. 1.

A 1/4" B+K microphone equipped with a nose cone to suppress aerodynamic self-noise was used. The microphone signal was analyzed by a real-time narrow band frequency analyzer set to a frequency range of 500 Hz and resolution of 1.25 Hz.

The photos Figs. 4 and 5 show the test setup and the cable with trip wire helix according to Fig. 3. The graphs Figs. 6 to 8 give typical examples of the narrow band spectra obtained.

Fig. 4 - Test setup in the wind tunnel
Left hand side : traversing unit with microphone boom
Right hand side : wind tunnel nozzle and vertically mounted cable

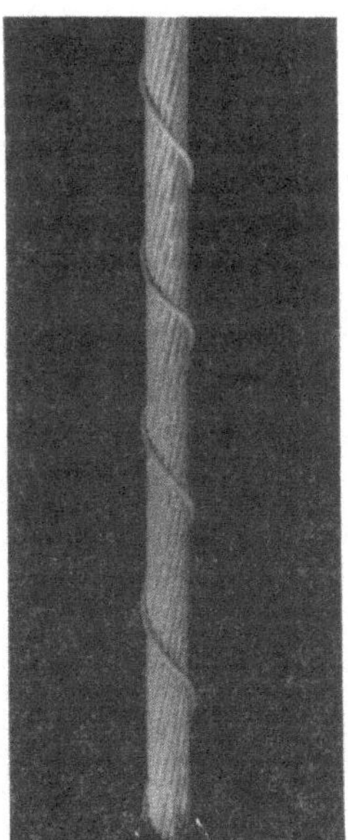

Fig. 5 - Cable with trip wire helix

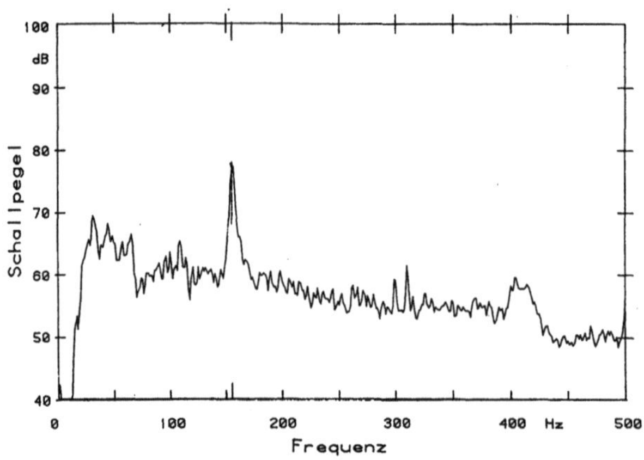

U_∞ = 17.5 m/s

Microphone position :

y = 1 m, x = 0

Fig. 6 - Sound spectrum of clean cable

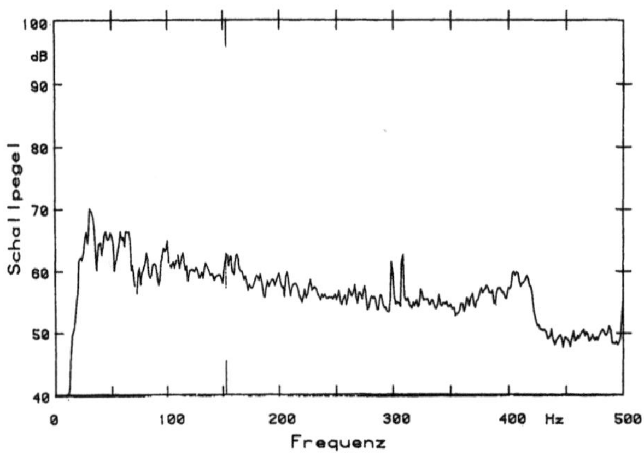

U_∞ = 17.5 m/s

Microphone position :

y = 1 m, x = 0

Fig. 7 - Sound spectrum of cable with trip wire helix

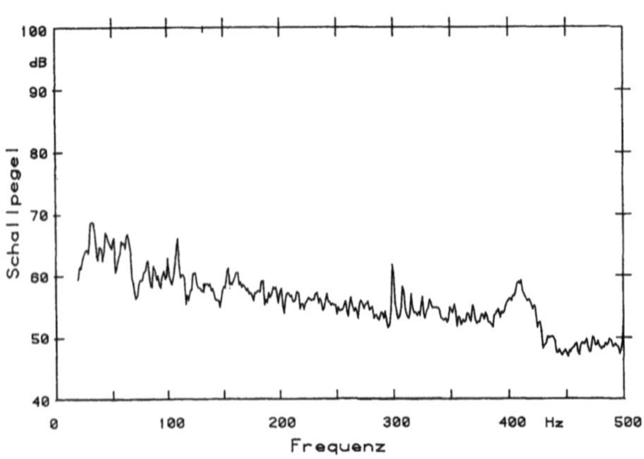

U_∞ = 17.5 m/s

Microphone position :

y = 1 m, x = 0

Fig. 8 - Sound spectrum of background noise without the model

3. RESULTS AND DISCUSSION

The spectrum Fig. 6 measured with the clean cable shows a prominent peak at f=155 Hz corresponding to a Strouhal number $Str=f\times D/U_\infty \cong 0.2$. The spectrum Fig. 7 of the cable with the trip wire exhibits only a slight hump at this frequency, it is almost identical with the spectrum of the background noise, Fig. 8. A similar behaviour of the spectra was observed at other flow velocities. We consider therefore the peak in the spectrum Fig. 6 as an aeolian tone which is strongly reduced in Fig. 7 by the effect of the trip wire.

The diagrams of Fig.9 present frequency and Strouhal number of the aeolian tone of the clean cable as functions of flow velocity. The Strouhal number is close to 0.2. It increases weakly with velocity between U_∞=10 m/s and U_∞=25 m/s. The flow velocity U_∞=15 m/s corresponds to a Reynolds number of $Re=2.2\times10^4$ (based on cable diameter).

The sound pressure level (SPL) of the aeolian tone is plotted versus flow velocity in Fig. 10 for the clean cable. The measured SPL are compared to theoretical estimates for smooth cylinders based on [4]. The predicted sound intensity is proportional to the axial correlation length s of the vortex

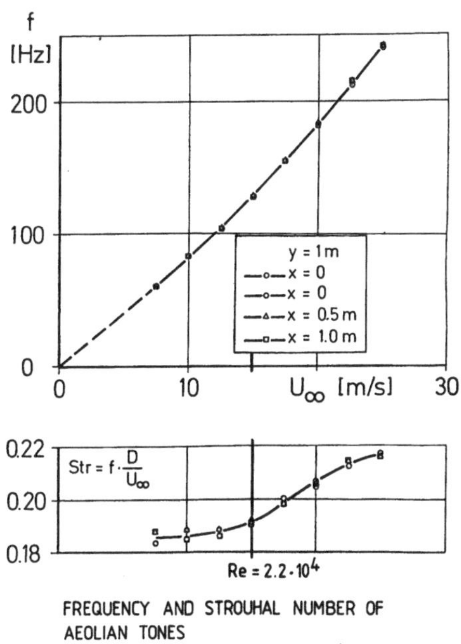

Fig. 9 - Frequency and Strouhal number of the clean cable

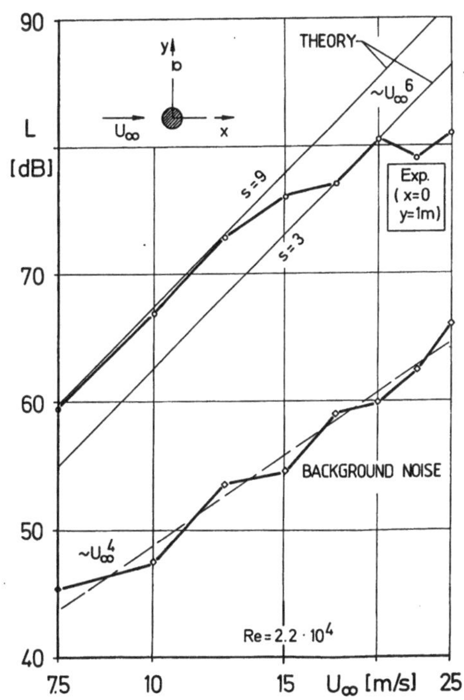

Fig. 10 - Sound pressure level of the clean cable, comparison with prediction shedding from the cylinder; for details see [5]. Two predictions with correlation lengths s=9 and s=3, s being normalized by the cylinder diameter D, are shown in Fig. 10. The value of s=9 was given in [6] for smooth cylinders at $Re < 5 \cdot 10^4$. The value of s=3 was proposed earlier in [4]. Measurements and the theoretical estimate with s=9 agree very well for $Re < 2 \cdot 10^4$. At $Re > 2 \cdot 10^4$, the measured levels deviate more and more from the prediction.

This is believed to be due to the surface roughness of the cable leading to boundary layer transition at lower Reynolds numbers than expected for smooth cylinders. If we consider the radius of the outer wires of the cable approximately as the roughness height, k_s=1.57 mm, the roughness parameter of the cable is $k_s/D \cong 0.07$.

According to experiments on periodic vortex shedding [3], the critical Reynolds number of rough surface cylinders varies with surface roughness and can be determined by the empirical relation $Re_{crit} = 6000/\sqrt{k_s/D}$.

Using this equation we find for the cable : $Re_{crit} \approx 2.3 \cdot 10^4$. This value agrees closely with the Reynolds number where the intensity of the aeolian tone begins to deviate from the estimate for smooth cylinders.

Fig. 11 (from [3]) shows the variation of the vortex shedding Strouhal number with Reynolds number for circular cylinders of different surface roughness. With $k_s/D \geqslant 0.003$, the Strouhal number increases only by about 10% to 20% with Reynolds number growing from subcritical to supercritical values. Also the

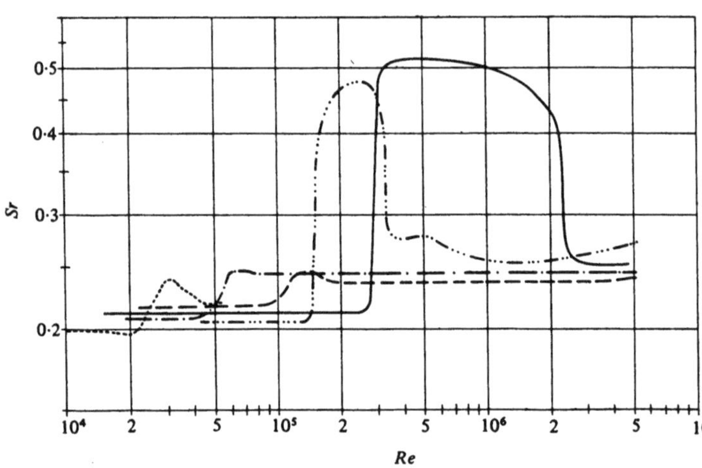

Fig.11 - Vortex shedding Strouhal number vs Reynolds number (after [3])
——smooth cylinder; —••—$k_s/D=7.5 \cdot 10^{-5}$;— — —$k_s/D=0.003$; —•—$k_s/D=0.009$; ---$k_s/D=0.03$

critical Reynolds number for $k_s/D=0.03$ is about $2.5\cdot10^4$. Our results, Fig. 9, agree well with these findings.

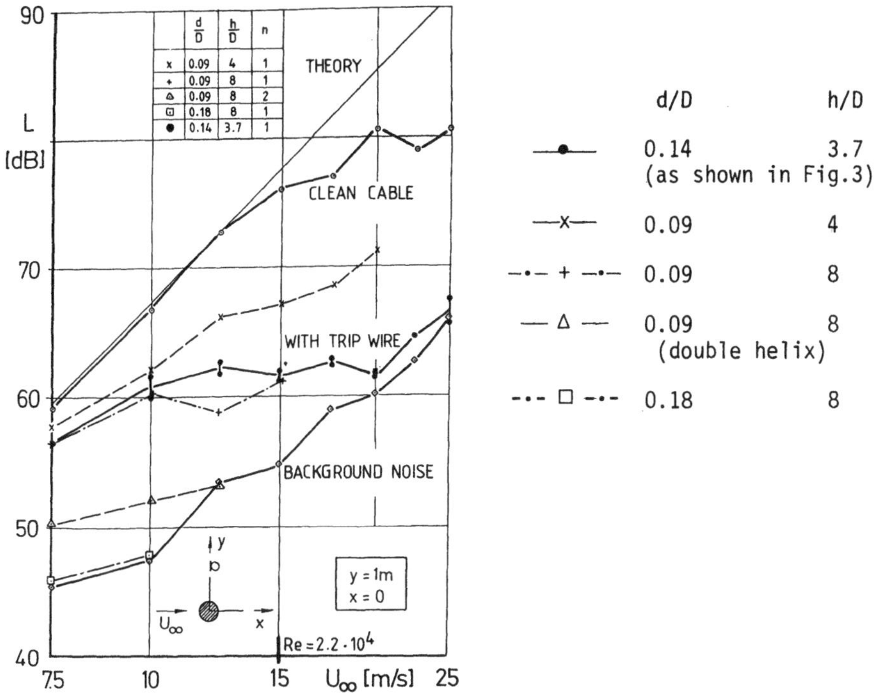

Fig. 12 - Sound pressure levels of the cable with and without a trip wire helix of wire diameter d and pitch h/D

The reduction of the aeolian tones by a trip wire helix is illustrated by Fig. 12. The full circles represent the sound pressure levels measured with the configuration shown in Figs. 3 and 5. The aeolian tones are reduced by up to 20 dB, the effect increases with Reynolds number, as expected.
Some results with different configurations are also given in Fig. 12. Within the range of parameters investigated, the effectiveness of the trip wire helix increases with the diameter d of the wire and with the pitch parameter h/D of the helix. Also, a double helix is significantly more effective than a single helix of the same wire diameter and pitch.

REFERENCES

1. STROUHAL, V.: Über eine besondere Art des Tonanregung. Annalen Phys. Chem., 5, 1978, pp 216-251.

2. BLEVINS, R.D.: Review of sound induced by vortex shedding from cylinders. J. Sound & Vibration, Vol. 92, 1984, pp 455-470.

3. ACHENBACH, E. & HEINECKE, E.: On vortex shedding from smooth and rough cylinders in the range of Reynolds numbers 6×10^3 to 5×10^6. J. Fluid Mechanics, Vol. 109, August 1981, pp 239-251.

4. PHILLIPS,O.M.: The intensity of aeolian tones. J. Fluid Mechanics, Vol. 1, Part 6, 1956, pp 607-624.

5. GROSCHE, F.-R., STIEWITT, H., HOEFER, H.: Windkanaluntersuchungen mit akustischen Messmethoden zur periodischen Strömungsablösung an Kreiszylindern. DFVLR IB 222-84A22, 1985.

6. SCHLINKER, R.H. & FINK, M.R.: Vortex noise from nonrotating cylinders and airfoils. AIAA P 76-81, 1976.

Characterizing Pressure Fluctuations on the Wall of a Circular Cylinder Perpendicular to a Flow

P. HOCQUET

Bassin d'Essais des Carènes
Paris, France.

B.E. FORESTIER

Institut de Mécanique des Fluides
Marseille, France.

Summary

The experimental results presented are included in a study on the pressure fluctuation field on the wall of bluff-bodies where the turbulent boundary layer shows significant favourable and adverse pressure gradients before separation. For these conditions, statistical pressure field studies were performed on walls of a rigid cylinder in the air and of a bluff-body under water. For this presentation, emphasis is placed on the results obtained on the cylinder in the frequency range $0,05 \leqslant \omega^* \leqslant 2,0$ ($\omega^* = \omega.\delta_1/U_e$, where ω is the pulse, δ_1 et U_e are respectively the displacement thickness and the mean velocity at the edge of the boundary layer). In this range, low wavenumber spectra estimation is presented and compared with the flat plate measurements obtained by Martin and Leehey /1/ and Jameson /2/.

1 - Introduction

Most studies on pressure fluctuations beneath turbulent boundary layer were carried out on flat plate, on body of revolution and on two-dimensional airfoil. The influences of significant pressure gradients, separation and curvature effect (case of bluff bodies), on the turbulent boundary layer pressure field, were investigated on a rigid circular cylinder in a wind tunnel with a low Mach number flow ($M = 0,03$) and a low turbulence intensity (0,4 %).

Some measurements were carried out in towing tank on the wall of a bluff body to compare the dimensionless wall pressure power spectral density with those on the cylinder.

2 - Results

For the cylinder, a regime simulation by artificial transition is needed to reach high Reynolds numbers with a low velocity flow (10 m/s) and consecutively a low background noise in a classical wind tunnel. This turbulent regime ($R_D = U_\infty.D/\nu \geqslant 10^6$) was reached by tripping the transition from

Aero- and Hydro-Acoustics
IUTAM Symposium Lyon 1985
Editors: G. Comte-Bellot and J. E. Ffowcs Williams

laminar to turbulent boundary layer with rough band and by fulfilling the correct simulation conditions as defined by Nakamura and Tomonari /3/. Mean characteristics of simulated boundary layer were determined from hot-wire anemometry.

Figure 1 shows the pressure spectral densities as measured in three positions : 1 , for $dP/dx < 0$; 2 , for $dP/dx = 0$; 3 , for $dP/dx > 0$. Results obtained for positions 1 and 2 show a fairly clear distinction between pressure fluctuations coming from the vortex shedding and the separation (phenomena observed also under laminar boundary layer) and those coming from the turbulent boundary layer ($\omega^* \geqslant 0,05$) where the $\omega^{*0,8}$ ($\omega^* <$ 0,3) increase has already been measured by Panton and al. /4/ on a glider for $dP/dx = 0$.

Cut-off frequencies, ω_c^*, for the three pressure gradients are the result of low Reynolds numbers, $U_\tau . \delta / \nu$, which are caused by the turbulent condition simulation (U_τ, friction velocity and δ thickness of boundary layer).

On position 2 , where the pressure gradient is zero, the power level at $\omega^* = 0,3$ is close to those reached on the flat plate.

The pressure fluctuation power measured on position 3 around $\omega^* = 0,3$ is located to that obtained in the towing tank on the bluff-body with an adverse pressure gradient and without tripping the transition.

On this same frequency band, the power levels measured on position 1 are relatively large when compared with those observed in the zero pressure gradient zone and it is difficult to confirm that the favourable pressure gradient increases power compared to case where $dP/dx = 0$, given the proximity of the rough band to this measurement position. In fact, the step effect created by the rough band may provide a power increase in the turbulent pressure fluctuations as noted by Farabee and Casarella /5/.

So as to predict the low wavenumber spectra in the frequency band, $0,05 \leqslant \omega^* \leqslant \omega_c^*$, through the use of Chase modelling /9/, the model constants were adjusted on the experimental results shown on figures 1, 2, and 3 around positions 1 and 3 . The figures show an universal coherence decrease following $\omega . r_1 / U_p$ for $\omega^* \geqslant 0,3$ (r_1 streamwise sensors separation and U_p measured phase velocity) and a frequency to frequency decrease for $0,05 \leqslant \omega^* < 0,3$ as was already noted by Brooks and Hodgson /6/.

The Chase model constants are :

Position 1 : $\mu = 0.12$; $r_T = 0.3$; $b_M = 0.7$; $b_T = 0.262$, $a_+ = 0.9$
Position 3 : $\mu = 0.7$; $r_T = 0.$; $b_M = 1.5$; $a_+ = 38.$

The phase velocities measured (fig. 4), U_p, show a variation in $\omega^{0.4}$ for $\omega^* < 0.3$ in position 1 . When connected with the spectra increase in $\omega^{0,8}$, this provides an increase in k^2 for the spectrum in wavenumber as predicted by the Kraichnan and Phillips theoretical studies. The spectral densities $P(k_1^*, 0, \omega^*)$ were calculated using Corcos and Chase /9/ models and are presented for $\omega^* = 0.3$, figure 5.

The two models provide identical convective peak levels which are higher by about 10 dB than those provided on flat plate with $dP/dx = 0$. Power levels for low wavenumbers (figure 6) were computed from the Chase model and, for the adverse pressure gradient, they are 30 dB above those measured by Martin and Leehey /1/ and Jameson /2/ on the flat plate ($dP/dx = 0$).

3 - Conclusion

For a circular cylinder perpendicular to the flow direction, dominant pressure fluctuations are at the same frequency as the vortex shedding. This provides lift fluctuations which are source of noise and induced vibration of the structure (see Blevins /7/). Beneath the turbulent boundary layer, the pressure fluctuations levels largely depend on the pressure gradient. In adverse pressure gradient, significant pressure fluctuations for low wavenumbers may generate wall vibrations at the rear part of a bluff-body.

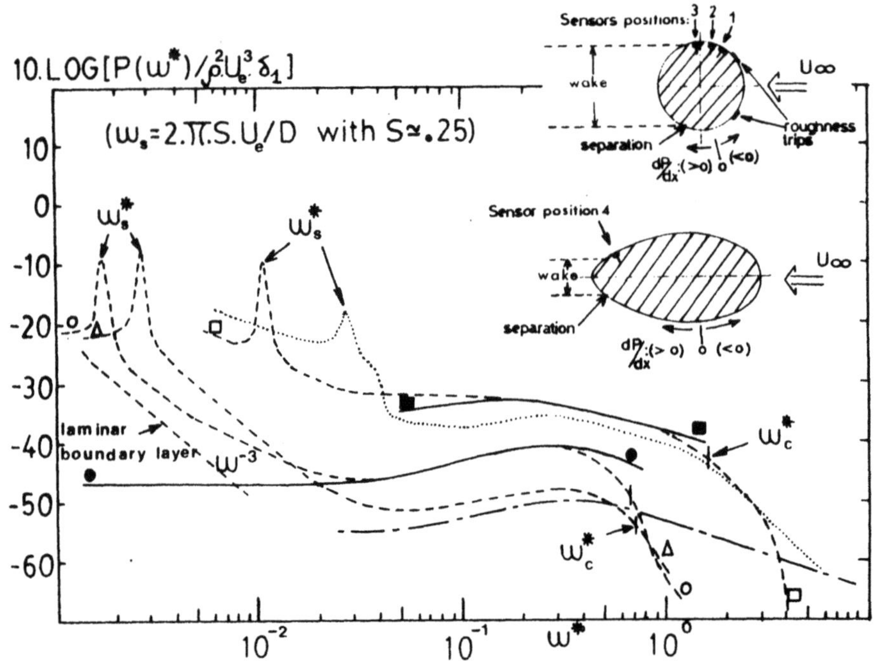

Fig.1. Dimensionless wall pressure power spectral density measured : Panton and al /4/ ;— - —; Position 1 ;o- - - -o; Position 2 ;Δ- - - - -Δ Position 3 ;□- - -□; Position 4 ;..........; and calculated from equation (74) /9/ : Position 1 ;—●; Position 3 ;—■.

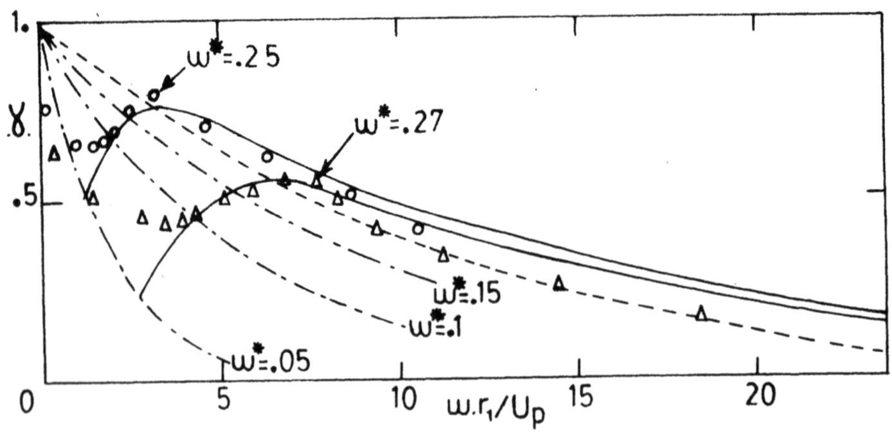

Fig. 2. Normalized longitudinal cross spectra measured : (0; $r_1/\delta = 1.55$, Δ ; $r_1/\delta = 3.4$) ; Position 1,- - - -,-;Schloemer (dP/dx < 0), and calculated from equation (71) /9/ : for the two dimensionless separations ;——, for different ω^*;— . —,(Position 1).

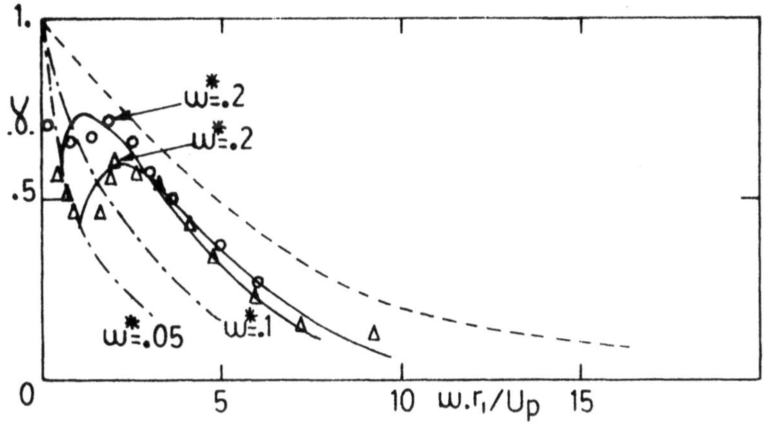

Fig. 3. Normalized longitudinal cross spectra measured : (0 ; r_1/δ = .45, Δ ; r_1/δ = .9) ; Position 3,----; Schloemer (dP/dx > 0) /8/, and calculated from equation (71) /9/ : for the two dimensionless separations ;——, for different ω^*,—·—, Position 3.

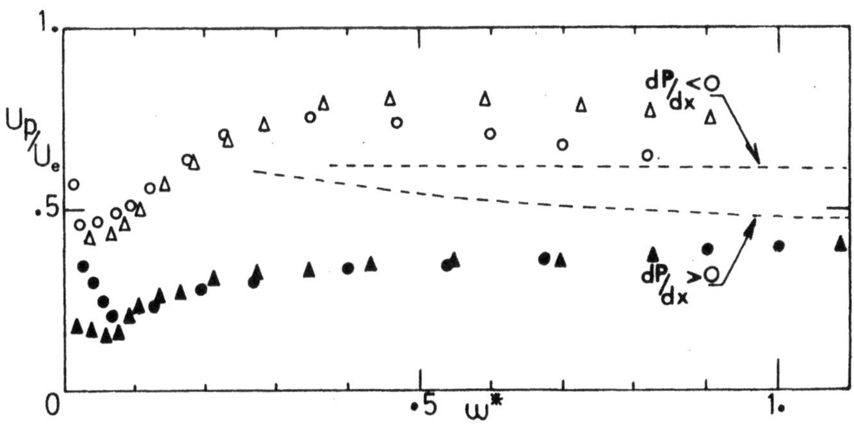

Fig. 4. Fractional phase convection speed for wall pressure vs. Dimensionless frequency measured : o , Δ ; Position 1 , ● , ▲ ; Position 3,---; Schloemer /8/.

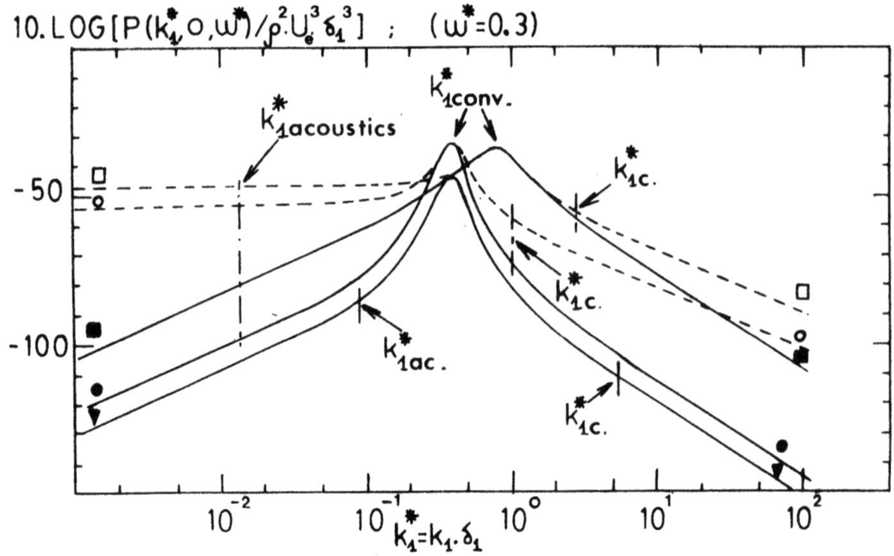

Fig.5. Dimensionless wavenumber spectra calculated (ω^*= 0,3) :----; Corcos model and ———; Chase model (equation (68) /9/).▼———▼; (Bull's data, dP/dx = 0 /9/),o----o and ●———●; Position 1,□---□ and ■———■; Position 3

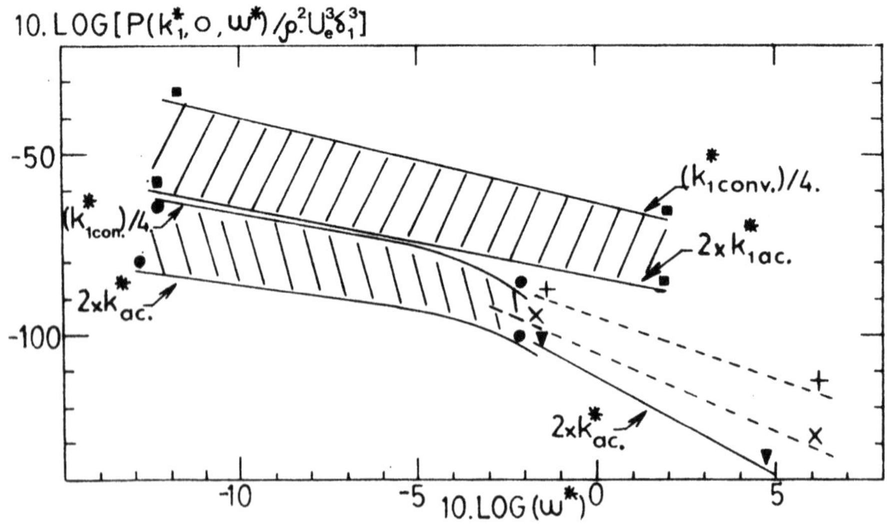

Fig. 6. Dimensionless low wave number spectra mesured :+----+; Martin and Leehey /1/ ,x---x; Jameson /2/ and calculated from equation (68) /9/ : ▽———▽; (Bull's data, dP/dx = 0 /9/),●———●; Position 1,■———■; Position 3.

References

1. Martin, N.C. and Leehey, P. : Low wave number wall pressure measurements using a rectangular membrane as a spatial filter. J.S.V. (1977), vol. 52 (1), pp. 95-120.

2. Jameson, P.W. : Measurement of the low wavenumber component of turbulent boundary layer pressure spectral density. Fourth Symposium on Turbulence in Liquids, University of Missouri, Rolla 192-200 (1975).

3. Nakamura, Y. and Tomonari, Y. : The effects of surface roughness on the flow past circular cylinders at high Reynolds numbers. J.F.M. (1982), vol. 23, pp. 363-378.

4. Panton, R.L. and al : Low-frequency pressure fluctuations in axisymmetric turbulent boundary layers. J.F.M. (1980), vol. 97, pp. 299-319.

5. Farabee, T. and Casarella, M. : Effects of surface irregularity on turbulent boundary layer wall pressure fluctuations. Symposium on Turbulence Induced Vibrations and Noise of Structures (1983), pp. 31-44.

6. Brooks, T.F. and Hodgson, T.H. : Trailing edge noise prediction from measured surface pressures. J.F.M. (1981), vol. 78, pp. 69-117.

7. Blevins, R.D. : Review of sound induced by vortex shedding from cylinders. J.S.V. (1984), vol. 92 (4), pp. 455-470.

8. Schloemer, H.H. : Effects of pressure gradients on turbulent boundary layer wall-pressure fluctuations. J.A.S.A. (1966), pp. 93-113.

9. Chase, D.M. : Modeling the wavevector-frequency spectrum of turbulent boundary layer wall pressure. J.S.V. (1980), vol. 70 (1), pp. 29-67.

Flow Effects on Measurement of the Modal Decomposition of Acoustic Field in a Hard Wall Cylindrical Duct

J.M. AUGER ; J.M. VILLE

UNIVERSITE DE TECHNOLOGIE DE COMPIEGNE
D.A.V.I. - B.P. 233 - 60206 COMPIEGNE CEDEX-FRANCE

Summary

A method of determining azimuthal and transverse wave modal decomposition in a hard wall cylindrical duct by a double Fourier-Lommel's transform of acoustic pressure in a cross section is experimented in a flow facility. It is shown that this method is suitable for flows with non uniform mean profile. But it is found that it is very sensitive to random noise which results from flow instabilities.

Introduction

To know the acoustic field inside a duct can be very useful in optimizing and guiding the choice of solutions to noise reduction problems. Therefore, the purpose of our work is to develop a method of measurement of the modal decomposition of the acoustic pressure inside a cylindrical duct. Already a method of determining directly spectral components of the transverse wave numbers based upon computation of the Lommel's transform of the acoustic radial pressure profile has been tested [1,2]. In order to improve this method and to extend its range of applications to an acoustic field with any azimuthal wave number spectrum, a double Fourier-Lomleks's transform is applied to the measured acoustic field in the duct cross section [3]. In the litterature similar works have been conducted [4,5] .

In this paper, this method is tested in a duct with an internal air flow. Assumptions of theory are limitating its applications to uniform and non time dependent flow. But interactions of flow with the acoustic field and with the immersed microphone are making them unrealistic . Hence, in order to set the range of validity of this method on the basis of criteria we are defining, experiments have been undertaken. Moreover, numerical simulation is used to study effects of flow boundary layer and of flow instabilities on modal decomposition.

Aero- and Hydro-Acoustics
IUTAM Symposium Lyon 1985
Editors: G. Comte-Bellot and J. E. Ffowcs Williams

Theoritical basis

The method presented is to determine transverse spectral components Amn and φmn of the acoustic field from the measurement of the pressure profile $P(r,\Theta)$ in a duct cross section and making the following assumptions.

a) Eigen functions of the duct are radially dependent Bessel functions of the first kind

b) Acoustic pressure is harmonically time dependent

As flow is supposed to be uniform and non time dependent, acoustic field can be expanded in series of the eigenfunctions $fmn(r,\Theta)$ where :

$$fmn(r,\Theta) = \exp(im\Theta)Jm(Bmn\ r/a) \tag{1}$$

Wavenumbers kmn, k, Bmn/a are connected by the dispersion relationship :

$$kmn^2 + (Bmn/a)^2 - (k-M\ kmn)^2 = o \tag{2}$$

From the hard wall duct boundary condition, Bmn are shown to be the zeros of the derivative of Jm.

If we applied the following double Fourier-Lommel's transform,

$$FL(P(r,\theta))(m,n) = 1/Nmn\langle P,fmn\rangle \tag{3}$$

$$Nmn = S\ (1-\ m^2/B^2mn)\ Jm^2\ (Bmn) \tag{4}$$

to the acoustic field $P(r,\theta)$ defined above, it can be demonstrated using the orthogonality properties of the eigenfunctions [6] that :

$$FL(P(r,\Theta))\ (m,n) = Amn\ \exp(i\varphi mn) \tag{5}$$

Experimental work

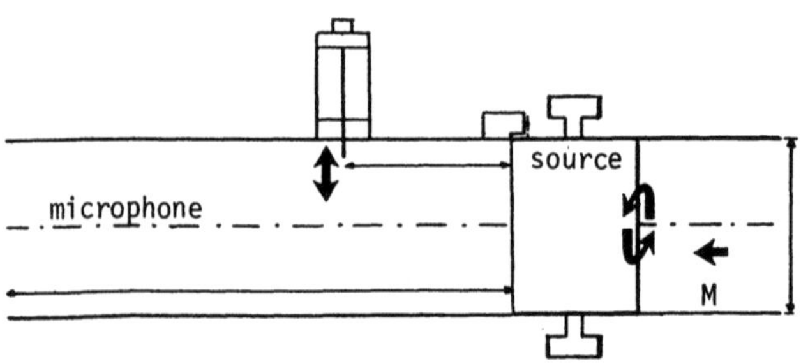

Fig. 1 - Experimental set up

Effects of flow on measurement of azimuthal and transverse wave number spectra are pointed out by means of experiments conducted with a set up described in Figure 1. A typical example is discussed with an acoustic field dominated by modes m=1, n=o and m=1, n=1 at an adimensionnal frequency ka = 5.44 in a flow at a mean Mach number M=0.3.

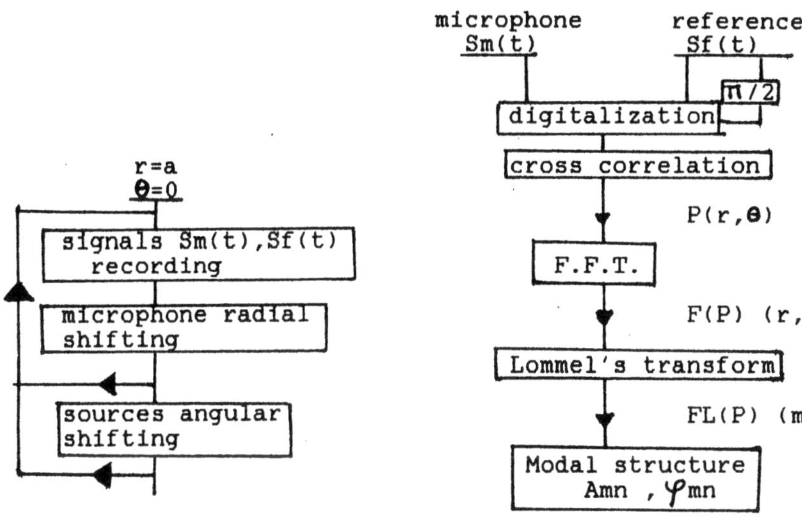

Fig. 2 - Measurement and treatement procedures

Measurement of P(r,Θ) is performed as shown in Figure (2). The microphone is traversed radially on 0.235 m radius and acoustic drivers are rotated azimuthally on 360°. Fifty values of the acoustic pressure are taken on a radius at sixteen angular locations. Four of these radial pressure profiles are plotted in Figure (3). The modulus of the Fourier-Lommel's transform of P(r,Θ) defined by equation (3) is plotted versus radial numbers Fig.4 n up to 9 with only the positive azimuthal wave number up to 3 considered. The acoustic field like for no flow case is dominated by modes m=1, n=0 and m=1, n=1 with amplitudes respectively 113dB and 128dB. However, it has to be pointed out that levels of the non propagative modes are high (up to 114dB) whatever the mode order is. Without flow, at the same frequency, propagative modes have levels 30dB higher than others. To show that this effect is due to flow instabilities radial profiles of acoustic pressure are computed and plotted after filtering all non propagative modes. Experimental curves are then smoothed (see Figure 3).

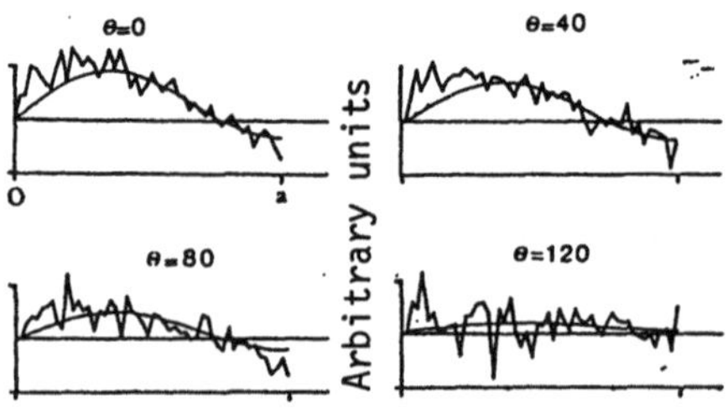

Fig. 3 - Experimental radial pressure profiles(with and without spatial filter)

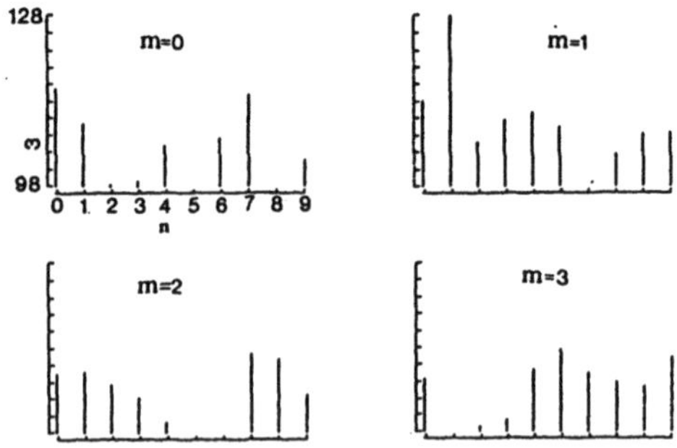

Fig. 4 - Experimental transverse wave number spectra

Numerical simulation in then carried out to study effect of noise addition on the signal and also to evaluate influence of a flow boundary layer on the results of this modal decomposition method. In order to quantify these effects, we defined the two following expressions :

$ET = 1/S <P,P>$ is total energy and,

$ETCP = 1/S \sum |Amn|^2 Nmn$ is total coherent energy of propagative modes with uniform flow (ETCPN is the same in a noisy case).

If we add to the acoustic pressure profile, as presented by smoothed curves in Figure 4, a random gaussian noise in both direction Θ and r with a standard deviation B, spectra shown in Figure 5 are obtained (spectra without noise is given by stars). These curves are similar to the experimental ones presented in Figure 4.

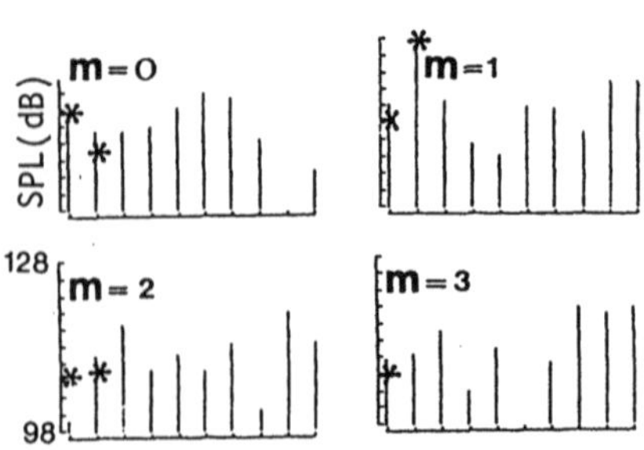

Fig. 5 - Simulated transverse wave number spectra with S/B ratio -3dB

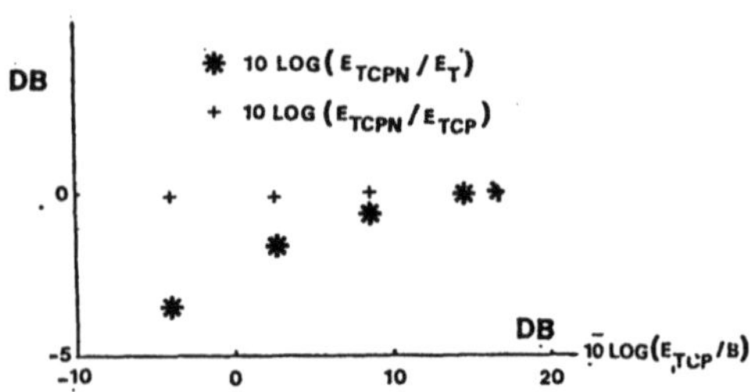

Fig. 6 - Noise effect on mode determination

Thus in order to know how precisely do amplitudes of the modes are mesured, variation of the ratio between coherent energy with and without noise versus the signal to noise ratio is plotted in Figure 6. It is shown that noise does not modify the overal energy carried out by propagative modes. But, when comparing uniform flow case (stars) and noisy case (solid lines) in Figure 5, we find that if the high level mode m=1, n=1 is not modified, other propagative modes with low level are changed up to 3dB.

Also plotted in Figure 6, is the variation of the ratio between coherent energy and total energy versus signal to noise ratio. This curve which emphasizes how the energy is spatially organized in propagative modes, can also be useful in determing the signal to noise ratio from experimental data.

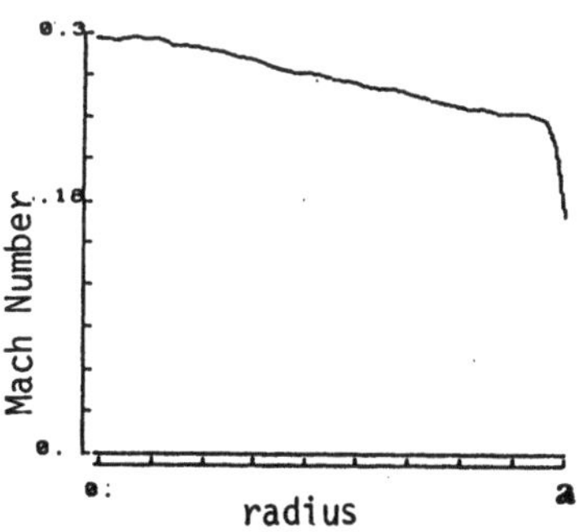

Fig. 7 : Experimental mean flow velocity profile

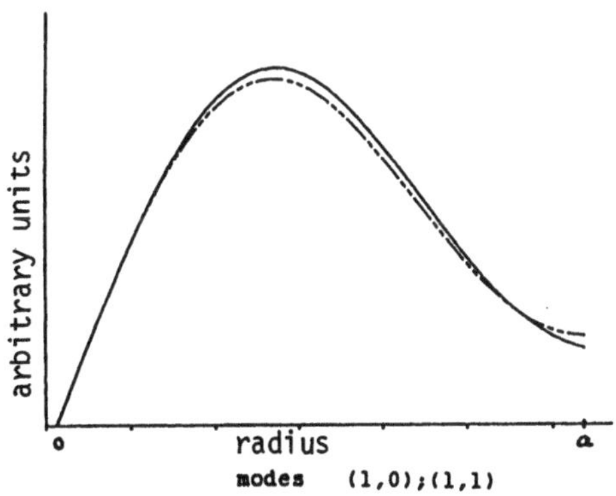

Fig. 8 : Radial acoustic pressure
—·-·— : uniform flow
——— : experimental flow

As shown in Figure 7, the experimental flow velocity profile was radially dependent with a boundary layer. Hence, in order to test the effect of a shear flow, the acoustic radial distribution function P(r) was numerically determined as in references [7,8]. A typical example of an acoustic pressure profile with the actual flow velocity profile is given in Figure 8. The effect of the shear flow is shown to be very samll.

Concluding remarks

This study has emphasized the effects of flow on a method of measurement of azimuthal and transverse wave numbers based upon a Fourier-Lommel's transform.

The influence of a non uniform flow velocity profile has been shown to be negligible. Flow instabilities due to flow interactions with the acoustic field and with the immersed microphone are simulated by a noise randomly distributed in space. They are found to reduce the possibilities of the method only in determining modes, the amplitude of which is much higher than noise. Improvement of Experimental set up should avoid this problem and makes this method available even for high velocity and non uniform flow cases.

ACKNOWLEGMENT

This work which has been conducted with the facilities of S.N.E.C.M.A. was supported by O.N.E.R.A. under contract.

Notations

M,c	=	Mach number, sound speed
f,a	=	frequency , duct radius
r,Θ,z	=	cylindrical coordinates
m	=	azimuthal wave number
Bmn,kmn	=	radial and axial wave numbers of the mode m,n
$\langle u,v \rangle$	=	$\int u\, v^* \, ds$

REFERENCES

1 J.P. PASQUALINI, "Développement d'une méthode de mesure de la structure modale du champ acoustique dans un conduit à partir d'une exploration radiale du profil de pression", Thèse de Docteur Ingénieur, Université de Compiègne, Janvier 1984

2 J.P. PASQUALINI, J.M. VILLE and J.F. de BELLEVAL, "Development of a method of determining the transverse wave structure in a rigid wallaxisymetric duct", J.A.S.A., 77(5), May 1985

3 J.M. AUGER, J.M. VILLE, "Détermination de la structure spatiale du champ acoustique par double transformée de Fourier-Lommel", 9ème Colloque d'Acoustique du G.A.L.F. Compiègne 14-16 Novembre 1984

4 R.J. SILCOX, H.C. LESTER, "Sound Propagation through a variable area duct : experiment and theory" A.I.A.A. Vol. 20, N° 10, October 1982

5 C.J. MOORE, "Measurement of radial and circumferential modes in annular and circular fan ducts" J.S.V. 1979, 62(2), 235-256

6 G.M. WATSON, "Theory of Bessel Functions", Cambridge University Press, London 1966

7 S.H. KO (1972) J.S.V. 22, 193-210 "Sound attenuation in acoustically lined circular ducts in the presence of uniform flow and shear flow"

8 M.K. BULL and N.K. AGARWAL, "Propagation of acoustic waves in fully-developed turbulent pipe flow" 11ème I.C.A. Paris 1983, Vol. 1, 295-298

The Effects of Density on Subsonic Jet Noise

D. J. SMITH and N. H. JOHANNESEN

Department of Aeronautical Engineering
University of Manchester
Manchester, England

Summary

The noise of impulsively started jets of mixtures of helium and argon ranging from pure helium to pure argon is compared with available data on the noise of hot air jets. It is concluded that there is strong evidence that the noise of subsonic jets in the rear arc depends on velocity and density only, and that the noise is independent of the way in which the density is achieved, whether by heating an air jet or by mixing cold light gases.

Introduction

During the early 1970s a series of experiments at SNECMA in France and NGTE in the United Kingdom established the effect of temperature on pure jet noise (Hoch et al. 1973). The results were presented in terms of the effect of density on total radiated power, and were incorporated in the SAE Aerospace Recommended Practice ARP 876 (latest edition 1983). Later an extensive series of experiments on the noise of hot jets were performed at Lockheed (1975).

In this paper we present a few selected results from an extensive investigation on the noise of cold jets of mixtures of helium and argon ranging from pure helium to pure argon. Results for subsonic jets in the rear arc are selected to demonstrate that the noise depends only on velocity and density, irrespective of whether the density is obtained by heating air or by mixing cold gases.

Experimental Technique

A 9.6m long 0.3m diameter very smooth cylindrical tube terminates in a 25mm diameter nozzle with an extended cylindrical

Aero- and Hydro-Acoustics
IUTAM Symposium Lyon 1985
Editors: G. Comte-Bellot and J. E. Ffowcs Williams

throat which incorporates a diaphragm station (Figure 1). The tube can be evacuated and filled with any gas or gas mixture to the desired pressure. The plastic diaphragm is trapped in contact with a mild steel ring embedded in insulating material. The ring is impulsively heated by a d.c. current and the diaphragm is blown clean out of the nozzle behind the starting shock wave. An expansion wave travels into the large diameter tube and establishes a region (1) in which conditions are constant until the reflected wave returns to the nozzle end. It is then again reflected and establishes a second region (2) of constant conditions. This process repeats and we thus get a succession of periods during which the stagnation conditions of the jet are constant. The duration of these periods depends on the gas, varying from 19ms for helium to 59ms for argon. The pressure step between each region of constant conditions is very small due to the large (144) geometric area ratio between the tube and nozzle. No evidence of jet noise forcing by this step has been found.

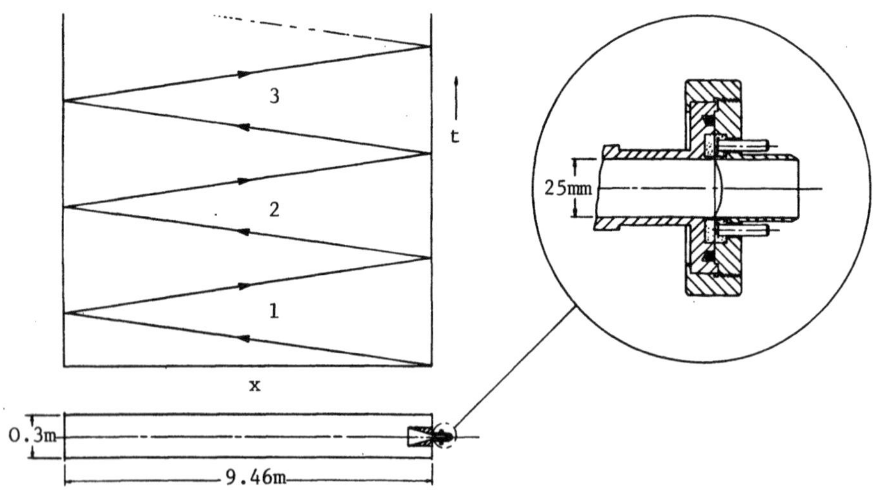

Fig. 1. Experimental rig with x,t diagram

During each run the noise of the jet which exhausts into an anechoic room is sampled digitally at microphones located at 1.25m from the nozzle exit at 15° intervals from 15° to 90°. Each sample contains 1024 points and the sampling rate is either 100k or 200k points per second. 10 or 5 samples are obtained from each run and they may be either from one or from several of the constant condition periods. Each sample gives an overall sound pressure level and is Fourier analysed using a standard FFT algorithm. Several samples from the same or repeated runs are averaged to give narrow band spectra and these are used to produce 1/3 octave spectra corrected for differences in band characteristics to give equivalent analogue spectra, so that a direct comparison with hot jet spectra is possible. All results presented were obtained at a room temperature of 295K and have been corrected to 760mm atmospheric pressure. All spectra have been corrected for microphone nonlinearity and atmospheric absorption. Noise measurements have been made on an air jet of Mach number 0.9 for varying time intervals after the impulsive start. It was found that the noise sources were sufficiently well established after only 5ms and that the far field noise was identical to that of a continuous air jet although the entire flow field including entrainment was unlikely to be fully established. No measurements are normally taken during the first region of constant conditions.

Results and Discussion

So far tests have been confined to the noise of subsonic jets in the rear arc using mixtures of argon and helium in the volume ratio 100/0, 75/25, 50/50, 25/75, and 0/100. Measurements were made at discrete values of $\log_{10}(V_J/a_A)$ where V_J is the jet velocity and a_A is the ambient speed of sound for velocities from 120m/s to sonic, although many measurements were taken at intermediate values.

In Figure 2 the power radiated in the rear arc is compared with the ARP prediction for discrete values of the logarithm (to base 10) of the ratio of the jet velocity to the ambient speed of sound. The prediction assumes that at any value of the velocity the power is proportional to the ratio of jet to ambient density to the power ω where ω depends on velocity only. Over the range of density for which the prediction can reasonably be used our results are in all cases within the quoted error band for the prediction. Our very much wider density range demonstrates that whereas the ω concept is useful for engineering purposes within

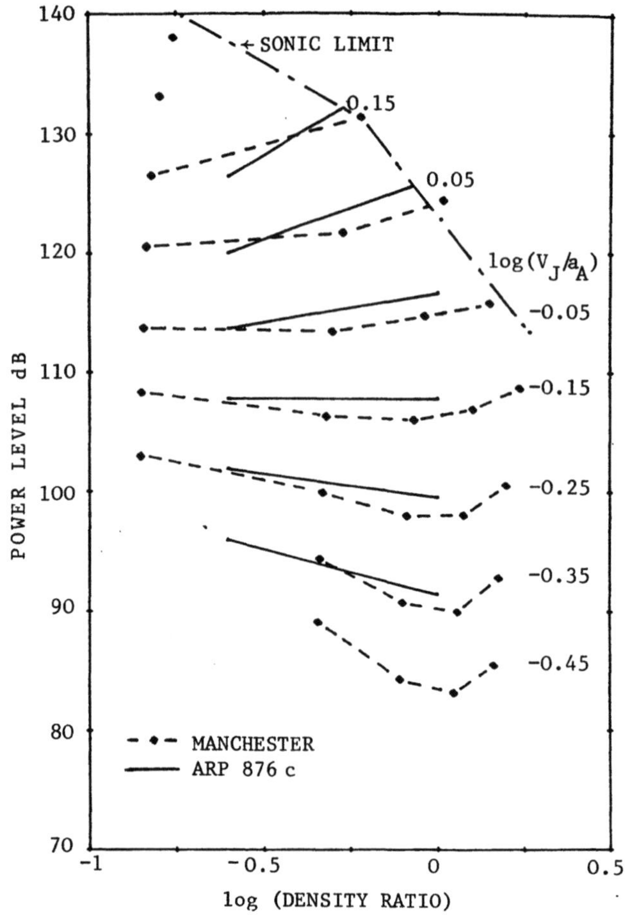

Fig. 2. Variation of rear arc power with density at different jet velocities.

the range covered by the prediction it is certainly not universally applicable and has no obvious physical basis.

The agreement between our results and those for hot jets is even more startling as demonstrated in Figure 3 which shows results at 30° to the jet axis plotted against the ratio of jet density to ambient density compared with the Lockheed hot jet results scaled and interpolated to our velocity values. This figure very strongly suggests that it is the density that matters, independently of how it is achieved. At very low densities and velocities we were unable to obtain reliable results because of

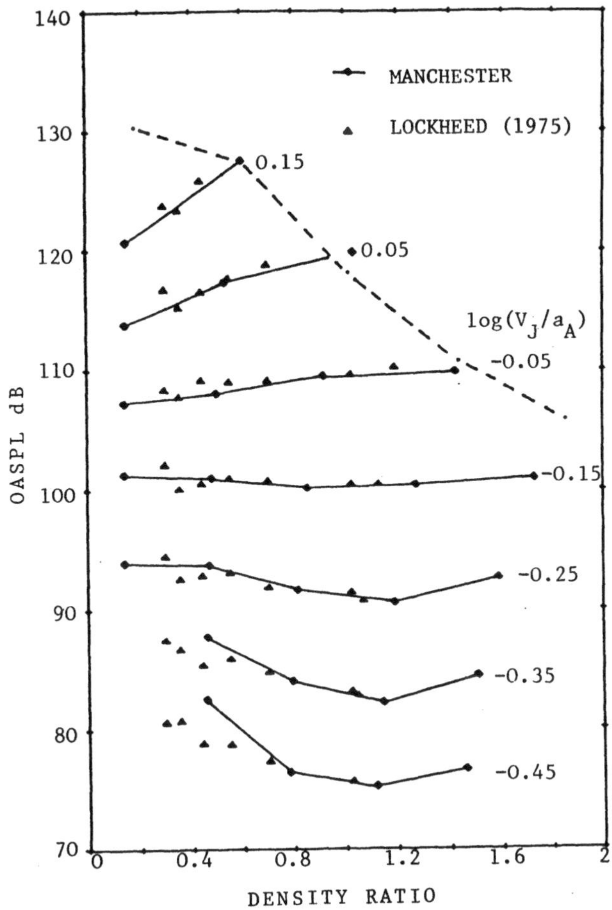

Fig. 3. Variation of overall sound pressure level at 30° to the jet axis with density at different jet velocities.

an oscillatory jet instability. In any case, in this region results appeared to depend on the precise nozzle geometry, and are, we believe, Reynolds number dependent. This effect is being further investigated.

A similar comparison of rear arc power again showed excellent agreement although the Lockheed results were now slightly higher than ours. A direct comparison of spectra is difficult because the Lockheed and our experimental points never completely overlap. It is nevertheless clear that whereas the spectra at low angles are in excellent agreement, the Lockheed results contain progressively more high frequency noise as we move from low angles to 90°. This was found to be a feature of nearly all the published results, and manifests itself in the overall sound pressure levels differing from our values by varying amounts but always being slightly higher.

There is nevertheless every indication that the simple and cheap technique which we have developed is perfectly adequate for the investigation of the effect of density on jet noise and that the modelling of temperature by modelling the density is fully justified.

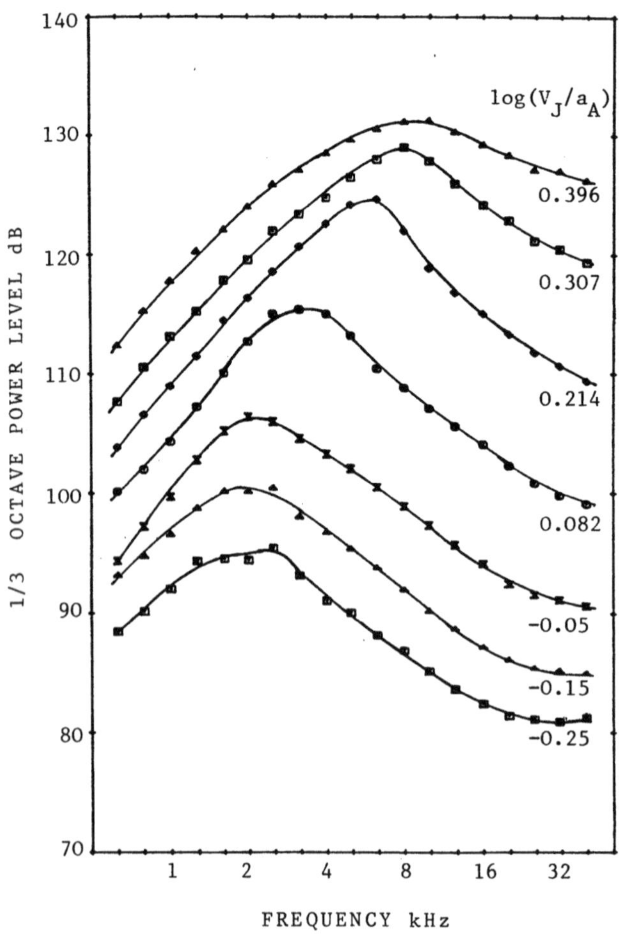

Fig. 4. Helium rear arc power spectra at different jet velocities.

This means that our results can be presented and interpreted in the same way as the many results in the literature, the main difference being that we cover a very much wider density range.

Of particular interest are the helium spectra because they correspond to air jets of a temperature of about 2000K. Figure 4 shows the rear arc power spectra over the whole velocity range. The spectra show the expected smooth increase of peak frequency with velocity. However, at low angles, as demonstrated by the spectra at 30° to the jet axis in Figure 5, the behaviour of the spectra at the higher velocities is highly irregular with sound

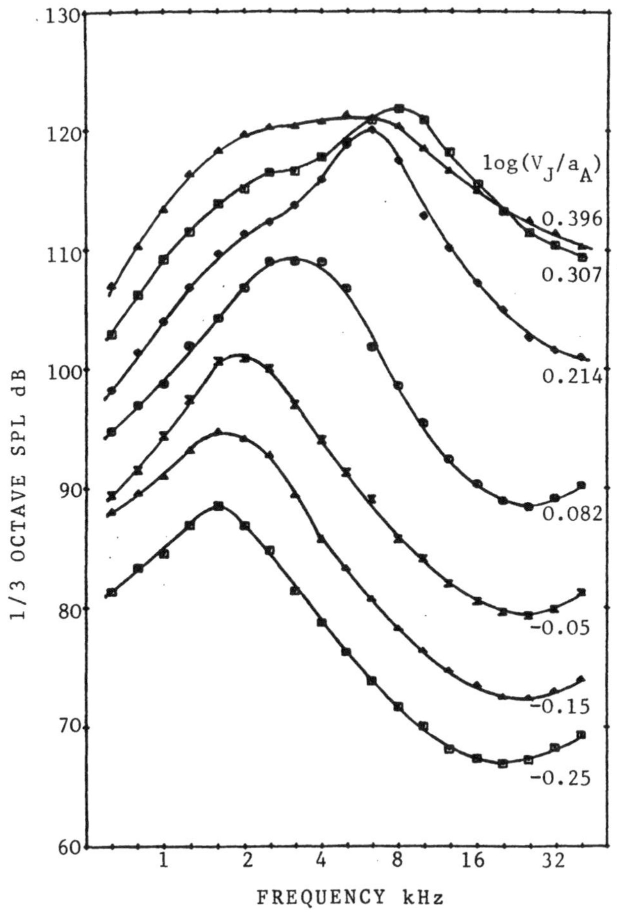

Fig. 5. Helium spectra at 30° to the jet axis at different jet velocities.

pressure level hardly increasing with velocity at the expected peak frequency but showing a large increase at frequencies below the peak.

In the past many investigators have attempted to analyse the noise at 90° at a particular density in terms of a sum of fourth, sixth, and eighth powers of velocity. A similar analysis of our data led us to two conclusions. Firstly, curve fitting techniques are unlikely to give definite answers about the relative importance of multipoles as they are extremely sensitive to small differences between data from different rigs. Nevertheless, we were able to conclude that the differences mentioned above between our own results and those of Lockheed at high angles were mainly due to the Lockheed results, in common with those of many other investigators, containing noticeably more noise proportional to the eighth power of velocity than did our results.

References

1. Hoch, R. G. et al. 1973: Journal of Sound and Vibration 28, 649-668.

2. Lockheed 1975. Turbulent mixing noise data, courtesy Dr. B. J. Tester.

Acoustics of Unstable Flows

Acoustics of Unstable Flows

D. G. CRIGHTON

Department of Applied Mathematical Studies
and Centre for Nonlinear Studies,
University of Leeds,
Leeds, England.

Summary

This review discusses much recent progress in understanding both the physical mechanisms and the mathematical description of the acoustical features of unstable flows. Attention is focussed on the receptivity problem for forced shear layers and the radiation fields accompanying the launching of an instability wave; on the evidence for significent direct radiation from the instability in both subsonic and supersonic jet flows; and on the role played by instabilities in aeroacoustic analogy theories. A number of important outstanding problems are identified.

Introduction

Flows which are in some sense "unstable" are commonplace in the scientific and technological areas covered by aero- and hydroacoustics. The word "instability" has in these fields two principal connotations. First, it suggests that there may be highly distinctive features in the acoustics and hydrodynamics of unstable flows; second, it suggests that there may be mathematical difficulties, with physical implications, in the theoretical modelling of the instabilities. Since the last IUTAM symposium on aeroacoustics, held in Göttingen in 1979 [1], there have been impressive and far-reaching developments in both of these aspects. Part of the purpose of this article is to review those developments. Significent issues remain, however, and it is equally our aim to record these, recognising (as I have argued at length elsewhere [2]) that aeroacoustics continues to act as a powerful stimulus for issues of fundamental importance in unsteady fluid mechanics.

The talk on which this article is based was centred on a wide-ranging comparison between various exact and asymptotic theories and experimental data from model rigs, using many figures taken from published papers. It would be inappropriate to include all these figures, and invidious to include only some. Consequently, this paper simply states briefly the circumstances for which good agreement is claimed, with enough detailed

Aero- and Hydro-Acoustics
IUTAM Symposium Lyon 1985
Editors: G. Comte-Bellot and J. E. Ffowcs Williams

reference to allow the interested reader to check for himself.

Acoustic-receptivity problems

In receptivity problems one studies the response of an unstable flow (most commonly a free shear layer, laminar or turbulent) to controlled excitation. The most intensively studied configurations are the plane mixing layer behind a thin splitter plate separating different uniform streams, and the circular jet emerging from a nozzle into co-flowing fluid of different velocity. Among the issues raised by these studies are the following:-

(i) how well are the hydrodynamic receptivity predictions borne out in experiment?

(ii) how well are the acoustic receptivity predictions borne out in experiment?

(iii) what evidence is there of direct radiation from the instability itself, and how does it compare with the radiation from the process in which the instability is launched from an edge or nozzle?

(iv) what role does the Kutta condition - or any other minimum singularity condition - play?

(v) what role does the causality principle have in determining the receptivity of a flow?

(vi) what is the relevance of the doubly-infinite model of the shear layers, in the absence of any edge or nozzle?

(vii) how does one deal with the multiple spatial instabilities of a given type, at a given frequency, that have been found in some flows?

(viii) does time-harmonic forcing lead to convective/spatial instability, or to absolute instability with exponential divergence everywhere with time?

(ix) how is an absolutely unstable flow controlled by the presence of a splitter plate, or other fixed boundary?

(x) what form does the nonlinear instability wave evolution take (for a convectively unstable flow)?

(xi) what are possible generalisations to other flows, other geometries, etc?

With regard to (i), theory and experiment are in excellent agreement for the plane shear layer/splitter plate configuration. The theory was first given in [3,4], and a form suitable for experimental test given in [5] together with experimental data. Theory is for a vortex sheet leaving a semi-infinite rigid plate under time-harmonic excitation by a prescribed source. The solution is unique (for zero flow on one side) with a Kutta condition imposed; this solution is also causal (i.e. would arise as the long-time limit of an initial value problem), but there are other causal solutions which do not satisfy the Kutta condition. What is predicted and compared with measurement is the downstream growth rate and the absolute amplitude of the shear layer velocity fluctuations as a function of downstream distance and of forcing strength and location. In the experiments [5] the shear layer thickness was controlled (over a 6:1 ratio) by boundary layer suction on the plate. Excellent agreement with the vortex sheet theory was found, with exponential growth over the range $0.5 \lesssim \omega x/U_\infty \lesssim 2.5$, with the exception of the thickest shear layer, for which vortex sheet theory predicted the correct growth rate but underpredicted the initial amplitude near the edge. Further experiments have since been carried out [6] and the predictions of the vortex sheet model (spatial instability, coupling to the splitter plate via a Kutta condition) verified in much further detail.

On (ii) there is extensive confirmation of the theory. A complete theory for cold jets, at any subsonic Mach number, under excitation by plane waves travelling down a circular hard-walled duct, at any forcing frequency, and with allowance for an infinite external co-flowing stream of different velocity, was given by Munt [7]. Various approximate theories have also been devised [8, 9, 10, 11], valid for low Mach number, or low Helmholtz number. All use a vortex sheet model for the shear layer; all impose a Kutta condition at the nozzle lip (or make an assumption equivalent to this) This guarantees a unique solution for the conditions so far studied, and it seems unnecessary to invoke causality - except implicitly to show that the flow is convectively unstable.

In [7] the directional distribution of the far-field intensity (i.e. the field shape) was compared very favourably with experiment for cold jets at Mach numbers up to 0.95, low to moderate Helmholtz numbers, and low Mach number external flow. Some calculations for hot jets at 830°K were also carried out, again with rather good agreement with experiment. The comparisons excluded a conical region around the jet axis in which the instability wave launched from the nozzle created a field varying exponentially with range. The field outside this cone may perhaps be thought of as generated by the coupling at the edge between the forcing plane waves and the vortical instability leaving the edge, rather than by the instability itself (which in linear parallel flow theory has no acoustic far field of the conventional kind). Of course, the acoustic and vortical fields are actually coupled everywhere outside the duct, but the receptivity of the latter to the former is dictated by the edge behaviour rather than by any local coupling elsewhere in the flow.

At the nozzle there is a conversion of almost all the acoustic energy into vortical fluctuations on the shear layer (which then amplify exponentially). If W_R is the total acoustic power radiated in the far field, W_T the net acoustic power incident on the nozzle (incident power - reflected power in the duct), it is found that $W_R/W_T << 1$ at low Helmholtz numbers. Howe [8] has shown that W_T is indeed essentially all transferred to the shear layer fluctuations; his theory is valid only for low Mach numbers, but gives excellent agreement with independent measurements of W_R/W_T by Moore [12] and Bechert [11]. The effect here is very large indeed; $10 \log_{10} W_R/W_T$ may be as large as (-25)dB at low frequency and low Mach number. It persists even to high Mach number at sufficiently low frequencies, as shown in experiment [11] and in an asymptotic low frequency theory due to Cargill [9] valid for all subsonic Mach numbers. This theory reasonably predicts W_R/W_T as a function of M, from M = 0.1 to 0.8, even for a Helmholtz number as high as $k_o a = 0.46$ (see Fig. 9 of [9]).

The pressure reflection coefficient in the duct has interesting features associated with the downstream instability and the nozzle Kutta condition. Ronneberger [13] first measured this coefficient in the presence of mean flow and showed that its magnitude increased above 1 as M increased, reaching a peak of around 1.3 at combinations of Mach number and Helmholtz number which yield Strouhal numbers in the range 0.3 - 0.5 corresponding to maximally amplified jet column instability modes. The increase above 1 does not violate energy conservation, but it is entirely associated with the edge coupling to downstream vortical fluctuations. The behaviour of the reflection coefficient is correctly predicted by Munt's theory; see Cargill [14] for details. Cargill [14] also provided an extension of his low frequency theory which gives analytical expressions for the reflection coefficient as a function of Helmholtz and Mach numbers and which satisfactorily predicts the reflection coefficient while firmly isolating the important contribution to it made by the shear layer motions.

While Munt's theory is general (within the vortex sheet/semi-infinite duct model) it needs numerical evaluation of the split functions arising from the Wiener-Hopf technique, and does not give analytical expressions from which trends and mechanisms can be seen. Such simple expressions are, however, provided by Cargill's [9, 14] low frequency theory, which is good for all subsonic Mach numbers. For example, the far field density fluctuation is

$$\rho - \rho_o = \frac{\rho_o A_J}{4\pi R c_o^2 (1 + M_o\cos\theta)(1 - M_R\cos\theta)^2} \left[\frac{\partial u_N}{\partial t}\right] \tag{1}$$

in terms of ambient density ρ_o, sound speed c_o, nozzle exit velocity fluctuations u_N, external flow Mach number M_o and relative jet Mach number M_R; θ is the angle from the exhaust, A_J the nozzle area and R the distance at the emission time indicated by [ ]. This expression can be interpreted in the light of the aeroacoustic analogies due to Lighthill and Curle [15, 16] and due to Dowling et al [17]. In the Lighthill-Curle analogy the dominant sources are monopoles and dipoles on the nozzle exit plane and quadrupoles in the unsteady downstream flow. Of these, the monopole and dipole strengths scale on the local jet density ρ_J, and one important function of the quadrupole terms is to remove that dependence and replace it with ρ_o. Thus the linearised quadrupoles make a major contribution to the sound field - and in this sense it may be said that Cargill's formula for $\rho - \rho_o$ does indeed include direct radiation from the instability wave. However the instability aspect of the vortex sheet fluctuations is absent in the low frequency limit; to leading order the fluctuations are simply convected at the jet speed, and in evaluating the quadrupole integrals Cargill assumes the fluctuations to vanish exponentially as $x \to +\infty$, rather than to amplify.

Interpretation in the terms of the Dowling-Ffowcs Williams-Goldstein analogy is also interesting. Here the quadrupole sources are nonlinear in the fluctuations and are negligible compared with the monopole and dipole terms. Expressions for these are provided by the Cargill-Munt theory on the basis of a causal Kutta condition solution for the fluctuations, and these expressions are then convolved with the bounded but non-causal Green's function which incorporates mean jet flow effects in the DFWG analogy. This recovery of the result (1) for $\rho - \rho_o$ from the two very different analogies provides an interesting test case for the DFWG analogy (non-causal Green's functions in particular).

We discuss point (iii) in more detail below. On (iv), there is now considerable direct and indirect evidence for the applicability of a Kutta condition at sharp trailing edges in unsteady subsonic flows. Some of the indirect evidence was mentioned above; the comparison between theory and experiment in the acoustic receptivity problems covers numerous facets of fields which, theoretically, are heavily dependent on satisfaction of a Kutta condition, and the comparison is generally very favourable. There is some direct evidence, in studies of pressure and velocity fields near the edges of oscillating aerofoils and near nozzle lips, but this is less conclusive, partly because of misconceptions as to what the Kutta condition represents. There is also now considerable theoretical support for the Kutta condition, in the form of studies, e.g., [18], which show that only that "outer" potential flow which satisfies a Kutta condition is, in general compatible with an "inner" multilayered viscous and nonlinear structure. Such studies also provide estimates of the frequency, amplitude and Reynolds number ranges over which the Kutta condition can be applied to the outer flow. For much further detail, see [19].

On (v) things are less clear. In the absence of any edge or boundaries the response of a convectively unstable flow to excitation is not uniquely determined unless one imposes a condition of (a) boundedness in the streamwise direction or (b) causality. The solution under (a) violates (b), and vice-versa. Dowling et al [17] argue for (a) for the Green's function of their acoustic analogy, while many authors (Tam [20] was perhaps the first in this context) have argued for (b) in deterministic calculations where it is believed that an exponential instability should be predicted. For trailing edge flows it now seems that in the simplest cases the causality condition is redundant, and superseded by the Kutta condition, which guarantees a causal solution containing a convective instability. In more complicated cases, however, this may no longer be the case, and causality may be a crucial principle (see discussion below).

On (vi), it can be shown [3] that the causal Kutta condition solution for the plane shear layer with a splitter plate and a time harmonic source contains two instability waves. One is confined to a wedge shaped region downstream of the source, and the solution in this region tends to that [21] for the doubly-infinite vortex sheet as the splitter plate is removed to upstream infinity. The second is confined to a wedge shaped region with the splitter plate edge as apex. Within this region - which ultimately includes every fixed point in space as the splitter plate recedes to upstream infinity - the field increases exponentially with distance from the edge. Therefore the "local" instability of the doubly-infinite model is overwhelmed by the instability launched from the edge, and increasingly so as the edge becomes more remote. This raises considerable doubts as to the relevance of the much-studied doubly-infinite vortex sheet - and indeed of all shear flows which are initiated at a sharp edge. The edge seems to be the primary mechanism for generating the instability (via the Kutta condition); the "local" source-induced instability (via the causality principle) is negligible. This point needs further study, because the doubly-infinite model is so widely adopted and so much simpler.

Point (vii) is very much bound up with the Kutta and causality principles, and absolute and convective instability, perhaps. Crow & Champagne [22, Figure 31] were the first to observe the possibility of more than one spatial instability of a given structure (axisymmetric) at a given frequency in their study of orderly large scale modes in turbulent jets. They were able to dismiss the second instability as probably unobservable, because of its very long wavelength. However, Michalke [23] also found two spatial axisymmetric instabilities in round jets, and showed that for hot jets the wavenumbers could coincide and lead to a complicated entangling of the two modes. Munt [7, Figure 2] gives an example of a low Mach number ($M = 0.2$), low frequency ($k_o a = 0.3$) cold jet calculation in which there are two axisymmetric instabilities (both of which have passed Munt's version of the causality criterion and both of which would therefore appear to be genuine candidates for instability waves, rather than merely complex zeros of some dispersion relation which might appear in space only in regions where they represented evanescent modes). It is not yet known how to deal with multiple modes of this kind, and no calculations of jet response to plane wave forcing have yet been carried out in parameter ranges where multiple modes occur. To each mode of instability there is an eigensolution of the jet-duct problem, and, in essence, where there is only one such eigensolution its amplitude can be fixed by a Kutta condition at the nozzle exit. How the amplitudes of two or more modes can be fixed if there is only one edge is not known. In the case of coaxial jets multiple modes arise again [24]. If the nozzle exits for the various streams are coplanar, then there is, as it were, one edge for each mode at the start of the flow region in which multiple modes can arise, and then the mode amplitudes can be fixed uniquely by Kutta conditions at each edge. If the exits are not coplanar the number of edges is less than the number of modes, and there seems to be a problem, as discussed in [24].

Point (viii) has been much discussed in the context of plasma instability, though there are at least three thorough studies [25, 26, 27] in the shear layer context. These set the standards by which instability of shear layers should be investigated in future (though the approaches here really need to be generalised as in [4, 7, 21] to allow vortex-sheet flows in which the growth rate increases linearly with frequency as the frequency becomes indefinitely large, because such flows are especially useful in yielding analytic dispersion relations and are almost the only flows for which solutions can be obtained when a boundary like a splitter plate is present). The question of whether a flow, in a given reference frame, is convectively unstable, or absolutely unstable (growth everywhere in time), or stable, is to be determined from considerations of the loci of roots in the complex k-plane (axial wavenumber) of the dispersion relation in which the complex frequency ω is allowed to vary arbitrarily in some entire half-plane. Crudely speaking, if roots move from one half of the k-plane, across the real axis integration path for a Fourier transform integral, and then remain in the other half, they correspond to convective instabilities, and time-harmonic response will follow time-harmonic forcing if the roots either behave this way or remain on one side or the other of the real k-axis. If, on the other hand, roots approach the integration path from opposite sides and "pinch" it when the frequency assumes some particular complex value, then the flow is absolutely unstable, with growth, everywhere in space, which is exponential in time, with that particular complex frequency.

In [27], Huerre & Monkewitz analyse the parallel mixing layer between two streams, showing it to be convectively unstable when the streams flow in the same direction and for small reverse flow of one stream, but to be absolutely unstable if the magnitude of the reverse flow exceeds a relatively small fraction of the other stream velocity. More interesting is the plane wake, shown by Koch [26] to be absolutely unstable over a narrow range of defect velocities - although there is no reverse flow anywhere in this case, stressing the point made in [27], that physical intuition alone cannot be used to determine the convective or absolute instability of a flow. Huerre

& Monkewitz also refer to work by Monkewitz on the hot jet, which is also shown to be absolutely unstable over some range of temperature ratios. Reference was made at the IUTAM Lyon Symposium to the possibility that this absolute instability may have led to the production of intense discrete tones in some circumstances in experiments, conducted by Professor N. H. Johannesen at Manchester, on the use of jets of helium to simulate the noise field of hot jets. The helium jets were not subjected to any deliberate forcing, and indeed the upstream conditions were quite unusually "clean". Further, the tone frequencies appeared not to correspond to any possible acoustic resonance. The link with the absolute instability is far from certain, but it needs to be pursued in detail, as it might show a clear acoustic effect of the absolute instability of a shear flow.

A more difficult point to resolve is (ix). Suppose we have a flow, like the hot jet, which is absolutely unstable when regarded as doubly-infinite. When we couple it to the duct and nozzle lip from which it exhausts, and subject it to time harmonic forcing by plane waves travelling down the duct, how does it respond? The conditions on the duct wall preclude exponential growth in time everywhere, but does such growth take place at each fixed position downstream of the exit plane? It should be possible to answer this question, with the vortex sheet model of the jet, by an extension of Munt's theory [7] and we hope to report on this in due course.

The nonlinear development of the instability downstream, point (x), (or perhaps also the linear development on a non-parallel flow) is more to do with the "direct" radiation from the instability than with the radiation occurring as the instability is launched. In experiment, however, the two cannot be distinguished in the far field (except possibly by source location techniques, though these will have poor spatial resolution in low speed flows). This should be borne in mind in considering discrepancies between experiment and the predictions of linear theories for parallel flows coupled to splitter plates or ducts. In the theories the instability wave field has an exponential character in all spatial directions, with no conventional acoustic far field, so that the predicted far field is entirely associated with the coupling between instability and boundary (this coupling including, of course, the sound sources at a nozzle exit that would be there even in the absence of mean flow).

On point (xi) we comment simply that the receptivity ideas, much studied for splitter plates and nozzle flows, can be extended to include shear layers resulting from flow separation at a leading edge and from smooth curved bodies. In the latter case, Goldstein [28] has argued that instabilities on the separation shear layer are coupled to externally imposed fluctuations through what amounts to a Kutta condition at the separation point (even though the body is smooth). This would presumably have consequences for the acoustic field, and one must expect that the acoustic field scattered or diffracted by a body with separated flow will differ substantially from that predicted on the assumption of attached potential flow.

It should be noted that all calculations referred to above deal with laminar flow and a highly idealised model of the shear layer (often a vortex sheet), while in the experiments the flow in the nozzle exit or trailing edge boundary layers may be laminar, transitional or turbulent. For many purposes - and in particular for receptivity problems involving large scale waves - the state of the boundary and shear layers is irrelevant; the turbulence averages itself, spatially and temporally, to a smooth mean with respect to the forcing and instability fields. In some circumstances, however, the turbulence and the broadband sound field are dramatically changed, even by very low levels of external forcing. The broadband turbulence and sound fields of a round jet can be amplified by as much as 8dB for a nozzle exit velocity fluctuation in the plane wave mode of as little as 0.1% of the mean velocity (see [12, 29]). Presumably the instability wave has a major role to play in this, but no mechanistic model has yet been put forward. Much remains to be done on this very important issue, on which little has been done since the author's review [2].

In summary, the hydrodynamic and acoustic consequences of controlled excitation of unstable shear layers and jets have been widely explored, and in the simplest cases there is extensive and very satisfactory agreement between experiment and theory. There are a number of directions for further work which emerge as pressing. Whether deliberate excitation offers a possible route to control of the noise of jets remains an open question; most attempts at control of high speed jets have led to an increase of the broadband components (though see [30]).

Direct radiation from instabilities

In most of the experiments cited above, the external forcing was relatively strong, producing a strong peak in the near and far field spectra (with relatively weak nonlinearly generated integral harmonics and often the broadband amplification) and often dominating the overall sound pressure level. In a few cases experiments have been carried out with very low levels of forcing, sufficient to trigger and phase-lock an instability but leading to much lower radiation fields associated with the coupling between instability and edge (according to the formula quoted earlier), and in such cases there is evidence for direct radiation from the instability. This radiation is entirely due to effects associated with nonlinearity and non-parallel flow if the phase speed of the instability is subsonic; if the phase speed is supersonic, Mach-wave radiation along the characteristics is possible, but nonlinearity and nonparallel flow effects are still important in determining the angular width of the Mach-wave beams.

The experiments of Laufer & Yen [31] on a very low speed jet flow are of particular interest, revealing quite unsuspected features of the radiation from instabilities - and possibly of low speed jet noise generally, if it is accepted that instabilities are a model for large-scale orderly structures which are now believed to be responsible for jet noise production around the spectral peak. The typical Mach number in these experiments was 0.1, diameter Reynolds number 10^5, and the frequency f_o of low level exit plane forcing was chosen to match that of maximal spatial amplification on the initial shear layer. Roll-up of the instability wave into a tightly wound vortex ring was found at a well-defined downstream location x_o, and consecutive such vortices "paired" first at x_1 to form a larger vortex with half the passage frequency, this being followed still further downstream by a second, and even third, vortex pairing, at distances x_2, x_3. All three vortex pairings are complete in less than one diameter downstream from the nozzle exit; at greater distances it is not possible to control the flow, which is then fully turbulent, by exit plane forcing. For $x \lesssim D$ the forcing is able to phase-lock the flow, much of the natural background turbulence being suppressed and most of the energy being contained in a few well-defined tones. At stations around x_o the main tone is at f_o, with weaker subharmonics at $f_1 = f_o/2$, $f_2 = f_o/4$ etc, and much weaker integral harmonics nf_o. Further downstream the fundamental f_o tone decays and f_1 grows, dominating the spectrum in the near field around x_1. This component then decays and gives way to f_2 around x_2, etc.

To the vortex pairings in (x,t) space corresponds a growth and decay in x of the various frequency components f_0, f_1, f_2 etc. Laufer & Yen found that the evolution of each of these was approximately linear (near field pressure and shear layer velocity fluctuations were linearly related, as was each to the forcing amplitude), and each had a phase structure with a virtually constant convection or phase speed (and hence wavelength). The amplitude distribution with x was found to take the form

$$\overline{p^2}(x, f_n) = \overline{p^2}_{max}(f_n) \exp\{-2 \left(\frac{x - x_n}{\lambda_n}\right)^2\} , \tag{2}$$

with x_n as defined earlier and λ_n the wavelength for frequency f_n (n = 0, 1, 2). The Gaussian form fitted to the data will be seen to be highly significant, though the data are confined to quite modest values of $|x - x_n|$ and the measured growth and decay of any one component covers a dynamic range of only about 10dB, so that the data could have been fitted also by a curve with algebraic variation.

The far field pressure spectra are dominated by tones at f_o, f_1, f_2 without any Doppler shift, indicating that sound is generated as the wave patterns amplify and decay through the fixed locations x_n. As a function of angle θ from the exhaust, and convection Mach number M_c, the directivity of each of these tones is extraordinary; Laufer & Yen fitted it, over the three tones and a range of forcing levels, by

$$\overline{p^2}(\theta, f_n) \sim \exp[90M_c \cos\theta] . \tag{3}$$

Crighton & Huerre [32] call this type of variation "superdirective". At the values of M_c concerned ($M_c \sim 0.05$) the field cannot be expanded in a multipole series of terms like $\cos^m\theta$. This goes against conventional low Mach number aeroacoustics, for the largest length scale in the "source" is the wave packet length, which from (2) can be taken as $2\lambda_n$, and the ratio of this to the acoustic wavelength is $2M_c$, which is close to the jet Mach number M, and so small here that the entire instability wave region must be regarded as "compact" in the usual sense. Lighthill theory [15] would then predict quadrupole directivity $\cos^4\theta$, possibly modulated by some function representing acoustic interference in the azimuthal direction around the thin annular shear layer (the jet diameter is not a significant dynamical scale for these high frequency instabilities, whose properties are essentially those of a plane flow scaling on the shear layer thickness). No trace of quadrupole behaviour or azimuthal interference is seen in the experiments of

[31].

The unsteady flow structure in the Laufer & Yen experiments is sufficiently simple that one could attempt either to predict it theoretically, or to curve-fit the data rather well, and from a functional description of the flow to predict the sound field from aeroacoustic theory. This was done in [33], where it was argued that around station x_n the mode f_n dominant there could be modelled as an almost neutrally stable mode with amplitude variation slow (compared with the phase variation) and controlled by mean flow divergence and nonlinear effects. A rational calculation allowing for linear shear layer spread gives the linear terms dA/dX + XA in the amplitude equation

$$\frac{dA}{dX} + XA - \Gamma|A|^2 A = 0 \tag{4}$$

while the nonlinear effects are simply modelled by the cubic term of familiar form (a rational deduction of this term will follow in future work [34]). Here nondimensional variables are used, the stream function being written as

$$\Psi(x, y, t) = \text{Re}\,\{A(X)\ \text{sech}\ y\ \exp i(x - t)\} \tag{5}$$

with x, X fast and slow variables in the axial direction, y normal to the shear layer, and where sech y is the transverse mode shape provided by Rayleigh's equation for a neutral mode on a tanh shear layer. The nonlinear equation (4) has a solution with many qualitative features seen in experiments [34]. The linear version was studied in [33] as appropriate to the Laufer-Yen experiments; it immediately predicts the Gaussian envelope of eq. (2), with a width which is the same fixed multiple of λ_n for each n. A standard estimate for the growth rate of a plane turbulent shear layer leads to a prediction identical to that of eq. (2), except that the coefficient 2 in exp is replaced by (2.2). In a private communication Laufer also reported that the mode shapes in y of the velocity fields derived from eq. (5) agreed well with hot wire data that are reported in [31].

Thus equations (4) and (5) seem to predict the flow structure quite well - but that is not essential for what follows; one can simply claim to have followed the thrust of all aeroacoustic work in now trying to calculate the sound field more or less directly from measurements of an unsteady flow field. The calculation has four stages. First, the individual quadrupole fields

T_{ij} are calculated from the approximation [15] $T_{ij} \underset{-}{\sim} \rho_o v_i v_j$ where v_i, v_j are the incompressible velocities derived from eq. (5). Appropriate quadrupole differentiations are carried out. Second, an instantaneous integration of T_{ij} normal to the jet shear layer is carried out; this is permissible because the flow field has no phase structure in the normal direction and the shear layer is very thin indeed on the acoustic scale. Third, T_{ij} is integrated in the axial direction, but allowance is made for retarded-time differences in this integration. This seems an unnecessary complication since, as pointed out earlier, the entire wave packet (as measured by its "standard deviation" width) has a scale of not more than one tenth of an acoustic wavelength - but the calculation reveals that this measure of "compactness" is inadequate here and that axial retarded-time differences are crucial. Finally, T_{ij} is integrated in the azimuthal direction, the shear layer forming a thin annulus whose diameter is comparable with the acoustic wavelength. There will therefore be substantial "azimuthal interference" between sources at different azimuthal locations, expressed of course by Bessel functions.

The upshot is a prediction

$$\overline{p^2}(\theta, f_n) \sim \exp(52 M_c \cos\theta)\,(\cos^2\theta\, J_o - \sin 2\theta\, J_1)^2 \qquad (6)$$

in which the Bessel functions have argument $(\omega_n a \sin\theta / c_o)$ with $\omega_n = 2\pi f_n$, a the nozzle radius and c_o the sound speed. In (6) the exponential is a superdirective "antenna factor", of the same form as in the measured eq. (3), but with too small a coefficient (52 in place of 90) despite the fact that the predicted wave packet length is very close to that measured. The functions $\cos^2\theta$, $\sin 2\theta$ correspond to longitudinal (x - x) and lateral (x - y) quadrupoles (the y - y quadrupole is much weaker) and the azimuthal interference factors J_0, J_1 have argument close to $(\sin\theta)$ and so cannot be approximated by their low frequency forms, 1 and 0 respectively.

The overall prediction is nothing like (3). Only some 16dB difference is predicted between $\theta = 0$ and $\theta = 90^o$, while 34dB is measured (in typical cases), while the combination of quadrupole and Bessel terms actually leads to a vanishing of $\overline{p^2}$ at about $\theta = 40^o$, with a large dip below the measurements over a wide range of angles around 40°. Apart from the overall problem of why the acoustic prediction is so poor, given, apparently, such a good description of the flow field, the following issues seem to be important:-

(i) Are theory and experiment describing the same thing?

(ii) Why is the superdirective effect underpredicted?

(iii) Why is there no trace of quadrupole or azimuthal interference functions in the experiments?

(iv) How can such a small ("compact") source have such a highly directional radiation pattern?

(v) Are these issues peculiar to this well-ordered low Mach number flow, or are they generally significant?

On the last point nothing is known; inasmuch as some of the difficulties reflect a delicate balance (e.g. the dip at 40° in the prediction) it is unlikely that they will arise in a high speed flow with high turbulence levels, but other difficulties may have wider implications.

On (i), recall that little difference is seen in [31] between the behaviours for components f_0, f_1 and f_2, though the only significant forcing was at frequency f_0. One might conclude that the radiation field did not contain any significant contribution from the forcing at the nozzle exit - and the theory leading to (6) excludes such effects. On the other hand, a sound field is generated by the coupling of an instability to an edge or nozzle even if there is no external forcing, and this field might be present in the experiment and ignored in the theory - but it is hard to see how such a nozzle-based source could have the kind of superdirective field observed in [31] and presumably generated by some spatially extended source with phase structure.

Point (ii) may perhaps be answered by saying that the prediction and near and far field measurements simply illustrate the fact that the level and directivity of the far field are extraordinarily sensitive to details of the source - and what matters, of course, is the Fourier component of the source at wavenumbers less than ω/c_o, generally equivalent to the space integral of the source. The measurements do not extend to sufficient distances to allow an accurate determination of this, and the fitting of the Gaussian expression (2) may not be good enough.

On (iii) we can see no reason why an expression of the form $(\cos^2\theta\, J_0 + \beta \sin 2\theta\, J_1)^2$ should not apply for an axisymmetric mode on a round jet. The choice $\beta = -1$ which leads to the undesirable dip depends on the mean velocity profile and on the shear layer divergence rate adopted. Monkewitz (private communication) has argued that the time mean profile is inappropriate and that any particular phase front of the instability sees a "mean" velocity which is piecewise uniform and of constant width, the sections being separated by the vortex pairing locations x_n at which the shear layer width abruptly doubles. The term XA for linear divergence in eq. (4) is then replaced by some g(X)A and the envelope is no longer a pure Gaussian. However, the effect in the far field is to destroy the cancellation dip and to give a directivity dominated by an antenna factor, with rather weaker features associated with quadrupole and azimuthal interference terms. Further detail on this will follow in [34]. Agreement with experiment is dramatically increased by this modification. Other physical mechanisms (e.g. departures from axisymmetry caused by random turbulent convection of the vortex rings or instability waves) seem not to be able to make such significant changes.

Point (iv) was considered in [32] on the basis of model problems which throw up a number of important points. Consider linear fluctuations in a two-dimensional flow. Then outside the shear layer the pressure fluctuations p(x, y, t) satisfy Laplace's equation, and the shear layer motions can be simulated by a boundary condition on p, on y = 0 say, in which the pressure has an amplitude-modulated travelling wave form. Thus we study

$$\frac{\partial^2 p}{\partial x^2} + \frac{\partial^2 p}{\partial y^2} = 0 \qquad (7)$$

in $y > 0$, $-\infty < x < \infty$, with

$$p(x, 0, t) = A(X) \exp(ix - it) \qquad (8)$$

and $X = \varepsilon x$. Near to $y = 0$ we have the well-known exponential decay of fluctuations in directions parallel to the wave crests,

$$p(x, y, t) \simeq A(0) \exp(ix - y - it). \qquad (9)$$

Far away we see a dipole field

$$p(r, \theta, t) \sim \frac{1}{\pi} \frac{\sin\theta}{r} F_0(t) \qquad (10)$$

whose strength

$$F_0(t) = \int_{-\infty}^{+\infty} A(X) \exp(ix - it)\, dx$$

is the net instantaneous force applied to the boundary y = 0. The question is "how, for $\varepsilon << 1$, does the transition take place, from the exponential decay of (9) in the near field to the algebraic decay of (10) in the (incompressible) far field, and how does the nature of the transition depend on properties of A(X)?" In [32] the question was answered in relation to three different A(X) for which all the calculations could be carried out explicitly.

For Case A: Exponential, $A(X) = \exp(-|X|)$, it was found that exponential decay takes place over $y \lesssim 1$ and is immediately followed by algebraic decay. For Case B: Algebraic, $A(X) = (1 + X^2)^{-1}$, exponential decay continues, with modification, out to $y = O(\varepsilon^{-1})$, where there is a thin "boundary layer" across which the decay changes to algebraic. For Case C: Gaussian, $A(X) = \exp(-X^2)$, exponential decay persists to $y = O(\varepsilon^{-2})$, where there is a rather wide transition region of scale ε^{-1}, beyond which algebraic decay follows. Thus the penetration depth is $O(1)$, $O(\varepsilon^{-1})$ and $O(\varepsilon^{-2})$ for the three cases, respectively, in units of the wavelength of the boundary pressure fluctuations. The significant aspect of the penetration depth is that it is the depth into the field over which the pressure retains some trace of the phase structure imposed at $y = 0$; beyond this depth that structure is lost and the field is characterised only by the coarse measure F_0.

To see how these features govern the acoustic directivity, consider the wave equation instead of (7), with the same boundary condition. There is a second small parameter, a compactness ratio δ, the ratio of the boundary wavelength (representing the scale of the hydrodynamic instability wave in a shear layer) to the acoustic wavelength at the same frequency. Since we are simulating a low Mach number flow, δ is small, of the order of the flow Mach number, but δ and ε (inverse of the wave packet scale) are independent. The limit $\delta \to 0$ enables the acoustic field to be found by matching the outer asymptotics of the inner incompressible field just described to the inner asymptotics of a wave solution to the Helmholtz equation.

For Case A, the incompressible field has algebraic decay for $y > O(1)$ and therefore matches a multipole wave field (dipole) whatever the relation of ε to δ; hence Case A cannot display superdirective features whatever δ/ε. For Case B the same applies provided the overlap region between inner and outer fields lies beyond the penetration depth, i.e. provided $\delta^{-1} >> \varepsilon^{-1}$, which states that the whole wave packet must be compact relative to the acoustic wavelength. If, however, $\varepsilon^{-1} \gtrsim \delta^{-1}$, the imprint of the boundary phase fluctuations is maintained in the overlap region, and then to infinity in the acoustic field. Thus for Case B, a superdirective field will result if $\varepsilon^{-1} \gtrsim \delta^{-1}$, i.e. if the wave packet is longer than one acoustic wavelength. For Case C it follows, similarly, that the acoustic field will be superdirective if $\varepsilon^{-2} \gtrsim \delta^{-1}$, but of dipole form if $\varepsilon^{-2} << \delta^{-1}$. Here, however, the wave packet can still be compact ($\varepsilon^{-1} << \delta^{-1}$) while satisfying the condition $\varepsilon^{-2} \gtrsim \delta^{-1}$ for a superdirective field.

Only Case B accords with normal intuition. If the wave packet is compact the field is dipole; if non-compact the field is superdirective. In Case A the field cannot be superdirective no matter how long the wave packet, while in Case C the wave packet can occupy an arbitrarily small fraction of an acoustic wavelength and still produce a superdirective field (e.g., take $\varepsilon^{-1} = \delta^{-\frac{1}{2}}$, $\varepsilon^{-1}/\delta^{-1} = \delta^{\frac{1}{2}} \to 0$).

If some unease is felt about these matching arguments, the reader can easily obtain the above results by writing down the exact solution to the Helmholtz equation

$$\frac{\partial^2 p}{\partial x^2} + \frac{\partial^2 p}{\partial y^2} + \delta^2 p = 0 , \tag{11}$$

with boundary value (8), in the form of a Fourier integral, and using stationary phase to obtain the acoustic far field, as was done also in [32].

These results indeed show that a small source of suitable structure can certainly have a superdirective far field. The Gaussian wave packet provides one example, and one that has been "seen" in experiments in which a small source did generate a superdirective field. In the Laufer-Yen experiments [31], the parameters ε^{-2} and δ^{-1} are close in numerical value, and the theoretical condition for superdirectivity of a Gaussian wave packet is satisfied. We hope to give more general conditions, in future work, for superdirectivity. For the present the message is first that these features are suppressed by approaches which ignore retarded-time variations on the usual criteria for compactness; and second, that approaches to the calculation of sound from instabilities on shear layers using matching arguments will be very difficult to execute correctly. The examples A, B, C show a tremendous variation in their near field structure and transition from exponential to algebraic decay. For an A(X) generated by some instability calculation (and possibly only available numerically) the near field structure would be very hard to determine, as would the penetration depth which, as we have seen, may be large compared with the total extent of A(X). Further, the actual levels of p around the penetration depth will be very small. In Cases B, C, these levels are of order $\exp(-1/\varepsilon)$ and $\exp(-1/4\varepsilon^2)$, respectively.

In summary, the experiments of [31] show strong radiation from instabilities in a low speed flow, with features not previously suspected (there are also some internal inconsistencies in the results reported in [31] which need further study - for example, a linear relation between shear layer velocity fluctuations and near field pressure, but a quadratically nonlinear one between those velocities and the far field pressure). The data pose severe difficulties for theory. An adequate prediction of them has not yet been obtained, but some theoretical conclusions have been reached which satisfactorily explain the most distinctive aspect of [31] - the superdirective field.

If the flow Mach number is not small it seems to be possible to avoid some of these pitfalls, at least in a formal sense, through the approach of Tam & Morris [35]. They begin, for a plane compressible shear layer with (small) growth rate measured by ε, by deriving a multiple scales expression for the pressure similar to eq. (5),

$$p \sim \sum_{n=0}^{\infty} \varepsilon^n \hat{p}_n (y,X) \exp(i\theta(x) - i\omega t) \qquad (12)$$

where $X = \varepsilon x$ and $\theta(x)$ is a complex phase function such that $d\theta/dx = \alpha(X)$ with α the wavenumber of locally parallel flow at x. All terms in this expansion decay exponentially with y (unless the instability wave has supersonic phase velocity) and there appears to be no radiation field. However, Tam & Morris show that $\hat{p}_n \sim y^{2n}$ as $y \to +\infty$ and so this multiple scales expansion is not uniformly valid for large y. Instead, a solution of the wave equation must be used there, and the solution will be fixed once p is given as a function of x, X on some fixed value of y, y = 0 say. The multiple scales solution can be used to provide this boundary data, in the form $A(X) \exp(i\theta(x) - i\omega t)$, and then an exact solution of the wave equation is written down in Fourier integral form, with the far field determined by stationary phase methods. Determination of A(X), $\theta(x)$ and the transform of $A(X) \exp i\theta$ at the acoustic wavenumber $(\omega/c_o)\cos\theta$ (θ the angle from the downstream direction) must all be carried out numerically, with the usual deformations into a complex y-plane for ranges of x in which the wave is decaying.

If the mean flow Mach number is not small, the acoustic and hydrodynamic scales are comparable, and a significant function of the "energy" in the transform

$$C(k) = \int_{-\infty}^{+\infty} A(X) \exp(i\theta(x) + ikx)dx$$

will lie in the radiating range, $|k| < \omega/c_o$, of components with supersonic phase velocity and, moreover, such components should be correctly predicted by the multiple scales analysis and the compressible flow instability eigenfunctions. Then the Tam & Morris approach seems to be viable, and was used in [35] to predict instability wave growth and decay and the directivity of the associated noise field for plane shear layers at Mach numbers 0.75, 1.25 and 1.5 and for three Strouhal numbers spread around that of maximal spatial amplification on the initial shear layer. Figure 15 of [35] shows, however, that even at M = 0.75 the spectral energy in the radiating range is only a small fraction of that elsewhere, and there are difficulties in this approach which presumably become rapidly greater as the Mach number decreases further. A comparison of the predictions of this approach with the data in [31] would be valuable, as would a comparison with the Lighthill-type prediction in [33]. No comparisons with experiment were given in [35].

For jets with supersonic perfectly expanded exhaust flow there is considerable experimental evidence - at low to moderate Reynolds number - for the dominance of instability wave modes in the flow structure and for significant radiation from them. Most of the evidence - indeed, the only convincing evidence involving measurements of the flow and the acoustic field on the same rig - is to be found in a series of publications from McLaughlin and colleagues, starting from low Reynolds numbers (by expanding the jet into a very low pressure chamber) and in the latest studies [36] dealing with jets at Mach numbers around 2 and Reynolds numbers just below 10^5. Acoustic data are restricted to the near field, but this is much less of a constraint than would be in the case at low Mach number. The near field data show a discrete tone structure when the jet is gently excited by a glow discharge at the nozzle exit, and azimuthal traverses reveal that only three azimuthal modes are significantly excited, with mode numbers n = 0, ± 1, and the relative phases of these components were also determined. In absolute levels the pressure fields radiated by these three modes are intense, and the typical overall level is comparable with that of jets of the same Mach number at much higher Reynolds number. McLaughlin and his collaborators argued that this fact, and the similarity in directivity of the tonal and broadband fields at low and high Reynolds numbers, respectively, indicate that the same mechanism is at work in both cases, and that increase of the Reynolds number, and of jet turbulence levels, simply broadens the spectrum associated with the instability waves, without any significant increase in the overall sound level.

Some comparisons with the experimental data of [36], both with regard to centreline velocity fluctuation and near field pressure contours, for Strouhal numbers 0.2 and 0.4 and a Mach number of 2.1, were made by Tam & Burton [37, 38]. In the theoretical part of this work the instability wave - sound field problem was tackled on a matched expansion basis. Linear theory is again taken, and the small parameter is the shear layer divergence rate ε. The Mach number must not be small - so that there are only two significant scales in the streamwise direction, the instability wavelength and the wave packet scale, giving the multiple axial scales x and $X = \varepsilon x$ as before. In the Tam & Morris work [35] it was recognised that the multiple scales expansion, equation (12), is not valid for large y, where in fact a wave equation must hold. Tam & Morris effectively used eq. (12) to provide boundary data, on some plane y = constant, whose Fourier transform in x could be used to find an exact solution of the wave equation. In [37], Tam & Burton modify this approach, arguing that if $Y = \varepsilon y$ is taken as an outer coordinate (and X is taken as the significant streamwise coordinate), then solutions to the wave equation in (X, Y) should be matched, rather than patched, to solutions of the multiple scales instability wave type in the inner variables (X, y). It was shown that the two approaches are equivalent provided the "supersonic damped wave" phenomenon does not arise - though of course the matching approach of [37] is superior, in that higher order terms can be calculated routinely, as indeed they were in [38], which deals with applications to the circular jet.

The comparisons made in [38] with the experimental data of [36] are generally favourable, though the flow structure comparisons are confined to the axial variation of centreline mass velocity fluctuation at two Strouhal numbers. Mean velocity profiles used in the instability wave calculations were taken as those measured in the presence of the weak forcing, rather than in the unforced case, a step argued to take out the most important nonlinear effect. There is good agreement between predicted and measured shapes of the near field pressure contours. Here there are three modes contributing, the phases between which are known, though the amplitude of the $n = \pm 1$ components relative to that for $n = 0$ is not. This relative amplitude is used to fix a single point on one of the theoretical constant pressure contours, and all are then fixed, giving measured and predicted field shapes and contour spacings in very close agreement. Unfortunately it was not possible to acquire genuine far field data on the rig used in [36], and the number of test conditions is extremely limited. However, there should be little difficulty in correctly inferring far field acoustic fields from this description of the flow and near field at Mach numbers around 2 (and cold flows - otherwise the problem of absolute instability may arise).

The "supersonic damped wave" problem of [37] deserves comment. If $\alpha = d\theta/dx$ is the local wavenumber at any station, then in the early stages of instability wave evolution downstream the wave is amplified ($\mathrm{Im}\ \alpha < 0$) and, in a very high speed flow, may well be supersonic relative to the ambient conditions ($0 < \mathrm{Re}\ \alpha < \omega/c_o$). As x increases, $\mathrm{Im}\ \alpha \to 0$ from below, and $\mathrm{Re}\ \alpha$ generally increases. If $\mathrm{Re}\ \alpha > \omega/c_o$ when $\mathrm{Im}\ \alpha$ reaches zero the wave is subsonic whenever it is damped, the instability wavenumber α stays away from the branch cut emanating from ω/c_o, and there is no problem (other than the usual problem for damped waves - see [35]) in following the whole growth and decay process. If $\mathrm{Re}\ \alpha < \omega/c_o$ as $\mathrm{Im}\ \alpha \to 0$, the wavenumber α runs into the cut - which means that then (and for greater x also) there is no instability wave eigensolution which either vanishes or represents outgoing waves at large transverse distances. Tam & Burton argue that this causes no concern, because the instability wave solution is irrelevant at large distances, where the outer solution applies and which can be made to satisfy appropriate conditions at infinity. They do not, however [37, § 7 "Resolution of the supersonic damped wave problem"], indicate what the inner and outer forms are at an x for which this occurs, and content themselves with asserting that in the integrals, which determine the amplitude variation A(X), the eigenvalue α and the associated eigenfunction are to be obtained simply by analytic continuation onto the second ("non-physical") sheet of the Riemann surface.

One is left wondering how suitable matching with an outer field can be achieved with an inner field which (on the second Riemann sheet) is exponentially increasing in the transverse direction. This, fortunately, is an issue which has arisen elsewhere - in the theory of fluid-loaded structures - and is not quite as new as claimed by Tam & Burton. It is discussed in [39], among other places, and has frequently given rise to misconceptions. There are really three points. First, exponential increase in y is easy to see. If the phase velocity is supersonic, the radiation can be thought of as Mach wave radiation in which, in a two-dimensional problem, the surface pressure say (at $y = 0$) is propagated unchanged into the ambient fluid along the characteristic lines. Consequently, the signal reaching an observer at an increased value of y originates on the surface at a smaller value of x, and in a wave damped in the x-direction the field increases exponentially as x decreases. Second, in any realistic physical situation the instability develops downstream of some edge, and as a result the acoustic far field associated with a wave of the type just described is contained within a downstream sector with the edge as apex. The field decreases exponentially in the x-direction, and in fact along any ray y/x = constant within the sector, but increases exponentially with y until the sector boundary is reached, beyond which the field vanishes. Such a field, which can develop in the coupled motion of a fluid and an elastic plate, for example, is called a "supersonic leaky wave". Conditions as $y \to \infty$ are obviously not violated by this wave as long as it is confined within what is effectively the Mach cone or wedge of some upstream edge, but it creates difficulties of interpretation in any doubly-infinite "free wave" problem. Third, one can argue that the location of the branch cuts in the complex wavenumber plane is not fixed in the way adopted by Tam & Burton, and so descent onto the second Riemann sheet can be avoided by re-routing the branch cut. Naturally the sum of the pole contribution and the branch line integral - i.e. the total response when some definite forcing is imposed - will be the same however these branch cuts are arranged, but it may not be possible to assign an unambigous meaning to the pole term in isolation, and that pole term (i.e. the instability eigenfunction) is all that is considered in an instability problem with no imposed forcing. These points could do with further consideration, because the flow speed does not have to be unrealistically high for this damped supersonic wave problem to arise; Tam & Burton give, as an example, the case of a cold jet at Mach number 1.5, where the problem certainly arises at Strouhal number 0.15 and mode number $n = 0$.

Instabilities and aeroacoustic analogy theories

In an aeroacoustic analogy, sound emission is calculated from a prescription in which certain linear and nonlinear terms are regarded as known source terms for some kind of forced wave equation with a linear wave operator incorporating certain aspects of the propagation process. What if the linear operator itself can sustain freely developing instability modes - as is the case in the Lilley theory, attempting to incorporate effects of mean velocity and mean shear in the source region (for a thorough recent review, see [40])? Then an arbitrary source will, in convolution with a causal Green's function, generate a convective instability, as predicted, for example, in [21] for the specific case of causal vortex sheet response to an impulsive localised source. But should this instability not already have been included in the source description, rather than as the unbounded response to a prescribed bounded source? This difficulty, the admission or suppression of instabilities in source-response theories incorporating mean shear, has arisen in one way or another in all such theories, and the position is still not clear.

One way round the difficulty is to argue, as in [17], that causality is not an appropriate requirement, and that a spatially bounded non-causal ("weakly causal" in [17]) Green's function should be used. The claim is that the turbulence is not the causal source of the sound, for the sound triggers instabilities, leading to further turbulence and a modification of the flow, to the extent that the flow is highly sensitive and responsive to its own sound field. On the other hand, in specific deterministic problems like the transmission of sound down a duct with flow, or the excitation of coupled waves in an acoustic liner or an elastic plate adjacent to flow, causality seems a proper requirement. Indeed, it is a necessary requirement, for a unique solution, in the flow of uniform fluid over an elastic plate, examined in [25], and probably generally, unless the configuration involves the production of a shear layer at an edge, where a Kutta condition seems to be the overriding requirement.

An illuminating discussion of these issues is given by Möhring & Rahman [41]. They point out that if the rearrangement of terms in an analogy is exact (as in Lighthill's, [15], for example) then, since in reality no unbounded exponential growth is actually observed, the source and causal Green's function must be orthogonal in an appropriate sense. Alternatively, the source must have zero spectral energy in the spectral regions corresponding to free modes of the linear operator. Thus Möhring & Rahman argue that if the source functions were known exactly it would be immaterial whether a bounded or a causal Green's function were used. This, however, is not a license to use one or the other at will, for it refers to the exact source functions, and in general these are known only in the very crudest way (scaling properties, typically). The authors of [41] emphasise that the orthogonality of source and Green's function may be easily destroyed by making approximations to the source, and then various choices of Green's function will lead to very different responses.

These issues really need to be examined in the context of a specific example. Reference has already been made to Cargill's [9] reconciliation of his prediction for plane wave radiation from a circular nozzle with flow with the Lighthill [15] and DFWG [17] analogies, though this is perhaps a degenerate case involving essentially neutral shear layer waves. Better would be examples where we have a nonlinear solution for the instability, and here there is a chance of development. A nonlinear version of the work of Brazier-Smith & Scott [25] on flow over an elastic plate or membrane seems possible, using typical multiple scaling arguments of weakly nonlinear instability theory, as do nonlinear versions of the work of [33, 34] on shear layer instabilities and jet noise and, third, of the Broadbent-Moore [42] vortex instability. These seem to offer the best prospect of resolving the current controversy in aeroacoustics on the role of instabilities and the choice between causal and bounded Green's functions.

References

1. Müller, E.-A. (ed.) Mechanics of sound generation in flows. (IUTAM/ICA/AIAA Joint Symposium, Göttingen 1979). Berlin, Heidelberg, New York: Springer-Verlag 1979.

2. Crighton, D. G. Acoustics as a branch of fluid mechanics. J. Fluid Mech. 106 (1981) 261-298.

3. Crighton, D.G.; Leppington, F. G. Radiation properties of the semi-infinite vortex sheet: the initial-value problem. J. Fluid Mech. 64 (1974) 393-414.

4. Morgan, J. D. The interaction of sound with a semi-infinite vortex sheet. Q. J. Mech. Appl. Math. 27 (1974) 465-487.

5. Bechert, D. W. A model of the excitation of large scale fluctuations in a shear layer. AIAA-83-0274 (1983).

6. Bechert, D. W.; Stahl, B. Shear layer excitation, experiment versus theory. DFVLR-FB-84-26 (1984) 1-84.

7. Munt, R. M. The interaction of sound with a subsonic jet issuing from a semi-infinite cylindrical pipe. J. Fluid Mech. 83 (1977) 609-640.

8. Howe, M. S. Attenuation of sound in a low Mach number nozzle flow. J. Fluid Mech. 91 (1979) 209-229.

9. Cargill, A. M. Low-frequency sound radiation and generation due to the interaction of unsteady flow with a jet pipe. J. Fluid Mech. 121 (1982) 59-105.

10. Rienstra, S. W. A small Strouhal number analysis for acoustic wave-jet flow-pipe interaction. J. Sound Vib. 86 (1983) 539-556.

11. Bechert, D. W. Sound absorption caused by vorticity shedding, demonstrated with a jet flow. J. Sound Vib. 70 (1980) 389-405.

12. Moore, C. J. The role of shear-layer instability waves in jet exhaust noise. J. Fluid Mech. 80 (1977) 321-367.

13. Ronneberger, D. Experimentelle Untersuchungen zum akustischen Reflexionsfaktor von unstetigen Querschnittsänderungen in einem luftdurchströmten Rohr. Acustica 19 (1967/68) 222-235.

14. Cargill, A. M. Low frequency acoustic radiation from a jet pipe - a second order theory. J. Sound Vib. 83 (1982) 339-354.

15. Lighthill, M. J. On sound generated aerodynamically. I. General Theory. Proc. R. Soc. Lond. A.211 (1952) 564-587.

16. Curle, N. S. The influence of solid boundaries upon aerodynamic sound. Proc. R. Soc. Lond. A.231 (1955) 505-514.

17. Dowling, A. P.; Ffowcs Williams, J. E.; Goldstein, M. E. Sound production in a moving stream. Phil. Trans. R. Soc. Lond. A288 (1978) 321-349.

18. Daniels, P. G. On the unsteady Kutta condition. Q. J. Mech. Appl. Math. 31 (1978) 49-75.

19. Crighton, D. G. The Kutta condition in unsteady flow. Ann. Rev. Fluid Mech. 17 (1985) 411-445.

20. Tam, C. K. W. Directional acoustic radiation from a supersonic jet generated by shear layer instability. J. Fluid Mech. 46 (1971) 757-768.

21. Jones, D. S.; Morgan, J. D. The instability of a vortex sheet on a subsonic stream under acoustic radiation. Proc. Camb. Phil. Soc. 72 (1972) 465-488.

22. Crow, S. C.; Champagne, F. H. Orderly structure in jet turbulence. J. Fluid Mech. 48 (1971) 547-591.

23. Michalke, A. Instabilität eines kompressiblen runden Freistrahls unter Berücksichtigung des Einflusses der Strahlgrenzschichtdicke. Z. Flugwiss. 19 (1971) 319-328.

24. Crighton, D. G.; Taylor, M. V. Acoustic and instability properties of coaxial jets. (In preparation).

25. Brazier-Smith, P. R.; Scott, J. F. Stability of fluid flow in the presence of a compliant surface. Wave Motion 6 (1984) 547-560.

26. Koch, W. Local instability characteristics and frequency determination of self-excited wake flows. J. Sound Vib. 99 (1985) 53-83.

27. Huerre, P.; Monkewitz, P. A. Absolute and convective instabilities in free shear layers. J. Fluid Mech. (1985). (In press).

28. Goldstein, M. E. Generation of instability waves in flows separating from smooth surfaces. J. Fluid Mech. 145 (1984) 71-94.

29. Bechert, D. W.; Pfizenmaier, E. On the amplification of broadband jet noise by a pure-tone excitation. J. Sound Vib. 43 (1975) 581-587.

30. Hussain. A. K. M. F.; Hasan, M. A. Z. Turbulence suppression in free turbulent shear flows under controlled excitation. Part 2. Jet-noise suppression. J. Fluid Mech. 150 (1985) 159-168.

31. Laufer, J.; Yen, T. Noise generation by a low -Mach-number jet. J. Fluid Mech. 134 (1983) 1-31.

32. Crighton, D. G.; Huerre, P. Model problems for the generation of superdirective acoustic fields by wave packets in shear layers. AIAA-84-2295 (1984).

33. Huerre, P.; Crighton, D. G. Sound generation by instability waves in a low Mach number jet. AIAA-83-0661 (1983).

34. Crighton, D. G.; Huerre, P.; Monkewitz, P. A. Instability and sound generation in low Mach number jet flows. (In preparation).

35. Tam, C. K. W.; Morris, P. J. The radiation of sound by the instability waves of a compressible plane turbulent shear layer. J. Fluid Mech. 98 (1980) 349-381.

36. Troutt, T. R.; McLaughlin, D. K. Experiments on the flow and acoustic properties of a moderate-Reynolds-number supersonic jet. J. Fluid Mech. 116 (1982) 123-156.

37. Tam, C. K. W.; Burton, D. E. Sound generated by instability waves of supersonic flows. Part 1. Two-dimensional mixing layers. J. Fluid Mech. 138 (1984) 249-271.

38. Tam, C. K. W.; Burton, D. E. Sound generated by instability waves of supersonic flows. Part 2. Axisymmetric jets. J. Fluid Mech. 138 (1984) 273-295.

39. Crighton, D. G. The free and forced waves on a fluid-loaded elastic plate. J. Sound Vib. 63 (1979) 225-235.

40. Goldstein, M. E. Aeroacoustics of turbulent shear flows. Ann. Rev. Fluid Mech. 16 (1984) 263-285.

41. Möhring, W. F.; Rahman, S. Sound generation in a flow near a compliant wall. J. Sound Vib. 66 (1979) 557-564.

42. Broadbent, E. G.; Moore, D. W. Acoustic destabilization of vortices. Phil. Trans. R. Soc. Lond. A290 (1979) 353-371.

Acknowledgement. This work was supported by the Office of Naval Research, and monitored by Dr. R. L. Sternberg, Code 425.

Hydrodynamic Instabilities and Surface Waves in a Flow over an Impedance Wall

S.W. Rienstra

National Aerospace Laboratory NLR
P.O. Box 90502, 1006 BM Amsterdam

Abstract

A 2-D analysis is made of the surface waves in a subsonic compressible uniform mean flow along an impedance wall. It is found that a maximum of four surface waves is possible. The regions of existence in complex impedance and wave number planes are given. One wave is found to travel upstream, while the others, one of which is an instability, propagate downstream. A Wiener-Hopf solution of plane wave scattering at a hard/soft discontinuity of the wall impedance is constructed, in which the instability wave provides the additional degree of freedom to satisfy the Kutta condition at the impedance jump. The present waves are limiting forms of certain irregular acoustic duct modes.

Introduction

It was recognized in Morse and Ingard [1] that 2 of the acoustic modes in a duct with an impedance wall sometimes show a certain irregular behaviour. In the presence of mean flow through the duct 2 additional modes appear (Tester [2], Koch and Möhring [3], Eversman [4]), also of unusual character since one of them is to be considered as an instability. A 2-D/semi-infinite analysis is described in the present paper to investigate these modes further; this is possible because these modes are not essentially duct modes, but only coupled to the impedance wall. So, the present analysis embodies a high-frequency limit of the duct problem. The present 2-D configuration allows more explicit and transparant results, revealing better the essential physics.

Model

Consider a perturbation potential ϕ and pressure p (dimensionless with mean density ρ_o, mean sound speed c_o, and some length) on a uniform mean flow with velocity $U_o = Mc_o$ $(0 < M < 1)$ in the space $y \geq o$, $-\infty < x < \infty$, satisfying

$$\nabla^2\phi - (i\omega + M\partial_x)^2\phi = o \tag{1}$$

$$p = -(i\omega + M\partial_x)\phi$$

Aero- and Hydro-Acoustics
IUTAM Symposium Lyon 1985
Editors: G. Comte-Bellot and J. E. Ffowcs Williams

where the time harmonic factor $\exp(i\omega t)$ is suppressed throughout. The boundary condition at y=o describes the effect of the complex impedance Z (Re Z ≥ o) modified by the (vanishing) mean flow boundary layer (Ingard [5]):

$$(i\omega + M\partial_x)^2\phi = i\omega Z\partial_y\phi \quad , \quad y=o \tag{2}$$

Furthermore:

$$\phi \to o \quad , \quad y\to\infty \tag{3}$$

Surface waves

Consider waves of the form

$$\phi = \exp(-i\alpha x - i\gamma y)$$

With the transformation $\omega = \beta\Omega$, $\sigma = \beta\alpha + M\Omega$, $\beta = (1-M^2)^{\frac{1}{2}}$, (1) and (2) yield

$$\gamma(\sigma) = (\Omega^2 - \sigma^2)^{\frac{1}{2}}$$

$$K(\sigma) = \beta^3\Omega Z\gamma + (\Omega - M\sigma)^2 = o \tag{4}$$

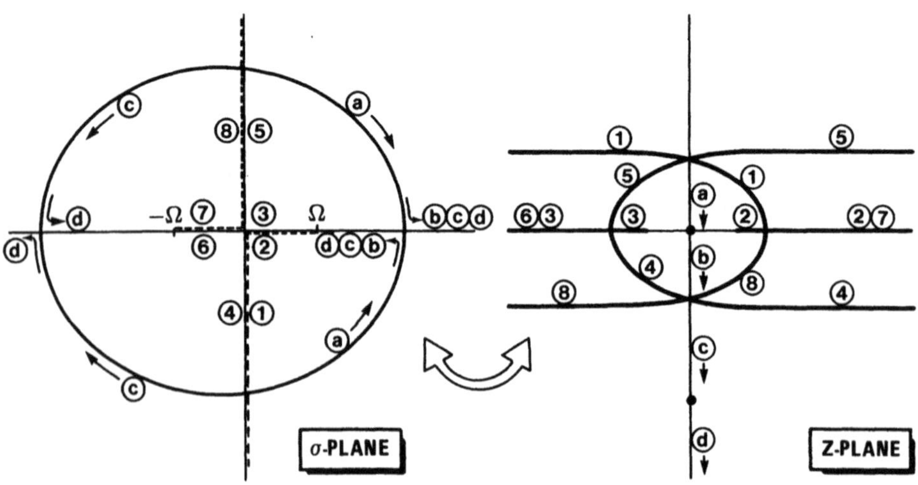

Fig. 1 Mapping $Z = -(\Omega - M\sigma)^2/\beta^3\Omega\gamma$
①,...,⑧ from σ to Z
ⓐ,...,ⓓ from Z to σ

The L-shaped branch cuts of γ along the lines Im γ = o, with the choice of the branch with Im $\gamma \leq$ o, garantees (3) (Fig. 3). The -at most 4- solutions σ of (4) correspond to the waves under consideration. Instead of solving σ, we start with considering equation (4) as defining $Z = Z(\sigma)$. Then, by mapping the branch cuts of $\gamma(\sigma)$, which are "lines of disappearance" of solutions σ, from the σ-plane to the Z-plane (Fig. 1) we obtain regions I to V in Z with different number of solutions of (4). Mapping back from Z to σ the line Re Z = 0 we obtain in the σ-plane the "egg-like" contour displaying the regions of existence of the various solutions. See figures 2 and 3. Note that the finer division II, III, IV was missed by previous authors ([4], [3]). For further details, see appendix A. From figures 2 and 3 it is seen that if $M \to$ o, ω fixed (no-flow limit), two solutions disappear to infinity, while the division of the Z-plane is only into two parts, along Im Z = o, which is in accordance with Wenzel [6]. Another interesting limit is $M \to$ o, $\omega \to$ o, with $S = \omega/M$ (Strouhal number) fixed (incompressible limit). If Z' (=Z/M) is the impedance dimensionless on $\rho_o U_o$, the solutions σ (if existing) are explicitly given by

$$\sigma = iS\left(\tfrac{1}{2}Z' - i \pm (1 + (\tfrac{1}{2}Z' - i)^2)^{\frac{1}{2}}\right),\ - iS\left(\tfrac{1}{2}Z' + i \pm (1 + (\tfrac{1}{2}Z' + i)^2)^{\frac{1}{2}}\right). \quad (5)$$

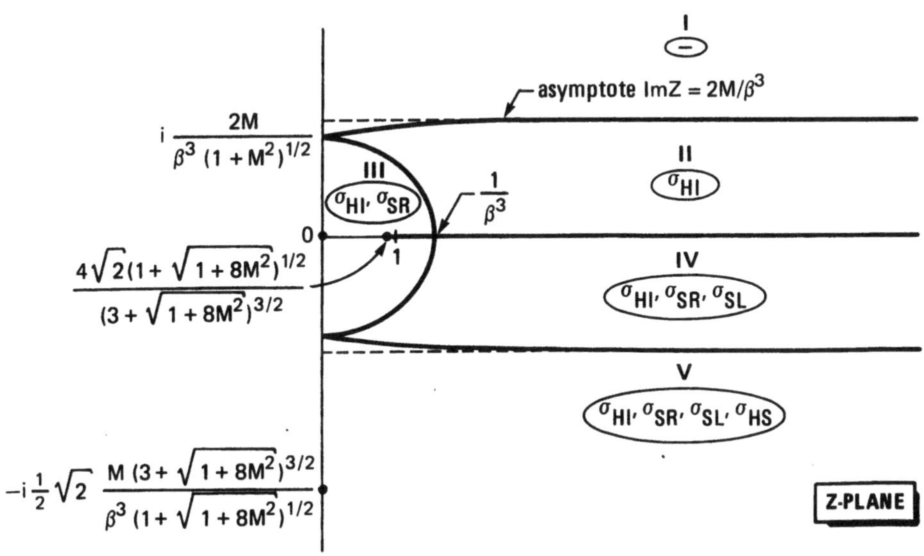

Fig. 2 Complex impedance Z-plane, divided into the regions with different number of solutions σ

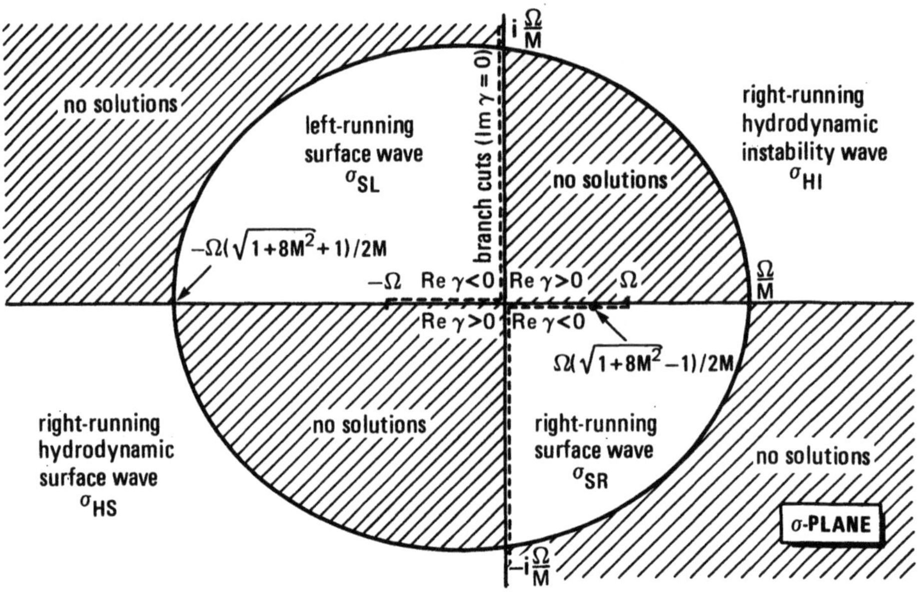

Fig. 3 Complex (transformed) wave number σ-plane, with regions of existence of the surface waves

Furthermore, $\gamma(\sigma) = -\ i\ \mathrm{sgn}(\mathrm{Re}\ \sigma)\ \sigma$, the contour in the σ-plane becomes a circle with radius S, and the division of the Z'-plane becomes threefold, viz. I: $\mathrm{Im}\ Z' > 2$, III: $-2 < \mathrm{Im}\ Z' < 2$, V: $\mathrm{Im}\ Z' < -2$.

<u>Scattering problem</u>

The scattered field generated by the plane waves

$$p = \tfrac{1}{2} \exp(-\ i\Omega\ (\cos\Theta_s - M)\ X - i\Omega\ \sin\Theta_s\ y)$$

($x = \beta X$, $-\pi < \Theta_s < o$), incident to a wall $y = o$, hard along $x < o$, and soft (→ (2)) along $x > o$ (Fig. 4), is found by a Wiener-Hopf procedure ([7], [3]):

$$p = \exp(-\ i\Omega\ (\cos\Theta_s - M)\ X)\ \cos(\Omega\ \sin\Theta_s\ y)$$

$$+\ \frac{1}{2\pi i} \oint_{-\infty}^{\infty} \left\{ \frac{K_-(\Omega_1)\ \gamma_-(\Omega_1)}{\sigma - \Omega_1} + E \right\} \frac{(\Omega - M\sigma)^2}{K_+(\sigma)\gamma_-(\sigma)}\ \exp(-i\gamma y - i\alpha x)\ d\sigma \qquad (6)$$

where $K(\sigma) = K_+(\sigma) / K_-(\sigma)$, $\gamma_\pm(\sigma) = (\Omega \mp \sigma)^{\frac{1}{2}}$, $\Omega_1 = \Omega \cos\Theta_s$, and E a constant (the amplitude of an eigensolution). See appendix B. Apart from an upward identation of the integration contour around Ω_1, the location of the contour around the poles, and thereby the split functions $K_\pm$, is not readily evident. The causality argument of Jones and Morgan [8] (a refinement of Briggs [9]) tells us to construct the correct solution by analytic continuation in ω from negative imaginary ($\omega = -i|\omega|$) to real values. This requires to adopt some frequency dependent $Z = Z(\omega)$, to deal with a physically realizable impedance. For the moment we have chosen a mass-spring-damper system: $Z = r + im\omega - ik/\omega$.

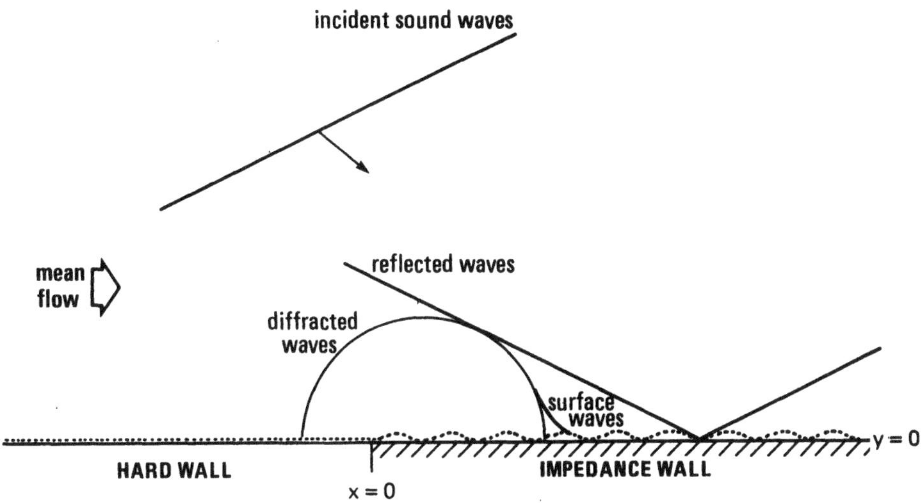

Fig. 4 Plane sound waves in an uniform mean flow, scattered by a wall impedance discontinuity

One zero of K, namely σ_{HI} (Fig. 3), appears for $\omega = -i|\omega|$ to move to the lower complex half-plane. For the general case a numerical search gave the evidence, but for the incompressible case (5) it can be proved analytically (App. C). As a consequence, $K_+(\sigma)$ contains the factor $\sigma - \sigma_{HI}$, and for real ω the contour has to be deformed such that it remains "above" the pole σ_{HI} (Fig. 5); thus σ_{HI} corresponds to a right-running instability. If $Z \in I$ (for real ω) σ_{HI} is in the second quadrant on the <u>other</u> Riemann sheet of γ, and the "causal" contour yields an unacceptable solution unbounded in y. In this case it is probably preferable to avoid the deformation; K_+ contains no fac-

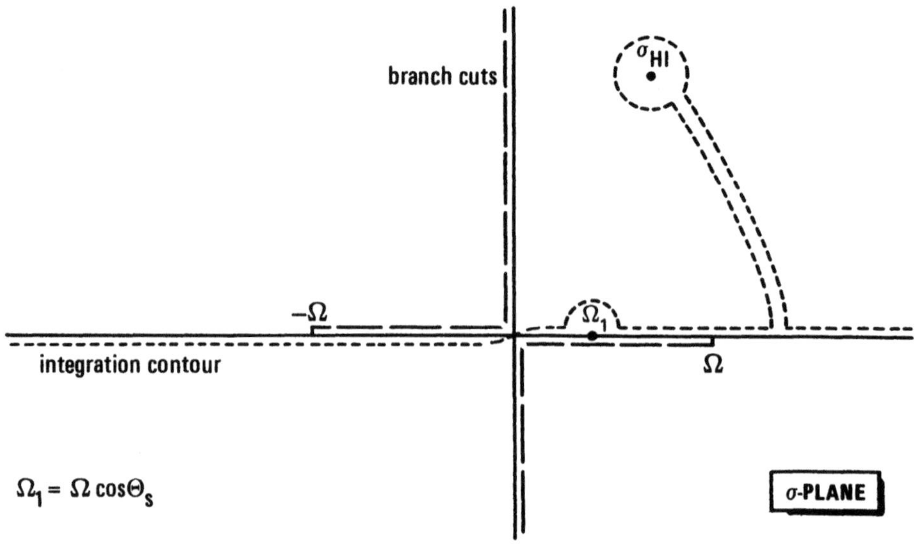

Fig. 5 Integration contour in σ-plane

tor $\sigma - \sigma_{HI}$, and E = o. If $Z \notin I$, the available eigensolution with amplitude E provides the possibility to satisfy the Kutta condition at x = y = o, with the choice E = o. In this case the sound couples to the generated surface waves in a way that acoustic energy may be exchanged with hydrodynamic (kinetic) energy, as was shown by Quinn and Howe [10] for the special case Z = o, where K can be split by inspection ($K_+ = K$, $K_- = 1$), and (6) can be reduced to Fresnel's integrals.

Relevance to duct modes

The axial wave numbers of the modes in a cylindrical duct of radius unity are given by

$$(\Omega - M\sigma)^2 - i\beta^3\Omega Z\gamma J_m'(\gamma)/J_m(\gamma) = o \tag{7}$$

which is the analogue of equation (4). J_m denotes the m-th order Bessel function. In the high frequency limit $\Omega \to \infty$ we can distinguish two cases: Im $\gamma = O(\Omega)$ in which case eq. (7) reduces to eq. (4), and Im $\gamma = O(1)$ which amounts to a boundary layer in the σ-plane enclosing the branch cuts of γ. The solutions inside this boundary layer correspond to genuine duct modes (their field covering the whole duct cross-section. The other possible four (or two, if $M = o$) with $\mathrm{Im}\,\gamma = O(\Omega)$ are surface waves, asymptotically described by the above analysis. If Z varies from one region to another, e.g. from I to II (Fig. 2), one of the genuine duct modes changes its character and turns into a surface wave, which is, being confined to the wall region, a physically different phenomenon.

References

1. Morse, P.M.; Ingard, K.U.: Theoretical Acoustics. McGraw-Hill, 1968 (509).
2. Tester, B.J.: JSV 28(2)(1973) 151-203.
3. Koch, W.; Möhring, W.: AIAA Jnl 21(2)(1983) 200-213.
4. Eversman, W.: JSV 50(1)(1977) 159-162.
5. Ingard, K.U.: JASA 31(7)(1959) 1035-1036.
6. Wenzel, A.R.: JASA 55(5)(1974) 956-963.
7. Noble, B.: The Wiener-Hopf Technique, Pergamon 1958.
8. Jones, D.S.; Morgan, J.D.: Proc. Roy. Soc. L. A 338 (1974) 17-41.
9. Briggs, R.J.: Res. Monogr. No. 29 MIT 1964.
10. Quinn, M.C.; Howe, M.S.: JSV 97(1)(1984) 1-9.

APPENDIX A

See figure 2. Impedances of special interest are (introduce $w = (1 + 8M^2)^{1/2}$):

$Z = 2iM/\beta^3(1+M^2)^{1/2}$ where $\sigma_{HI} = i\Omega/M$, $\sigma_{SR} = -i\Omega/M$.

$Z = -2iM/\beta^3(1+M^2)^{1/2}$ where $\sigma_{SL} = i\Omega/M$, $\sigma_{HS} = -i\Omega/M$.

$Z = 1/\beta^3$ where (in the upper half of III) $\sigma_{SR} = o$, or (in IV) $\sigma_{SL} = o$

$Z = o$ where σ_{HI} and σ_{SR} coincide: $\sigma_{HI} = \sigma_{SR} = \Omega/M$

$Z = -i\tfrac{1}{2}\sqrt{2}\, M(3+w)^{3/2}/\beta^3(1+w)^{1/2}$ where σ_{HS} and σ_{SL} coincide:

$$\sigma_{HS} = \sigma_{SL} = -\Omega(w+1)/2M$$

When Z, varying in region III, crosses the real interval ($4\sqrt{2}(1+w)^{1/2}/(3+w)^{3/2}$, $1/\beta^3$) with, say, increasing imaginary part, then in the σ-plane σ_{SR} disappears somewhere along (o , $\Omega(w-1)/2M$), but appears at the same time somewhere along ($\Omega(w-1)/2M$, Ω), thus leaving the number of solutions the same. When $Z \to 4\sqrt{2}(1+w)^{1/2}/(3+w)^{3/2}$ the two representations of σ_{SR} coincide (K becomes quadratic) at $\sigma_{SR} = \Omega(w-1)/2M$.

If Re $Z = o$ we have (see Fig.1) along ⓐ two complex conjugate solutions, which become real along ⓑ ; along ⓒ we have two complex conjugate and two real solutions, and along ⓓ all four are real.

APPENDIX B

It is convenient to introduce the position of the wall streamline $y=h(x)\exp(i\omega t)$, and to write instead of (2) ([3],[5]):

$\partial_y\phi = (i\omega + M\partial_x)h$, $p = -i\omega Zh$ for y=o, x>o. Then, if

$$\phi = \psi + \frac{i\beta \cos(\Omega \sin\Theta_s\, y)}{\Omega(1 - M\cos\Theta_s)} \exp(-i\Omega\,(\cos\Theta_s - M)\,X)$$

$$\hat{\psi}(\alpha,y) = \int_{-\infty}^{\infty} \psi(x,y)\exp(i\alpha x)\,dx \quad , \quad H_+(\alpha) = \int_{o}^{\infty} h(x)\exp(i\alpha x)\,dx$$

$$P_-(\alpha) = \int_{-\infty}^{o} (i\omega + M\partial_x)\psi(x,o)\exp(i\alpha x)\,dx$$

we obtain (assume for the moment $\cos\Theta_s > o$, $\mathrm{Im}\,\Omega < o$)

$$\hat{\psi} = A\exp(-i\gamma y), \quad A = -(\Omega - M\sigma)H_+/\beta\gamma,$$

$$P_- + i\beta/(\sigma - \Omega_1) = -iKH_+/\beta^2\gamma.$$

In the usual way ([7]) we rewrite this equation into a left hand side, analytic in σ in a lower half plane (-), and a right hand side, analytic in an upper half plane (+), both then equal to one entire function $E(\sigma)$. As explained in the main text, we start for causality with $\omega = -i|\omega|$. Then σ_{HI} belongs to the lower half plane, and we have the estimates for $|\sigma| \to \infty$: $K_+(\sigma) = O(\sigma^2)$, $K_-(\sigma) = O(1)$. In addition, $H_+(\sigma) = O(\sigma^{-3/2})$, $P_-(\sigma) = O(\sigma^{-\frac{1}{2}})$, as results from $h(x) = O(x^{\frac{1}{2}})$, $p(x,o) = O(x^{-\frac{1}{2}})$ for $x \downarrow o$. It turns out then that E is a constant. Collecting the various terms, a Fourier transform back into the x- domain, and rotating ω from the negative imaginary axis to the positive real axis (preserving analyticity in ω by deforming the σ-integration contour, as explained in the main text) then gives eq. (6).

APPENDIX C

Since $\sigma_{HI} = iS\exp(\mathrm{arsinh}\,(\frac{1}{2}Z'-i))$ we have $\arg\sigma_{HI} = \frac{1}{2}\pi + \arg S + \mathrm{Im}(\mathrm{arsinh}(\frac{1}{2}Z'-i))$. If $S \to -i|S|$, Z' becomes real and $\mathrm{Im}(\mathrm{arsinh}(\frac{1}{2}Z'-i)) < o$, and hence $\arg\sigma_{HI} \in (-\frac{1}{2}\pi\ , o)$.

Wake-Induced Shear Layer Excitation in a Basic Annular Jet

K. M. LAM and N. W. M. KO

Department of Mechanical Engineering
University of Hong Kong
Hong Kong

ABSTRACT

The excitation of the outer shear layer by the disturbances associated with the wake structures shedding from the central body in a basic annular jet is studied with the conditional sampling technique. The educed velocity perturbations show that at locations upstream of the rollup of the first thin initial vortices, the outer shear layer is already under a slight oscillation at the wake-shedding frequency. This instability which triggers the formation of the wake-induced structures in the outer shear layer and the geometrical structure of these structures seem to be governed largely by the large-scale lateral velocity fluctuations of the quasi-two-dimensional orderly pattern of the shedding wake structures in the inner region.

1. INTRODUCTION

Various investigations have shown that when a jet flow is subjected to an internal or external perturbation in pressure or velocity at an appropriate frequency, instability waves grow on the shear layer (for example, Moore [1], Bechert & Pfizenmaier [2]). In a basic annular jet, in which the central body has a blunt downstream end or interface, periodic wake structures are shed in the inner region and the associated disturbances excite the outer shear layer, resulting in a train of wake-induced structures which have their quasi-passing frequency being the same as the wake-shedding frequency (Ko & Chan [3]). Using the mode expansion technique, Ko & Lam [4] has shown that both the wake-induced structures and the wake structures contain a dominating first azimuthal mode constituent. This feature of the wake-induced structures is different from the axisymmetric nature of the structures formed in the axisymmetric mixing layers of unexcited circular jets and of circular jets planar acoustic excitations. In this paper, some eduction results on the velo-

Aero- and Hydro-Acoustics
IUTAM Symposium Lyon 1985
Editors: G. Comte-Bellot and J. E. Ffowcs Williams

city perturbations within the outer shear layer of the annular jet will be presented.

2. EXPERIMENTAL PROCEDURES

The annular jet has an outer diameter D_o of 6.6 cm and an inner diameter of 3.0 cm. The jet exit velocity $\overline{U}_o$ was 50 m/s giving a Reynolds number $\overline{U}_o D_o/\nu$ of 2 x 10^5. The exit conditions of both the outer and inner shear layers were laminar. A single hot wire which was used as the triggering probe was placed at a fixed location in the potential core near the inner mixing region. The location chosen was $x/D_o = 0.5$ and $y/D_o = 0.275$. The signal was dominated by the contribution of the shedding wakes at a frequency of about 415 Hz or a quasi-period of about 2.4 msec. A cross-wire probe was used to recovery the velocity fluctuations along $\eta = (y - D_o/2)/x \simeq -0.06$. The eduction was achieved with triggering on a threshold level of $+2\sigma_u$, where σ_u was the r.m.s. value of the triggering hot-wire signal (Figure 1).

3. RESULTS AND DISCUSSIONS

Figures 2a and b show the recovered velocity perturbations u_R and v_R in the outer shear layer. Rollup of the first initial vortices from the shear-layer mode instability occurs at x/D_o = 0.15. Upstream of this, the recovered traces at $x/D_o = 0.075$ and 0.1 already show the oscillations of the shear layer at the wake-shedding frequency. The magnitudes of these oscillations are very small and this wake-induced instability amplifies and saturates at $x/D_o \simeq 0.4$. The recovered traces suggest that the wake-induced instability and the natural shear-layer instability seem to exhibit some mutual interactions, which may be responsible for the modulated signals at x/D_o = 0.15 to 0.3.

The most strongly educed u_R peaks in Figure 2a and v_R peaks in Figure 2b are both convecting downstream. The u_R peaks are coincident with the v_R peaks at all axial locations. Hence the wake-induced structures are associated with radial momentum transports and Reynolds stress production.

Figure 3 shows the u_R traces educed in the outer shear layer on the side opposite to the triggering signal across the jet. By comparing the traces with those in Figure 2a, especially at the three most upstream locations, it is obvious that the wake-induced instabilities on the two opposite sides of the jet are out-of-phase.

Velocity signals have been recovered at different lateral locations on the diametrical plane of the jet containing the triggering probe at x/D_o = 0.5 and 1. The traces were bandpassed at 100 to 1600 Hz to emphasize the large-scale motions and iso-contour maps were constructed from the traces. The contour map of v_R at x/D_o = 0.5 (Figure 4) shows the quasi-two-dimensional vortex street pattern of the wake structures in the inner mixing region. It has been reported by Fuchs et al. [5] and Roberts [6] that the wake behind a circular disc bears a close resemblance to the two-dimensional Karman vortex street behind a cylinder in cross flow, except for the random oscillation of the plane of the quasi-two-dimensional orderly pattern about the axis of symmetry. In the outer shear layer, a number of small flow structures seem to be in stages of merging forming the wake-induced structures. At this axial location, the lateral velocity perturbations due to the wake and the wake-induced structures are uniformly phase-aligned at all different lateral locations on the whole diametrical plane.

Further downstream at x/D_o = 1 (Figure 5), where the wake-induced structures become well-developed, the phase alignment of the v_R perturbations is very obvious. Correspondingly, the wake and the wake-induced structures are orientated in a neat staggered quasi-two-dimensional orderly pattern. Hence the wake-induced instability in the outer shear layer which triggers the formation of the wake-induced structures is likely to be affected largely by the lateral velocity perturbation pattern of the quasi-two-dimensional wake structures in the inner region.

Finally, Figure 6 shows the pressure traces recovered at different azimuthal angles at x/D_o = 1 and r/D_o = 0.4. The results were obtained in another experimental setup involving the conditional sampling of pressure fluctuations which has been discussed in Ko & Lam [4]. The wake-induced structures do not assume the geometry of axisymmetric toroidal structures but span about half the circumference of the outer mixing region with successive structures alternating on opposite sides. This provides further evidence to the influence of the wake pattern on the wake-induced instability.

4. CONCLUSIONS

The eduction results show that the quasi-two-dimensional vortex street orderly pattern of the shedding wakes from the interface induces a wake-induced instability in the outer shear layer. This instability is governed largely by the quasi-two-dimensional and alternating features of the wake pattern. These characteristics are also found in the wake-induced structures whose formation is locked onto the instability.

In the circular jet flow, higher-mode excitation has been rare. Bechert & Pfizemaier [7] injected lateral velocity fluctuations about a transverse diametrical plane at the exit of a circular jet to achieve a first mode excitation. They observed far-field noise generation different from the excitation in the planar mode. Hence the study of the flow structures under higher mode excitation may be useful in the noise generation of real jet engines.

ACKNOWLEDGMENTS

The authors would like to thank the Croucher Foundation in awarding a Croucher Foundation Studentship to K.M. Lam. The work was partly supported by a research grant from University of Hong Kong awarded to N.W.M. Ko.

REFERENCES

1. Moore, C.J.: JFM 80 (1977) 321-367.
2. Bechert, D. and Pfizenmaier, E.: JSV 43 (1975) 581-587.
3. Ko, N.W.M. and Chan, W.T.: JFM 93 (1979) 549-584.
4. Ko, N.W.M. and Lam, K.M.: AIAA Journal (1985) in press.

5. Fuchs, H.V.; Mercker, E. and Michel, U.: JFM 93 (1979) 185-207.

6. Roberts, J.B.: AIAA Journal 11 (1973) 1569-1571.

7. Bechert, D. and Pfizenmaier, E.: AIAA Journal 15 (1977) 1268-1271.

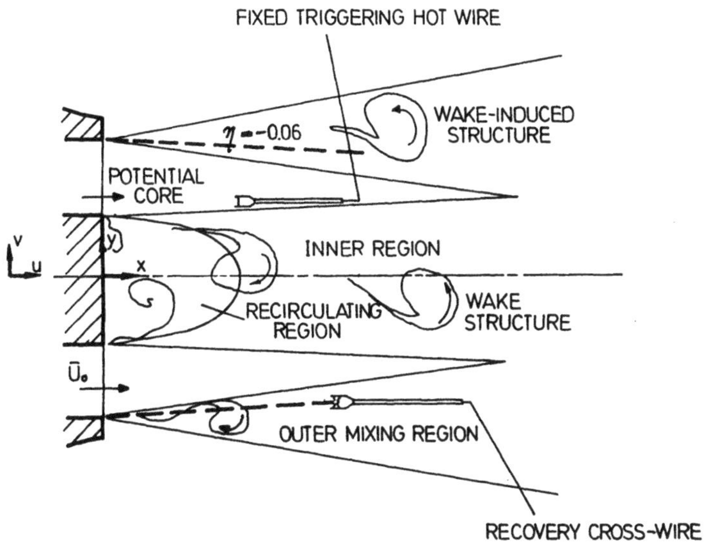

Figure 1. Schematic diagram of the flow and experimental arrangement.

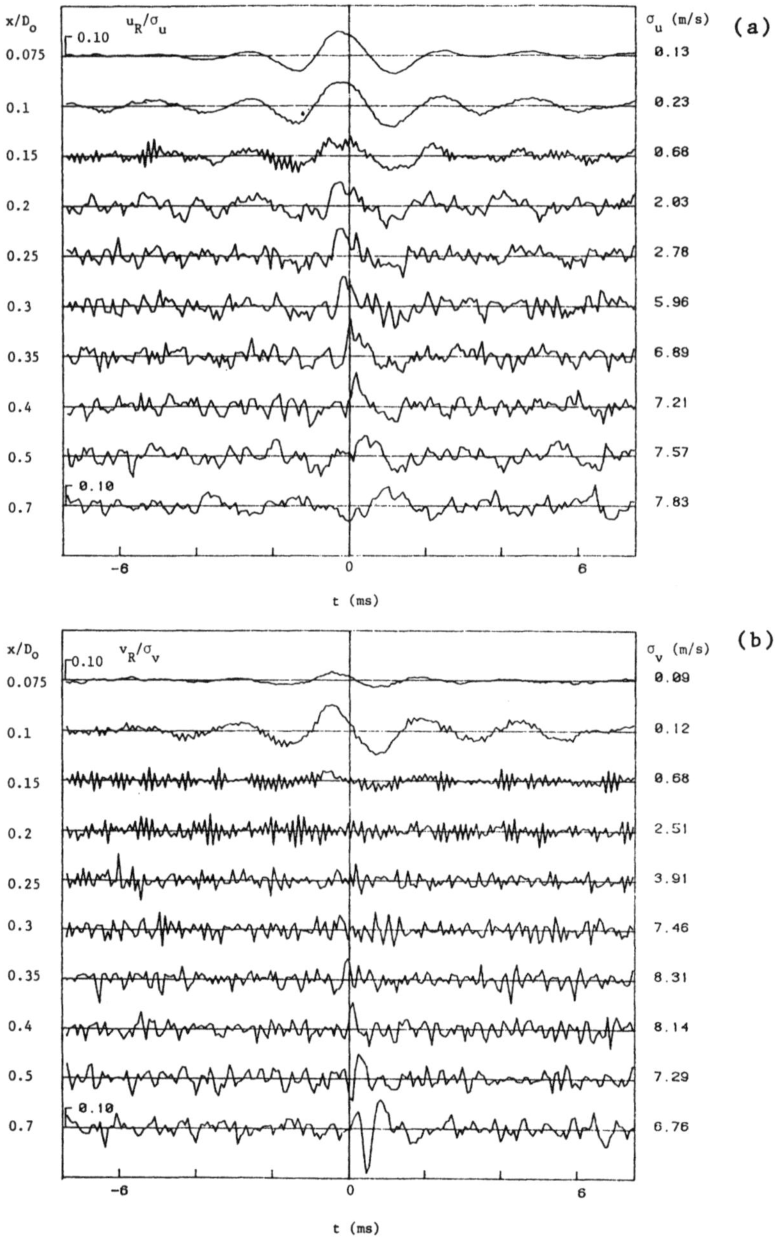

Figure 2. Streamwise evolutions of recovered velocity perturbations in the outer shear layer. t is the time relative to the triggering instant.
(a) $u_R(t)$; (b) $v_R(t)$.

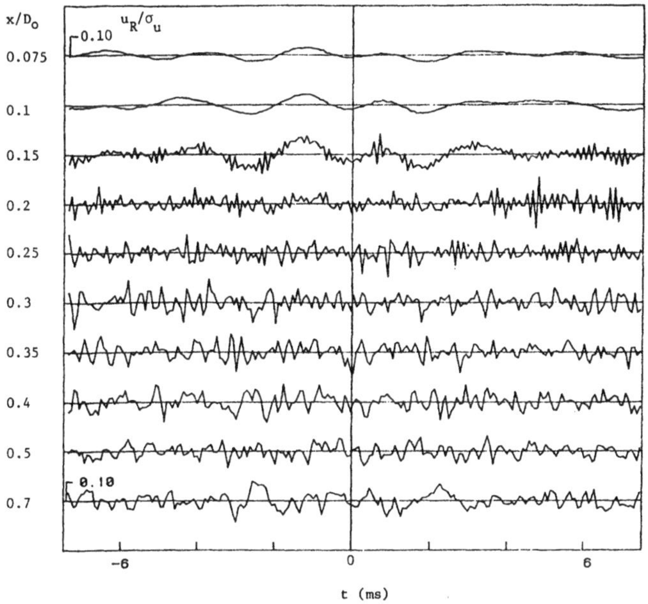

Figure 3. Streamwise evolution of recovered $u_R(t)$ in the outer shear layer on the aximuthal plane opposite to the triggering probe.

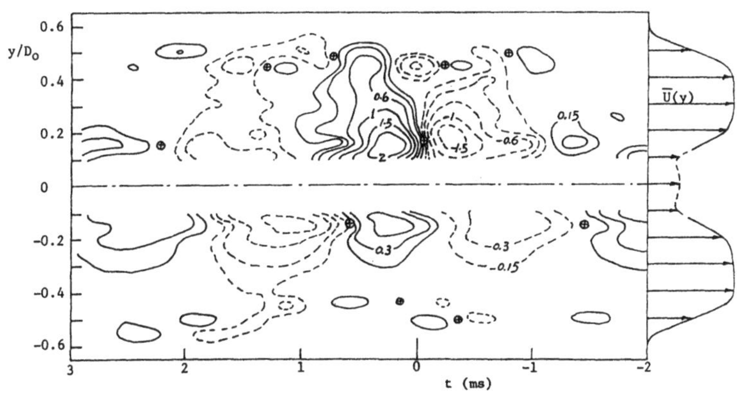

Figure 4. Contour map of recovered $v_R(t,y)$ at x/D_o=0.5. Contour levels in m/s. The locations of $-0.15<y/D_o<0.15$ are inside the recirculating region. The approximate centre locations of flow structures as deduced from induced v_R fluctuations are marked by the symbol ⊕.

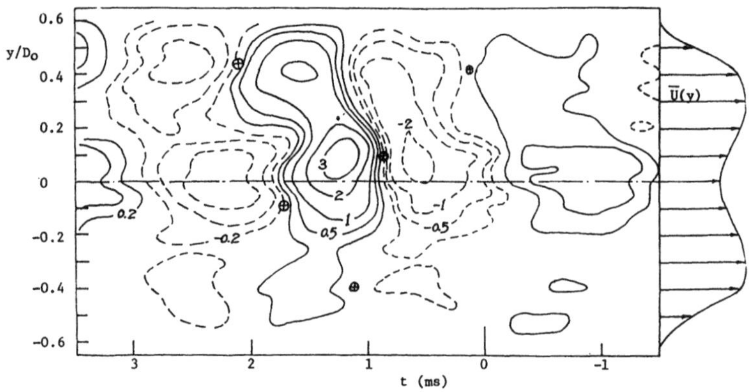

Figure 5. Contour map of recovered $v_R(t,y)$ at x/D_o=1. Contour levels in m/s.

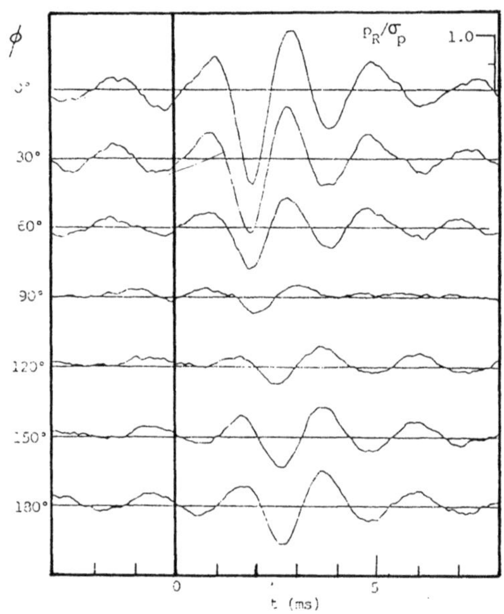

Figure 6. Pressure perturbations $p_R(t)$ recovered at different azimuthal angles at x/D_o = 1 and r/D_o = 0.4.

Transient Vortex Modelling of Flow Induced Acoustic Resonances Near Cavities or Obstructions in Ducts

M.C. WELSH and A.N. STOKES*

Commonwealth Scientific and Industrial Organization
Division of Energy Technology, Victoria, 3190, Australia

*Division of Mathematics and Statistics

Summary

When flow in a duct passes a cavity or an obstruction and generates a shear layer, an acoustic resonance may be excited which interacts with that shear layer. The result is a sustained resonance and changed flow pattern with generally two distinct vortices forming during one cycle of the acoustic resonant frequency. This effect occurs over discrete ranges of flow velocity. To explain the fluid mechanics of the process Howe's theory of aerodynamic sound is used to calculate the energy exchanged, during one acoustic cycle, between the sound field and the flow. The flow velocities corresponding to sustained resonance correlate well with velocities where the net flow of energy from the flow to the sound field is calculated to be greatest.

Introduction

There has been much interest recently in the problem of flow-excited acoustic resonance in enclosed and partly enclosed spaces. Analysis of the characteristic frequencies associated with the flow has been based on the response of a detached shear layer to perturbations (Rockwell and Schachenmann [1]), Bechert [2]) or alternatively on a momentum balance equation due to Howe [3]. This latter theory was used by Welsh, Stokes and Parker [4] to explain the resonant process in ducts containing plates. Nelson, Halliwell and Doak [5] used a similar theory to describe resonance in cavities.

The aerodynamic acoustic resonant process has four components [4]:

(a) a shear layer, which is excited by sound, generating coherent large scale vortex structures in the flow,

Aero- and Hydro-Acoustics
IUTAM Symposium Lyon 1985
Editors: G. Comte-Bellot and J. E. Ffowcs Williams

(b) sound generated by those vortex structures as they pass through the resonant sound field,

(c) feedback via the resonant sound field of a sufficient perturbation of the shear layer to establish, at the sound frequency, coherent large scale vortex structures, and

(d) the transfer of acoustic energy away from the source region at the rate at which it is being generated by the large scale vortex structures.

To describe the resonant process we have developed a mathematical model using an adaption of the "discrete vortex" technique which is briefly reviewed by Rockwell and Naudascher [7]. The model shows the energy transfer between the flow and sound fields, and identifies the regions of net gain and loss. It has so far been applied to flows past plates in ducts [4,8], but experience with the modelling process suggests it has application to other enclosed flows where resonances occur. The aim of this paper is to briefly describe the model and show its potential for wider application.

Discussion of Acoustic Sources

The power transferred from the flow to the sound field is given by a momentum balance equation due to Howe [3]:

$$P = 2\pi\rho \int \underline{\omega} \cdot (\underline{v} \times \underline{u})\, dV \qquad (1)$$

where ρ = density of fluid

$\underline{\omega}$ = curl $\underline{v}$ = vorticity

$\underline{v}$ = total velocity

$\underline{u}$ = acoustic particle velocity

dV = volume element.

The following qualitative deductions can be made about the likely strength of a region of flow as a source of resonant sound:

(i) $\underline{\omega}$, $\underline{v}$ and $\underline{u}$ must all have significant magnitude. This implies that an acoustic source is associated with vorticity which cannot be at an acoustic particle velocity node or where the flow velocity is small.

(ii) there must be a substantial angle between the vectors of vortex velocity and acoustic velocity. In particular no acoustic source can be expected where the sound velocity is approximately parallel to the vortex paths.

(iii) even with the two previous conditions met, the instantaneous power will be oscillatory due to the oscillatory nature of $\underline{u}$. Consequently, the instantaneous powers will tend to cancel when integrated over a full cycle. In a Lagrangian frame of reference, the magnitudes or the directions of the vectors in the triple product (equation 1) must therefore change during a cycle to prevent total cancellation. This is called "imbalance" during a cycle.

(a)

(b)

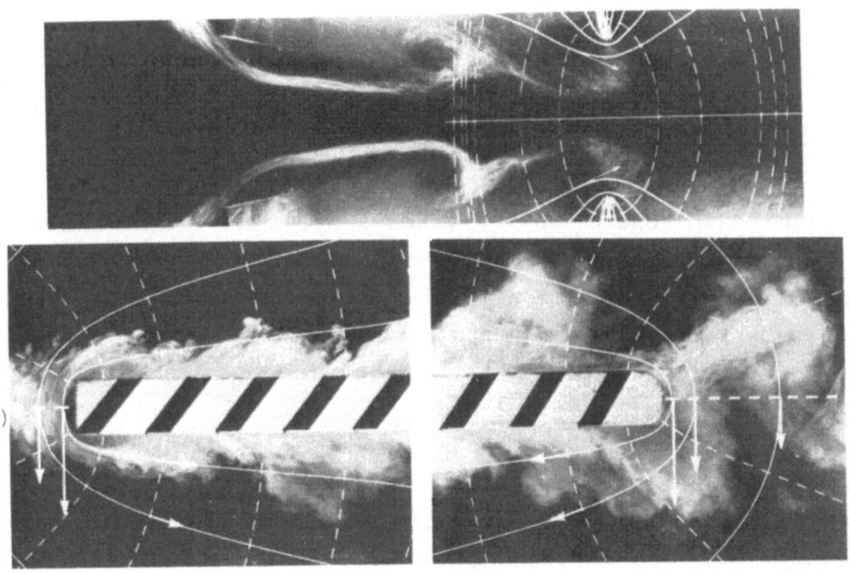

Fig. 1. Flows and acoustic fields at an instant during resonance; ——→ , instantaneous acoustic velocity; ----, contour of RMS acoustic pressure; —— , lines tangential to acoustic velocities; (a), baffles in a duct; (b), plate in a duct.

Figures 1(a) and 1(b) show two resonant acoustic fields superimposed on photographs of the flow. Figure 1(a) shows a flow which was first photographed by Nomoto et al. [6]. There are two pairs of baffles placed in a duct and away from the baffles the dominant acoustic mode is a plane wave with a wave length approximately equal to the length of the duct. The angle between vortex motion and sound velocity is small when the vortex is about midway between the baffles, increasing as the second pair of baffles is approached and then becoming zero again as the vortices pass through the plane of the baffles. When the vortices are approaching the second pair of baffles the triple vector product gives a positive instantaneous value for P. When the vortices move downstream from the baffles the angle between the vortex motion and the sound velocity increases again but now has opposite sign which leads to a negative instantanteous value for P. However, the sound field also oscillates in time and if the acoustic velocities change sign as the vortices pass through the plane of the baffles (which has been observed experimentally), then the triple vector product gives a positive instantaneous value of P; consequently P is positive throughout a full acoustic cycle giving a strong net sound source.

In Figure 1(b) there is a Parker β-mode [9] surrounding a rectangular plate centrally located in a duct. Qualitative features of the source are described in the next section and verified by calculations from a mathematical model.

Verification by Mathematical Model

In the case of the rectangular plate shown in Figure 1b, the flow separates with vorticity being shed from the leading edge corner. A time-varying potential flow which is two-dimensional, inviscid, incompressible and isentropic was used to model this situation. Discrete vortices, originating at the apparent point of shedding, were allowed to increase in circulation over time. Two vortices (one above and one below the plate centre line) were formed during each acoustic cycle. A Kutta condition was applied at the corners and this determined the rate of vortex growth. The vortex paths and circulations could then be determined. The sound velocity was also modelled by an oscillating potential flow [3] and the power transferred from the flow to the sound field was computed from (1) where with discrete vorticity, the integral is replaced by a discrete sum.

The flow region along the plate fails as a sound source because the sound velocity is either approximately parallel to the flow or tending to zero midway between the leading and trailing edges. However, near the ends, particularly the trailing edge, all the conditions for a source region are met. The sound velocity is close to maximum, changing rapidly, and becoming orthogonal to the flow. The triple product of (1) is large and has marked imbalance over a cycle.

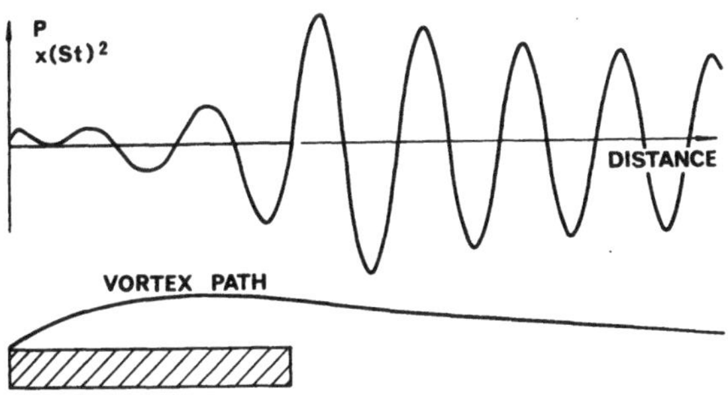

Fig. 2 Instantaneous acoustic power generated by a vortex moving through a β - mode acoustic field; Strouhal number (St) based on plate chord and upstream flow velocity = 1.7

Figure 2 shows the instantaneous contributions of a single vortex to the power integral as it follows the flow for a Strouhal number (St) of 1.7. The symmetries are such that all other vortices make the same contribution (beginning at different times). Since the vortices are shed periodically, the total energy they contribute over a lifetime, which is the integral of the power over time, is proportional to the acoustic power generated. These results illustrate points (i) and (ii) above but must be interpreted in the light of (iii) when evaluating a flow region as a potential sound source. Downstream of the plate the instantaneous powers remain large but become virtually sinusoidal, with little imbalance and hence approximately zero contribution over a cycle.

To illustrate the effects of imbalance we have applied a smoothing filter to the contributions which is equivalent to taking two successive moving averages over a cycle. This particularly suppresses effects at the sound frequency. The resulting curve (Figure 3) has the same total area underneath, but is much less oscillatory and identifies the major acoustic source and sink regions located near the edges, with the trailing edge dominating. The result at the trailing edge is particularly dependent on the phase of arrival of the vortices there (relative to the sound field) and in Figure 3, when the St is increased to 2.1, there is a dominant sink at the trailing edge and no net energy is available for resonance. Correspondence with experimental data is shown in Figure 4, where the computed power of the net source due to all the vortices over a range of St's is compared with experimentally recorded data.

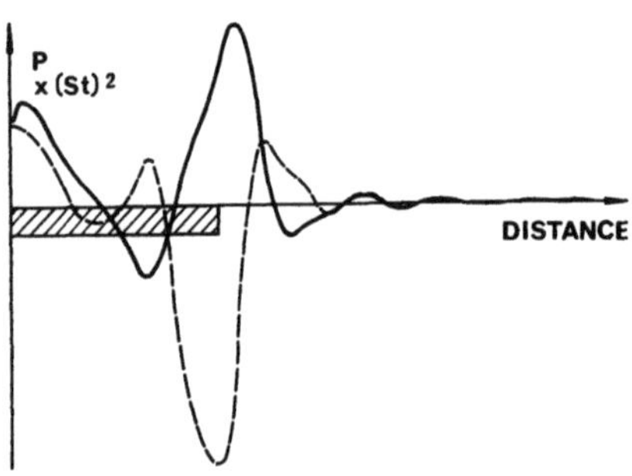

Fig. 3. Averaged acoustic power of one vortex over a full cycle; ——, St = 1.7; ----, St = 2.1.

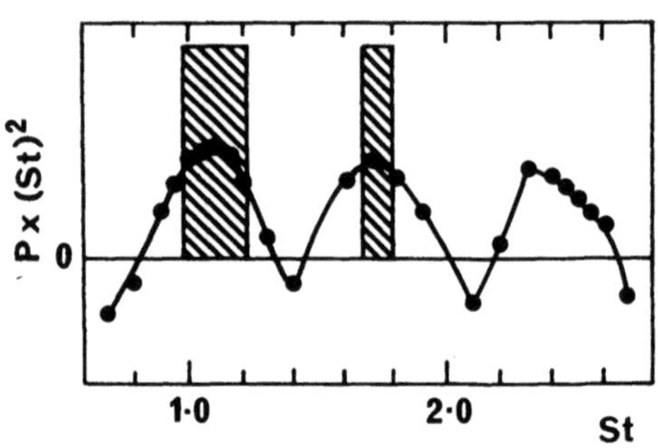

Fig. 4. Comparison between predicted total acoustic power, —•——•—, and experimentally observed resonance, [hatched box].

Conclusion

The evaluation of a momentum balance integral due to Howe with a time varying potential flow model provides a qualitative explanation of the fluid mechanics of the flow induced acoustic resonance process for a rectangular plate installed in a duct. The principles of this mathematical model also explains the fluid mechanics of the flow induced acoustic resonance generated with two pairs of baffles installed in a duct.

References

1. Rockwell, D. and Schachenmann, E. (1982), J.Fluid Mech., 425-441.
2. Bechert, D.W. and Pfizenmaier, E. (1975), J.Sound Vib., 581-587.
3. Howe, M.S. (1984), I.M.A. J.Appl. Math., 187-209.
4. Welsh, M.C., Stokes, A.N. and Parker, R. (1984), J.Sound Vib., 95(3),
305-323.
5. Nelson,P., Halliwell, N.A. and Doak, P.E. (1983), J.Sound Vib., 375-402.
6. Nomoto,H. and Culick, F.E.C. (1982), J.Sound Vib., 84(2), 247-252.
7. Rockwell, D. and Naudascher, E. (1979), Annual Rev. Fluid Mech., 67-94.
8. Stokes, A.N. and Welsh, M.C, (1985), J.Sound Vib., (in press).
9. R. Parker (1967), J.Sound Vib., 332-343.

Etude du Couplage Acoustique-Instabilités Aérodynamiques sur une Maquette de Statoréacteur

D. BIRON*, P. HEBRARD*, S. PAUZIN*, A. LAVERDANT**

ONERA-CERT/DERMES*, 2 Avenue Edouard Belin - 31055 TOULOUSE CEDEX
ONERA PALAISEAU OE**, Fort de Palaiseau - 91120 PALAISEAU

Résumé

Les résultats d'une étude expérimentale et numérique des phénomènes aérodynamique et acoustique apparaissant, lors du fonctionnement sans combustion d'une maquette à échelle réduite de statoréacteur de géométrie simplifiée sont présentés. Un ensemble de spectres de fluctuations de vitesse ou de pression permet de mettre en évidence des fréquences fixes, liées à l'acoustique de l'ensemble, et une fréquence moyenne fonction linéaire de la vitesse, associée au détachement tourbillonnaire périodique en aval de divers obstacles. Afin de préciser le couplage pouvant exister entre ces deux phénomènes, on étudie l'amplification des instabilités aérodynamiques résultant de la superposition d'une excitation acoustique proche de la fréquence naturelle. Des calculs, par la méthode des éléments finis, permettent de retrouver les différents modes acoustiques de l'ensemble.

Introduction

Au cours d'essais de mise au point de foyers de statoréacteur, la présence d'instabilités est observée soit dans le domaine des basses fréquences, quelques 100 Hz, soit dans une gamme plus élevée, quelques KHz. Jusqu'à présent le choix des solutions technologiques portant remède à ce problème est rendu délicat par l'absence de connaissances de base sur les phénomènes en présence. Les résultats présentés ici se rapportent à une étude /1/ ayant double objectif : d'une part, acquérir une description détaillée des mécanismes mis en jeu, et d'autre part, préciser l'influence des différents paramètres. Les conclusions de cette étude doivent pouvoir être utilisées sur des statoréacteurs à combustible liquide ou gazeux.

1. Description de l'installation et moyens de mesure

L'installation d'essais présentée (Figure 1) fonctionne en aspiration entre l'atmosphère et un circuit de vide assuré par trois pompes à anneaux liquides. Ce montage comporte de l'amont à l'aval : un caisson d'aspiration de 1 m^3, deux manches sur lesquelles sont montés en dérivation deux baffles permettant de

Aero- and Hydro-Acoustics
IUTAM Symposium Lyon 1985
Editors: G. Comte-Bellot and J. E. Ffowcs Williams

superposer à l'écoulement une excitation acoustique à fréquence et déphasage variables, le corps du stato (cylindre + sphère) muni d'un système porte sonde à déplacement programmable (3 directions). Les mesures effectuées, soit dans l'écoulement, soit en paroi utilisent une chaine anémomètrique DISA ou un microphone quart de pouce équipé de son préamplificateur. Les sorties de ces capteurs sont enregistrées sur un magnétique deux pistes puis dépouillées à l'aide d'un analyseur bicanal type 2032 B & K.

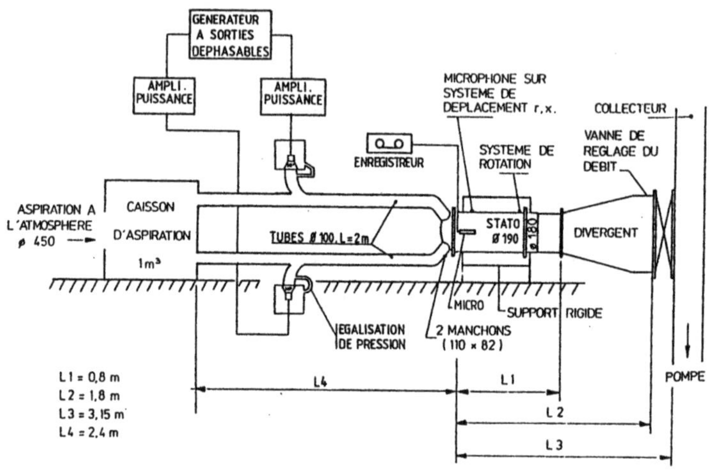

Figure 1 : Schéma de l'installation

2. Résultats : Essais et interprétations

2.1. Qualification acoustique

Après avoir évalué le niveau de bruit de fond de la veine pour plusieurs conditions aux limites (avec ou sans diffuseurs et tuyère), les mesures de pression pariétale ou dans l'écoulement nous ont permis de distinguer un certain nombre de pics à fréquences discrètes. Ces pics sont mis en évidence soit par une excitation acoustique large bande de l'ensemble sans écoulement, soit par l'écoulement lui-même et de façon d'autant plus nette que la vitesse est faible. Un cacul monodimensionnel permet d'identifier ces fréquences (fonction des conditions aux limites et des dimensions) comme associées aux premiers modes longitudinaux. A titre d'exemple, on présente :

. Les spectres (Figure 2) correspondant à une configuration

sans tuyère ni diffuseurs

$$L_1 = 0,8 \text{ m} \qquad f_1 = \frac{C}{2L_1} = 215 \text{ Hz} \qquad f_{exp} = 225 \text{ Hz}$$

$$L_2 = 1,8 \text{ m} \qquad f_2 = \frac{C}{2L_2} = 95,5 \text{ Hz} \qquad f_{exp} = 102 \text{ Hz}$$

$$L_3 = 3,15 \text{ m} \qquad f_3 = \frac{C}{2L_3} = 54,6 \text{ Hz} \qquad f_{exp} = 49 \text{ Hz}$$

$$L_4 = 2,4 \text{ m} \qquad f_4 = \frac{C}{2L_4} = 71,6 \text{ Hz} \qquad f_{exp} = 78 \text{ Hz}$$

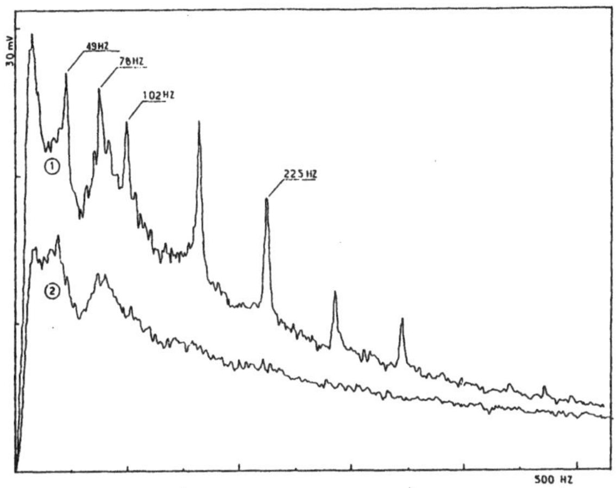

Figure 2 : Modes longitudinaux
1 microphone - 2 fil chaud.

. L'évolution axiale de l'amplitude des raies correspond , en présence de diffuseurs amont et tuyère aval, aux trois modes longitudinaux suivants : fondamental dans la chambre (317 Hz), fondamental (72 Hz) et premier harmonique (150 Hz) dans les manches (cf. Fig. 5). Ces résultats expérimentaux sont en parfait accord avec les résultats numériques présentés au § 3.

2.2. Instabilités aérodynamiques

Des mesures semblables effectuées à l'aide du fil chaud et du microphone muni de son ogive permettent de remarquer l'existence d'une fréquence (d'autant plus nette que la vitesse est faible) dont la valeur croît linéairement avec la vitesse (Figure 3). Ce phénomène, non détecté en paroi, mais particulièrement net dans le plan de sortie des manches est caractéristique d'instabilités aérodynamiques associées au détachement périodique de tourbillons alternés dans le sillage de différents obstacles. Par exemple, on obtient pour le sillage de pontets de forme triangulaire un nombre de Strouhal $St = \frac{fL}{V}$ (L base pontet, V vitesse dans les manches) de l'ordre de 0,12.

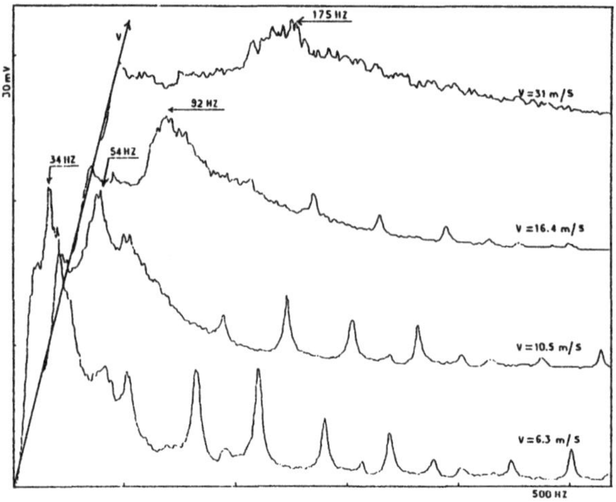

Figure 3 : Evolution de la fréquence F (vitesse en Amont du Pontet)

2.3. <u>Couplage acoustique - aérodynamique</u>

L'origine, des instabilités à basse fréquence, signalée plus haut pouvant trouver son explication dans un mécanisme d'amplification de ce dernier phénomène lorsque fréquences aérodynamique et acoustique sont voisines, il nous a paru intéressant d'étudier ce couplage en superposant une excitation acoustique à l'écoulement. Les résultats de la figure 4 correspondant à un cas d'écoulement V = 11,4 m/s et $f_{naturelle}$ = 70 Hz conduisent aux remarques suivantes : Une excitation à fréquence éloignée de la fréquence aérodynamique ne se traduit par une aucune amplification. Par contre, une excitation à même fréquence entraîne une amplification (R_A) de la pression acoustique d'autant plus importante que l'excitation (R_{Ej}) dans les deux manches se fait à niveau plus important et déphasage plus faible. Cette amplification (cf. Figure 4) qui présente un seuil et une valeur asymptotique traduit, comme l'ont montré d'autres études (Réf. 3, 4) L'augmentation de cohérence du mécanisme de détachement tourbillonnaire résultant du "pilotage" de ces instabilités par une fréquence voisine de la fréquence naturelle.

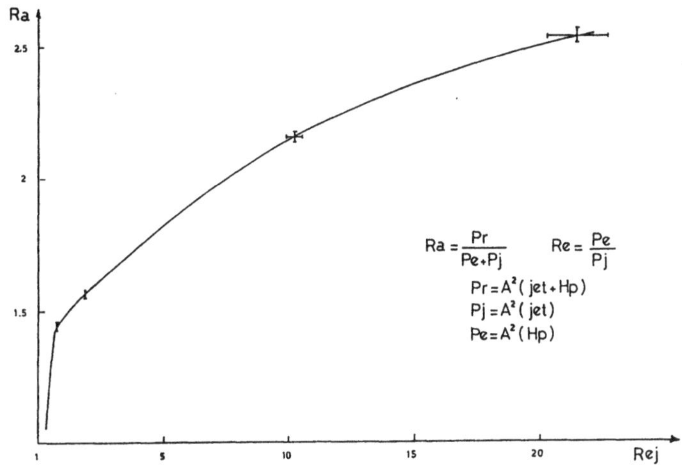

Figure 4

3. Etude des modes acoustiques

L'analyse modale est effectuée à partir de l'équation d'Helmholtz établie pour un milieu homogène au repos. Cette équation vérifie une condition de Neumann sur les parois de la cavité, le plan d'entrée de la tuyère (col amorcé) et sur l'axe de la chambre (mode symètrique), ou une condition de Dirichlet à l'entrée de la prise d'air, sur l'axe de la chambre (mode antisymétrique) et le plan d'entrée de la tuyère (absence d'écoulement).
Une formulation par éléments finis de type Galerkine peut être établie à l'aide du théorème de Green :

$$- \int_S \nabla \tilde{p} \cdot \nabla \phi \, dS + \frac{\omega^2}{c^2} \int_S \tilde{p} \; \phi \, dS = 0$$

où ϕ est une fonction test qui vérifie les conditions aux limites.
L'analyse modale est ensuite réalisée en divisant la surface de la cavité en éléments triangulaires où la pression est décrite par une approximation linéaire. Le problème de valeurs propres obtenu est résolu à l'aide d'un algorithme de Jacobi /5/.
Les calculs ont été effectués pour une chambre de longueur 0,605 m, une célérité du son de 344 m/sec et un rapport de sections entrée d'air/chambre identique à celui de l'installation.
Le deuxième mode longitudinal de la cavité composée de l'entrée d'air et de la chambre possède une fréquence de 149 Hz (Figure 5) tandis que le premier mode longitudinal de la chambre est couplé

avec un mode longitudinal de l'entrée d'air et présente une fréquence de 329 Hz (Figure 5). Des calculs effectués pour une cavité ouverte à chaque extrêmité donnent une fréquence de 69 Hz pour le premier mode longitudinal (Figure 5).

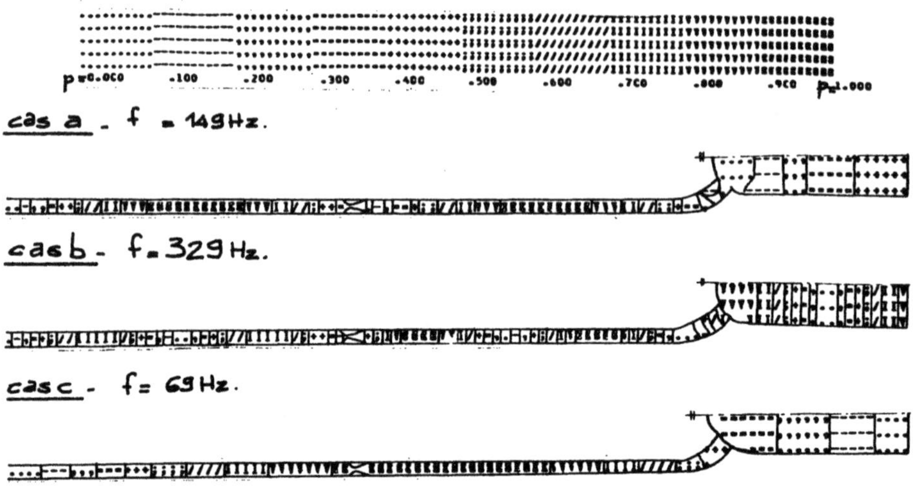

Figure 5 : Structure spatiale de la pression acoustique

a, b : cavité ouverte à l'amont et fermée à l'aval

c : cavité ouverte aux deux extrêmités.

L'examen des résultats numériques montre que l'identification des modes et des fréquences propres est en bon accord avec les résultats expérimentaux présentés en 2.1. (Cf. Figure 7 comparaison dans les manches et cf. Figure 6 comparaison dans le corps stato).

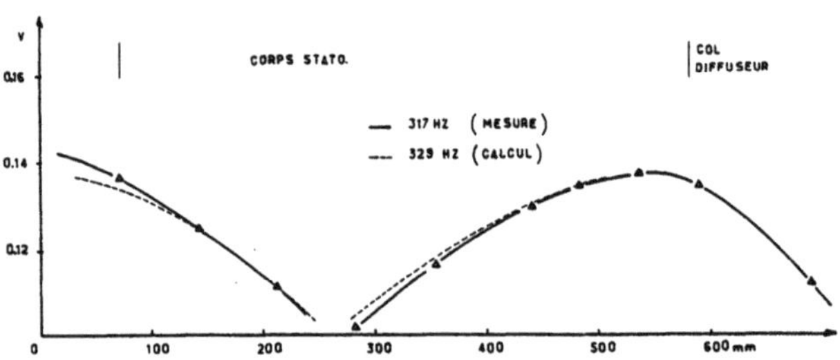

Figure 6 : Evolution du fondamental dans le corps.

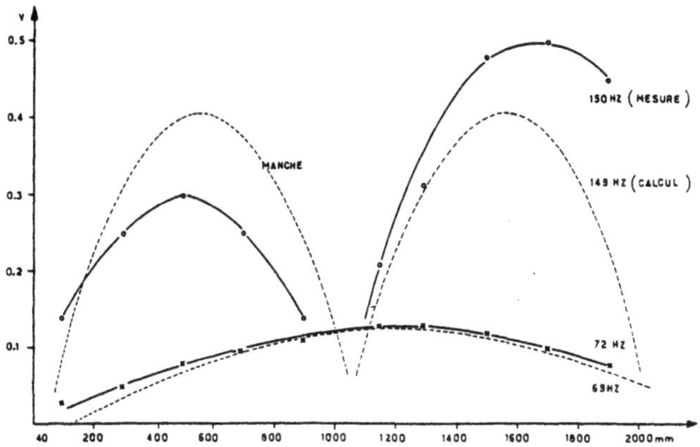

Figure 7 : Evolution du fondamental et premier harmonique dans les manches.

4. Conclusion

Devant la complexité des phénomènes intervenant dans la génération d'instabilités au cours du fonctionnement de statoréac - teurs, la présente étude fournit un élément de réponse permettant déjà d'expliquer certains résultats obtenus en combustion. Le nombre important de paramètres mis en jeu, nous a conduit depuis à orienter une partie de cette recherche vers une étude plus fondamentale, effectuée sur géomètrie simplifiée, sur l'influence des conditions aux limites et de la position de l'obstacle dans le champ acoustique sur ce couplage acoustique-instabilités d'écoulement.

Bibliographie

1 - BIRON, D.; HEBRARD, P.; PAUZIN, S.; : Rapport 2157/CERT-MES Février 1983.

2 . BIRON, D.; HEBRARD, P.; PAUZIN, S.; : Rapport 2182/CERT-MES Février 1984.

3 . BERGER, C.; GIOVANNINI, A.; HEBRARD, P.; MATHE, J.M., : Rapport 4/2029/CERT-MES. Novembre 1975.

4. BLEVINS, R.D.; Ed. VAN NOSTRAN : Flow induced vibration, 1977.

5. DHATT, G. ; TOUZOT, G.; : Une présentation de la méthode des éléments finis. Ed. MALOINE, PARIS, 1981.

Flow-Acoustics Interaction

On the Influence of Opening Geometry on Noise from Air Jets

PETER NØHR LARSEN & LEIF BJØRNØ

Industrial Acoustics Laboratory
Institute of Manufacturing Engineering
Technical University of Denmark
DK-2800 Lyngby, Denmark.

Summary

This paper presents a theoretical and experimental investigation of flow and noise data arising from a wide range of compressed air nozzles. The study includes narrow slits of different geometries, quadratic openings and combinations of 2 and more circular openings with the same or with different diameters. The experimental noise data are compared with theoretical predictions, and good agreements are found.

Introduction

Due to the extensive use of compressed air in industry free turbulent jets occur in many factories and workshops, where they among other things are used for cooling, cleaning, and transportation purposes. These jets are everywhere recognized as extremely powerful noise sources, with sound pressure levels frequently exceeding 100 dB(A).

Some attempts have been made to develop quiet outlet systems. But so far no systematic investigation of the influence of the geometry and dimensions of the nozzles on the emitted noise has been carried out.

Theoretical background

Using Lighthill's inhomogenous wave-equation /1/ and Batchelor's theory of axisymmetric turbulence /2/ it is shown /3/ and /4/ that the noise intensity received at a distance X from a subsonic air jet issued from a nozzle with opening area A and with the volumetric mean velocity U may be expressed by:

$$I(X) = \frac{A\rho}{X^2}\cdot\left(\frac{U^8\cdot G_1(\Theta)}{c^5\cdot(1-M_c\cos\Theta)^5} + \frac{U^4\cdot G_2(\Theta)}{c\cdot(1-M_c\cos\Theta)^3}\right) \qquad (1)$$

Where ρ is the ambient air density, c is the ambient sound ve-

Aero- and Hydro-Acoustics
IUTAM Symposium Lyon 1985
Editors: G. Comte-Bellot and J. E. Ffowcs Williams

locity, M_c is the eddy convection Mach-number, Θ is the angle of observation to the jet axis. G_1 and G_2 are dimensionless functions only being dependent on Θ, and they are properties of the nozzle. The first term on the right hand side of (1) will dominate by air jets in good flow conditions (e.g. laboratory jets), because this term represents the noise generated by the free turbulent mixing of the jet in the atmosphere (quadrupole noise). The last term will frequently dominate by commercial jets having poor flow conditions, because the term represents the noise generated by turbulence (and other disturbances of the flow) arising from positions up-stream the nozzle (monopole noise).

The frequency spectrum of the noise intensity emitted from an air jet may be shown to be expressed by a dimensionless function /3/. At right angle to the jet axis, this function only depends on the Strouhal's number, St.= f·a/U. f is the frequency, and a is a characteristic dimension of the nozzle. For a circular nozzle a is the nozzle diameter, and for a slit nozzle a is the smallest dimension of the slit. The frequency spectrum emitted at right angle to the jet axis thus becomes:

$$I_{90^\circ}(X,f) = K(St.)\,\frac{Aa\cdot\rho}{X^2}\cdot\left(\frac{G_1}{c^5}\cdot U^7 + \frac{G_2}{c}\cdot U^3\right)/(G_1 + G_2) \qquad (2)$$

where K(St.) is the dimensionless function.

Experimental set-up

Noise and flow data arising from 22 different nozzles have been studied experimentally. The nozzles are shown in figure 1. The noise data, i.e. sound power levels, frequency spectra, and sound pressure directivity patterns, are experimentally determined under anechoic conditions according to ISO 3745 as functions of volumetric mean velocity of the nozzle flow and pressure drop across the various nozzles.

Experimental results

In figure 2 the sound power level as a function of velocity is experimentally determined for the 4 slits tested. As may be seen a U^8-variation of sound power level is found for all 4

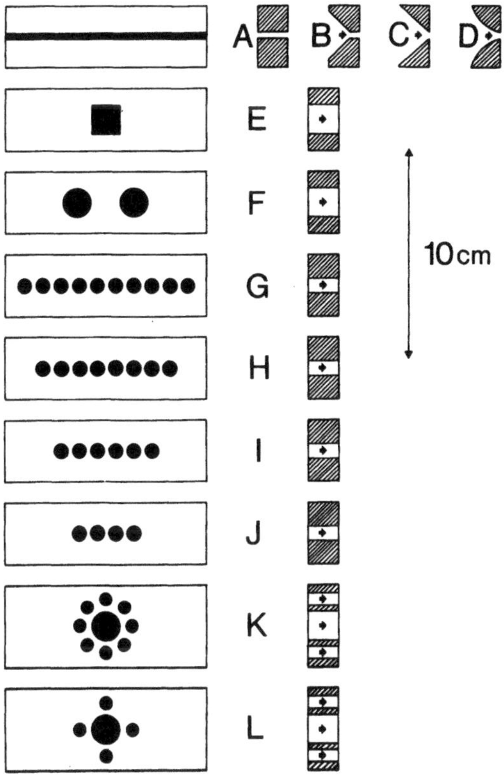

Figure 1. Geometries of the nozzles tested. Cross sections are shown with inducation of flow direction.

slits. This result corresponds very well with theory by (G_2=0). If sound power levels from jets issued from slit A are compared with the levels associated with slit B, a reduction of about 1.5 dB is observed. This is due to the convergent inlet section of slit B. In addition, if a smooth inlet section to the slit is used (slit D), an additional reduction in sound power level of about 1 dB may be obtained. Slit C represents the worst case, due to the sharp edge outlet of the slit. Its sound power level is 6 dB above the level of slit D, and it is most probably caused by increased turbulence. The frequency spectrum of the noise from slit jets is measured at an angle of 84.3° (angle according to ISO 3745) to the jet axis (predominant self noise). In order to be able to compare

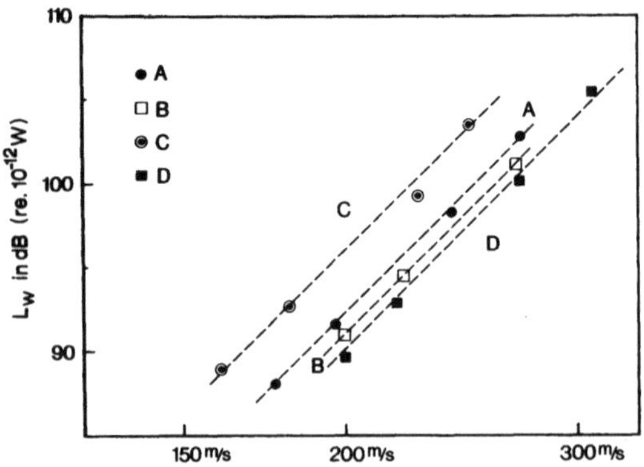

Figure 2. Measured sound power level emittted from nozzles A, B, C, and D as a function of volumetric mean velocity.

noise spectra from jets with different velocities and nozzle geometries, the measured spectrum is normalized according to Eq.(2) leading to a determination of K(St.). In figure 3 the normalized spectrum of the noise from a jet out of slit A at 4 different velocities may be seen. A good agreement between measured spectra is seen normalizing the data using Eq.(2).

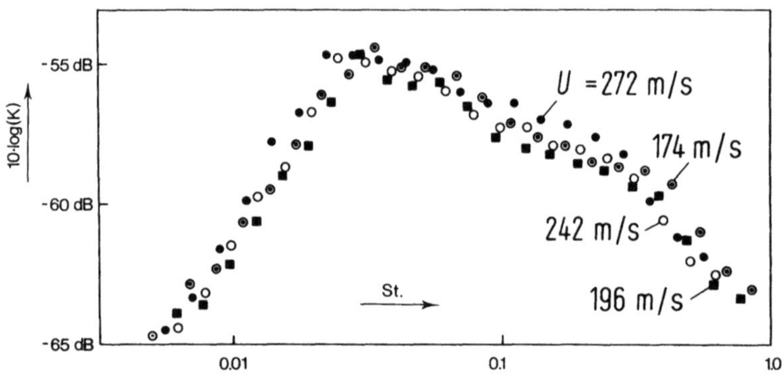

Figure 3. Normalized frequency spectra obtained from slit A with opening height a=1.5mm. Data from 4 different velocities are shown.

Apart from level the noise spectra from slits B, C, and D are found only to differ slightly from that of slit A.
In figure 4 the emitted sound power level from the quadratic opening, nozzle E, is shown as a function of volumetric mean velocity of the jet. The dimensions of the opening are 14mm x 14mm. Nozzles E may be regarded as a slit with aspect ratio

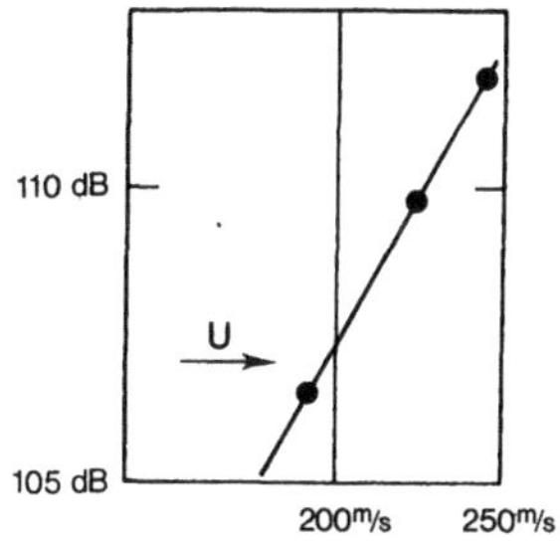

Figure 4. Measured sound power level emitted from nozzle E as a function of volumetric mean velocity. a=14 mm.

equal to one. If this is done, the results may be compared with those of nozzle A with opening height a=2 mm (equal opening area). Doing this by using figure 2 and Eq.(1) it is found that the quadratic opening is far noisier than the slit opening. This is due to the fact that the flows in the 4 corners of the quadratic nozzle constitute a greater part of the total flow through the nozzle, than they do in the case of a slit nozzle.

As a basis for the multi-jet nozzles, measurements are performed on a twin-jet nozzle with two 14 mm diameter outlets separated by the distance S, (measured between the two jet axes), nozzle F in figure 1. The emitted sound power level is found as a function of volumetric mean velocity for 4 different S-values, and are shown in figure 5. To be able to perform a direct comparison between the sound power levels of the twin-jets and of two similar but individual jets, the sum of sound power levels from two single jets (D=14 mm) is also shown in figure 5. This doubled sound power level is referred to as S = O. As can be seen from the figure, the sound power level from the twin-jet nozzles and for the sum of the two jets are essentially the same. This means that no reduction in the sound power level can be gained by placing two jets closely together. However, the main effect of placing two jets closely together is in the directivity pattern of the emitted noise, where a shielding effect based upon mutual sound refraction in the two jets is observed. In figure 6 the azimuthal directivity for nozzle F is shown. It is seen, that the greatest noise reduction is gained around $\varphi = 0°$ (i.e. in the plane of the

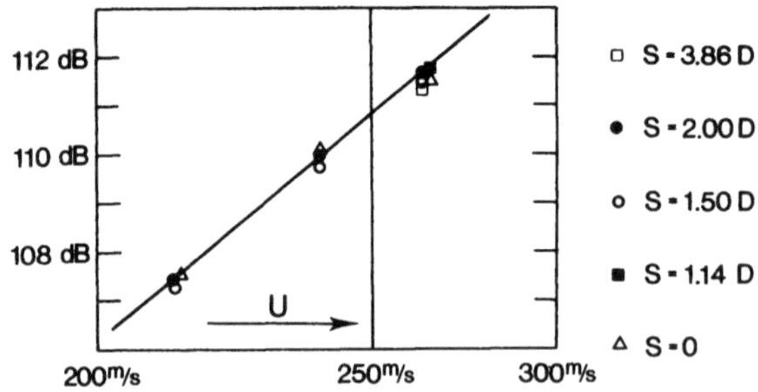

Figure 5. Measured sound power level emitted from nozzle F as a function of volumetric mean velocity. S is the distance between the two jet axes.

two jet axes), and an increase in noise level is observed at azimuth angles above approximately 50°. The azimuthal directivity of the nozzle with the largest jet separation distance deviates most from that of the single-jet nozzle.

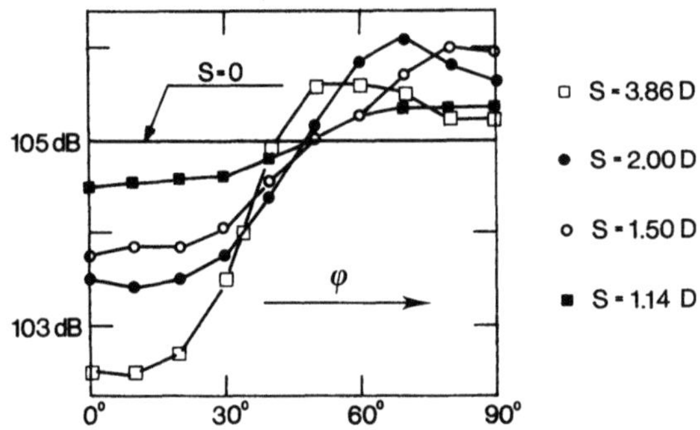

Figure 6. φ-directivity in sound pressure level measured from nozzle F in the plane Θ = 45°. U = 260 M/S. φ=0° corresponds to the plane of the 2 jet axes.

The results obtained from the twin-jet nozzle, nozzle F, may be applied for prediction of the noise data arising from nozzles consisting of more than two jets. As shown in figure 5, the emitted sound power level will not be influenced by placing two jets closely together. This means that the total sound power level arising from a multi-jet nozzle may be predicted if the sound power level from one single jet is known. In figure 7 the sound power level from a single jet (D=7mm) has been used to predict the sound power levels from nozzles consisting of 2-4-6-8 and 10 similar jets in line. And as can be seen, a very fine agreement exists between predicted and measured levels. In figure 8 the directivity in sound pressure level of a twin-jet nozzle has been used to predict the azimuthal directivity of two multi-jet nozzles. The one nozzle consists of 4 similar jets in line (nozzle J.) and the other nozzle consists of 8 similar jets in line (nozzle H). As can be seen, an overall good agreement exists between the predicted and measured levels.

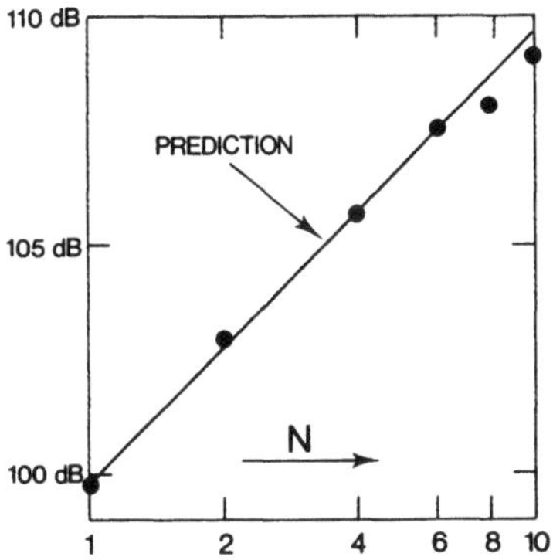

Figure 7. Measured and predic- sound power levels in dB emitted from multi-jet nozzles. N is the number of jets in line.

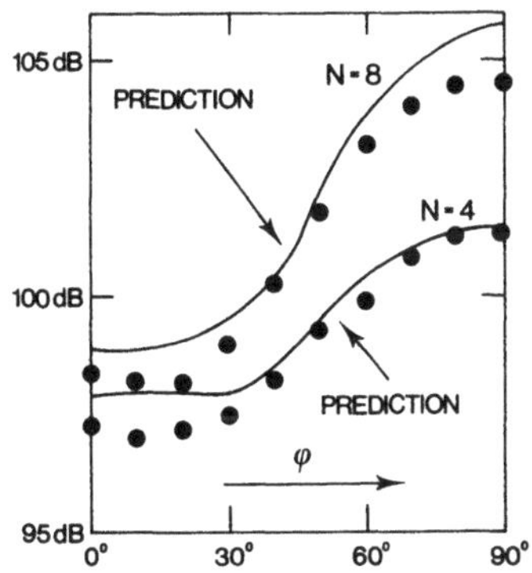

Figure 8. Measured and predicted φ-directivity in sound pressure level in dB emitted from multi-jet nozzles.

As seen from figure 8 a reduction in sound pressure level of approximately 7 dB in the most quiet direction may be gained by using a nozzle consisting of 8 jets in line.

Conclusions

It is found that the sound power level arising from a jet issued from a narrow slit with a smooth inlet section (nozzle D) is about 2.5 dB below the level of a jet from a slit with no inlet section (nozzle A). The experimental data from slit jets are found to be well represented by the theoretical model. Regarding the noise data arising from twin-jet nozzles it is found that the total emitted sound power level and the frequency spectrum of the noise will not be influenced by the separation distance between the 2 jets. The only effect of using

a nozzle consisting of 2 closely situated jets is in the directivity pattern of the emitted noise. Here sound refraction will lead to a noise reduction in the plane of the 2 jet axes and a consequent noise increase in the directions perpendicular to the plane. For the twin-jet nozzle a maximum reduction in SPL of about 2 dB is found. Using a nozzle consisting of 8 similar jets in line a further reduction in SPL of approximately 7 dB is found in the plane of the jet axes. But increasing the number of jets beyond 8 does not lead to a significant further reduction. The noise from the multi-jet nozzles may be predicted using noise data from twin-jet nozzles.
By using a nozzle consisting of one large outlet surrounded by an array of smaller outlets (nozzles K and L), it is found (data not shown in this paper) that the jets issued from the smaller outlets are able to deflect the noise from the central large jet. But due to the noise generated by the smaller jets themselves, the effect is not pronounced.
The results obtained here demonstrate that it is possible to design nozzles, which are more quiet and which will give noise reductions in preferred directions. But more research is still necessary in order to be able to design a nozzle having optimum noise qualities.

Acknowledgment

The financial support received from The Danish Council for Scientific and Industrial Research, from Laurits Andersens Fund, and from Fisker & Nielsens Fund is very gratefully acknowledged.

References

1. Lighthill, M.J., 'On sound generated aerodynamically, I (General theory)' . Proc. Roy. Soc., London (1952) A 211, pp. 564-587.

2. Batchelor, G.K., 'Theory of axisymmetric turbulence' . Proc. Roy. Soc., London (1946) A 186, pp. 480-502.

3. Larsen, P.N. 'Noise generated by air jets from a rectangular slit' . Ph.D. thesis, Technical University of Denmark, Dec. 1983.

4. Bjørnø, L. and Larsen, P.N., 'Noise of air jets from rectangular slits' . Acustica (1984), Vol. 54, pp. 247-255.

Response of a Flat Plate Turbulent Boundary Layer to Intense Sound Waves Moving in the Upstream Direction

E. BROCHER, E. DUPORT and M.P. DEVIC

Institut de Mécanique des Fluides
1, rue Honnorat, 13003 Marseille

Abstract

An experimental study has been undertaken to determine how a turbulent boundary layer developing on a flat plate responds to plane sound waves moving in the upstream direction. These waves are generated by a resonator located downstream of the flat plate. The resonator produces a nearly pure tone with a fundamental frequency of 200 Hz and an intensity slightly higher than 140 dB. Measurements made with hot wires across the boundary layer show that the velocity amplitude is much larger within the boundary layer than outside. Also, the relative amplitude of harmonics increases in the vicinity of the plate.

Introduction

In fluid mechanics, there are several configurations in which strong flow oscillations spontaneously occur. These configurations include : diffusors, sudden enlargments, turbomachines, spiked bodies and Hartmann - Sprenger tubes. The origin of these oscillations is not well known but what appears to be common to these flows is a periodic boundary layer separation when the oscillations have reached their full amplitude. This separation could be provoked by positive pressure waves moving in the upstream direction. In Hartmann-Sprenger tubes, for instance, it has been observed that very stable and strong oscillations can be obtained in the 3 configurations depicted in Fig. 1, even if the jet is subsonic (and not an underexpanded supersonic jet as is the case for the classical Hartmann - Sprenger tube).

In Savory's experiments (Fig.1a), plates were placed near the jet boundary. It was found [1] that if the plates were moved perpendicularly to the jet axis, oscillations would occur in the cavity as soon as the plates would touch the jet boundary. For the needle configuration [3], it has been demonstrated that the needle diameter needs only to be a few percent of the nozzle diameter to obtain oscillations of optimal amplitude. In

Aero- and Hydro-Acoustics
IUTAM Symposium Lyon 1985
Editors: G. Comte-Bellot and J. E. Ffowcs Williams

these 3 configurations, the boundary layer developing on the plates (Savory), on the stem (Hartmann) and on the needle (Brocher) is submitted to waves reflected by the end wall of the cavity and separates periodically.

In view of the important flow configurations listed above for which periodic boundary layer separation occurs, it appears worthwhile to undertake a fundamental study on the response of a turbulent boundary layer to sound waves propagating in the upstream direction.

In recent years, a large amount of papers [4] has been devoted to the study of unsteady boundary layers but they were dealing with an external flow obeying the relation

$$u\ (x,t) = u_o + u_1 \sin \omega t$$

Patel [5,6] was the first, and presumably the only one, to study both experimentally and theoretically the response of a boundary layer with a free stream flow of the type

$$u\ (x,t) = u_o + u_1 \sin \omega(t - \frac{x}{Q})$$

where Q is the wave propagation velocity. Patel theoretically showed that the external velocity fluctuation can be amplified within the boundary layer, the amplification depending on the traveling wave velocity ratio $K = u_0/Q$ and on the frequency parameter $\bar{\omega} = \omega x/u_0$. In Patel 's experiment, the velocity fluctuations were produced by travelling vortices and K was always positive whereas in the experiments reported here K is negative.

Experimental set-up

A schematic view of the experimental set up is shown in Fig. 2.

Wind tunnel

Experiments have been carried out in the new subsonic, open-circuit, wind-tunnel S_4 of the I.M.F.M. The test section is 50 cm x 50 cm and 200 cm in length. All the test reported here were run with a free stream velocity of 50 m/s. At this speed, the turbulence level in the test section is about 2.5 %.

Flat plate

The flat plate is 150 cm in length, has the same width as the test section and is 15 mm thick. Its leading edge has an elliptical shape with an axis ratio of 5. Machining of the leading edge creates small steps which have been left on purpose so as to trigger turbulence on the flat plate. Pressure gauges (Endevco 8510 B-5) are inserted in the plate every 10 cm along the plate.

Resonator

A resonator is placed downstream of the flat plate. Its width is the same as that of the test section and its inside height is 39 mm. Its depth is 397 mm. The resonator mouth is located 14 cm downstream of the plate trailing edge. The depth L of the resonator determines the frequency f of the acoustic waves ($f = a/4L$, where a represents the speed of sound). This frequency is nearly equal to 200 Hz with the length L choosen.

The resonator is triggered by a small wedge located upstream of its mouth. The performances of the resonator were optimised by varying the wedge cross-section and the distance ℓ between the wedge apex and the resonator. The pressure oscillations were recorded with Kistler gauges mounted on the end plate of the resonator.

The criteria retained for the optimisation were

- stability of the oscillations
- largest pressure amplitude possible
- low harmonic content
- absence of transversal mode within the resonator.

The best results were obtained with a wedge 7.6 mm in height and 9 mm in width located at ℓ = 59 mm. With this configuration the pressure amplitude was 245 mbar and the harmonics amplitudes were less than 2 % of that of the fundamental frequency.

It was previously shown [7] that the maximum pressure amplitude in resonators of the Hartmann - Sprenger type is equal to $\Delta p_{max} \cong 2 \gamma M p_a$ where γ represents the ratio of specific heats, M the Mach number of the driving flow and p_a the ambient pressure. For air at 50 m/s this gives $\Delta p_{max} \cong 410$ mbar. Thus, the largest amplitude obtained represents about 60 % of the maximum achievable amplitude. This ratio is less than what has been measured in H-S tubes with square cross-section (up to nearly 100 %), but the difficulty with the resonator used in this experiment is caused by the very large width-to-height ratio (12.5). It has been found necessary to insert 2 dividing plates in the resonator to guide the flow in the axial direction so as to avoid transverse modes. Moreover, to synchronize the oscillations in the 3 cavities so created, a spacing of 50 mm was left between the downstream edge of these plates and the end plate of the resonator.

The mechanism of excitation of the resonator by the wedge is not fully understood at the present. Presumably, the vortices created by the wedge impinge on the lips of the resonator and the acoustic waves moving up and down the resonator interact with the flow around the lips. It is likely that there exists some analogy with the physics of organ pipes. It is planned to visualise the flow with the Schlieren technique or with smoke to better understand the oscillation mechanism.

Hot-wire support

Hot-wire measurements have been made in the boundary layer developing over the flat plate. Since the sound level is very high (more than 140 dB) great care was taken to avoid spurious readings due to hot-wire vibrations. A special displacement mechanism was designed. Streamlining of the vertical support was made. The hot-wire ceramic was glued into a metallic tube which was then firmly clamped into the support. Hot-wire signals were sent on a FFT analyser (HP 3522 A).

Results

It was first verified that, without the sound waves, the velocity profiles corresponded to a well developped turbulent boundary layer on the flat plate.

Pressure measurements on the flat plate gave peak-to-peak amplitude of 7 mb with no significant attenuation of the waves moving in the upstream direction. This corresponds to a sound intensity slightly higher than 140 dB. The wave propagation speed was verified to be very nearly equal to $-(a - u_\infty)$.

Typical frequency spectra are shown on Fig. 3. Outside of the boundary layer ($u/u_\infty = 1$), the turbulence is very low so that velocity fluctuations due to the acoustic field dominate. The harmonics (408 and 612 Hz) can be seen but they are much less intense than the fundamental frequency (204 Hz). At some intermediate position within the boundary layer ($u/u_\infty = 0.52$), the turbulence level is high and the only peak which can be seen corresponds to the fundamental frequency. Nearer to the plate ($u/u_\infty = 0.29$), turbulence level is lower and harmonics can be seen again.

Amplification of the fundamental frequency across the boundary layer is shown on Fig. 4. It is observed that the amplification becomes very high near the plate. Such high amplification does not appear to be predicted by Patel's theory and the reason for this discrepancy will be investigated.

Amplification of the fundamental frequency along the flat plate at $u/u_\infty = 0.5$ is shown on Fig. 5. It is seen to vary substantially with the frequency parameter $\bar{\omega}$.

Future work

The results presented in this paper are of exploratory nature and a very intensive research programm is now planned. Instead of sending the signal directly on the FFT analyser, ensemble average will be performed on the data so as to get more significant results. In particular, amplification and phase shift will be determined on ensemble average values.

After the measurements with the 0-spatial pressure gradient have been completed, a diffusor effect ($\partial p/\partial x > 0$) will be imposed to the flat plate boundary layer. The response of the boundary layer should be even larger in this case and acoustic waves should eventually lead to unsteady separation.

Aknowledgments

This work has been performed under D.R.E.T.-Contrat n° 82/490.

References

1. Savory, L.E. : Experiments with the Hartmann acoustic generator. Engineering 170 (August 1950) 99, 100, 136, 137, 138.
2. Hartmann, J.; Trudso, E. : Synchronisation of air-jet generators with an appendix on the stem-generator. Dan. Mat. Fys. Medd. 26 (1950).
3. Brocher, E.; Maresca, C. : Condition de fonctionnement d'un tube de résonance alimenté par un jet subsonique. C.R. Acad. Sci. (Paris) 268(A) (1969), 749-752.
4. Michel, R.; Cousteix, J.; Houdeville, R. (eds) : Unsteady turbulent shear flows. Springer-Verlag, Berlin, Heidelberg, New-York, 1981.
5. Patel, M.H. : On laminar boundary layers in oscillatory flow. Proc. R. Soc. Lond. A, 347 (1975) 99-123.
6. Patel, M.H. : On turbulent boundary layers in oscillatory flow. Proc. R. Soc. Lond. A, 353 (1977) 121-144.
7. Brocher, E.; Maresca, C. : Recherches sur les tubes de résonance excités par un jet subsonique. J. Mécanique 8 (1969), 21-39.

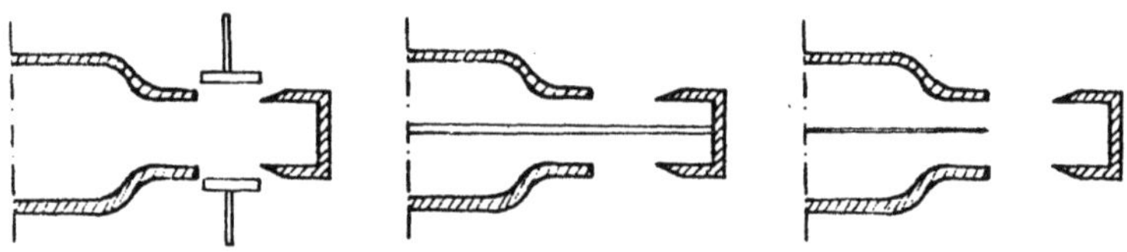

a) Savory [1] b) Stem-generator Hartmann [2] c) Needle-generator Brocher et al. [3]

Fig. 1. Hartmann-Sprenger tubes configurations

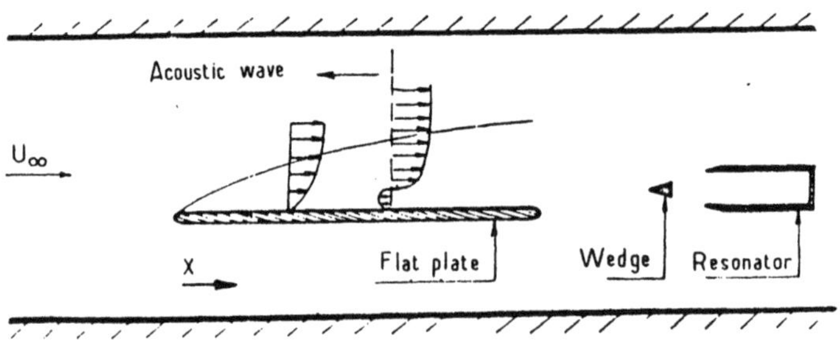

Fig. 2. Experimental set-up (schematic)

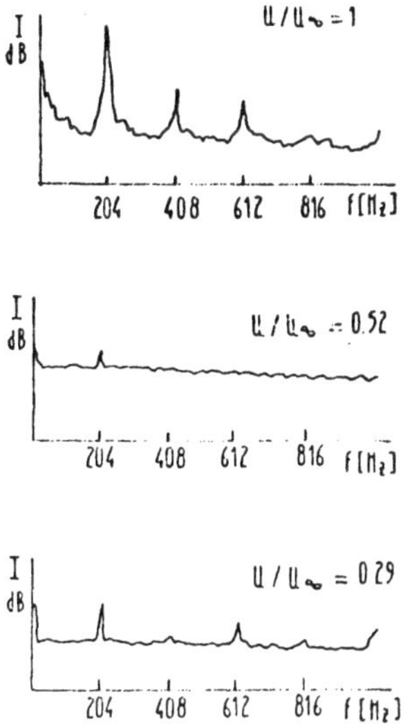

Fig. 3. Typical velocity fluctuations spectra

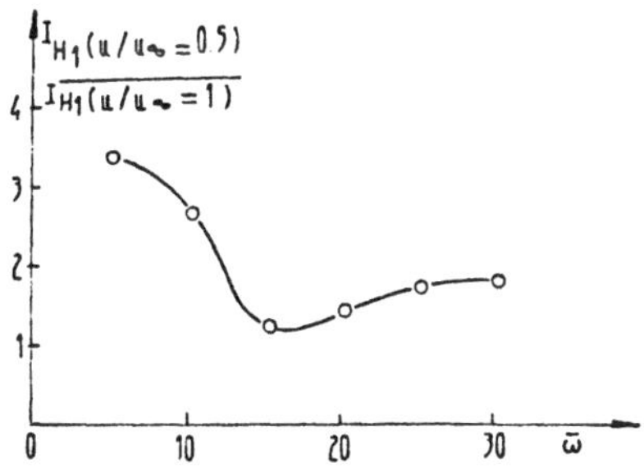

Fig. 5. Amplification of fundamental frequency along the plate

Fig.4. Amplification of fundamental frequency H_1 across boundary layer

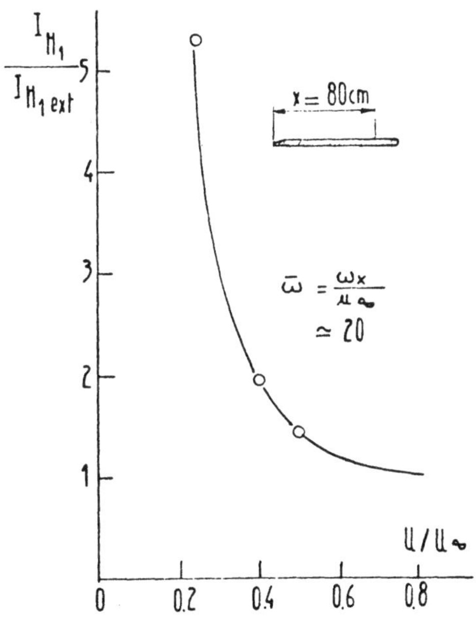

An Experiment on Sound Turbulence Interaction by 'Impulsive Sound Tube'

H.YOSHIDA, H.HONMA, A.KAWAGUCHI* and K.UZAWA

Department of Mechanical Engineering,
Chiba University, Chiba 260, Japan
(* Mitsubishi Heavy Industries Co.)

Summary

The 'Impulsive Sound Tube (IST)' was constructed to study sound turbulence interaction mechanisms and, through the present experiment, it was found to be a useful tool for such a purpose. The IST is a kind of suction wind tunnel composed of a long circular tube with a sound source by spark discharge and a vacuum tank. Sound waves with amplitude below 100Pa and duration of 2-3ms are generated in fully-developed turbulent flows in the tube. Downstream propagation is observed (flow Mach number M<0.1). It is found that the flow speed has a remarkable influence on the waveform of the impulsive sound. The net attenuation of acoustic energy, which is evaluated from integration over the frequency range from 0 to 10kHz, is particularly noticeable for velocities U_m>20m/s (R_e>41,200). For these velocities considerable attenuation occurs in low frequencies below 2kHz. The frequency shift or spectral broadening of the impulsive wave due to turbulent flow is not clearly confirmed in the present experiment.

Introduction

The 'Impulsive Sound Tube (IST)' was constructed to facilitate studies on the interaction mechanisms between sound waves and turbulent flows[1]. Use of the impulse technique enables us to separate the incident wave from the reflected wave. Moreover, since each of the pulses contains spectral information over wider frequency range, the propagation characteristics can be computed from a single test. The most important matter is that, by using the impulsive waves, a precise investigation on substantial (or net) change of the acoustic energy due to turbulent flow is possible. To observe such a change of acoustic energy caused by the turbulent flow is the main subject of the present study.

The sound propagation through turbulent pipe flow has been investigated by many authors[2-4], mostly by using the standing waves with fixed frequencies. For certain flow conditions and

Aero- and Hydro-Acoustics
IUTAM Symposium Lyon 1985
Editors: G. Comte-Bellot and J. E. Ffowcs Williams

frequencies, the extra attenuation of sound due to flow was observed[3,4]. However, most of the results concern about the attenuation for fixed, individual frequencies because of the use of the standing wave technique. By using this technique, it seems to be difficult to examine the sound turbulence interactions in which the frequency shift or spectral broadening of sound wave and also the net acoustic energy change occur. These phenomena have not yet been thoroughly understood.

In this paper results of the impulsive waves propagating through the turbulent pipe flow are presented. Based on these results the net change of the acoustic energy and spectral broadening effect due to turbulent flow are examined.

Impulsive Sound Tube (IST)

In Fig.1 is shown a schematic diagram of the IST. The IST is a kind of suction type wind tunnel, which is composed of an impulsive sound-source by spark discharge, a vacuum tank ($2m^3$) and a 8.5m long, 31mm I.D. and 3.6mm thick circular tube made of polyvinyl chloride. The tube inlet is open to the atmosphere.

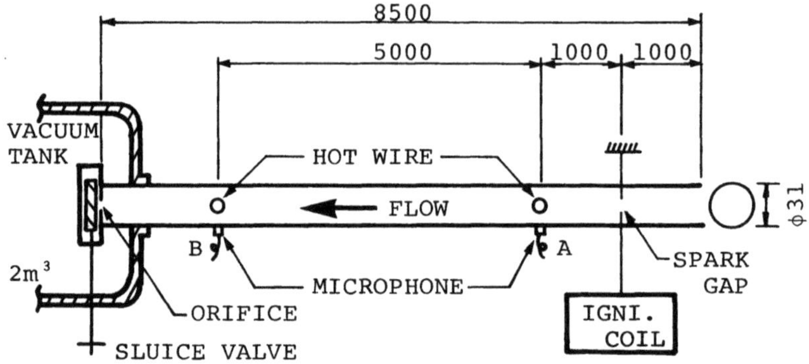

Fig.1 A schematic diagram of 'Impulsive Sound Tube'

First the vacuum tank is evacuated below 133Pa(1mmHg) by a rotary pump, and then the sluice valve is opened to produce in the tube a quasi-steady, turbulent flow with low background noise. The steady, quiet flow is established within a few seconds after opening the valve and it continues as long as the flow at the orifice (placed at the tube outlet) is choked: the duration of the steady flow is 30-100s for orifices of 1-15mm. The rotary pump is stopped during the acoustic measurements. While the flow is steady, an impulsive wave is emitted by spark discharge. The acoustic signals are perceived by two microphones(B&K 4135, 6mm O.D.) at the locations A and B. The outputs of them are first recorded by the transient memories(KAWASAKI TM-1410) in digital forms(5-10μs sampling rate, 1kword) and later analyzed by a mini-computer. The signal averaging technique[5] is used to increase the S/N ratio of the acoustic data. Note that in the IST system one does not need any special devices such as anechoic rooms or muffler sections.

Waveforms for Various Flow Velocities

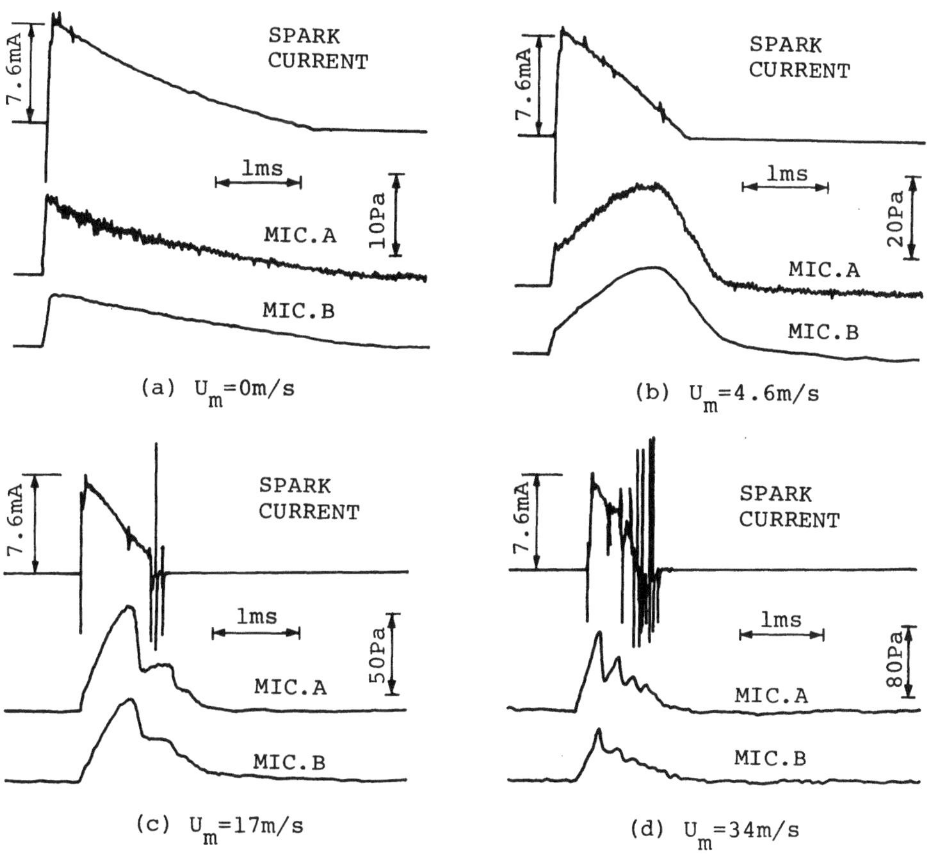

Fig.2 Spark currents and acoustic pressures observed in IST.

Fig.2 shows the discharge currents of the source and the corresponding microphone output signals at the locations A and B. Profiles of the discharge currents, which are measured by inserting a 10Ω resistance into the earth-side of the electrode, is basically triangular shape with an initial negative pulse irrespective of the flow velocity. For higher velocities, the basic (triangular) current profile is disturbed by secondary small pulses. The duration of the discharge decreases from about 3ms to 1ms with increase of the mean flow velocity U_m, while the magnitude of the first positive peak current is almost constant(≃7.6mA) for all cases. Regarding the acoustic waves produced by the spark discharge, their waveform is strongly affected by the flow. That is, for U_m= 0m/s(Fig.2(a))it is, similar to the discharge current, triangular and has a shock-like discontinuous front. Moreover, high-frequency pressure fluctuations at A disappear at B, and also the peak is rounded due to the viscous effect of the fluid. For U_m=4.6m/s (R_e=9,500, Fig.2(b)) and 17m/s(R_e=35,100, Fig.2(c)) the discontinuous fronts are followed by a gradual increase in acoustic pressure, which reaches its maximum in the latter half of the wave. For U_m=34m/s(R_e=70,200 , Fig.2(d)), a remarkable oscillation of pressure appears, which seems to be related to the corresponding discharge characteristic. The oscillation mentioned above decays significantly during the propagation. Moreover, in this case the discontinuous front is merged in the main pulse and cannot be distinguished.

Attenuation Spectra of the Impulsive Waves

In Fig.3 the measured attenuation spectra for several flow velocities are plotted with the theoretical static value[6]. For the case of U_m=0m/s(Fig.3(a)), over most frequencies up to about 5kHz, the measured value is slightly larger than the theoretical one. However, both values show quite similar frequency dependence. For higher frequencies above 5kHz the measured value fluctuates appreciably. This is considered to be due to the effect of the higher-order mode attenuations:the first circumferential mode is estimated to appear around 6.5kHz. As shown in Figs.3(a)-(c), effect of the flow on the attenuation, which appears as a deviation from the static attenuation, starts to appear from higher

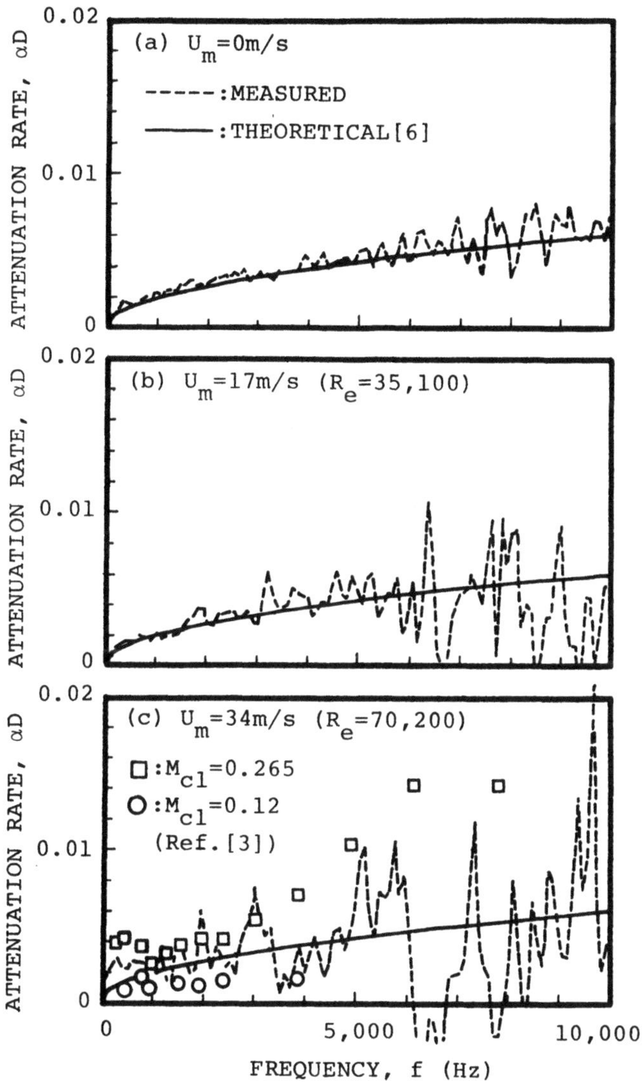

Fig.3 Attenuation spectra of the impulsive sound waves.

frequencies to lower frequencies with increase of U_m. That is, for the highest velocity case of U_m=34m/s(R_e=70,200 ,Fig.3(c)), increase of the attenuation becomes noticeable in the lower frequencies up to about 3kHz. Although the flow conditions are somewhat different from the present ones, results obtained by the standing waves[3] are reproduced in Fig.3(c) as a comparison.

Net Attenuation of Acoustic Energy due to Flow

We introduce a dimensionless parameter defined as

$$\Delta E_{f_1,f_2} = (E^A_{f_1,f_2} - E^B_{f_1,f_2})/(E^A_{f_1,f_2} - E^{B'}_{f_1,f_2}),$$

where $E^i_{f_1,f_2} = \int_{f_1}^{f_2} |P^i(f)|^2 df$, (i = A or B)

$$E^{B'}_{f_1,f_2} = \int_{f_1}^{f_2} \left(|P^A(f)|\exp[-\alpha_0(f)L]\right)^2 df,$$

$|P(f)|$ is the absolute value of the Fourier amplitude, the superscripts A and B indicate values at the locations A and B, respectively, $\alpha_0(f)$ is the theoretical static attenuation rate[6] and L is the distance between A and B. Therefore, ΔE gives us a measure of the net acoustic energy change due to flow (not to the viscous effect of the fluid). In Fig.4 the ΔE for three frequency ranges (0-2kHz, 0-4kHz and 0-10kHz) are plotted as a function of U_m. For U_m<20m/s (R_e< 41,200) the extra attenuation due to flow does not appear in the considered frequency ranges, while for U_m> 20m/s a marked increase in the extra attenuation occurs. Particularly, the acoustic energy of the lower frequencies (0-2kHz) is most attenuated. It is said that the appreciable extra attenuation

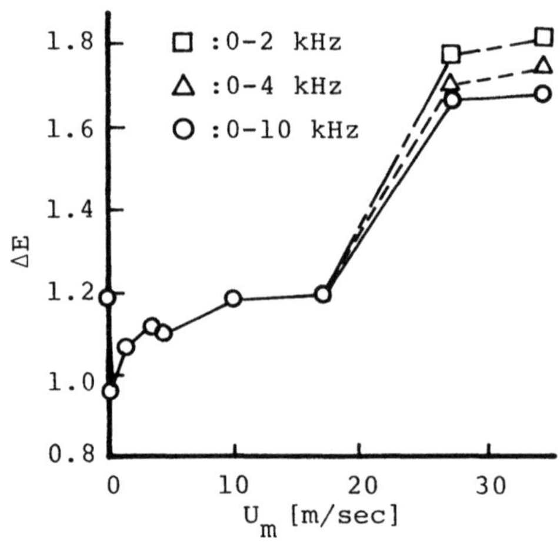

Fig.4 Net attenuation of the impulsive waves vs. U_m.

occurs when the acoustic boundary layer thickness, δ_a, becomes comparable with the viscous sublayer, δ_v, [3]. For the present turbulent flows ($1.6m/s < U_m < 34m/s$ and measured pipe friction factor $\lesssim 0.01$), δ_a becomes comparable with δ_v in the frequency range from about 100Hz to 2,000Hz. In other words, in this frequency range, the extra attenuation is expected to appear for the velocity range considered. This seems to be true for velocities $U_m > 20m/s$ but not for $U_m < 20m/s$. This fact probably implies that the attenuation of the impulsive wave depends not only on the geometric parameter such as δ_a/δ_v but also on the U_m or the turbulent velocity, U_t, itself. Note that in the present impulsive waves large amplitude components concentrate in the frequencies up to about 2-3kHz (see Fig.5). Therefore, the sound turbulence interactions which occur within these frequencies are expected to be easily perceived by the present technique.

In Fig.5 the measured amplitude $P^B(f)$ is compared with the modified one $P^{B'}(f)$ $\left(= P^A(f)\exp[-\alpha_0(f)L]\right)$ for $U_m = 34m/s$. The result in this figure also shows noticeable extra attenuation in lower frequencies mentioned previously. On the other hand, under the present flow conditions ($R_e < 70,200$ and $U_t < 1.1m/s$), the frequency shift or spectral broadening due to turbulence, on which Sakao [7] confirmed by different method, is not clearly confirmed at least within the plane mode frequencies (cf. $U_t = 3.2m/s$ in [7]).

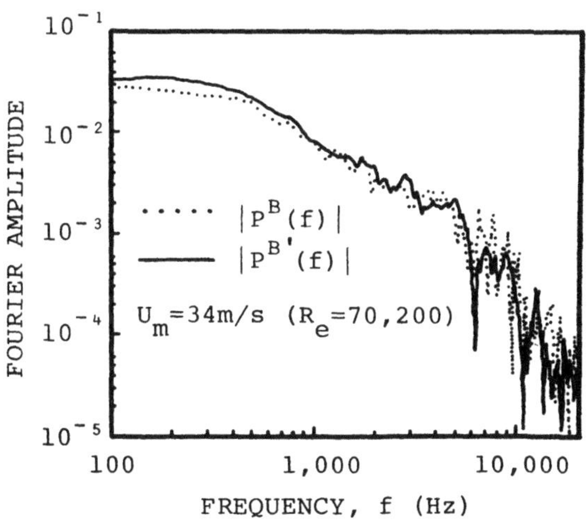

Fig.5 Fourier amplitude of the impulsive wave at B.

Concluding Remarks

An experiment on sound turbulence interaction has been made by using the Impulsive Sound Tube. Downstream propagation for the mean flow Mach numbers below 0.1 is observed. The main results can be summarized as follows:

The waveform of the impulsive wave generated by the spark discharge is strongly affected by the flow. That is, for example, its duration noticeably decreases with increase of U_m. The influence the turbulent flow on the attenuation spectrum appears in higher frequencies for low speed flows. Its range extends from higher to lower frequencies as U_m or U_t increases. The net attenuation of the acoustic energy, which is evaluated from integration over the frequencies from 0 to 10kHz, is particularly noticeable for U_m>20m/s (R_e>41,200). In this flow range, the energy of 0-2kHz frequency range is most attenuated. The net attenuation seems to depend not only on δ_a/δ_v but also on the U_m or U_t itself. The frequency shift or spectral broadening of the impulsive wave is not clearly confirmed in the present experiment.

Acknowledgment

The authors wish to thank Prof. T.Ueno for his helpful advice on the signal processing technique and Messrs. Y.Taketani, Y.Tsujino and T.Yano for their assistance in preparing the figures in this paper. This work was supported by the Ministry of Education under Grant-in-Aid for Scientific Research.

References

1. Honma,H.,Yoshida,H. and Kawaguchi,A., Proc. 14th Int. Symp. on Shock Tubes & Waves, 247-254 (1984).
2. Comparin,R.J.,Rapp,J.R. and Singh,R., J.Acoust.Soc. Am 72,7-12 (1982).
3. Kuhn,F. and Morfey,C.L., J. Sound Vib. 44, 525-529 (1976).
4. Ahrens,C. and Ronneberger,D., ACUSTICA 25, 150-157 (1971).
5. Salikuddin,M.,Ahuja,K.K. and Broun,W.H.,J.Sound Vib. 94,33-61 (1984).
6. Lord Rayleigh, 1949 Theory of Sound Vol.2,New York,Dover Pub.
7. Sakao,F., NAGARE J.Japn.Soc.Fluid Mech. 1,360-372 (1982).

Disturbance Induced by a Boundary Layer on Sound Sources Location. Application to the Case of a Vehicle.

H. ARBEY*, M. CHIOLLAZ**, P. DUFOURCQ***, B. ESCUDIE**

* Institut National de Recherche et de Sécurité, 54501 Vandoeuvre

** Institut de Chimie et Physique Industrielles, 69288 Lyon

*** Ecole Centrale de Lyon, 69131 Ecully

Introduction

Interfering Acoustic Imaging Technique in far field describes the distribution of acoustic intensity radiated by a sound source as a function of angle and frequency. Such a technique gives a description of the different sources responsible for noise level observed at a given measurement point. This process can be seen as a practical "tool" to describe and analyse the different noise sources from a vehicle. However the mean flow due to vehicle motion induces disturbances which can modify sources radiation [1], [2], [3]. Therefore it is necessary to estimate the disturbance influence on Acoustic Imaging Technique.

For this purpose two experiments have been made. The first one concerns the analysis of the disturbance induced by a mean flow when locating sound sources. The other intends to apply Acoustic Imaging Technique to a vehicle (High Speed Train) moving with a velocity of U = 70 m/s.

Disturbance induced by a boundary layer on sound sources location

Experimental conditions :

The experimental set-up was placed in the anechoïc room of the Ecole Centrale de Lyon. It consisted in a half-opened wind tunnel as described by Dufourcq [4], (fig. 1). A plane plate (2 m long and 0.20 m wide) was fixed along the smallest lip of the rectangular mozzle exit. This plane plate was flanked by two walls to obtain a two-dimensionnal mean flow along the spanwise direction x_3.

The velocity field characteristics (mean velocity $U(x_2)$ and fluctuating velocity $u(x)_2$) were measured with a hot-wire probe and a constant tempe

Aero- and Hydro-Acoustics
IUTAM Symposium Lyon 1985
Editors: G. Comte-Bellot and J. E. Ffowcs Williams

rature anemometer. The fluctuating pressure field characteristics induced on the plane plate was measured by condenser microphones with a linear frequency response in the range 2 Hz - 10 kHz. These microphones were fixed downstream to the plate at the following coordinates :
π_1 : x_1 = 0.2 m ; π_2 : x_1 = 0.4 m ; π_3 : x_1 = 0.6 m.

The sound sources referred as E_1, E_2, E_3 and E_4 consisted in piezoelectric disks embedded downstream in the plane plate. Their coordinates were : E_1 : x_1 = 0.2 m ; E_2 : x_1 = 0.4 m ; E_3 : x_1 = 0.6 m ; E_4 : x_1 = 0.8 m. The characteristics of these sound sources have been given in [4]. Their frequency response was quasi-linear in the range 2 kHz - 5 kHz with a resonant peak beyound 4 kHz. The in-phase lines (measured without flow) of the radiated sound field were almost spherical. Finally, the noise level radiated in the range 2 kHz - 4 kHz was higher than 70 dB at 3 m from the sound source.

Sources location was carried out using an acoustic antenna as described in [4]. This latter was placed at 3 m from the source and oriented in such a way that its axis coïncided with the source center (fig. 1). Note that the Acoustic Imaging Technique gives the angle α = SOS' where S is the actual sound source, O the observing point (center of antenna) and S' the "apparent" source [1], [5].

Measurements and experimental results :

The characteristics of the velocity field developping on the plane plate were measured under four mean flow velocity conditions and at three locations on the plate. In each case, boundary layer displacement thickness δ_1 and momentum thickness δ_2 were determined. These measurements point out the different flow regimes. The boundary layer remained laminar up to the Reynolds number $Rx_1' = 4.10^5$ (where x_1' was the coordinate measured from the fictitious origin of the boundary layer), as shown by the variation of the factor H_{12} ($H_{1/2} = \delta_1/\delta_2$) (fig. 2). The transition zone was located by measuring the pressure fluctuation induced on the plate by the boundary layer. Pressure fluctuation level grew exponentially from $Rx_1' = 4.10^5$ to 6.10^5, stabilised between 7.10^5 and 9.10^5 and then decreased to $\Pi'/q = 0.008$ for values of Rx_1' higher than 12.10^5. This value $\Pi'/q = 0.008$ stands in the typical range for a turbulent boundary layer.

The location of the sound sources were performed for 3 emitting points (E_1, E_2 and E_3) and at 4 frequencies (fe = 2, 3, 3.5 and 4 kHz). A typical result is shown in fig. 3. In each case the "apparent" source moved downstream ($\alpha < o$) with an increasing velocity. However this motion was not linear. For small values of U_∞ the angle $|\alpha|$ increased slowly. Then $|\alpha|$ increased strongly for $U_\infty \simeq 20$ m/s and finally remained constant $(|\alpha| = |\alpha_{max}|)$ for values of U_∞ in the range 30 m/s - 50 m/s. The maximum angle value $|\alpha_{max}|$ varied from 0.9° to 2.5° depending on both excitation frequency fe and point source location. In the same way, the velocity range associated with an important increasing displacement depended at the same time on frequency fe and source location. This range moved to lower values of the flow velocities when both the emission point moved downstream and when frequency fe decreases.

To determine with accuracy this velocity range, a velocity value noted U* (fig. 3) for which $|\alpha| = |\alpha_{max}/2|$ was defined for each measurement. At this value, the curve shows an inflexion point corresponding to a maximum shift rate of the source. This velocity U* depends on x'_1 and fe simultaneously. From our experimental results we obtained the following expression :

$$U^* = k\, x_1'^{-0.3}\, fe \quad ; \quad \text{where } k \simeq 162\ m^{1.3}$$

The important dependence of U* with exciting frequency fe lead us to seek for a relationship with a fluctuating parameter of the flow field. For that purpose, it was noted that U* (fe, x'_1) was laying within the flow velocity range in which occured the transition of the boundary layer. Therefore we were conducted to relate the movement of the "apparent" source to the transition process and then to the characteristics of the fluctuating pressure field induced by the flow.

From pressure fluctuation spectra measurements at x'_1 = 0.6, 0.8 and 1.0 m, we obtained the Strouhal number :

$$Stc = 2\Pi\ fc\ (U_\infty, x'_1)\ \delta_1\ (U_\infty, x'_1)\ /\ U_\infty$$

where fc is the cut-off frequency of the pressure spectrum and δ_1 is the boundary layer displacement thickness. This Strouhal number can be compared to that of characteristic sound emission :

$$Ste = 2\Pi\ fe\ \delta_1\ /\ U^*$$

When source displacement rate was maximum $(U = U^*\ (x'_1, fe))$, these two numbers were nearly equals, as shown in figure 4. Therefore, the flow velocity at which shifting occurred was the same than the velocity at which cut- off frequency of pressure spectrum coïncided with the excitation frequency.

In the same step, the maximum deviation angle had to be related to the pressure fluctuation rate Π'/q. As a matter of fact, this ratio increased during boundary layer transition (fig. 2) and then stabilised in the turbulent regime. The same variation occured for α. Then $|\alpha_{max}/2|$ value was compared with pressure fluctuation Π'/q (measured under the same flow velocity condition). This result (given in [4]) confirmed the dependence of α with pressure fluctuation on the plate. However, at this stage, no final conclusion can be given as Π'/q variations were very small.

Application to the case of a vehicle improvement of space frequency acoustic imaging by the Wigner-Ville joint representation

Classical imaging by spectral analysis techniques

Acoustic Imaging displays the incoming intensity as a function of frequency ν (Hz) and space frequency ($f = \nu \sin\alpha / C_o$). Such a display is a spectral analysis under the classical constraints : [6], [7] : - far filed conditions, - received signals are stationnar ones as time functions.

The main constraints in Acoustic Imaging is the "length" constraints on antennas. If in optics the ratio $d/\lambda \simeq 1\,000$, d being the radius of a given lens, such a ratio is very near 10 in Acoustical experiments [7]. Such a constraints is a severe drawback for angular resolution : [7], [6].

$$\Delta(\sin\alpha) = \nu_M / (p.\nu) \ ; \ \nu_m \leqslant \nu \leqslant \nu_M \ ; \ x_M = p\,\lambda m/2 \tag{1}$$

if x_m is the width of the receiving antenna.

Attempt to detect and measure the dowstream displacement described above, requires a high resolving power on the space frequency axis. Such an attempt appears a very difficult one when trying to identify noise sources on the High Speed Train, the speed of which is very near 70 m/s. Our last attempt failed to detect such displacements under the following conditions [8] : d = 150 m, $x_M = p\,\lambda m/2$, p = 12 (fig. 5). Then for 1 kHz octave band : $707 \leqslant \nu \leqslant 1414$ Hz ; $\Delta(\sin\alpha) \geqslant 1/p = 8{,}33.10^{-2}$; $\Delta\alpha \geqslant 4.8°$ x_M d.

To get stationnary conditions the observation length or duration Δt is defined by the "blurrless conditions" :

$$U.\Delta t/d \leqslant 1/(p.m) \ ; \ m \simeq 3 \text{ to } 5 \tag{2}$$

So as to get an angle $\Delta\alpha$ less than the resolution defined by $\Delta(\sin\alpha)$. Under the present constraints Δt is defined as :

$$\Delta t \leqslant d/(U.m.p) \tag{3}$$

with m = 4 ; d = 150 m ; U = 70 m/s; $\Delta t \leqslant 40$ m/s.

The instantaneous frequency $\nu_1(t)$ of the received signal, when emission is a pure tone the frequency of which is ν_o, is expressed by the Doppler modulation law. Under the previous constraint on Δt the received signal may be interpretated as a continuous wave with a constant frequency. Such an assumption may be controlled by the modulation parameter :

$$\sqrt{Bd}\,\Delta t = \sqrt{|d\nu_1/dt|}\,.\,\Delta t \simeq 1$$

when using the values defined by equation (3). All these constraints are not sufficient ones to get a true detection of the downstream displacement because of the law resolution power ($\Delta\alpha \simeq 4°$) and the value $U\,\Delta t \simeq 3$ m. The main consequence is to avoid assumptions such as stationnary ones and get a valid information on angular motion.

<u>Imaging by cross Wigner-Ville representation</u> :

To avoid the stationnary assumption a possible way is to describe the non-stationnary signals by the Wigner-Ville joint representation [9]. This representation displays the variance of a non-stationnary process as a function of canonically conjugated variables as t and ν or x and f. Taking into account the definition of such a representation :

$$f_z\,(t,\,\nu\;;\,1) = \int_R Z(t+\frac{\tau}{2})\;Z^*(t-\frac{\tau}{2})\;e^{-2i\pi\nu\tau}\,d\tau \qquad (4)$$

one derived easily the main invariance properties under translation and scaling transformation [10]. These properties are used to described translation of the array along the x axis and the z axis (fig. 5). Acoustic image is defined as the cross Wigner-Ville representation to two sensors outputs, the distance between them is x :

$$I(t,\,\nu,\,f) = \int_{R^2} Z_r(t+\frac{\tau}{2},\,-\frac{x}{2})\;Z^*(t-\frac{\tau}{2},\,\frac{x}{2})\;G(x)e^{-2i\pi(\nu\tau+xf)}\,d\tau dx \qquad (5)$$

G(x) indicates the spatial weighting due to the finite length array. Zr is the analytic received signal dealing with pure tone signal, as those radiated by vehicles the derivation indicates :

$$I(t,\,\nu,\,f) = \Pi_{\Delta t}(t)\;\omega\,\underset{\substack{\uparrow\\ \text{frequency}\\ \text{modulation}\\ \text{of } Zr(t)}}{(\nu-\nu_1(t)}\;\;g\underset{\substack{\uparrow\\ \text{space frequency}\\ \text{modulation}\\ \text{related to source movement}}}{\big(f-f_1(t)\big)},\quad f = -\frac{\nu}{Co}\sin\alpha \qquad (6)$$

under a quadratic phase assumption. This assumption indicates that f_1 (t) and ν_1(t) are linear frequency modulated functions. Under these conditions no blurr effect is present and the maximum values of I are located along the curve $[\nu_1(t), f_1(t)]$ in the (t, ν, sin α) space. This property is due to the fact that with LFM signals the Wigner-Ville analysis is the Fourier transform of a bilinear form using parity transformation [11]. Figure 6 display such a property for time or space modulated signals. The duration Δt may be now computed from the constraint on LFM signals. This means near the origin $\alpha = 0$ that $\sin\alpha \simeq tg\alpha \simeq \alpha$. This result may be expressed as [11] :

$$(U.\Delta t'/d_1) \leqslant (\Pi/N)rd \ ; \ N \simeq 5 \text{ to } 10 \text{ or } \Delta t' \leqslant (d_1.\Pi)/(U.N) \qquad (7)$$

When comparing with previous values under the stationnary assumption we get :

$$\Delta t'/\Delta t = (d_1.\Pi.mp) : (d.N) \text{ if distance } d_1 \neq d.$$

The present constraint (7) is a weak one. One may decrease d_1 and get a better linear resolution along the cars of the train passing by. In an High Speed Train experiment the parameters may be now : d_1 = 35 m, U = 70 m/s, N = 8, $\Delta t' \leqslant$ 200 ms and $U.\Delta t \simeq 14$ m $\simeq d_1/2$ without "blurr effect" due to the bilinear structure of the Wigner-Ville image representation.

These results indicate that a better experiment can now be planned for High Speed Train experiment. With a better space location of wheels and trucks along the track by infrared transmetters we can get accurate images and try to detect the possible downstream displacements, due to the boundary layer, of the sound sources that are the rolling and vibrating parts of the cars.

Conclusion

The experimental analysis of the perturbation introduced by a mean flow, surrounding an acoustic source, when source location is obtained using an Acoustic Imaging Technique, was made for various flow conditions and for different sound frequencies. For low values of the Reynolds number ($Re \leq 7.10^5$) there was a downstream displacement of the "apparent" sound source well predicted by Amiet's theory. However the source displacement was greater than predicted values when the boundary layer became turbulent and when the Strouhal number Stc (associated to the pressure fluctuations measured at the wall) was equal to the Strouhal number Ste (associated to the sound emission). The maximum value of that displacement was proportional to the pressure fluctuation amplitude measured at the wall. A second experiment was carried out to tempted to point out similar phenomena occuring during sound sources location from a vehicle. For this purpose, an analysis using an Acoustic Imaging Technique was made on a High-Speed Train with a 70 m/s velocity. However at this stage, no conclusion can be given on this effect, because of technical limitation in Acoustic Imaging Technique (angular resolution $\Delta\alpha \simeq 4°$) and in available accuracy for High Speed Train location by optical way.

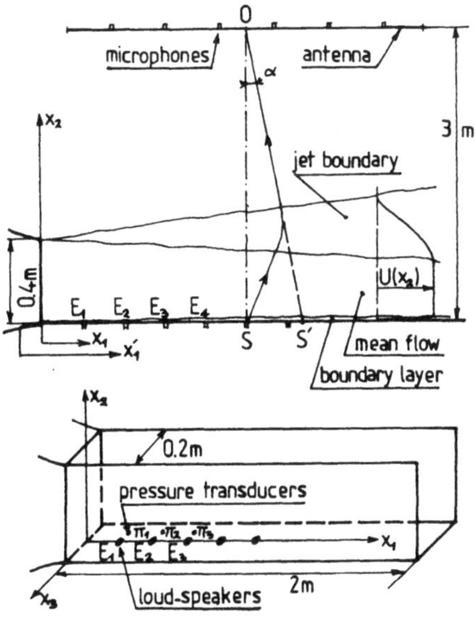

Fig.1 - Experimental set-up.

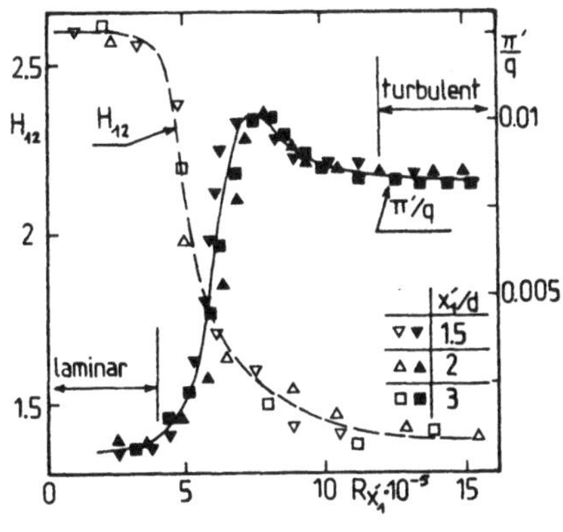

Fig.2 - Evolution of H_{12} and Π'/q during the transition boundary layer.

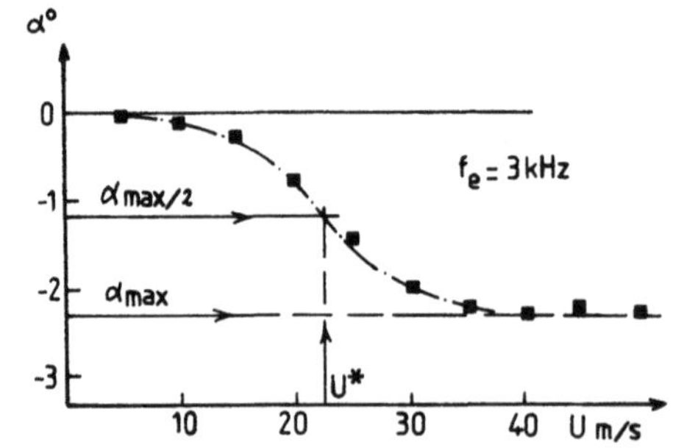

Fig. 3 - Evolution, with the flow velocity, of the angle α (fe = 3 kHz, antenna located at 3 m from the source).

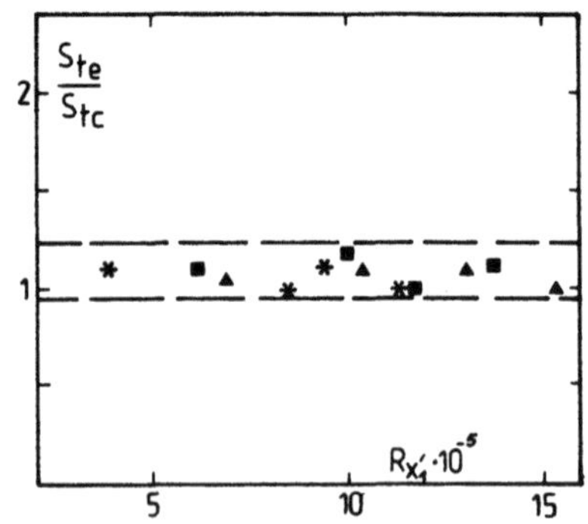

Fig. 4 - Evolution, with the Reynolds number, of the ratio Ste/Stc for various sources locations (E_1 : * ; E_2 : ▲ ; E_3 : ■) and various excitation frequencies (fe = 2 ; 3 ; 3,5 ; 4 kHz).

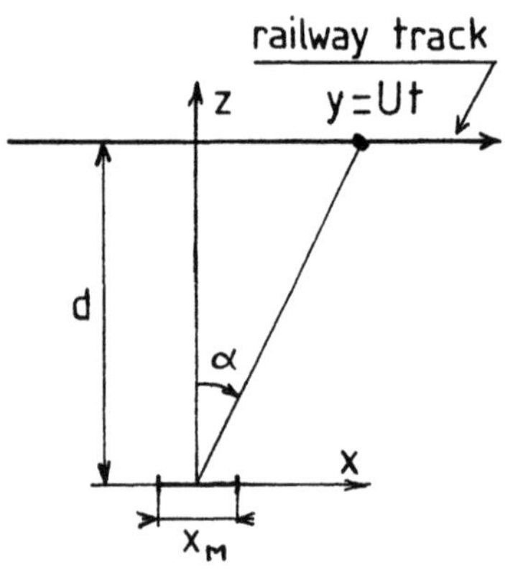

Fig. 5 - Experimental conditions.

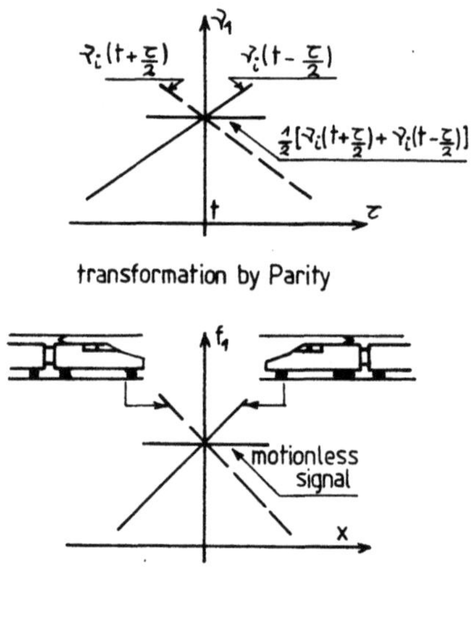

Fig. 6 - Structure of Wigner-Ville representation for frequency modulated signals.

REFERENCES

[1] ARBEY H., HELLION A., ESCUDIE B. - 4e Cong. Français Méca., Nancy 1979

[2] ARBEY H., BATAILLE J. - J. Fluid Mech, 1973, 134, p. 33

[3] AMIET R.K. - J. Sound and Vib., 1978, 58 (4), p. 467

[4] DUFOURCQ P. - - Thèse Docteur Ingénieur, Lyon, 1984

[5] DUFOURCQ P., ARBEY H., ESCUDIE B.
11th Int. Congr. Acoustic, Paris, 1983, 6, p. 181

[6] BIRAUD Y., ESCUDIE B., HELLION A., MUNIER J.
Proceeding IAU 49, 1979, Vol 76, p.299 - C. Van Schoonereld (Ed.)

[7] CHIOLLAZ M., ESCUDIE B., BIRAUD Y., PACHIAUDI G.
I.C.A.S.S.P. Meeting IEEE, Paris, mai 1982

[8] ARBEY H. et al. - Compte Rendu DGRST. PO/18, Contrat 81/5/0867, 1982

[9] FLANDRIN P., ESCUDIE B. - Signal Processing - Vol. 6 p. 27-36, 1984

[10] FLANDRIN P. - Thèse de Docteur-Ingénieur, INP Grenoble, 1982

[11] ESCUDIE B., CHIOLLAZ M. - 10e Colloque Nat. Trait. Signal, GRETSI,
Nice, 1985, pp 313-317

MIX
Papier aus verantwortungsvollen Quellen
Paper from responsible sources
FSC® C105338

If you have any concerns about our products,
you can contact us on
ProductSafety@springernature.com

In case Publisher is established outside the EU,
the EU authorized representative is:
Springer Nature Customer Service Center GmbH
Europaplatz 3, 69115 Heidelberg, Germany

Printed by Libri Plureos GmbH
in Hamburg, Germany